Communication of Innovations:

A CROSS-CULTURAL APPROACH

Communication

A CROSS-CULTURAL APPROACH

EVERETT M. ROGERS

With F. FLOYD SHOEMAKER

of ***I****nnovations*

SECOND EDITION

THE FREE PRESS, NEW YORK

COLLIER-MACMILLAN LTD. LONDON

To David

The Free Press
A Division of The Macmillan Company
866 Third Avenue, New York, New York 10022

Collier-Macmillan Canada Ltd., Toronto, Ontario

Library of Congress Catalog Card Number: 78–122276

1 2 3 4 5 6 7 8 9 10

Contents

v

2 Merging Diffusion Research Traditions: The Middle Range Analysis

3 The Innovation-Decision Process

4 Perceived Attributes of Innovations and Their Rate of Adoption

5 Adopter Categories

6   Opinion Leadership and the Multi-Step Flow of Ideas

7   The Change Agent

8   Communication Channels

9 Collective Innovation-Decisions

11 Consequences of Innovations

Preface

The first edition of this book, *Diffusion of Innovations*, was published in 1962. It ended with the remark:

This book is actually the first of two volumes. The second volume can perhaps be written in ten or fifteen years after the leads for research suggested here have been followed up and expanded upon.

The tremendous increase in research on the diffusion of innovations during the 1960s makes the second edition a necessary reality in about half the time originally estimated. There are about three times as many publications on the subject than there were eight years ago, which means more diffusion research has been done in the past few years than in the previous thirty years. Publications now arrive at the Diffusion Documents Center at Michigan State University at the rate of several a day. In writing the first volume, the senior author conveniently kept all 405 reports then available in a cardboard box under his writing table. In order to make full use of the currently available methodologies and findings, we have now adopted a computerized information retrieval system. Its utility is illustrated in the present volume in the form of generalizations about diffusion, which we derived from a content analysis of past research.

Not only have the number of publications increased, but the nature of diffusion studies has become much more varied. The first edition referred to only a few investigations that had been conducted in less developed countries. As a result, we could only hypothesize about the cross-cultural validity of diffusion generalizations derived from investigations in the United States. Today, we can offer definite support that the diffusion models of 1962 are not entirely culture-bound. However, to account for the low levels of literacy, lack of mass media exposure, and the relatively traditional style of life among peasants in less developed countries, appropriate modifications must be made in many generalizations derived from the United States. In the present edition we shall attempt to highlight cross-cultural similarities, as well as differences, in the diffusion of innovations. Hence, the subtitle of this volume.

In the present edition we have, to a far greater extent, integrated diffusion research with the scientific study of human communication. Diffusion researchers have long been aware that they were investigating a special type of communication behavior. In this book we stress communication concepts and frameworks in our analysis of the diffusion process. We feel this provides an advantage of conceptual clarity as well as ease of wide expression. Our adoption of the communication viewpoint is reflected in the addition of several chapters, the complete reorganization of all chapters, and the frequent inclusion of new concepts.

The two editions are similar. Both utilize a series of generalizations as the basic structure for organization of the chapters. This book is directed to social scientists with an academic interest in the microanalysis of communication and change, and to change agents whose purpose is to diffuse innovations. We feel that generalization from diffusion research offers one route toward a more general theory of social change. Likewise, the corpus of empirical research results about diffusion offers a basis for synthesizing strategies of change.

The need for improved communication among diffusion researchers continues. The various diffusion research traditions, described eight years ago, still operate separately. One objective of our book is to show that diffusion research results have a great deal in common, even though the innovation studied and the academic discipline of the investigator, makes some difference.

The senior author must acknowledge the Agricultural Development Council, Programa Interamericano de Información Popular, UNESCO, the Ford Foundation, and the U.S. Agency for International Development for their sponsorship of his diffusion research in Colombia, India, Thailand, Nigeria, and Brazil. This internationalization has undoubtedly caused the present generalizations to become more cosmopolite than their 1962 counterparts. Another pressure in the same direction is the senior author's experience in teaching courses on diffusion in the Faculty of Sociology at the National University of Colombia, Bogotá, in 1963–1964 as a Fulbright lecturer, and at CIESPAL (Centro International de Estudios Superiores de Periodismo para America Latina) in Quito, Ecuador, in 1968, where mimeographed chapters from the present volume were pretested. Similarly, feedback from readers of the various foreign language translations of the 1962 book have been useful, especially the Japanese, Arabic, and Spanish translations.* Chapters of this book were written and revised in such varied locales as Hyderabad, Enugu, East Lansing, Belo Horizonte, Berlin, Buffalo, Bogotá, Quito, Ft. Collins, and Bangkok, as well as during lengthy transcontinental flights. The contents of this book and the approach used in writing it are fully international.

The most constructive critics of the present manuscript have been Professor A. W. van den Ban at the Agriculture University, Wageningen, Netherlands, and Dr. Nan Lin of the Johns Hopkins University. To these helpful colleagues, who served as intellectual mirrors for our minds, as well as to our students at Michigan State University and Colorado State University who were subjected to earlier versions of this work, we acknowledge our gratitude. The present work is, therefore, the product of a computer and an editorial committee of our peers, as well as our pens.

We acknowledge the editorial and theoretical contribution of our former colleague, Lynne Svenning, especially to the early chapters of this book. She added clarity where it was needed, in our expression and in the conceptual framework.

Greatest thanks go to our 1,500 unknowing and unintended "co-authors" of this work. They, like we, hope the present volume is a useful synthesis for research and action.

Lansing, Michigan   E.M.R.
Ft. Collins, Colorado   F.F.S.

---

*As well as being a translation, the Spanish edition is actually a revision of the original 1962 book with special application to Latin America. It is published by Ediciones Tercer Mundo and the Facultad de Sociología, Universidad Nacional de Colombia, Bogotá. The Japanese translation is published by Baifukan, Tokyo, and the Arabic translation by Aalam Alkotob, Cairo.

1   *Elements of*

. .—.. . .— —— . —. . — ...   —— ..—.   —.. .. ..—. ..—. ..— ... .. —

If every human group had been left to climb upward by its own unaided efforts, progress would have been so slow that it is doubtful whether any society by now could have advanced beyond the level of the Old Stone Age.

RALPH LINTON (1936, p. 324)*

There is nothing new under the sun.   ECCLESIASTES, 1:9

*All references cited are listed in the bibliography (Appendix B) at the end of this book.

Diffusion:

AN OVERVIEW

T HIS is a book about *communication,* a special type of communication, the diffusion of new ideas, new practices—*innovations.* The phenomenal rate at which innovations are being invented, developed, and spread makes it important to look at how these new ideas affect (or fail to affect) the existing social order. An eminent social scientist once observed that the two central tasks for students of society are to find out: (1) how social systems perpetuate themselves by maintaining their structure, and (2) how social systems change their structural form (Radcliffe-Brown, 1957). The second of these tasks is our main concern in the present book: To explore how social systems are changed through the diffusion of new ideas.

Although it is true that we live more than ever before in an era of change, prevailing social structures often serve to hamper the diffusion of innovations. Our activities in education, agriculture, medicine, industry, and the like are often without the benefit of the most current research knowledge. The gap between what is known and what is effectively put to use needs to be closed. To bridge this gap we must understand how new ideas spread from their source to potential receivers and understand the factors affecting the adoption of such innovations. We need to learn why, if 100 different innovations are conceived simultaneously, ten will spread while ninety will be forgotten (Tarde, 1903, p. 140).

If the reader is not yet completely convinced that the diffusion of innovations is a crucial problem in contemporary life, the following

1

case illustration provides insight into some difficulties facing agents of change.

Water-Boiling in a Peruvian Village:
*An Example of Innovation That Failed**

The public health service in Peru attempts to introduce innovations to villagers to improve their health and lengthen their lives. The change agency enjoys a reputation throughout Latin America as efficient. It encourages people to install pit latrines, burn garbage daily, control house flies, report suspected cases of communicable disease, and boil drinking water. These innovations involve major changes in thinking and behavior for Peruvian villagers, who have little knowledge of the relationship between sanitation and illness.

Water-boiling is a necessary method of preventive medicine for these people. Unless they boil drinking water, patients who are "cured" of infectious diseases in village medical clinics often return within the month to be treated for the same disease.

A two-year water-boiling campaign conducted in Los Molinos, a peasant village of 200 families in the coastal region of Peru, persuaded only eleven housewives, who are the key decision makers in the family, to boil water. From the viewpoint of the health agency, the local hygiene worker, Nelida, had a simple task: To persuade the housewives of Los Molinos to add water-boiling to their pattern of existing behavior. Even with the aid of a medical doctor, who gave public talks on water-boiling, and fifteen village housewives who were already boiling water before the campaign, Nelida's program of directed change failed. To understand why, we need to take a closer look at the culture, the local environment, and the individuals.

The Village

Most residents of Los Molinos are peasants who work as field hands on local plantations. Water is carried directly from stream or well by can, pail,

*This case illustration is adapted from Wellin (1955, pp. 71–103), and is used by permission.

gourd, or cask. Children are the usual water carriers; it is not considered appropriate for teenagers of courtship age or for adult men to carry water, and they seldom do. The three sources of water in Los Molinos include a seasonal irrigation ditch close by the village, a spring more than a mile from the village, and a public well whose water the villagers dislike. All three are subject to pollution at all times and show contamination whenever tested. Of the three sources, the irrigation ditch is most commonly used. It is closer to most homes and children can be sent to fetch the water; it has the advantage of being running water, rather than stagnant, and the villagers like its taste.

Although it is not feasible for the village to install a sanitary water system, the incidence of typhoid and other water-borne diseases could be reduced by boiling the water before consumption. During her two-year residence in Los Molinos, Nelida paid several visits to every home in the village but devoted especially intensive efforts to twenty-one families. She visited each of these selected families between fifteen and twenty-five times; eleven of these families now boil their water regularly.

What kinds of persons do these numbers represent? By describing three village housewives—one who boils water to obey custom, one who was persuaded to boil water by the health worker, and one of the many who rejected the innovation—we may add further insight into the process of planned diffusion.

Mrs. A : Custom-Oriented

Mrs. A is about forty and suffers from sinus infection. She is labeled by the Los Molinos villagers as a "sickly one." Each morning, Mrs. A boils a potful of water and uses it throughout the day. She has no understanding of germ theory, as explained by Nelida; her motivation for water-boiling is a complex local custom of hot and cold distinctions. The basic principle of this belief system is that all foods, liquids, medicines, and other objects are inherently hot or cold, quite apart from their actual temperature. In essence hot–cold distinctions serve as a series of avoidances and approaches in such behavior as pregnancy and child rearing, food habits, and the entire health–illness system.

Boiled water and illness are closely linked in the folkways of Los Molinos; by custom, only the ill use cooked, or "hot" water. Once an individual becomes ill, it is unthinkable for him to eat pork (very cold) or to drink brandy (very hot). Extremes of hot and cold must be avoided by the sick; therefore, raw water, which is perceived to be very cold, must be boiled to overcome the extreme temperature.

Villagers learn from childhood to dislike boiled water. Most can tolerate cooked water only if flavoring, such as sugar, cinnamon, lemon, or herbs, is added. Mrs. A likes a dash of cinnamon in her drinking water. At no point in the village belief system is the notion of bacteriological contamination of water involved. By tradition, boiling is aimed at eliminating the innate "cold" quality of unboiled water, not the harmful bacteria. Mrs. A drinks boiled water in obedience to local custom; she is ill.

Mrs. B: Persuaded

The B family came to Los Molinos a generation ago, but they are still strongly oriented toward their birthplace, located among the peaks of the high Andes. Mrs. B worries about lowland diseases which she feels infest the village. It is partly because of this anxiety that the change agent, Nelida, was able to convince Mrs. B to boil water.

Nelida is a friendly authority to Mrs. B (rather than a "dirt inspector," as she is seen by most housewives), who imparts knowledge and brings protection. Mrs. B not only boils water but also has installed a latrine and has sent her youngest child to the health center for an inspection.

Mrs. B is marked as an outsider in the community by her highland hairdo and stumbling Spanish. She will never achieve more than marginal social acceptance in the village. Because the community is not an important reference group to her, Mrs. B deviates from group norms on innovation. Having nothing to lose socially, Mrs. B gains in personal security by heeding Nelida's friendly advice. Mrs. B's practice of boiling water has no effect on her marginal status. She is grateful to Nelida for teaching her how to neutralize the danger of contaminated water, a lowland peril.

Mrs. C: Rejector

This housewife represents the majority of Los Molinos families who were not persuaded by the efforts of the change agent during the two-year health campaign. Mrs. C does not understand germ theory, in spite of Nelida's repeated explanations. How, she argues, can microbes survive in water which would drown people? Are they fish? If germs are so small that they cannot be seen or felt, how can they hurt a grown person? There are enough real threats in the world to worry about—poverty and hunger—without bothering with tiny animals one cannot see, hear, touch, or smell. Mrs. C's allegiance to traditional customs are at odds with the boiling of water. A firm believer in the hot–cold superstition, she feels that only the sick must drink boiled water.

Several housewives, particularly those of the lower social class, are rejectors because they have neither the time nor the means to boil water, even if they were convinced of its value. These women lack time to gather firewood and to boil water. The poor cannot afford the cost of fuel for water-boiling and the wives often work as field laborers beside their husbands, leaving them less time to boil water for their families.

Understanding Why Water-Boiling Failed

This intensive two-year campaign by a public health worker in a Peruvian village of 200 families, aimed at persuading housewives to boil drinking water, was largely unsuccessful. Nelida was able to encourage only about 5 percent of the population, eleven families, to adopt the innovation. In contrast, change agents in other Peruvian villages were able to convince 15 to 20 percent of the housewives. Reasons for the relative failure of the campaign in Los Molinos can be traced partly to the cultural beliefs of the villagers. Local tradition links hot foods with illness. Boiling water makes it less "cold," and hence, appropriate only for the sick. But if a person is not ill, he is prohibited by cultural norms from drinking boiled water. Only the least integrated individuals risk defying community norms on water-boiling. An important factor affecting the adoption rate of any innovation is its compatibility with the cultural beliefs of the social system.

Nelida's failure demonstrates the importance of reference group influences in the adoption and rejection of an innovation. Mrs. B was socially an outsider, "marginal" in the Los Molinos community, although she had lived there for years. Nelida was a more important referent for Mrs. B than were her neighbors, who shunned her. Anxious to secure social acceptance from the more cosmopolite Nelida, Mrs. B adopted water-boiling, not because she understood the correct health reasons, but to obtain Nelida's approval. The marginal Mrs. B adhered to the standards of a cosmopolite reference group outside the local community.

Nelida worked with the wrong housewives if she wanted to launch a self-generating diffusion process in Los Molinos. She concentrated her efforts on village women like Mrs. A and Mrs. B. Unfortunately, they were perceived as a sickly one and a social outsider and were not respected as models of water-boiling behavior by the other women. The village opinion leaders, who could have been a handle to prime the pump of change, were ignored by Nelida.

The way that potential adopters view the change agent affects their willingness to adopt his ideas. In Los Molinos Nelida was seen differently by lower and middle status housewives. Most poor families saw the health worker as a "snooper" sent to Los Molinos to pry for dirt and to press already harassed housewives into keeping cleaner homes. Because the lower status housewives had less free time, they were not likely to initiate visits with Nelida about water-boiling. Their contacts outside the community were limited, and as a result, they saw the cosmopolite Nelida with eyes bound by the social horizons and cultural beliefs of Los Molinos. They distrusted this outsider, who they perceived as a social stranger. Further, Nelida, who was middle class by Los Molinos standards, was able to secure more positive results from housewives whose socioeconomic level and cultural background were more similar to hers. This tendency for effective communication to occur with those who are more similar is a common experience of change agents in most diffusion campaigns.

In general Nelida was much more "innovation-oriented" than "client-oriented." Unable to put herself in the role of the village housewives, her attempts at persuasion failed to reach her clients because the message was not suited to their needs. Nelida did not begin where the villagers were; instead she talked to them about germ theory, which they could not, and did not need to, understand.

We have cited only some of the factors that produced the innovation failure with which Nelida is charged. Understanding the water-boiling case will be much easier, once the remainder of this book has been read.

Social Change*

The theme to be developed throughout this book is: *Communication is essential for social change.*** In this section we focus on the nature of social

---

*A number of the central ideas in this section are adapted from Rogers with Svenning (1969).
**Although one might think of some rare exceptions, such as when social change is caused by a natural event like a volcanic eruption.

change. The case of water-boiling in Los Molinos illustrates change that occurred as a result of new ideas diffusing through a social system. The process of social change consists of three sequential steps:* (1) invention, (2) diffusion, and (3) consequences.** *Invention* is the process by which new ideas are created or developed. *Diffusion* is the process by which these new ideas are communicated to the members of a social system. *Consequences* are the changes that occur within a social system as a result of the adoption or rejection of the innovation. Change occurs when a new idea's use or rejection has an effect.*** Social change is therefore an effect of communication.

What Is Social Change?

Social change is the process by which alteration occurs in the structure and function of a social system. National revolution, invention of a new manufacturing technique, founding of a village improvement council, adoption of birth control methods by a family—all are examples of social change. Alteration in both the structure and function of a social system occurs as a result of such actions. The structure of the social system is provided by the various individual and group statuses which compose it.**** The functioning element within this structure of statuses is the role or actual behavior of an individual in a given status. Status and role reciprocally affect one another.

*Some observers, especially anthropologists, specify two additional steps in this sequential process. One of these is an innovation's *development*, which occurs after invention and before diffusion. It is the process of putting the new idea in a form that meets the needs of an intended audience of receivers. We do not discuss this step because it does not always need to occur, as when the invention is already in a useful form. The last stage, occurring after consequences, is the discontinuance and decline of the innovation. We discuss this step in Chapter 3 but do not consider it as a major part of the change process in the present chapter. Further, some authors distinguish a specific sequence that usually occurs in (our) consequences stage, that of *integration*, the process by which a new idea is incorporated into the continuing operations and way of life of the members of a social system. Some anthropologists also conceive of the process of *reinterpretation*, which occurs when the receivers use an innovation for different purposes than when it was invented or diffused to them.
**An interesting illustration of these three processes is in Lamb's (1948, pp. 385–386) fictional account of the invention of roast pig in China.
***Failure to adopt a new idea can produce change in individuals or social systems when the rejection of the innovation causes an alteration. For example, the failure of many Indians to adopt birth control methods will eventually change the structure and function of their social order as a result of population increase.
****A status or position in a social system may be formal, like school teacher, or informal, like neighborhood opinion leader. Both types carry an implicit set of expectations about the behavior of the individual operating in that status or position.

The status of school teacher carries with it expectations concerning the behavior of the individual in that position, influencing that individual's actual behavior. Should the individual deviate too far from the prescribed set of behaviors, his status is likely to be changed. Social function and social structure are closely linked and reciprocally affect each other. In the process of social change, as one is altered, so is the other. The founding of a new campus organization affects the university social structure by defining a new set of statuses. As individuals begin to function in these new positions, they are likely to affect the overall functioning of the university, as in the case of student protests and demonstrations.

Immanent and Contact Change: Categories of Social Change

One of the more useful ways of viewing social change is to focus on the source of change. When the source is from within the social system under analysis, it is *immanent* change; and when the source of the new ideas is outside the social system, it is *contact* change (Table 1-1).

Table 1-1 Paradigm of Types of Social Change

Recognition of the Need for Change	ORIGIN OF THE NEW IDEA	
	Internal to the Social System	External to the Social System
Internal: Recognition is by members of the social system	I Immanent change	II Selective contact change
External: Recognition may be by change agents outside the social system	III Induced immanent change[a]	IV Directed contact change

[a]Although this situation might be improbable, it is not impossible. A missionary in a peasant village may recognize a need or problem and bring it to the attention of the villagers but not offer suggestions on how to change the situation. Once the problem has been called to their attention, the villagers proceed to invent their own solution.

1. *Immanent change* occurs when members of a social system with little or no external influence create and develop a new idea (that is, *invent* it), which then spreads within the system. A farmer in the senior author's home community in Iowa invented a simple hand tool to clear cornpickers that were plugged with damp cornstalks. The invention was easy to make and a great time-saver. In a short time, most of the inventor's neighbors were using it. Immanent change, then, is a "within-system" phenomenon.

2. The other type of social change, *contact change*, occurs when sources

external to the social system introduce a new idea. Contact change is a "between-system" phenomenon. It may be either *selective* or *directed*, depending on whether the recognition of the need for change is internal or external.

Selective contact change results when members of a social system are exposed to external influences and adopt or reject a new idea from that source on the basis of their needs.* The exposure to innovations is spontaneous or accidental; the receivers are left to choose, interpret, and adopt or reject the new ideas. An illustration of selective contact change occurs when school teachers visit a neighboring school that is especially innovative. They may return to their own classrooms with a new teaching method, but with no pressure from school administrators to seek and adopt such innovations.

Directed contact change, or planned change, is caused by outsiders who, on their own or as representatives of change agencies, intentionally seek to introduce new ideas in order to achieve goals they have defined.** The water-boiling compaign in Peru is an example of directed contact change. The innovation, as well as the recognition of the need for the change, originates outside the social system in the case of directed change. The many government-sponsored development programs designed to introduce technological innovations in agriculture, education, health, and industry are examples of contemporary directed change.

The grand theorists of social change disagree on the merits of directed change. August Comte, for example, advocated directed social change in contrast to Herbert Spencer (noted for his "social Darwinism"), who argued for complete *laissez faire* and an evolutionary survival of the fittest. In the present era most national governments show clear preference for the Comtean approach.*** These governments want higher levels of living for their people,

*A few authors restrict use of the term "diffusion" to unplanned communication of new ideas (selective contact change), as opposed to the concept of "dissemination," which they define as planned communication (directed contact change). However, we use diffusion and dissemination interchangeably to denote both selective and directed contact change, essentially in a sense synonymous with the communication of new ideas. Probably most diffusion today comes about as a result of planned change, that is, as a result of the efforts of change agents.

**Bennis and others (1962, p. 154) define directed (or planned) change in essentially similar terms to ours, except that they stress the mutuality of client and change agent in planning the change. We feel this mutuality is not always necessary for directed change to occur, although it facilitates its likelihood of success. Our notion of directed change bears a close resemblance to what Kelman (1961) terms "compliance," an alteration of the receiver's beliefs or behavior in a direction the source of communication feels is desirable.

***Methods of planned change and diffusion are themselves changing. An example of a recent improvement in methods of diffusion to peasants is the low-cost transistor radio, which has invaded almost all villages in less developed nations.

a goal that can be gained effectively only through massive programs of directed change. Programs of planned change are largely the result of dissatisfaction with the rate of change that results from immanent and selective contact change.*

In the long range we might wish that the major type of change will be spontaneous, rather than directed. As people become more technically expert and sophisticated in diagnosing their needs, selective change can occur more rapidly and can be effected more efficiently. In this sense change agents may eventually work themselves out of a job, or at least into a different role. Instead of diagnosing needs and then promoting innovations to meet them, the change agent would answer requests for innovations from his clients. While the current thrust is upon directed change, if change agents work to improve their clients' capabilities and competence to analyze needs, future emphasis may be to make immanent and selective change easier.

The prevailing enthusiasm for planned change has not always been matched by overwhelming success. The desire for rapid change has outrun the application of scientific know-how about how to introduce innovations. Most studies about the diffusion of innovations which we shall synthesize in this text are investigations of planned social change. As communication research is conducted on the spread of new ideas and as the results are accumulated in a meaningful way, we shall be able to use these findings to design more effective programs of planned change.

Individual and Social System Change: Levels at Which Change Occurs

We have been looking at social change from the viewpoint of the innovation's origin. Another perspective is provided by the nature of the *unit* that adopts or rejects the new ideas.

1. Many changes occur at the *individual* level; that is, the individual is the adopter or rejector of the innovation. Change at this level has variously been referred to as diffusion, adoption, modernization,** acculturation, learning, or

*In the present volume we shall not deal much with the ethical considerations of when and why man is entitled to change his fellow man, although we touch on this issue in Chapters 7 and 11.

**Modernization is defined as the process by which individuals change from a traditional way of life to a more complex, technologically advanced, rapidly changing style of life (Rogers with Svenning, 1969).

socialization. We might term this the microanalytic approach to change analysis in that it focuses on an individual's change behavior.

2. Change also occurs at the *social system* level where it has been diversely termed development,* specialization, integration, or adaptation. Here our attention is centered on the change process at the social system level and is thus macroanalytic in approach.

Of course, change at these two levels is closely interrelated. If we regard a school as a social system, then the school system's adoption of team teaching will lead to individual teachers' decisions to change their teaching methods. Similarly, the aggregation of a multitude of individual changes produces a system-level alteration. Farmers' decisions to adopt a more productive coffee variety in an African nation may eventually result in changes affecting the country's international balance of trade.

Perhaps all analyses of social change must ultimately center primary attention upon communication processes. In fact, all explanations of human behavior directly stem from an examination of how individuals acquire and modify ideas through communication with others. The learning process, the diffusion process, the change process, and so on all basically involve the communication of new ideas.

Communication and Social Change

Communication is the process by which messages are transferred from a source to a receiver. We might think of the communication process in terms of the oversimplified but useful S–M–C–R model. A *source* (S) sends a *message* (M) via certain *channels* (C) to the *receiving* individual (R). One can easily see how communication factors are vitally involved in many aspects of the decision processes which together make up social change: A farmer's decision to move to the city or to participate in a government program, an industrialist's adoption of a new manufacturing technique, or the decision of a husband and wife to engage in family planning. In each of these instances, a message (M) is conveyed to individuals (R) via communication channels (C) from a source individual (S), which causes the receivers to change an existing behavior pattern.

*Development is a type of social change in which new ideas are introduced into a social system in order to produce higher per capita incomes and levels of living through more modern production methods and improved social organization. Development is modernization at the social system level.

Although communication and social change are not synonymous, communication is an important element throughout the social change process. Essentially, the concept of social change includes, in addition to the communication process, the societal and individual consequences that result from the adoption or rejection of an innovation. When examining social change, our concern is with alteration in the structure and function of a social system, as well as the process through which such alteration occurs.

Communication and Diffusion

Diffusion is a special type of communication. *Diffusion* is the process by which innovations spread to the members of a social system. Diffusion studies are concerned with messages that are new ideas, whereas communication studies encompass all types of messages (Figure 1-1). As the messages are new in the case of diffusion, a degree of risk for the receiver is present. This

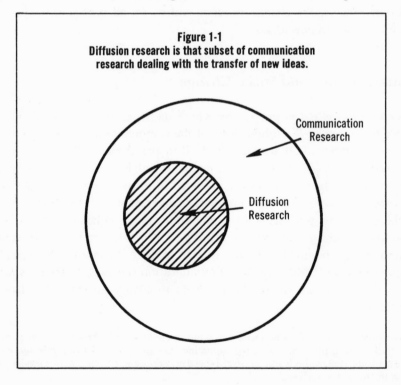

Figure 1-1
**Diffusion research is that subset of communication
research dealing with the transfer of new ideas.**

Communication
Research

Diffusion
Research

leads to somewhat different behavior on his part in the case of innovations than if he were receiving messages about routine ideas.

There is often a further difference between the nature of diffusion research versus other types of communication research. In the latter, we often focus on attempts to bring about changes in knowledge or attitudes by altering the makeup of the source, message, channels, or receivers in the communication process. For instance, we may seek to present the source as more credible to the receiver, because communication studies indicate that when this is done there is likely to be greater persuasion or attitude change on the part of the receivers. But in diffusion research we usually focus on bringing about *overt behavior change*, that is, adoption or rejection of new ideas, rather than just changes in knowledge or attitudes.* The knowledge and persuasion effects of diffusion campaigns are considered mainly as intermediate steps in an individual's decision-making process leading eventually to overt behavior change.

The focus on new ideas by diffusion researchers has led to a more thorough understanding of the communication process. The conception of the flow of communication as a multi-step process lacked clear conceptual development until it was probed by researchers studying the diffusion of innovations.** They found that new ideas usually spread from a source to an audience of receivers via a series of sequential transmissions, rather than in the over-simplified two steps that had been originally postulated.*** By tracing communication patterns over time, diffusion researchers expanded the conceptual repertoire of communication researchers. Until students of diffusion began studying the flow of communication, consideration of the role of different communication channels at various stages in the innovation-decision process was masked (van den Ban, 1964b).**** Specifically, it was learned that mass media channels are often more important at creating awareness-knowledge of a new idea, whereas interpersonal channels are more important in changing attitudes toward innovations.

*This emphasis upon behavior change in diffusion research is rather important, because we know that knowledge change and persuasion do not always lead directly and immediately to behavior change (as will be discussed in Chapter 3).
**For example, Menzel and Katz (1955) concluded: "We found it necessary to propose amendments for the model of the two-step communications by considering the possibility of multi-step rather than two-step flow." This idea of the multi-step flow model emerged from one of the first tests of the two-step flow in the diffusion of an innovation.
***By Lazarsfeld and others (1944) in their investigation of mass media channels and opinion leaders in a U.S. presidential election campaign.
****Many other illustrations could be cited (and will be in future chapters of this book) where the diffusion approach has uncovered additional dimensions of the communication process by its attention to the time variable as well as by its focus on overt behavior change.

Communication research on concepts such as credibility, persuasion, perceptions, selective exposure, and message variables has also provided further insights for diffusion scholars. We hope to bring these two closely allied fields more closely together in the present volume, because we feel that the communication approach* offers fruitful potential for the analysis of the diffusion of innovations.

Heterophily and Diffusion

One of the obvious principles of human communication is that the transfer of ideas occurs most frequently between a source and a receiver who are alike, similar, homophilous. *Homophily*** is the degree to which pairs of individuals who interact are similar in certain attributes, such as beliefs, values, education, social status, and the like. In a free-choice situation, when a source can interact with any one of a number of receivers, there is a strong tendency for him to select a receiver who is most like himself.

There are many reasons for this homophily principle. Similar individuals are likely to belong to the same groups, to live near each other, to be drawn by the same interests. This physical and social propinquity makes homophilous communication more likely. "Birds of a feather flock together," as the old adage goes.

But in many situations, propinquity explains only a part of homophilous tendencies. *More effective communication occurs when source and receiver are homophilous*.*** When they share common meanings, a mutual subcultural language,

*The communication approach is utilized by social scientists in a number of disciplines such as sociology, social psychology, anthropology, political science, and economics. The viewpoint is essentially interdisciplinary and multidisciplinary and perhaps is best expressed in its most integrated aspects by communication researchers in the modest number of departments and institutes of communication which have sprung up largely since the end of World War II, both in the U.S. and abroad.

**This concept and its opposite, heterophily, were first called to scientific attention by Lazarsfeld and Merton (1964, p. 23). *Heterophily*, the mirror opposite of homophily, is defined as the degree to which pairs of individuals who interact are different in certain attributes. The term homophily derives from the Greek word "homoios," meaning alike or equal. Thus, homophily literally means affiliation or communication with a similar person.

***A further refinement of this proposition includes the concept of *empathy*, defined as the ability of an individual to project himself into the role of another: *More effective communication occurs when source and receiver are homophilous, unless the source and the receiver have high empathy*. Heterophilous individuals who have high empathy are, in a social psychological sense, really homophilous.

The proposition about effective communication and homophily can also be reversed:

and are alike in personal and social characteristics, the communication of ideas is likely to have greater effects in terms of knowledge gain, attitude formation and change, and overt behavior change. When homophily is present in communication, therefore, interaction is likely to be more rewarding to both source and receiver. As they become gradually conditioned to homophily, the choice of other homophilous interaction partners is made even more likely.

Many examples could be cited to support the proposition about homophily and effective communication. In everyday life most of us interact with others who are quite similar in social status, education, and beliefs. And when we occasionally seek to communicate with those of a much lower social status, many problems of ineffective communication arise. Consider the middle class teacher who seeks to communicate with slum children, the social worker who tries to change the behavior of her lower class or foreign born clients, the technical assistance worker overseas who attempts to introduce innovations to peasants.

One of the most distinctive problems in the communication of innovations is that the source is usually quite heterophilous to the receiver. The extension agent, for instance, is much more technically competent than his peasant clients. This frequently leads to ineffective communication. They simply do not talk the same language. In fact, when source and receiver are identical regarding their technical grasp of the innovation, no diffusion can occur. Therefore, the very nature of diffusion demands that at least some degree of heterophily be present between source and receiver.* Ideally, they are homophilous on all other variables (education, social status, and the like) even though heterophilous regarding the innovation. In actuality, source and receiver are usually heterophilous on all of these variables because innovation competence, education, social status, and so on are highly interrelated.

This heterophily gap in diffusion is much wider when source and receiver do not share a common culture, as we shall see when we focus upon the transfer of technological innovations from more to less developed countries. Thus, one of the special communication problems for analysis in this book is that of

---

Effective communication between source and receiver leads to greater homophily in knowledge, beliefs, and overt behavior. The proverb just cited would read: "When birds flock together, they become more of a feather."

*We shall see in later chapters that receivers often seek sources that are slightly more technically competent about innovations than themselves, but not *too much* so. For instance, opinion leaders who are sought for information about innovations are usually somewhat more innovative in adopting new ideas than their followers, yet the leaders are seldom innovators, the very first to adopt. This suggests that there is an optimal degree of heterophily for effective diffusion.

cross-cultural heterophily. And an important and recurring theme of this book is the special problems of ineffective communication that come about because source and receiver are usually heterophilous in diffusion.

Time Lags in Diffusion

Evidence that diffusion is not a simple, easy process is the time that it requires. Change takes time, much time. Despite generally favorable attitudes toward change in nations like the United States, a considerable time lag exists from the introduction of a new idea to its widespread adoption. This is true even when the economic benefits of the innovation are obvious.

1   A forty-year time lag existed between the first success of the tunnel oven in the English pottery industry and its general use (Carter and Williams, 1957).
2   More than fourteen years were required for hybrid seed corn to reach complete adoption in Iowa (Ryan and Gross, 1943).
3   U.S. public schools required fifty years to adopt the idea of the kindergarten in the 1930s and 1940s (Ross, 1958), and more recently, about five or six years to adopt modern math in the 1960s (Carlson, 1965).*

One of the goals of diffusion research is to shorten this time lag. Although this goal is obviously important in nations like the United States, it is more important in less developed nations, because the time lags in such nations may be even longer. For example, more than thirty years passed from first use to complete adoption of chemical fertilizer in a small Colombian village (Deutschmann and Fals Borda, 1962a).

It is clear that research alone is not enough to solve most problems; the results of the research must be diffused and utilized before their advantages can be realized.** Even diffusion research findings must be diffused before

---

*The difference in the time lag for kindergartens versus modern math may reflect a different climate for educational change in the post-Sputnik era or simply the fact that massive funds were spent in "packaging" and promoting modern math.
**This viewpoint has been widely recognized in a number of U.S. government agencies that: have recently created research utilization sections. Examples are the Technology Utilization Branch of NASA, the Research Utilization Branch of the U.S. Office of Education, the Applied Research Branch of NIMH, the Research Utilization Branch of the U.S. Office of Economic Opportunity's Community Action Program, and the Research Utilization Branch of the U.S. Vocational Rehabilitation Service. Another indication of the importance of keeping abreast of research results in one's professional field is illustrated by a recent

their benefits can be derived; one goal of the present text is to facilitate the widespread use of what we know about diffusion.

In spite of the fact that the communication of most innovations involves a considerable time lag, there is a certain inevitability in their diffusion. Most attempts to prevent innovation diffusion over an extended period of time have failed. For instance, the Chinese were unsuccessful in their attempt to maintain sole knowledge of gunpowder. And today, a growing number of nations share the secret of the nuclear bomb with the United States.* Similar are university administration and campus police attempts to prevent the widespread adoption of marijuana smoking among students.

Consequences of Innovations

The consequences of innovations are a third part of the social change process, following invention and diffusion. Because consequences have an obvious interface with diffusion (for example, the selection of diffusion strategies affects the consequences that accrue), we shall present a brief discussion of consequences here and an expanded treatment in our last chapter.

Consequences are the changes that occur within a social system as a result of the adoption or rejection of an innovation. There are at least three classifications of consequences:

1 *Functional* versus *dysfunctional* consequences, depending on whether the effects of an innovation in a social system are desirable or undesirable.
2 *Direct* versus *indirect* consequences, depending on whether the changes in a social system occur in immediate response to an innovation or as a result of the direct consequences of an innovation.
3 *Manifest* versus *latent* consequences, depending on whether the changes are recognized and intended by the members of a social system or not.

Change agents usually introduce into a client system innovations that they expect will be functional, direct, and manifest. But often such innovations result in at least some latent consequences that are indirect and dysfunctional for the system's members. An illustration is the case of the steel ax introduced

lawsuit in which a medical doctor was held liable for failing to utilize a new curative technique, with the consequence that the patient died. The basis for this legal decision was that the physician had a responsibility for knowing about major innovations stemming from research in his field.
*A discussion of the international diffusion of nuclear weapons is found in Beaton and Maddox (1962).

by missionaries to an Australian aborigine tribe (Sharp, 1952, pp. 69–72). The change agents intended that the new tool should raise levels of living and material comfort for the tribe. But the new technology also led to breakdown of the family structure, the rise of prostitution, and "misuse" of the innovation itself. Change agents can often anticipate and predict the innovation's *form*, the directly observable physical appearance of the innovation, and perhaps its *function*, the contribution of the idea to the way of life of the system's members. But seldom are change agents able to predict another aspect of an innovation's consequences, its *meaning*, the subjective perception of the innovation by the clients.

Elements in the Diffusion of Innovations

Crucial elements in the diffusion of new ideas are (1) the *innovation* (2) which is *communicated* through certain *channels* (3) over *time* (4) among the members of a *social system*.* It is the element of time which distinguishes diffusion from other types of communication research. As pointed out previously in this chapter, diffusion research deals only with messages that are *new* ideas.

The four elements of diffusion differ only in nomenclature from the essential elements of most general communication models. For example, Aristotle proposed a very simple model of oral communication consisting of the speaker, the speech, and the listener. Laswell described all communication as dealing with "*who* says *what,* through *what channels* of communication, to *whom* with what . . . *results*" (Smith and others, 1946, p. 212). The S–M–C–R model, cited earlier, consists of (1) source, (2) message, (3) channel, and (4) receivers, to which we might add (5) the *effects* of communication. Obviously, this S–M–C–R–E communication model (Berlo, 1960) corresponds closely to the elements

---

*These four elements are similar to those listed by Katz and others (1963) as essential in any diffusion study: (1) the *acceptance*, (2) over *time*, (3) of some specific *item*—an idea or practice, (4) by individuals, groups, or other *adopting units,* linked to (5) specific *channels* of communication, (6) to a *social structure*, and (7) to a given system of values or *culture*. We do not include element 1 as a separate item among our four, as we see acceptance or adoption as an effect of communication (our element 2). We collapse Katz and others' (1963) elements 4, 6, and 7 in our fourth element, because they make up various aspects of the social system. No single chapter in the present volume deals solely with the effects of culture on diffusion, but in various chapters we shall discuss specific interfaces of culture and diffusion. Examples are cultural consequences of diffusion (Chapter 11), how an innovation's cultural compatibility affects its rate of adoption (Chapter 4), and the influence of traditional and modern norms on diffusion (Chapter 1).

of diffusion (Figure 1-2): (1) the receivers are members of a social system, (2) the channels* are the means by which the innovation spreads, (3) the message is a new idea, (4) the source is the origin of the innovation (an inventor, scientist, change agent, opinion leader, and the like), and (5) the effects are changes in knowledge, attitude, and overt behavior (adoption or rejection) regarding the innovation.

The Innovation

An *innovation* is an idea, practice, or object perceived as new by an individual. It matters little, so far as human behavior is concerned, whether or not an idea is "objectively" new as measured by the lapse of time since its first use or discovery. It is the perceived or subjective newness of the idea for the individual that determines his reaction to it. If the idea seems new to the individual, it is an innovation.

"New" in an innovative idea need not be simply new knowledge. An innovation might be known by an individual for some time (that is, he is aware of the idea**), but he has not yet developed a favorable or unfavorable attitude toward it, nor has he adopted or rejected it. The "newness" aspect of an innovation may be expressed in knowledge, in attitude, or regarding a decision to use it.

Every idea has been an innovation sometime. Any list of innovations must change with the times. Black Panthers, computers, micro-teaching, birth control pills, chemical weed sprays, LSD, heart transplants, and laser beams might still be considered innovative ideas at this writing, but the reader in North America will probably find many of these items adopted or even discontinued at the time of reading. This list also illustrates the great variety

*The source and channel are seldom distinguished in most diffusion research, which is usually conducted by obtaining recall data from receivers via personal interviews. These receivers can report the communication channels from which they directly obtained information about the innovation, but they may be unaware of the original source of the new idea (in the case of technological innovations, this original source is ultimately a scientist). Some receivers may recognize the immediate source of information about the innovation when the channels are interpersonal, as in the case of a farmer-receiver obtaining information via interpersonal channels from a change agent, such as an agricultural extension agent. But generally the equivalent of the communication channel element in diffusion research is both the source and channel in the S–M–C–R–E model.

**There is, of course, more to knowledge about an innovation than simply awareness of it. Also important as a basis for effective decision making (adoption or rejection) about an innovation is the degree of knowledge about how properly to use the idea.

Figure 1-2
**Elements in the diffusion of innovations
and the S-M-C-R-E communication model are similar.**

Source	—	Message	—	Channel	—	Receiver	—	Effects
Inventors, scientists, change agents, or opinion leaders		Innovation (Perceived attributes, such as relative advantage, compatibility, etc.)		Communication channels (Mass media or interpersonal)		Members of a social system		Consequences over time 1. Knowledge 2. Attitude change (persuasion) 3. Behavioral change (adoption or rejection)

Elements in
the S-M-C-R-E
Model:

Corresponding
elements in
the diffusion
of innovations:

of material products, ideological beliefs, social movements, and so on that qualify as innovations. Among the wide range of innovations that have been analyzed by diffusion researchers are a new speech form among oil drillers (Boone, 1949), nuclear warfare among nations (Beaton and Maddox, 1962), a rumor aboard a submarine (Allingham, 1964), and snowmobilies among the Lapps (Pelto and others, 1969).

It should not be assumed that the diffusion and adoption of all innovations is necessarily desirable. In fact, we shall review studies of harmful and uneconomical innovations* that are generally not desirable for either the individual or his social system.**

Symbolic Versus Action Innovation-Decisions

Most but not all of the new ideas analyzed in this book are of a material, technological variety,*** consisting of an object as well as an idea. An innovation may have two components: (1) an *idea* component, and (2) an *object* component (that is, the material or physical product aspect of the idea). All innovations must have the ideational component, of course, but many do not have a physical referent. One criterion for classifying innovations is whether or not the innovation has an object component associated with it.

Innovations with only an idea component cannot be adopted in a sense that can be physically observed. Adoption is essentially a *symbolic* decision in this case. In contrast, innovations that also have an object component invoke an *action* adoption.**** Examples of innovations requiring symbolic decisions are Communist ideology, news events, and rumors.

*We have relatively few such studies in comparison to the number of inquiries about "desirable" innovations. One illustration is Francis' (1960) investigation of the adoption of a bogus and costly item of farm equipment, the oats incubator. There are also a few studies of the *discontinuance* of "undesirable" innovations, like Graham and Gibson's (1967) study of the cessation of cigarette smoking for health reasons.

**Further, the same innovation may be desirable for one adopter in his situation but undesirable for another potential adopter in a different situation.

***When a more restrictive definition of innovation is needed, it can be preceded by an appropriate adjective such as "technological" or "material." Actually, most innovations studied in past research have been both technological and material. We do not limit our meaning of technological to material objects only. For example, the idea of assembly line organization in a factory is an example of social technology. Technology can be broadly defined as a design for instrumental action.

****Further, some innovations cannot (or should not) be adopted immediately. Change agents are seeking, in these cases, only to persuade their clients to be willing (and perhaps able) to adopt without actually doing so. An example is the efforts by the U.S. Office of Civil Defense to inform the U.S. public about the nature of fallout shelters and to convince them

Characteristics of Innovations

It should not be assumed, as it often has in the past, that all innovations are equivalent units of analysis. This is a gross oversimplification. Whereas it may take an innovation like modern math only five or six years to reach complete adoption, another new idea such as team teaching may require several decades to reach widespread use. The several characteristics of innovations, as sensed by the receivers, contribute to their different rate of adoption.

1. *Relative advantage* is the degree to which an innovation is perceived as better than the idea it supersedes. The degree of relative advantage may be measured in economic terms, but often social prestige factors, convenience, and satisfaction are also important components. It matters little whether the innovation has a great deal of "objective" advantage. What does matter is whether the individual *perceives* the innovation as being advantageous. The greater the perceived relative advantage of an innovation, the more rapid its rate of adoption.

2. *Compatibility* is the degree to which an innovation is perceived as being consistent with the existing values, past experiences, and needs of the receivers. An idea that is not compatible with the prevalent values and norms of the social system will not be adopted as rapidly as an innovation that is compatible. The adoption of an incompatible innovation often requires the prior adoption of a new value system. An example of an incompatible innovation is the use of the IUCD (intra-uterine contraceptive device) in countries where religious beliefs discourage use of birth control techniques.

3. *Complexity* is the degree to which an innovation is perceived as difficult to understand and use. Some innovations are readily understood by most members of a social system; others are not and will be adopted more slowly. For example, the rhythm method of family planning is relatively complex for most peasant housewives to comprehend because it requires understanding human reproduction and the monthly cycle of ovulation. For this reason, attempts to introduce the rhythm method in village India have been much less successful than campaigns to diffuse the loop, a type of IUCD, which is a much less complex idea in the eyes of the receiver. In general those new ideas requiring little additional learning investment on the part of the receiver will be

to use the shelters if the need should ever occur. But there is no intent by the Civil Defense officials to bring about actual use of the shelters. A similar illustration of an *anticipatory innovation-decision* was that made in Iowa in 1968 where the extension service conducted an active campaign to inform farmers about how to control the corn pest that was currently a major problem in the neighboring state of Nebraska, in anticipation of the future need for such knowledge.

adopted more rapidly than innovations requiring the adopter to develop new skills and understandings.

4. *Trialability** is the degree to which an innovation may be experimented with on a limited basis. New ideas which can be tried on the installment plan will generally be adopted more quickly than innovations which are not divisible. Ryan and Gross (1943) found that not one of their Iowa farmer respondents adopted hybrid seed corn without first trying it on a partial basis. If the new seed could not have been sampled experimentally, its rate of adoption would have been much slower. Essentially, an innovation that is trialable represents less risk to the individual who is considering it.

5. *Observability*** is the degree to which the results of an innovation are visible to others. The easier it is for an individual to see the results of an innovation, the more likely he is to adopt. For example, a technical assistance agency in Bolivia introduced a new corn variety in one town. Within two years the local demand for the seed far exceeded the supply. The farmers were mostly illiterate, but they could easily observe the spectacular results achieved with the new corn and were thus persuaded to adopt. In the United States a rat poison that killed rats in their holes diffused very slowly among farmers because its results were not visible.

The five attributes just described are not a complete list, but they are the most important characteristics of innovations, past research indicates, in explaining rate of adoption.

Given that an innovation exists and that it has certain attributes, communication between the source and the receivers must take place if the innovation is to spread beyond its inventor. Now we turn our attention to this second element in diffusion.

Communication Channels

Communication is the process by which messages are transmitted from a source to a receiver. In other words communication is the transfer of ideas

---

*This attribute of an innovation was referred to as "divisibility" in an earlier version of this book (Rogers, 1962b, p. 131). We prefer the convention of "trialability" because it implies a somewhat broader meaning, including the notion of psychological trial (see Chapter 4 of this book).

**This attribute was referred to as "communicability" in the earlier edition of this book (Rogers, 1962b, p. 132), but we prefer the present convention because of its more precise meaning (see Chapter 4).

from a source with a viewpoint of modifying the behavior of receivers. A communication *channel* is the means by which the message gets from the source to the receiver.

As we stated earlier, diffusion is a subset of communication research that is concerned with new ideas. The essence of the diffusion process is the human interaction by which one person communicates a new idea to one or several other persons. At its most elementary level, the diffusion process consists of (1) a new idea, (2) individual A who has knowledge of the innovation, (3) individual B who is not yet aware of the new idea, and (4) some sort of communication channel connecting the two individuals. The nature of the social relationships between A and B determines the conditions under which A will or will not tell B about the innovation, and further, it influences the effect that the telling has on individual B.

The communication channel by which the new idea reaches B is also important in determining B's decision to adopt or reject the innovation. Usually the choice of communication channel lies with A, the source, and should be made in light of (1) the purpose of the communication act, and (2) the audience to whom the message is being sent. If A wishes simply to inform B about the innovation, *mass media channels** are often the most rapid and efficient, especially if the number of Bs in the audience is large. On the other hand, if A's objective is to persuade B to form a favorable attitude toward the innovation, an *interpersonal channel*** is more effective.

Therefore, the source should choose between mass media and interpersonal channels on the basis of the receiver's stage in the innovation–decision process, whether at the knowledge or persuasion stage. This brings us to discussion of a third element in diffusion, time.

Over Time

Time is an important consideration in the process of diffusion.*** The time dimension is involved (1) in the innovation-decision process by which an individual passes from first knowledge of the innovation through its adoption

---

*Mass media channels are all those means of transmitting messages that involve a mass medium, such as radio, television, film, newspapers, magazines, and so on, which enable a source of one or a few individuals to reach an audience of many.
**Interpersonal channels are those that involve a face-to-face exchange between two or more individuals.
***"Time is the key to diffusion research" (Katz and others, 1963).

or rejection, (2) in the innovativeness of the individual, that is, the relative earliness-lateness with which an individual adopts an innovation when compared with other members of his social system, and (3) in the innovation's rate of adoption in a social system, usually measured as the number of members of the system that adopt the innovation in a given time period.

The Innovation-Decision Process

The *innovation-decision process* is the mental process through which an individual passes from first knowledge of an innovation to a decision to adopt or reject and to confirmation of this decision. Many diffusion researchers have conceptualized a cumulative series of five stages in the process: (1) from awareness (first knowledge of the new idea), (2) to interest (gaining further knowledge about the innovation), (3) to evaluation (gaining a favorable or unfavorable attitude toward the innovation), (4) to small-scale trial, (5) to an adoption or rejection decision.* We prefer to conceptualize four main functions or steps in the process: (1) knowledge, (2) persuasion, (3) decision, and (4) confirmation.** The *knowledge function* occurs when the individual is exposed to the innovation's existence and gains some understanding of how it functions. The *persuasion function* occurs when the individual forms a favorable or unfavorable attitude toward the innovation. The *decision function* occurs when the individual engages in activities which lead to a choice to adopt or reject the innovation. The *confirmation function* occurs when the individual seeks reinforcement for the innovation-decision he has made, but he may reverse his previous decision if exposed to conflicting messages about the innovation.

An example is presented in Figure 1-3 to clarify the meaning of the innovation-decision process and to show the importance of the time dimension. Mr. Skeptic, an Iowa farmer, first learned of hybrid seed corn from an agricultural extension agent in 1935 (the knowledge function). However, he was not convinced to plant hybrid corn on his own farm until 1937, after he had discussed the innovation with several neighbors (the persuasion function).

*These five stages were utilized in the previous version of this book (Rogers, 1962b).
**We feel this four-stage process is an improvement over the traditional "adoption process," which it replaces, because our present model of the innovation-decision process makes provision for rejection as well as adoption decisions and allows for post-decision communication behavior which usually reinforces the original decision, but may lead to its reversal. The present model is conceptually linked to the notions of decision making, the learning process, and dissonance reduction.

Skeptic purchased a small sack of hybrid seed in 1937 and by 1939 was planting 100 percent of his corn acreage in hybrids. When did he adopt hybrid corn?

Skeptic adopted in 1939 when he decided to continue full scale use of the innovation (decision function). *Adoption* is a decision to make full use of a new idea as the best course of action available.* The *innovation-decision period* is the length of time required to pass through the innovation-decision process; in

---

Figure 1-3
The innovation-decision period
from knowledge to decision and confirmation.

The innovation-decision process is the mental process through which an individual passes from first knowledge of an innovation to a decision to adopt or reject and to later confirmation of this decision. There are four functions in this process: (1) knowledge, (2) persuasion (attitude formation and change), (3) decision (adoption or rejection), and (4) confirmation. For the farmer depicted in this diagram, confirmation of his adoption decision continued until 1941, when conflicting messages reached him, and he decided to discontinue the innovation. The innovation-decision period is the length of time required to pass through the innovation-decision process.

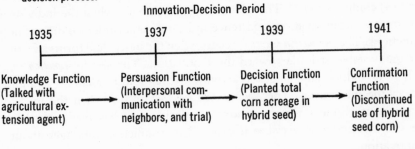

Innovation-Decision Period

| 1935 | 1937 | 1939 | 1941 |

| Knowledge Function (Talked with agricultural extension agent) | Persuasion Function (Interpersonal communication with neighbors, and trial) | Decision Function (Planted total corn acreage in hybrid seed) | Confirmation Function (Discontinued use of hybrid seed corn) |

---

the present instance it lasted four years.** The innovation decision can also take a negative turn; that is, the final decision can be *rejection*, a decision not to adopt an innovation.

The last function in the innovation-decision process is confirmation, a stage at which the receiver seeks reinforcement for the adoption or rejection

*This definition, based upon Zaltman (1964, p. 24), differs slightly in wording from that utilized in the previous edition of the present book (Rogers, 1962b, p. 17), where adoption was defined as a decision to continue full-scale use of an innovation.

**For practical reasons, the length of the innovation-decision period is usually measured from first knowledge until the decision to adopt (or reject), although in a strict sense it should perhaps be measured to the time of confirmation. The trouble with this latter procedure is that the confirmation function may continue over an indefinite time period.

decision he has made. Occasionally, however, conflicting and contradictory messages reach the receiver about the innovation, and this may lead to discontinuance on one hand or later adoption (after rejection) on the other. In the case of Mr. Skeptic, a decision was made to *discontinue* use of the innovation after previously adopting it. Farmer Skeptic became dissatisfied with hybrid seed and discontinued its use in 1941, when he again planted all of his corn acreage in open-pollinated seed. Discontinuances occur for many other reasons, including replacement of the innovation with an improved idea. Discontinuances occur only after the individual has fully adopted the idea.

Innovativeness and Adopter Categories

If Skeptic adopted hybrid seed in 1939 and the average farmer in his community adopted in 1936, Skeptic is less innovative than the average member of his system. *Innovativeness* is the degree to which an individual is relatively earlier in adopting new ideas than the other members of his system.* Rather than describing Mr. Skeptic as "less innovative than the average member of his social system," it is handier and more efficient to refer to him as being in the "late majority" adopter category. This shorthand notation saves words and contributes to clearer understanding, for diffusion research shows clearly that each of the adopter categories has a great deal in common. If Skeptic is like most others in the late majority category, he is below average in social status, makes little use of mass media channels, and secures most of his new ideas from peers via interpersonal channels. In a similar manner, we shall present a concise word-picture of each of the other four adopter categories in this book (Chapter 5). *Adopter categories* are the classifications of members of a social system on the basis of innovativeness. The five adopter categories used here are: (1) innovators, (2) early adopters, (3) early majority, (4) late majority, and (5) laggards.

Obviously, the measure of innovativeness and the classification of the system's members into adopter categories are based upon the relative time at which an innovation is adopted.

Rate of Adoption

There is a third specific way in which the time dimension is involved in the diffusion of innovations. *Rate of adoption* is the relative speed with which an

*By "relatively earlier" we mean earlier in terms of actual time of adoption, rather than whether the individual *perceives* he adopted the innovation relatively earlier than others in his system.

innovation is adopted by members of a social system. This rate of adoption is usually measured by the length of time required for a certain percentage of the members of a system to adopt an innovation. Therefore, we see that rate of adoption is measured using an innovation or a system, rather than an individual, as the unit of analysis. Innovations that are perceived by receivers as possessing greater relative advantage, compatibility, and the like have a more rapid rate of adoption (as we pointed out in a previous section of this chapter).

There are also differences in the rate of adoption for the same innovation in different social systems. Generally, diffusion research shows that systems typified by modern, rather than traditional, norms will have a faster rate of adoption. How do we classify systems as to modern or traditional? What is a social system?

Among Members of a Social System

A *social system* is defined as a collectivity of units which are functionally differentiated and engaged in joint problem solving with respect to a common goal. The members or units of a social system may be individuals, informal groups, complex organizations, or subsystems. The social system analyzed in a diffusion study may consist of all the peasants in a Latin American village, students at a university, high schools in Thailand, medical doctors in a large city, or members of an aborigine tribe. Each unit in a social system can be functionally differentiated from every other member. All members cooperate at least to the extent of seeking to solve a common problem or to reach a mutual goal. It is this sharing of a common objective that binds the system together.

It is important to remember that diffusion occurs within a social system, because the social structure of the system affects the innovation's diffusion patterns in several ways. The social system constitutes a set of boundaries within which innovations diffuse. In this section we shall deal with the following topics: How the social structure affects diffusion, the effect of traditional and modern norms on diffusion, the roles of opinion leaders and change agents, and types of innovation-decisions. All these issues involve interfaces between the social system and the diffusion process that occurs within it.

Social Structure and Diffusion

To the extent that the members in a social system are differentiated, structure then exists within the system. Social structure develops through the arrangement (such as in an hierarchical fashion) of the statuses or positions in a system. A formal organization such as a government agency has a well-developed formal social structure consisting of titled positions, giving those in a higher ranked status the right to give orders to those of lesser rank and to expect the orders to be carried out. Even an informal grouping has some degree of structure inherent in the interpersonal relationships among its members, determining who interacts with whom and under what circumstances. Naturally, both formal and informal social structures have an effect on human behavior and how it changes in response to communication stimuli.

Diffusion and social structure are complexly interrelated.

1. *The social structure acts to impede or facilitate the rate of diffusion and adoption of new ideas through what are called "system effects."** The basic notion of system effects is that the norms, social statuses, hierarchy, and so on of a social system influence the behavior of individual members of that system. *System effects* are the influences of the system's social structure on the behavior of the individual members of the social system.

In the case of innovation diffusion, one can conceptualize an individual's innovation behavior as explained by two types of variables: (1) the *individual's* personality, communication behavior, attitudes, and so on, and (2) the nature of his *social system*. Which type is more important?

Several investigations point out the importance of system influences. Van den Ban (1960) studied the effects of traditional and modern norms (for a sample of Wisconsin townships) on the innovativeness of farmers. Although such individual characteristics as a farmer's education, size of farm, and net worth were positively related to his innovativeness, the township norms were even better predictors of farmer innovativeness. Van den Ban concluded that a farmer with a high level of education, on a large farm, and with a high net worth, but residing in a township with traditional norms, adopted fewer farm innovations than if he had a lower level of education and a smaller farm in a township where the norms were modern.

Qadir (1966) conducted a somewhat parallel inquiry in twenty-six Filipino rural neighborhoods. He found that in modern systems with a social climate favorable to the adoption of innovations, even individuals lacking much education, mass media exposure, or modern attitudes, acted in an innovative manner. This finding suggests Generalization 1-1: *System effects* (such as

---

*Sometimes system effects are synonymously referred to as "compositional effects" (Davis and others, 1961), as "contextual effects," or as "structural effects" (Blau, 1957 and 1960; Campbell and Alexander, 1965; and Tannenbaum and Bachman, 1964). None of these studies, however, looked at system effects on diffusion behavior.

system norms, the composite educational level of one's peers, and the like) *may be as important in explaining individual innovativeness as such individual characteristics as education, cosmopoliteness, and so on.**

2. *Diffusion may also change the social structure of a system.* Some new ideas are "restructuring" innovations in that they change the structure of the social system itself. The adoption of a village development council changes the village social structure by adding a new set of statuses. The initiation of a research and development unit within an industrial firm and the departmentalization of a public school are also restructuring innovations. In many instances the restructuring affects the rate of future innovation diffusion within the system.**

A system's social structure and the way in which innovations diffuse in that system bear a close but subtly intertwining relationship. As Katz (1961) remarked, "It is as unthinkable to study diffusion without some knowledge of the social structures in which potential adopters are located as it is to study blood circulation without adequate knowledge of the structure of veins and arteries." Therefore, a system's social structure affects diffusion, and vice versa.

System Norms and Diffusion

We have just pointed out that a system's norms affect an individual's innovation-adoption behavior. *Norms* are the established behavior patterns

---

*Since system effects have only recently been recognized as particularly important, we need carefully designed investigations of the role of these effects in the diffusion of innovations. Further support for this proposition comes from Saxena's (1968) study of system effects on farmer innovativeness in Indian villages, and from Davis' (1968) companion investigation in Nigeria. In both studies, system (village) variables explained a portion of the variance in individual innovativeness not also explained by individual variables (e.g., literacy, mass media exposure, and so on). The specific research studies which support or do not support each of the generalizations in this book are listed in Appendix A.

**In fact one strategy for speeding the rate of diffusion of new ideas is to restructure the social system. Such an approach is generally more appropriate and perhaps easier to accomplish when the system is a formal organization, than when it is a relatively more informal system such as a peasant village. Changing the norms of such a system might be a very difficult feat for a change agent, at least in the short range.

Those of highest status and power in a system, the elite, are obviously in a position to serve as "gatekeepers" in controlling the flow of innovations into the system from external sources. They are more likely to favor the introduction of functioning innovations, which do not threaten to disturb the status quo of the system's social structure, than restructuring innovations. Perhaps an illustration of this point is the oligarchic leaders of Latin American nations who promote technological innovations in agriculture and marketing, but oppose such restructuring innovations as land reform and overhauling of the tax system. This viewpoint will be discussed further in Chapter 11.

for the members of a given social system.* They define a range of tolerable behavior and serve as a guide or a standard for the members of a social system.

A system's norms can be a barrier to change, as was shown in our example of water-boiling in a Peruvian community. Such resistance to new ideas is often found in norms relating to food. In India, for example, sacred cows roam the countryside while millions of people are undernourished. Pork cannot be consumed by Moslems. Polished rice is eaten in most of Asia and the United States, even though whole rice is more nutritious.

In additonal to influencing the original adoption or rejection of an innovation, norms also influence the manner in which an innovation will be integrated into the existing way of life of the receivers, that is, the consequences of the innovation. When horses were introduced into the Shoshone culture, the Indians readily accepted them for they had prior experience with horses, which they had stolen from pioneers for food (Harris, 1940). Although U.S. Bureau of Indian Affairs agents had intended that the horses be used for transportation, the Indians ate them!

Most of the researches which provide evidence of the influence of norms on diffusion have suffered from methodological shortcomings such as the limitations of their case study nature. Only one or two social systems** are usually included in these analyses. Needed are investigations of a large sample of comparable systems, so that we can have some assurance of the generalizability of the results. In such investigations it is necessary to conceptualize and operationalize the traditional-modern dimension of system norms, which is itself a serious methodological task.

Traditional and Modern Norms

We conceptualize system norms that are most relevant for innovation diffusion as either traditional or modern. These two kinds of norms are *ideal*

---

*There have been two schools of thought among sociologists as to the meaning of norm. The neo-positivists (such as Lundberg, Chapin, and Dodd) defined norms as a *standard behavior* represented by such measures of central tendency in a distribution as a mean, median, or mode. The social actionists defined norm as a group *expectation* for a certain type of behavior. This argument between the "what is" versus the "what ought to be" meanings of norm has subsided recently with a tendency toward a more operational definition of norm that may reflect either a standard or an expectation for behavior, or both.

**Illustrative of such investigations in which a pair of contrasting social systems (such as rural communities, schools, or colleges), one with modern and one with traditional norms, are compared in terms of their social structure and the communication behavior of their members are: Eibler (1965), Leuthold (1965), Lionberger and Chang (1965), Campbell and Holik (1960), van den Ban (1963b), Davis (1965), Rogers and van Es (1964), and Rogers with Svenning (1969).

types, conceptualizations based on observations of reality and designed to facilitate comparisons. Ideal types do not necessarily exist empirically but may be constructed by abstracting to a logical extreme the characteristics of the behavior under analysis. Developed purely for methodological reasons, ideal types provide a framework for anlysis. They are ideal not in the sense that they describe what ought to be, but rather in the sense that they logically accentuate some dimension of analysis. Since early times social scientists have conceptualized opposing pairs of ideal types, which are called "polar types," for the purposes of analyzing behavior occurring within social systems. Our conception of traditional and modern norms is a synthesis, at least in part, of various aspects of these previous typologies.*

A number of synonyms may be used to describe modern norms: Innovative progressive, developed, scientific, rational, and so on. The crucial dimension is that individuals in social systems with modern norms view change favorably, predisposing them to adopt new ideas more rapidly than individuals in traditional systems. Traditional social systems can be characterized by:

1   Lack of favorable orientation to change.
2   A less developed or "simpler" technology.
3   A relatively low level of literacy, education, and understanding of the scientific method.
4   A social enforcement of the status quo in the social system, facilitated by affective personal relationships, such as friendliness and hospitality, which are highly valued as ends in themselves.
5   Little communication by members of the social system with outsiders. Lack of transportation facilities and communication with the larger society reinforces the tendency of individuals in a traditional system to remain relatively isolated.
6   Lack of ability to empathize or to see oneself in others' roles, particularly the roles of outsiders to the system. An individual member in a system with traditional norms is not likely to recognize or learn new social relationships involving himself; he usually plays only one role and never learns others.

In contrast, a modern social system is typified by:

1   A generally positive attitude toward change.

*These ideal types include, for example, the *Gemeinschaft* and *Gesellschaft* of Töennies, Weber's rational and traditional types, Merton's local and cosmopolitan, and the sacred and secular types of Becker. The traditional and modern ideal types are based most directly on the work of Parsons (1951, pp. 101ff), Parsons and Shils (1951, pp. 80ff), Redfield (1956), Weber (1947, pp. 115–116), Wolf (1955), and particularly Lerner (1958).

2   A well developed technology with a complex division of labor.

3   A high value on education and science.

4   Rational and businesslike social relationships rather than emotional and affective.

5   Cosmopolite perspectives, in that members of the system often interact with outsiders, facilitating the entrance of new ideas into the social system.

6   Empathic ability on the part of the system's members, who are able to see themselves in roles quite different from their own.

In summary, a social system with modern norms is more change oriented, technologically developed, scientific, rational, cosmopolite, and empathic. A traditional system embodies the opposite characteristics.*

There is one danger in attempting to fit our thinking into the framework of ideal types: There is a tendency to overemphasize the extent of the differences. Traditional and modern ideal types are actually the end points of a continuum on which actual social system norms may range. We should not forget that the norms of most systems are distributed between the extremes that we have described.

One should not conclude from our discussion that traditional norms are necessarily undesirable. In many cases, tradition lends stability to a social system that is undergoing rapid change and is in danger of disorganization.** Modern systems have their own unique drawbacks, including slums, pollution of water and air, alienation, neuroses, and an almost endless list of social problems rooted in the consequences of "progress."

The reader should remember that it is possible for an individual to be a

---

*Three main methods have been utilized to measure modern and traditional norms in diffusion research: (1) the average innovativeness of the system's members, (2) their attitudes toward innovators, and (3) key informants' ratings (Rogers, 1962b, pp. 67–70). None of these measures of system norms is above methodological criticism, and future research efforts should be directed toward developing improved measures. Nevertheless, existing measures provide a rough indication of a system's norms, in that it is usually possible to say that the norms of one social system are relatively more traditional or modern than those of another system (or systems).

**Perhaps an ideal rate of change is somewhat less than that which would lead to disequilibrium within the system. *Disequilibrium* occurs when the rate of change is too rapid to permit the system to adjust. A system is said to be in a state of *dynamic equilibrium* when change is occurring at a rate commensurate with the system's ability to cope with it. *Stable equilibrium* occurs when there is almost no change in the structure or functioning of the social system. The latter state would occur on a traffic circle without vehicles; a dynamic equilibrium has cars moving around it. Disequilibrium would occur if there were one car too many and all traffic stopped. For a more detailed discussion of ideal rates of change, see Chapter 11.

member of more than one social system and that the norms of these different systems may vary on the traditional-modern continuum. And if the norms of the systems to which the individual belongs are widely divergent, he is likely to experience cross-pressures in making innovation decisions. For instance, a school teacher who is continuing his part-time graduate education in a university where new ideas are constantly discussed is likely to experience conflict when he attempts to introduce these innovations into the traditional school system where he teaches.

Not only is the traditionalism-modernism of a social system's norms important in predicting individual diffusion behavior, but also the *commitment* of the individual to the social system affects his conformity to its norms. An innovative teacher in a traditional school may be relatively unaffected by the norms because the local school is not important as a reference group to him. Thus, an individual's integration into a social system,* as well as the nature of the system's norms, need to be studied in order to fully explain his adoption behavior.**

Opinion Leaders and Change Agents

In this section we have discussed the influence of the social structure on the members' diffusion behavior. Now we turn to the different roles that individuals play in a social system and the effect of these roles on diffusion patterns. Specifically, we shall look at two roles: Opinion leaders and change agents.

Very often the most innovative member of a system is perceived as a deviant from the social system, and he is accorded a somewhat dubious status of low credibility by the average members of the system. His role in diffusion (especially in persuading others about the innovation) is therefore likely to be limited. On the other hand there are members of the system who function in

*Both Yadav (1967) and Guimarães (1968) found that a high degree of communication integration characterized more modern peasant villages. *Communication integration* is the degree to which the units in a social system are interconnected by interpersonal communication channels. It seems logical that more integrated systems should be more modern (and their members relatively more innovative) because once a new idea enters these systems, it will spread quickly to all members of the systems.

**Jamias (1964) found that Michigan dairy farmers who were more dogmatic (that is, whose belief systems were more rigidly compartmentalized) conformed more closely to social system norms than did farmers who were less dogmatic. This finding suggests that perhaps personality variables such as dogmatism may affect the degree to which system norms influence individual behavior.

the role of opinion leader. They provide information and advice about innovations to many others in the system.

Opinion leadership is the degree to which an individual is able to informally influence other individuals' attitudes or overt behavior in a desired way with relative frequency.* It is a type of informal leadership, rather than being a function of the individual's formal position or status in the system. Opinion leadership is earned and maintained by the individual's technical competence, social accessibility, and conformity to the system's norms.** Several researches indicate that when the social system is modern, the opinion leaders are quite innovative; but when the norms are traditional, the leaders also reflect this norm in their behavior. By their close conformity to the system's norms, the opinion leaders serve as an apt model for the innovation behavior of their followers.

In any system, naturally, there may be both innovative and also more traditional opinion leaders. These influential persons can lead in the promotion of new ideas, or they can head an active opposition. In general, when opinion leaders are compared with their followers, we find that they (1) are more exposed to all forms of external communication, (2) are more cosmopolite, (3) have higher social status, and (4) are more innovative (although the exact degree of innovativeness depends, in part, on the system's norms).

Opinion leaders are usually members of the social system in which they exert their influence. In some instances individuals with influence in the social system are professionals representing change agencies external to the system. A *change agent* is a professional who influences innovation-decisions in a direction deemed desirable by a change agency. He usually seeks to obtain the adoption of new ideas, but he may also attempt to slow down diffusion and prevent the adoption of what he believes are undesirable innovations. Change agents often use opinion leaders within a given social system as lieutenants in

---

*Thus, our definition of opinion leadership implies a leadership-followership *relation* between two or more people, rather than an abstract attribute of an individual leader. We follow Merton (1957, p. 415) in the notion that opinion leadership has not occurred unless the opinions or overt behavior of the followers is different from what it would have been if the opinion leader had not interacted with others in the system. Our present definition of opinion leadership also differs somewhat from that expressed in the previous edition of the present book (Rogers, 1962b, p. 16), which we now feel did not adequately convey the notion that opinion leaders may be either active or passive (or both) in influencing their followers. That is, opinion leaders may be sought by, or they may seek, their followers.
**These criteria for opinion leadership correspond roughly to those suggested by Katz (1967).

their campaigns of planned change. There is research evidence that opinion leaders can be "worn out" by change agents who overuse them.* Opinion leaders may be perceived by their peers as too much like the change agents; thus, the opinion leaders lose their credibility with their former followers.

Types of Innovation-Decisions

The social system has yet another important kind of influence on the diffusion of new ideas. Innovations can be adopted or rejected by individual members of a system or by the entire social system. The relationships between the social system and the decision to adopt innovations may be described in the following manner:

1. *Optional decisions* are made by an individual regardless of the decisions of other members of the system. Even in this case, the individual's decision is undoubtedly influenced by the norms of his social system and his need to conform to group pressures. The decision of an individual to begin wearing contact lenses instead of eye glasses, an Iowa farmer's decision to adopt hybrid corn, and a housewife's adoption of birth control pills are examples of optional decisions.

2. *Collective decisions* are those which individuals in the social system agree to make by consensus. All must conform to the system's decision once it is made. An example is fluoridation of a city's drinking water. Once the community decision is made, the individual has little practical choice but to adopt fluoridated water. It does indeed "take two to tango" (Katz, 1962), once the partners have agreed to dance.

3. *Authority decisions* are those forced upon an individual by someone in a superordinate power position, such as a supervisor in a bureaucratic organization. The individual's attitude toward the innovation is not the prime factor in his adoption or rejection; he is simply told of and expected to comply with the innovation-decision which was made by an authority. Few research studies have yet been conducted of this type of innovation-decision, which must be very common in an organizational society such as the U.S. today. In all authority decisions we must distinguish between (1) the decision maker, who is one (or more) individual(s), and (2) the adopter or adopters, who carry out the decision. In the case of optional and collective decisions these two roles (of deciding and adopting) are performed by the same individual(s).

*This point will be discussed in detail in Chapters 6 and 7.

These three types of innovation-decisions range on a continuum from optional decisions (where the adopting individual has almost complete responsibility for the decision), through collective decisions (where the adopter has some influence in the decision), to authority decisions (where the adopting individual has no influence in the innovation decision). Collective and authority decisions are probably much more common than optional decisions in formal organizations,* such as factories, public schools, or labor unions, in comparison with other fields like agriculture and medicine where innovation-decisions are usually optional.

Generally, the fastest rate of adoption of innovations results from authority decisions (depending, of course, on whether the authorities are traditional or modern). In turn, optional decisions can be made more rapidly than the collective type. Although made most rapidly, authority decisions are more likely to be circumvented and may eventually lead to a high rate of discontinuance of the innovation.** Where change depends upon compliance under public surveillance, it is not likely to continue once the surveillance is removed.

The type of innovation-decision for a given idea may change or be changed over time. Automobile seat belts, during the early years of their use, were installed in private autos largely as optional decisions. Then in the 1960s many states began to require by law installation of seat belts in all new cars. In 1968 a federal law was passed to this effect. An optional innovation-decision then became a collective decision.***

There is yet a fourth type of innovation-decision which is essentially a sequential combination of two or more of the three types**** we have just discussed. *Contingent decisions* are a choice to adopt or reject which can be made only after a prior innovation-decision. An individual member of a social system is free to adopt or not to adopt a new idea only after his system's innovation-decision. A teacher cannot adopt or reject the use of an overhead

*We shall focus on collective innovation-decisions in Chapter 9 and on authority decisions in Chapter 10.
**Therefore, authority innovation-decisions often result in a rapid rate of adoption but in a relatively low quality decision, that cannot effectively be put into action, at least over an extended period of time.
***But in another sense an optional decision was still required by the automobile driver or passenger to use the belts, that is, to fasten them when getting in the seat. The collective decision in 1968 led to a rapid (100 percent) installation of the belts but not to a parallel increase in the use of the safety devices.
****Of course there are also many innovation-decisions which are difficult to categorize in that they fall between these three types; nevertheless, the three types are heuristically distinct.

projector in his classroom until the school system has decided to purchase one; at that point the teacher can decide to use or reject the overhead projector. In the Punjab State of India hybrid corn adoption is a contingent decision because hybrid corn requires a growing season two weeks longer than open-pollinated varieties, and villagers release their cattle to roam for forage across the unfenced fields once their corn is harvested. One can readily imagine the difficulty of making an optional decision to adopt hybrid corn in the Punjab without a prior collective decision by the entire village. One can also imagine contingent decisions in which the first decision is of an authority sort followed by optional or collective decisions. The distinctive aspect of contingent decision making is that two (or more) tandem decisions are required; either of the decisions may be optional, collective, or authority.

Summary

A main theme in this book is that communication is essential for social change.

Social change is the process by which alteration occurs in the structure and function of a social system. We suggest three sequential stages in the process of social change: (1) *invention*, the process by which new ideas are created or developed, (2) *diffusion*, the process by which these new ideas are communicated to the members of a given social system, and (3) *consequences*, the changes that occur within the social system as a result of the adoption or rejection of the innovation.

Change is either immanent or contact. *Immanent change* occurs when members of a social system with little or no external influence create and develop a new idea (that is, invent it), and then it spreads within the system. *Contact change* occurs when sources external to the social system introduce a new idea; contact change may be either selective or directed. *Selective contact change* results when members of a social system are exposed to external influences and adopt or reject a new idea from that source on the basis of their needs. *Directed contact change,* or planned change, is caused by outsiders who, on their own or as representatives of change agencies, intentionally seek to introduce new ideas in order to achieve goals they have defined. Much change that occurs today is directed, and this variety is the main concern of the present book.

Communication is the process by which messages are transferred from a source to a receiver. Essential elements in the communication process are the source, message, channels, and receivers. Diffusion is essentially a special type of communication concerned with the spread of messages that are *new* ideas. A certain degree of risk is usually associated with the reception of innovations, and this leads to somewhat different behaviors on the part of the individual than if he were receiving routine ideas.

Another distinctive aspect of diffusion as a subfield of communication is that heterophily is most often present between source and receiver. *Heterophily* is the degree to which pairs of individuals who interact are different in certain attributes, such as beliefs, values, education, social status, and the like. The opposite of heterophily is *homophily*, the degree to which pairs of individuals who interact are similar in certain attributes. Generally, most human communication takes place between individuals who are homophilous, a situation that leads to more effective communication. Therefore, the extent of source-receiver heterophily, which is often present in the diffusion of innovations, leads to special problems in securing effective communication.

The main elements in the diffusion of new ideas are: (1) the *innovation*, (2) which is *communicated* through certain *channels*, (3) *over time*, (4) among the members of a *social system*. An *innovation* is an idea, practice, or object perceived as new by an individual. The characteristics of an innovation, as perceived by the members of a social system, determine its rate of adoption. Five attributes of innovations are: (1) relative advantage, (2) compatibility, (3) complexity, (4) trialability, and (5) observability.

Communication channels are the means by which a message gets from a source to a receiver. Mass media channels are more effective in creating knowledge of innovations, whereas interpersonal channels are more effective in forming and changing attitudes toward the new idea.

Time is involved in diffusion in (1) the innovation-decision process, (2) innovativeness, and (3) an innovation's rate of adoption. The *innovation-decision process* is the mental process through which an individual passes from first knowledge of an innovation to a decision to adopt or reject, and to later confirmation of this decision. We conceptualize four functions in this process: (1) knowledge, (2) persuasion, (3) decision, and (4) confirmation. *Adoption* is a decision to make full use of a new idea as the best course of action. *Rejection* is a decision not to adopt an innovation. *Discontinuance* is a decision to cease use of an innovation after previously adopting it. Discontinuance, then, is essentially adoption of an innovation, followed by rejection.

Innovativeness is the degree to which an individual is relatively earlier in adopting new ideas than other members of his social system. We specify five *adopter categories*, classifications of the members of a social system on the basis of innovativeness: (1) innovators, (2) early adopters, (3) early majority, (4) late majority, and (5) laggards. *Rate of adoption* is the relative speed with which an innovation is adopted by members of a social system.

A *social system* is a collectivity of units which are functionally differentiated and engaged in joint problem solving with respect to a common goal. It is important to remember that diffusion occurs within a social system, because the system's social structure can have an important influence on the spread of new ideas.

Social structure consists of the statuses or positions in a social system and how these statuses are arranged, such as in an hierarchical fashion. The social structure of a system acts to impede or facilitate the rate of diffusion and adoption of new ideas through what are called "system effects." The norms, social statuses, hierarchy, and so on of a social system influence the behavior of individual members of that system. Evidence from several researches indicates that the system effects may be as important in explaining individual innovativeness as education, cosmopoliteness, and the like. Diffusion may also change the social structure of a system in the sense that many innovations are of a restructuring nature.

Norms are the established behavior patterns for the members of a social system. Two ideal types of norms can be distinguished: Traditional and modern. A system with modern norms is more change oriented, technologically developed, scientific, rational, cosmopolite, and empathic, whereas a traditional system is the opposite of these.

Opinion leadership is the degree to which an individual is able to informally influence other individuals' attitudes or overt behavior in a desired way with relative frequency. A *change agent* is a professional person who attempts to influence innovation-decisions in a direction that he feels is desirable.

We consider three main types of innovation-decisions: (1) *optional decisions,* which are made by an individual regardless of the decisions of other members of the system; (2) *collective* decisions, which individuals in the social system agree to make by consensus, and (3) *authority* decisions, which are forced upon an individual by someone in a superordinate power position. These three types of decisions may, of course, occur in a sequential order so that an optional decision cannot be made until a prior collective decision has been made; these *contingent innovation-decisions* are a choice to adopt or reject which can be made only after a prior innovation-decision.

Purpose of This Book

In one sense this first chapter has been a concise picture of the contents of the entire book. The reader should now have a fairly good idea of the main concepts, methods, and themes of the chapters that are to follow.

The primary purpose of this book is to synthesize a series of generalizations from research on the diffusion of innovations. Each of these generalizations represents the relationship found between two or more concepts.* This book is essentially, then, a distillation of the results of more than 1,500 diffusion publications.** Our objective in deriving these generalizations is to facilitate understanding of the diffusion process by change agents and by social scientists.

DeFleur (1966) argues that the twin needs of diffusion research are: (1) conceptual standardization, and (2) providing linkages with more general social science theory. Hopefully, this present volume is a step in the direction of fulfilling both needs. It relates the diffusion process to communication concepts and theories and seeks to relate empirical understandings about the diffusion of ideas to a theory of social change. Further, our approach is heavily cross-cultural. Throughout the discussions that follow, we shall test the cross-cultural similarities, as well as contrasts, in diffusion generalizations. Our chief focus is upon determining the comparability of diffusion understandings from research in more developed countries versus less developed countries. This cross-cultural theme is facilitated by the great increase in diffusion research in less developed nations, mainly completed in the 1960s. Prior to that time most diffusion inquiry was conducted in countries such as the United States.

A secondary purpose of this book is to prevent the unnecessary duplication of research effort which often results when a corpus of research literature is not thoroughly analyzed and synthesized. Every research topic reaches a point where greater returns come from a synthesis of findings already present than from investing resources in an additional research study. Therefore, areas of needed research are suggested throughout the present volume. It is thus a

*For a complete list of these generalizations and each of the studies that supports (or does not support) them, see Appendix A, "Generalizations about the Diffusion of Innovations."
**All the diffusion publications cited in this book are held in the Diffusion Documents Center, an information storage and retrieval system in operation in the Department of Communication at Michigan State University. Each publication reporting empirical research results is content analyzed in terms of its methods and the generalizations that are reported in it. Further details on the Diffusion Documents Center and our approach to synthesis of its materials will be found in Chapter 2, Appendix A, and Appendix B, a complete bibliography of diffusion publications.

conceptual inventory as well as a propositional index, in that we call for needed research as well as synthesize past findings.*

*Our approach in this book is somewhat parallel to McGrath and Altman's (1966) synthesis of 250 small group research studies. They also use a propositional inventory as a basic method of synthesis, and their system of classification of small group variables bears striking similarity to our Diffusion Documents Center's system, although both were developed independently. March and Simon (1958) utilized a somewhat similar procedure in the field of research on formal organizations.

2 Merging
Diffusion Research

Research on mass communications and on the acceptance of new farm practices may be characterized as an interest in campaigns to gain acceptance of change. Despite their shared problems, these two fields have shown no interest in each other.

ELIHU KATZ (1960)

A science without a theory is blind because it lacks that element which alone is able to organize facts and give direction to research. It is necessary to have a theory . . . which is empirical and not speculative. This means that theory and facts must be closely related to each other. KURT LEWIN (1936, p. 4)

The studies of the diffusion of innovations, including the part played by mass communication, promise to provide an empirical and quantitative basis for developing more rigorous approaches to theories of social change. MELVIN L. DeFLEUR (1966, p. 138)

Traditions:

THE MIDDLE RANGE
ANALYSIS . .... . ._ ._. _._. .... _ ._. ._ _.. .. _ .. ___ -

I~N~ 1955, the senior author discovered an educational journal devoted to reviewing research studies on the diffusion of educational innovations. This was the first convergence between two research traditions that had both been investigating the diffusion of new ideas for over seventeen years! And the two fields, we soon found, had uncovered remarkably similar findings: Both had determined that the diffusion of an innovation follows an S-shaped curve over time, that innovators were wealthier and more cosmopolite than later adopters, and so on.

Unfortunately, there has been a definite lack of diffusion of diffusion research.* One reason for this poor interchange lies in the existence of diffusion research traditions. A *research tradition* is a series of investigations on a similar topic in which successive studies are influenced by preceding inquiries. One effect of this academic partitioning into diffusion research traditions has been an inadequate flow of research findings among diffusion researchers. The result has often been unnecessary duplication and unwanted replication.

But in the mid-1960s a gradual breakdown occurred in the formerly

---

*"Ironically, it almost seems as if diffusion research in the various research traditions can be said to have been independently invented! Indeed, diffusion researchers in the several traditions which we have examined scarcely know of each other's existence" (Katz and others, 1963).

impermeable boundaries between the diffusion research traditions,* even though the traditions continued to exist as distinct communities of scholars. Figure 2-1 shows evidence of this increasing breakdown of the "paper curtains" between the members of the diffusion research

*We can only speculate about the specific techniques that were instrumental in causing greater interdisciplinary awareness in the diffusion field. Among the most important are probably the availability of courses on diffusion at various universities, the appearance of a general textbook on the subject (that was not tradition-bound), and the activities of such scholars as Elihu Katz at the University of Chicago, Wilbur Schramm at Stanford University, and the late Paul Deutschmann at Michigan State University, all of whom promoted a more cosmopolite, view of diffusion research to their colleagues, both in their writings and in their professional activities.

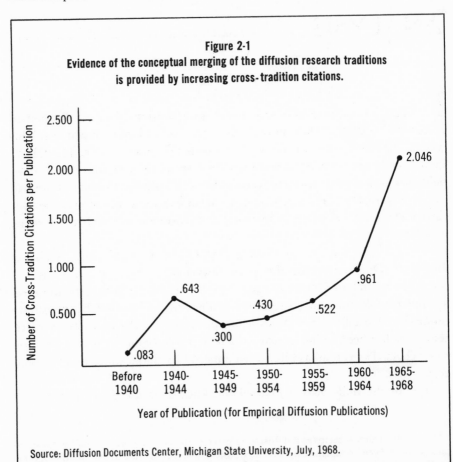

Figure 2-1
Evidence of the conceptual merging of the diffusion research traditions
is provided by increasing cross-tradition citations.

Source: Diffusion Documents Center, Michigan State University, July, 1968.

traditions.* As time goes on, the typical diffusion researcher is more likely to be aware of parallel methodologies and results in other research traditions. And his own investigations, we feel, will profit from this broader perspective. *Diffusion research is thus emerging as a single, integrated body of concepts and generalizations, even though the investigations are conducted by researchers in several scientific disciplines.*

The merging of the diffusion research traditions is a major theme of this chapter, in which we shall present a brief picture of the research approaches used in each tradition. Then we shall argue that a procedure of theory construction and testing known as *middle range analysis* should be used to obtain greater conceptual significance from existing diffusion research findings. This approach essentially consists of accumulating and synthesizing a series of middle range generalizations from empirical findings on the diffusion of innovations. These propositions are at the so-called middle range because they are midway in specificity-generality between empirical data and grand theory. It is assumed that more general theories of social change can be derived most effectively by accumulating in the direction of generality from empirical data to more abstract theory.** The problem in the past has been the lack of rapprochement between research and theory. We see middle range analysis as a way to bridge this gap, to cross the lacunae.

Middle range analysis may be a route toward not only a theory of social change but also toward a more complete integration of various diffusion research traditions. The second theme of this chapter is the potential intellectual profitability of a middle range analysis of diffusion, a point which we hope to demonstrate in the remaining chapters by a series of generalizations (derived from middle range analysis of diffusion findings) which synthesize this huge corpus of empirical research results.

After discussing each of the major diffusion research traditions, roughly in their historical sequence, we turn our attention to a typology of main types of diffusion research, to the contributions and shortcomings

*The index of cross-tradition citations per publication (Figure 2-1) is computed on the basis of the research traditions represented in the footnotes and bibliography of each empirical diffusion publication. Even though there is evidence of greater cross-tradition citing, especially in more recent years, the typical diffusion report cited only about two other traditions (from the seven main traditions) in the mid-1960s.

**The alternative approach, obviously, is to proceed deductively from the grand theories to middle range propositions to gathering empirical data to test these hypotheses. Like Merton (1957), however, we believe a more effective approach to be by induction. Popper (1961, p. 59) also expressed his preference for this direction in the process: "Theories are nets cast to catch what we call 'the world': to rationalize, to explain, and to master it. We endeavor to make the mesh ever finer and finer."

of diffusion research, and close with a description of middle range analysis.

Traditions of Research on Diffusion

In this section we trace the intellectual ancestry of each of the seven major diffusion research traditions.* Our major criterion for the delineation of these traditions is the disciplinary affiliation of the researcher, modified somewhat by the nature of the innovation studied (farm ideas, educational methods, and the like). Table 2-1 shows for each tradition the main types of innovations studied, methods of data gathering and analysis, unit of analysis, and the types of findings. This overview and comparison of the diffusion research traditions is complemented by the narrative description of each tradition which follows. For each tradition we generally present some detail about one classic research study.

Anthropology

The anthropology diffusion tradition is the oldest of the seven traditions. It has had great influence on the early sociology, rural sociology, and medical sociology fields but only limited impact on the other traditions.

In the early days of the anthropology tradition, a major argument raged as to whether diffusion or parallel invention was more important in social change. The debate centered on whether ideas were independently invented in two different cultures or whether they were invented in one culture and diffused to the other.**

*The exact number of major diffusion research traditions is, of course, somewhat arbitrary. We chose these seven because they represent the relatively greatest number of empirical diffusion publications (an exception is the early sociology tradition which is included because of its considerable influence on most of the other traditions which develop later). The seven traditions represent a total of 857 empirical publications available in mid-1968. This figure represents 79 percent of the total empirical reports then available.

**Diffusionism* is the point of view in anthropology that explains change in a society as a result of the introduction of innovations from another society. Some diffusionists claim that all innovations spread from one original source, which, of course, argues against the existence of parallel invention. There are two main schools of diffusionists in anthropology, the German-Austrian and the British. Neither today has much of a following, perhaps because of their extreme view that all change could be explained by diffusion alone. Their contribu-

As the field developed, a great number of anthropological studies were concerned with investigating the introduction of modern Western ideas to primitive societies. Compared to other diffusion traditions, anthropology has generally been more concerned with the exchange of ideas between societies than with the spread of an idea within a society. It has tended to emphasize the social consequences of innovation more than any other diffusion tradition, as exemplified by Sharp's (1952) analysis of the effects of the steel ax on a tribe of Australian aborigines.*

More generally, anthropologists have centered their research concerns upon the connections between culture** and social change. They have shown how the consequences of innovations affect a culture. On the other hand, they have demonstrated that the receivers' culture has much influence upon the decision to adopt an innovation. If the new idea is compatible with existing cultural values, its likelihood of adoption is much greater, and its rate of adoption is more rapid.*** In recent years the anthropology tradition has turned its attention somewhat to analyses of cross-cultural programs of technical assistance. In many of these research reports anthropologists show that the technical assistance planners failed to take fully into account the cultural values of the target audience. Of those change programs analyzed by anthropologists the number that failed probably outnumbers those that succeeded.****

The anthropology tradition contains many investigations of the specific particulars of diffusion in a small social system, often a single peasant village. This is because much anthropological research is a one-man operation, and the investigator is therefore limited to what he himself can observe in a limited setting. The results of such inquiry provide valuable insight into the details of social change at a microcosmic level. But there has been little concern by anthropologists with *generalization*, that is, determining the degree to which their descriptions of change in single villages are applicable to other villages,

tion was chiefly that of calling the importance of diffusion to the attention of other social scientists (Kroeber, 1937, pp. 137–142). As was pointed out in Chapter 1, we see social change as caused by *both* invention and diffusion, which often occur sequentially.
*A brief description of this study appears later in this chapter, and further detail is provided in Chapter 11.
***Culture* consists of material and nonmaterial aspects of a way of life, shared and transmitted among the members of a society.
***See Chapter 4 for a detailed review of evidence for the proposition that an innovation's perceived compatibility is positively related to its rate of adoption. Many of the diffusion studies bearing on this statement are by anthropologists.
****Examples of this type of anthropological report are Dobyns (1951), Bliss (1952), Mead (1955), Erasmus (1961), and Niehoff (1966a).

Table 2-1   Comparison of the Major Diffusion Research Traditions

DIFFUSION RESEARCH TRADITION	NUMBER OF EMPIRICAL PUBLICATIONS AVAILABLE	TYPICAL INNOVATIONS STUDIED	METHOD OF DATA GATHERING AND ANALYSIS	MAIN UNIT OF ANALYSIS	MAJOR TYPES OF FINDINGS
1 Anthropology	69	Technological ideas (steel ax, the horse, water-boiling, e.g.)	Participant and non-participant observation and the case study approach	Tribal or peasant villages	Consequences of innovations; relative success of change agents
2 Early sociology	10	City manager government, postage stamps, ham radios	Data from secondary sources and statistical analysis	Communities or individuals	S-shaped adopter distribution; characteristics of adopter categories
3 Rural sociology[a]	480	Agricultural ideas (weed sprays, hybrid seed, fertilizers, e.g.) and health ideas (vaccinations, latrines, e.g.)	Survey interviews and statistical analysis	Individual farmers in rural communities	S-shaped adopter distribution; characteristics of adopter categories; perceived attributes of innovations and their rate of adoption; communication channels by stages in the innovation-decision process; characteristics of opinion leaders
4 Education	71	Kindergartens, driver training, modern math, programmed instruction	Mailed questionnaire, survey interviews, and statistical analysis	School systems or teachers	S-shaped adopter distribution; characteristics of adopter categories

Table 2-1 Comparison of the Major Diffusion Research Traditions—continued

DIFFUSION RESEARCH TRADITION	NUMBER OF EMPIRICAL PUBLICATIONS AVAILABLE	TYPICAL INNOVATIONS STUDIED	METHOD OF DATA GATHERING AND ANALYSIS	MAIN UNIT OF ANALYSIS	MAJOR TYPES OF FINDINGS
5 Medical sociology	76	Medical drugs, vaccinations, family planning methods	Survey interviews and statistical analysis	Individuals	Opinion leadership in diffusion; characteristics of adopter categories; communication channels by stages in the innovation-decision process
6 Communication	87	News events, agricultural innovations	Survey interviews and statistical analysis	Individuals	Communication channels by stages in the innovation-decision process; characteristics of early and late knowers, of adopter categories, and of opinion leaders
7 Marketing	64	New products (a coffee brand, the touch-tone telephone, clothing fashions, e.g.)	Survey interviews and statistical analysis	Individual consumers	Characteristics of adopter categories; opinion leadership in diffusion
8 Other traditions[b]	227	—	—	—	—
Total	1,084				

[a] The rural sociology tradition actually includes 70 publications that are, strictly speaking, in the subtradition of extension education.
[b] Includes general sociology, agricultural economics, psychology, general economics, geography, industrial engineering, and several others.
Source: Diffusion Documents Center, Michigan State University, July, 1968.

states, or nations. The administrator of a national change agency in a less developed country needs to know how to diffuse new ideas throughout his nation. The anthropological studies tell us much about a little but not much about a lot.

Recently, Niehoff and his associates* attempted to determine more general patterns in the cross-cultural introduction of innovations. Niehoff's approach was to gather and content analyze the main variables in a large number of case studies of change (many of these case descriptions are village studies by other anthropologists). Then, Niehoff determined which variables were most important in explaining the success or failure of these diffusion campaigns. His evidence suggests that the way in which the change agent communicates the new idea to his clients is especially important in determining whether they will adopt it. Niehoff sees his efforts as both a route toward general conclusions from past anthropological case studies and as a guide for planning future studies of change.

Early Sociology

The intellectual tradition that we refer to as "early sociology" traces its ancestry to a French sociologist, Gabriel Tarde (1903), but most of the research publications in this tradition appeared from the late 1920s to the early 1940s. The true significance of the field lies not in its volume of investigations nor in the sophistication of its research methods but in the considerable influence of early sociologists upon later diffusion researchers.

Tarde (1903) proposed several novel notions for testing by later diffusion investigators. He was among the first to suggest that the adoption of a new idea follows a normal, S-shaped distribution over time.** He also argued that the greater cosmopoliteness of innovators is one reason for their early adoption of new ideas. Probably Tarde's greatest contribution was his insight into the process by which the behavior of opinion leaders is imitated by other individuals.

Most early sociologists traced the diffusion of a single innovation over a

---

*Their work is reported in Arensberg and Niehoff (1964), Niehoff (1964a, b; 1966a, b; 1967), Niehoff and Anderson (1964a, b, c), and Niehoff and Niehoff (1966).

**Figure 2-2 is an illustration of these S-shaped adopter distributions. First a small number of individuals adopt the innovation; then there is a rapid rate of adoption followed by a slower rate as the few remaining individuals in the system adopt. A detailed discussion of S-shaped adopter distributions is found in Chapter 5.

geographical area.* The motivating interest of the early sociologists was primarily in the diffusion of innovations which contributed to contemporary social changes. With the exception of Bowers (1937 and 1938), who investigated the diffusion of ham radio sets, early sociologists did not emphasize the innovation-decision process nor did they concentrate upon the process by which opinion leaders influence others in their system to adopt or reject new ideas.

Bowers' (1937 and 1938) investigation was probably the first study in the early sociology tradition to utilize primary data from respondents, in addition to those from available government records. He contacted a sample of 312 ham radio operators in the United States by mailed questionnaire in order to determine the influences that led to their adoption of radios. Bowers (1938) was the first researcher to find that interpersonal channels are more important than mass media channels for later adopters than for earlier adopters.

The number of amateur radio operators in the United States had increased sharply from about 3,000 in 1914 to 46,000 in 1935. Bowers attempted to determine whether this adopter distribution followed a normal shape, and he concluded that it was generally an S-shaped normal curve (except for a plateau near the middle of the distribution). In a fashion similar to others in the early sociology tradition, Bowers traced the relationship of such ecological factors as city size and region to the rate of adoption of ham radios.

Rural Sociology

The research tradition which boasts the largest and most enduring concern with diffusion is rural sociology. The tradition dates from the 1920s when administrators in the U.S. Department of Agriculture launched a series of evaluation studies of diffusion campaigns that had been conducted by state extension services. One handy measure of a campaign's success was the adoption of innovations recommended and promoted by county extension agents. As early as 1925 Wilson and his associates were studying the ratio of innovations adopted to the relative cost of their diffusion (Wilson and Gallup, 1955.)**

*One exception to this is McVoy (1940), who constructed an "index of progressiveness" composed of twelve new ideas in state government. This index was an attempt to measure a general innovativeness dimension; it set a methodological precedent for later researchers, particularly those in rural sociology.
**Wilson's studies in the 1920s certainly influenced later studies in the rural sociology tradition, but actually he was neither trained in rural sociology nor associated with a department of rural sociology at the time of his diffusion studies. In general it is somewhat difficult to distinguish clearly between the rural sociology tradition and the subtradition of extension, as we show later.

After a relatively slow start, the number of studies in the rural sociology tradition mushroomed in the 1950s and 1960s. Ryan and Gross' (1943) analysis of the diffusion of hybrid seed corn led directly to investigations of correlates of innovativeness and to the roles of various communication channels by functions of the innovation-decision process. The hybrid corn study is undoubtedly the most widely known rural sociological inquiry of all time, and even today it ranks as a classic study of diffusion. For this reason we shall review its methods and findings in some detail.

The Hybrid Seed Corn Study

The Ryan and Gross (1943) investigation of the diffusion of hybrid seed corn, more than any other study, influenced the methods, findings, and interpretations of later students in the rural sociology tradition. A total of 259 farmers were interviewed* in two small Iowa communities, Grand Junction and Scranton (Gross, 1942). The innovation studied was the result of years of intensive research by agricultural scientists. The hybrid vigor of the seed did not continue in the second generation; farmers had to purchase hybrid seed each year, whereas previously they had selected their own open-pollinated seed. The major advantage of the innovation was a 20 percent increase in yield. Hybrid seed was not available until about 1928, but it was almost completely adopted by 1941. The diffusion of the new idea was aided by the Iowa Extension Service and the lively commercial interests of seed corn salesmen.

The major findings** from the hybrid corn study are:

1. First use of hybrid seed followed a bell-shaped (but not exactly normal) distribution when plotted over time (Ryan and Gross, 1943). Four adopter categories were classified on the basis of first use of hybrid seed (Gross, 1942).

---

*At the time of the inquiry Neal Gross was a young graduate student in rural sociology at Iowa State University. He actually knew little about rural people and nothing about securing information from them in research interviews. His first research task, Ryan told him upon his arrival in Ames, would be to interview farmers concerning their adoption of hybrid corn. Someone advised Gross that Iowa farmers arose at 6 A.M., and so the next morning he was waiting at daybreak in the barnyard of his first respondent. The story goes that by that evening, he had completed twenty-one research interviews. In fact, Gross averaged about fourteen interviews a day during the data-gathering period in 1941! This is particularly amazing when one considers that a rate of four interviews a day is considered average by modern survey researchers.

**Publications from the study are Gross (1942 and 1949), Ryan and Gross (1943 and 1950), Gross and Taves (1952), and Rohwer (1949). These reports are supplemented through private communication with both Professor Ryan and Professor Gross by the senior author.

The social characteristics, such as age, social status, and cosmopoliteness,*
of both the earliest and the latest adopters were then determined.

2. The innovation-decision period from first knowledge to the adoption
decision averaged about nine years for all respondents, a finding which led to
a clearer realization that the innovation-decision process involved considerable
deliberation by most adopters, even in the case of an innovation with spectacular results.

3. The typical farmer first heard of hybrid seed from a salesman, but
neighbors were the most frequent channel leading to persuasion. Salesmen
were more important channels for earlier adopters, and neighbors were more
important for later adopters.

In the hybrid seed corn investigation no analysis of opinion leadership of
diffusion of the idea was attempted, although the sample design, which consisted of a complete enumeration in two communities, would have made the
use of sociometric questions easily possible. "Information was simply collected
from all community members as if they were unrelated respondents in a
random sample" (Katz and Levin, 1959).

In spite of this possible shortcoming, however, the scope and depth of the
hybrid corn study is impressive for its time. A number of subsequent researchers made important advances on the basis of leads set forth by Ryan and
Gross (Katz, 1961). However, a great number of later rural sociological
studies have followed an unimaginative "variables-related-to-innovative-
ness" approach. The results add very little in many cases to present knowledge
of how new ideas diffuse except further verification of previous findings,
which is often unnecessary.

Later Developments in the Rural Sociology Tradition

Since the mid-1950s there has been a great proliferation of research
studies by rural sociologists.** Most of these studies have been financed by
state agricultural experiment stations or the U.S. Department of Agriculture
(and also in recent years by commercial agricultural companies). Federal and

---

*In fact, the hybrid corn study was one of the first to establish the relationship between
cosmopoliteness and innovativeness. For an up-to-date report on the research evidence for
this generalization, see Chapter 5 and Appendix A.

**Crane (1968c) explains this spurt in the rural sociology diffusion tradition on the basis of
her analysis of sociometric data from a sample of rural sociologists: "A few individuals
developed a high degree of commitment to it [diffusion] and as a result were able to direct
the activities of others in the field and to make it viable as a research area." See Crane (1969)
for further detail on the growth of the rural sociology diffusion tradition.

state agencies spend sizable sums for research on agricultural technology. Their administrators are convinced of the value of sociological inquiry to trace the diffusion of these research results to farm people. Most rural sociologists are employed by state agricultural universities, and the proximity of these sociologists to state extension services has affected the tradition by leading to a strong interest in applied research.

In the 1960s several U.S. rural sociologists completed research studies in foreign countries using essentially similar methods to those used in the U.S. The contemporary international outlook in the rural sociological tradition provides a cross-cultural test of the propositions generated by research in the U.S. (and abroad). The primary focus of rural sociologists upon *agricultural* innovations has been especially appropriate in less developed nations, where problems of agricultural development are usually paramount.

The trend to internationalization of the tradition in the 1960s has also contributed to increased attention by rural sociologists to the diffusion of health and family planning innovations. This type of inquiry is facilitated by the fact that the respondents in these overseas studies are villagers, as is true with the agricultural diffusion research in less developed nations.

One general criticism of the rural sociology tradition which has been voiced by rural sociologists themselves* is its lack of attention to sociological theory. There is a noticeable tendency for many rural sociology diffusion studies to approach raw empiricism, with little emphasis upon the sociological significance of the findings.

The rural sociology tradition probably has received more favorable evaluation from outsiders than from its own members.** For instance Katz (1960) states that rural sociology ". . . is an island of communications research deriving from a sociological tradition which has taken account of the fact that 'farmers talk to other farmers' and that such interaction has consequences for the response of individuals and groups. . . ."

Rural sociologists have been especially active diffusers of the diffusion approach to other fields, such as education, public health, and marketing. One

---

*An example is Lionberger (1952 and 1960), who stated that rural sociological diffusion research is "for the most part characterized by a dearth of sociological theory."

**In an otherwise generally caustic review of the sociological contributions of rural sociology, Anderson (1959, pp. 370–371) evaluated diffusion research in favorable terms: "In one area, rural sociologists are breaking new ground today; diffusion of technical practices. . . . This may well be the most important topic that rural sociologists have ever studied. It is 'realistic' and appeals to farmers, administrators, extension workers, and technologists." Loomis and Loomis (1967) also see diffusion research as one of the main contributions of rural sociology.

result is that the specific research methods satisfactory for the study of farmers' adoption of hybrid corn have occasionally been utilized without complete adaptation in other research contexts.* Much of the total body of diffusion research, especially such matters as the innovation-decision studied, the reliance upon recall data from personal interviews, and the individual variables correlated with innovativeness, bears the indelible stamp of its intellectual ancestry in rural sociology. And this academic inheritance is frequently inappropriate for the varied settings in which diffusion research is conducted today.

Extension Education

An important and partially distinct subtradition of rural sociology has appeared in recent years. The representatives of this subtradition are usually found in agricultural universities, on the staffs of agricultural extension services, or else in departments of extension education. This is true in almost every state in the U.S. and overseas, especially in countries like India, where there are such departments in almost every agricultural college. The primary concern of extension educators is the training of extension change agents. To make this training more appropriate, as well as to improve the efficiency of extension diffusion campaigns, members of this subtradition rather naturally become involved in diffusion research. Their studies have usually been designed to evaluate the effects of extension diffusion efforts.

The extension education subtradition is heavily influenced by rural sociology, so much so that we find it impossible clearly to separate the two. Most extension educators have received graduate training in rural sociology, and their research methods, concepts, and even the type of innovations and respondents that are studied are similar. There are seventy empirical publications in the extension education subtradition, which is 7 percent of the total in the larger rural sociology tradition.

Education

One of the larger traditions in terms of the number of studies,** education is one of the lesser traditions in terms of its contributions to understanding the

---

*This and other closely-related biases in diffusion research are discussed in a later section of the present chapter.

**The number of empirical diffusion publications is especially large if one also includes the 150 studies listed in a review of the field by Ross (1958).

diffusion of innovations or to a theory of social change. Carlson (1968) concurs in this rather low evaluation of educational diffusion: "Data collection on acceptance has not been characterized by rigor. . . . Given this weak base, it is rather difficult to count on what is known about the diffusion of educational innovations." But there are certain novel aspects of educational diffusion (for instance, that it occurs in formal organizations) that mark its *potential* contribution as exciting.

The Teachers College Studies

A majority of early educational diffusion studies were completed at one institution, Columbia University's Teachers College, and under the direction of one man, Paul Mort. The roots of this tradition trace to early research (in the 1920s and 1930s) by Mort and others on local control over school financial decisions, and on whether this local control led to school innovativeness. Mort and Cornell (1938) state: "To operate schools today in terms of the understandings of a half century ago is to waste school funds and school time. Adaptability, or the capacity to meet new needs by taking on new purposes and new practices, is indispensable to the effective functioning of any school system." This statement, and the contents of any of the Columbia University education diffusion publications,* shows that innovativeness in schools was welcomed as desirable.**

The data were most often gathered by mailed questionnaires from school superintendents or principals.*** The unit of analysis was the school system in almost all these investigations. A number of central findings have emerged from the Columbia University diffusion studies which may be summarized as follows:

---

*Most of the Columbia University research studies were published as doctoral theses or as Teachers College reports.

**As was local school control. Mort and Cornell (1938) state: "For more than a decade the senior author of this book has cast his lot with those seeking to maintain local initiative in the control and financial structure of states moving toward central financing in education. . . . He took this position in the faith that local initiative contributes to adaptability and that it should therefore not be destroyed without a demonstrably effective substitute."

***It is interesting to note that the financial support for these studies came largely from the public schools being studied! The total research budget for the Columbia University education diffusion investigations was over one-quarter million dollars in 1959. Most of this financial support was donated by an annual fee from each member of the Metropolitan School Study Council, mostly public schools in the New York City area, and member schools in the Associated Public School Systems, located throughout the United States.

1. The best single predictor of school innovativeness is educational cost per pupil. The wealth factor almost appears to be a necessary prerequisite for innovativeness among public schools.*

2. A considerable time lag is required for the widespread adoption of new educational ideas. "The average American school lags 25 years behind the best practice" (Mort, 1953, pp. 199–200). Why is the diffusion of educational ideas slower than that of farm innovations or medical drugs? It may be because of (1) the absence of a scientific source of innovations in education, (2) the lack of change agents to promote new educational ideas, and (3) the lack of economic incentive to adopt.

There is, of course, a wide range in the rate of adoption of educational ideas, as is shown in Figure 2-2. These data raise the question of why modern math diffused so much more rapidly than kindergartens or even driver training. Likely reasons include (1) the aftermath of Sputnik, which caused widespread dissatisfaction with U.S. public education and has led since 1958 to massive federal programs of financial assistance to encourage educational innovations; (2) the active promotion of modern math (and driver training, to a lesser degree) by a prestigious and powerful change agency (thus, modern math is an example of directed change, whereas kindergartens illustrate selective contact change); and (3) the contemporary value placed by the public on change in education, which was not so strong prior to the late 1950s.

3. The pattern of adoption of an educational idea over time approaches an S-shaped curve.** At first only a few innovator schools adopt the idea, then the majority decide the new idea is desirable, and finally the adopter curve levels off as the last remaining schools adopt (Figure 2-2).

Later Studies on Educational Diffusion

After Paul Mort's death in 1959, Teachers College of Columbia University lost its monopoly control on educational diffusion. Recent studies have focused (1) upon teachers as respondents, rather than simply on administrators, (2) on within-school, as well as school-to-school diffusion, and (3) on educational change in less developed nations. An illustration of the latter trend is an

---

*Although this relationship of wealth to innovativeness was not supported by the results of Carlson's (1964) study of the diffusion of modern math, perhaps because much of the cost of adoption was provided by change agencies rather than the adopting schools.
**Studies supporting this statement are Farnsworth (1940), Mort and Cornell (1941), Cocking (1941), Barrington (1953), Lovos (1955), and Adler (1955).

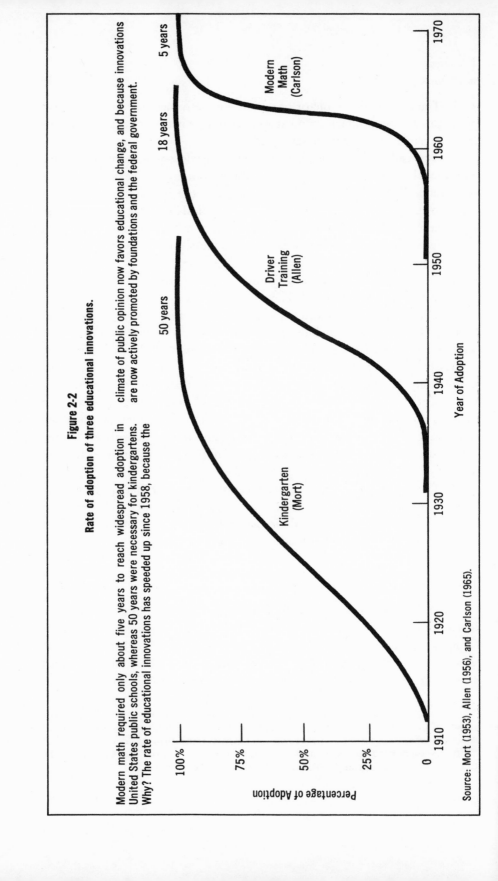

Figure 2-2

Rate of adoption of three educational innovations.

Modern math required only about five years to reach widespread adoption in United States public schools, whereas 50 years were necessary for kindergartens. Why? The rate of educational innovations has speeded up since 1958, because the climate of public opinion now favors educational change, and because innovations are now actively promoted by foundations and the federal government.

Source: Mort (1953), Allen (1956), and Carlson (1965).

investigation of the diffusion of educational innovations in Thailand by Rogers and others (1968). The Minister of Education in Thailand gives orders for adoption of new educational ideas, such as objective testing, classroom discussion (rather than lecture), school libraries, and audio-visual methods. These mandates are transmitted via a hierarchy of officials to local school teachers. Thus, the diffusion researchers were led, on the basis of the Ministry's organizational chart and how it supposedly functioned, to expect a downward flow of innovations. Their research, however, showed that most of the new ideas flowed *upward* in the bureaucracy. The reason, apparently, is that most teachers learned of the innovations from seeing them used by their professors in Thai colleges of education. These professors had received graduate training in the U.S. or Europe where they were exposed to the new ideas. The innovations then flowed into local school systems from the young, newly trained teachers to their older colleagues and to some of the administrators. Therefore, the diffusion patterns went upward rather than downward.

Probably the best piece of educational diffusion research is Carlson's (1965) analysis of the communication of modern math among school administrators in Pennsylvania and West Virginia. He studied the opinion leadership patterns in the diffusion of modern math among school superintendents, variables correlated to innovativeness, perceived characteristics of innovations and their rate of adoption, and the consequences of one educational innovation.*

One of the distinctive aspects of educational diffusion is that it often occurs within bureaucratic structures.** Many more of the innovation-decisions are authority or collective decisions rather than optional innovation-decisions, but most of the past research has treated educational innovations as if they were individually adopted, even though many are not. A further shortcoming of the education tradition is that researchers have largely ignored (1) considera-tion of communication channels,*** and (2) how the social structure acts to impede or facilitate diffusion. These conceptual shortcomings need to be overcome in future studies of educational diffusion.

---

*Further detail on the Carlson study is found in Chapters 6 and 11.

**Carlson (1968), in his critique of educational diffusion, notes: "The startling fact is that even though research has taken the school systems as the adopting unit, very limited attention has been paid to concepts related to organizational theory. . . . The fact that school systems are organizations has been generally overlooked."

***Carlson (1968) agrees: "Communication has been neglected in adoption studies of educational innovations. . . . Overall, the neglect of communication is rather awesome."

Medical Sociology

This diffusion tradition began in the 1950s, about the same time that medical sociology began to be recognized as a field of sociological specialization.* The innovations studied have consisted of (1) either new drugs or medical techniques, where the adopters are doctors, or (2) polio vaccine, family planning methods, or other medical innovations, where the adopters are clients or patients.

Columbia University Drug Study

The classic study in this tradition was completed by three sociologists, Elihu Katz, Herbert Menzel, and James Coleman, then of Columbia University's Bureau of Applied Social Research.** The significance of this investigation, hereafter referred to as the "drug study," is almost comparable to that of the Ryan and Gross analysis of hybrid corn, in terms of its contribution to our knowledge of the diffusion of new ideas.

One of the Bureau's alumni was the director of market research for a large pharmaceutical firm, Charles Pfizer and Company, in New York. The company provided a grant of about $40,000 to the Bureau for the project, which began in 1954. A pilot study of the spread of a new drug among thirty-three doctors in a New England town was carried out in May, 1954 (Menzel and Katz, 1955). The main investigation was conducted, after methodological techniques had been pretested in the pilot study, in four cities in Illinois in late 1954.***

The drug study analyzed the diffusion of a new antibiotic that had appeared in late 1953. The innovation was referred to by the Bureau researchers in most of their published reports by a pseudonym, "gammanyn." The drug had been

*It is important to note that many diffusion researchers in the medical sociology tradition do not necessarily identify themselves as "medical sociologists", and some are affiliated with university schools of public health, rather than departments of sociology.

**This research center is one of the most noted for research on mass communication. Sociologists like Paul Lazarsfeld and Robert Merton pioneered early studies of the two-step flow of communication and opinion leadership, local and cosmopolitan leadership patterns, and the importance of interpersonal communication channels in consumer decisions. The Bureau was fertile ground for conduct of the drug study, and even though the project was not designed with much knowledge of existing diffusion frameworks, it did capitalize on the academic tradition of communication research expertise at the Bureau.

***The Illinois data are reported in Menzel and others (1959), Coleman and others (1957, 1959, and 1966), Katz (1956a, 1957, and 1961), Katz and Levin (1959), and Menzel (1957, 1959, and 1960). The present discussion features data mainly from the four Illinois cities, rather than from the pilot study in New England.

tried at least once by 87 percent of the Illinois doctors, who had been making extensive use of two other closely related "miracle" drugs belonging to the same antibiotic family as gammanyn. The new drug superseded an existing idea just as hybrid corn replaced open-pollinated seed.

It is, of course, the patient and not the doctor who pays for a costly new drug, although it is the doctor who makes the innovation-decision. The Bureau sociologists interviewed 125 general practitioners, internists, and pediatricians in the four Illinois cities. These were 85 percent of the doctors practicing in specialities where "the new drug was of major potential significance" (Coleman and others, 1957). These 125 respondents sociometrically designated in their cities at least 103 additional doctors in other specialities who were also interviewed. Whereas many of the findings from the drug study are based upon the sample of 125 physicians, the sociometric analyses of opinion leadership come from the responses of the total sample of 228 doctors,* which constituted 64 percent of all doctors in active private practice in the four cities (Coleman and others, 1957).

One of the neat methodological twists of the drug study was the use of an *objective measure* of time of adoption from the written record of drugstore prescriptions. This is the only study where the researchers were not forced to depend upon recall-type data on innovativeness.** There was, in fact, a marked tendency for many doctors to report having adopted the drug earlier than the prescription records indicated (Menzel, 1957).

Hawkins (1959a) criticized the drug study for its inadequate review of existing literature*** in the drug diffusion field. The Bureau investigators certainly were not aware of other research traditions on diffusion at the time the gammanyn data were gathered. The researchers make no secret of their surprise upon discovery of the hybrid seed study. Katz (1961) states "The

---

*There appears to be some ambiguity as to the total number of doctors interviewed. The sample size is variously reported as 228 by Coleman and others (1957) and 216 by Katz (1956, p. 185). These differences have never been explained by the Bureau researchers, but they may be the result of discarding certain interview schedules for various reasons, such as incomplete data.

**But the drug study methodology, although eliminating one source of inaccuracy in adoption dates of gammanyn, introduced another. The three-day sampling period out of each month leaves something to be desired. What of the doctor who prescribed the drug on one of the twenty-seven days each month not sampled but did not prescribe it during the three days that were sampled? Nevertheless, the use of prescription records was probably a major improvement in accuracy over recall data. It is unfortunate that the researchers used only a three-day sample per month, but the volume of effort involved in a complete enumeration of prescriptions evidently prohibited a complete audit of the drug records.

***Such as Caplow (1952), and Caplow and Raymond (1954).

drug study was completed only a few years ago without any real awareness of
its many similarities to the study that had been undertaken by Ryan and Gross
almost fifteen years before."

Three findings from the drug study with greatest significance to the present
book are:

1   The detailed sociometric data secured from the physicians allowed an
    analysis in depth of the patterns of opinion leadership through which
    gammanyn spread in the medical community. A positive relationship
    between opinion leadership and innovativeness was established.

2   Variables correlated to innovativeness (such as cosmopoliteness, social
    status, size of operation, and communication behavior) were studied
    for a different kind of respondent (physicians) than had been previously
    investigated.

3   The third major contribution of the drug study is the methodological
    technique of determining the date of doctors' use of gammanyn from
    prescription records, rather than only from recall.

Later Studies in the Medical Sociology Tradition

Since the classic investigation of drug diffusion, a considerable number of
other studies have been completed in the medical sociology tradition. These
include inquiries on the adoption of polio vaccine and other health practices
by the public and researches on the diffusion of family planning methods. The
latter type of study has often been conducted in less developed nations, where
population pressures on food are especially serious and much emphasis is given
to family planning campaigns. The number of family planning communication
studies has perhaps increased faster than any other type of diffusion research
in the 1960s, but the intellectual contribution of these studies has been less
spectacular. In fact most of the independent variables related to knowledge or
adoption of family planning innovations are such demographic dimensions as
age, family size, education, and the like. Social-psychological factors, communication, and such economic variables as financial incentives for adoption
have been neglected.*

One exception to this generally dreary picture is a field experiment on the
diffusion of family planning ideas in Taiwan, conducted by Berelson and

*Rogers and Bettinghaus (1966) are highly critical of the past contribution of family
planning diffusion research (when compared to its potential). They advocate more adequate
use of diffusion concepts and propositions other than those of family planning in designing
future research on the communication of contraceptives.

Freedman (1964).* The design of this study is especially noteworthy because it is a field experiment, that is, an experiment conducted in the "real world," rather than the laboratory. In a field experiment data are gathered from a sample of respondents at two points in time by means of a benchmark and a follow-up survey. Soon after the benchmark survey, a treatment (or treatments) is applied to the sample. The effects of the treatment can be determined by measuring the change in some variable (for instance, adoption of innovations) between the benchmark and the follow-up survey. The main advantage of field experimental designs is that they allow the researcher to determine the *time-order* of his independent (treatment) variable on the dependent variable. There has been a trend toward the field experimental approach in diffusion research in very recent years,** and the Berelson-Freedman study in Taiwan is one of the best, as well as one of the biggest: "This effort . . . is one of the most extensive and elaborate social science experiments ever carried out in a natural setting" (Berelson and Freedman, 1964).

The researchers applied four different communication treatments to approximately 2,400 neighborhoods (each of 20 to 30 families) in one city in Taiwan:

1   Neighborhood meetings with leaders and distribution of posters about family planning.
2   Posters and meetings, plus mailed information about the innovation to newlyweds and to couples with two or more children (who were most likely to be interested in adopting family planning methods).
3   Posters and meetings plus a personal visit to the home of all married women aged twenty to thirty-nine by a trained nurse–midwife who provided contraceptives, answered questions and sought to persuade the women to adopt.
4   Posters and meetings but with personal visits by the change agents to both husbands and wives.

The innovations promoted in this experiment included the usual types of contraceptive techniques (foam tablets, diaphragm, condoms, rhythm, and

*A number of other publications report details on this research: Freedman (1964), Freedman and others (1964), Freedman and Takeshita (1965 and 1969), Gillespie (1965), Takeshita (1964 and 1966), and Takeshita and others (1964).
**Of 1,084 empirical diffusion investigations in the Diffusion Documents Center at Michigan State University, sixty-five were field experiments and sixty-seven were panel studies over time. The latter can also provide data indicating the time-order of variables. Forty-six (71 percent) of the sixty-five field experiments were completed after 1960.

oral pills), plus a then-new method, the intrauterine device (IUCD), a plastic loop or coil inserted into the woman's uterus.

The results of this diffusion experiment were truly spectacular: 40 percent of the eligible audience of about 10,000 women adopted some form of family planning. Pregnancy rates immediately decreased by about 20 percent. Most of the innovations (78 percent) adopted were IUCDs. Did the adopters later discontinue? A check-up six months later indicated that about 20 percent of the IUCD adopters had removed the innovations (or else the IUCDs had been involuntarily expelled from the woman's body).

The results of the Taiwan diffusion experiment provided reason for optimism among change agents responsible for nationwide population control programs which were being initiated in countries like Korea, India, Turkey, and Pakistan. In the years since the Berelson-Freedman study, however, we have come to realize that it is another matter to secure comparable results to those achieved in Taiwan. This is because the Taiwan experiment involved very intensive efforts (such as the change agents' visits to families), which require more resources than can usually be mustered on a national basis. Further, negative rumors about the IUCD,* which were rare in the Taiwan city, are common in India, Costa Rica, and many other nations, and a high rate of discontinuance of the IUCD usually results.

Communication

A recent diffusion research tradition, which has been recognized only since 1962, is communication. There are about eighty departments or institutes of communication in U.S. universities today, and a number are springing up in other nations. They have arisen largely in the past twenty years in response to a need for applied social science research on human communication problems. Departments of journalism, speech, theater, audio-visual education, advertising, and television and radio broadcasting have, of course, existed for some years. These departments mainly teach applied communication skills to undergraduates; they are primarily concerned with producing professional communicators. Professors in these academic settings realized that the training which they provide would be much more useful and more academically respectable if it were based on scientific research results. Departments and

---

*The exact nature of these rumors is somewhat different in each locale, but they usually involve an (untrue) anecdote about a husband and wife who are stuck together during intercourse because of the IUCD. In some versions, the husband's organ must be amputated; in others the wife dies. In any case, the result of the rumor is a high rate of discontinuance of the innovation.

institutes of communication were created to conduct such inquiry and to train communication researchers. Such communication departments typically engage in both experimental laboratory research on interpersonal communication, and in survey research of a more sociological nature. These field investigations typically deal with the interfaces of mass media and interpersonal communication channels in changing human behavior.

One of the important concerns of these communication researchers is the diffusion of news events carried by the mass media. About thirty-five such studies have been completed, dealing with such news as Alaskan statehood, Russia's launching of Sputnik,* and President Kennedy's assassination.** The results show that news events diffuse in a generally similar fashion to technological innovations: The distribution of knowers over time follows an S-shaped curve, interpersonal and mass media channels play comparable roles, and so on. One difference from technological innovation diffusion is that news events spread much more rapidly; for example, 68 percent of the adult U.S. public was aware of the events in Dallas within thirty minutes of the shot that felled President Kennedy (Sheatsley and Feldman, 1964).

Since the early 1960s, communication researchers have investigated the transmission of technological ideas, especially agricultural, health, educational and family planning innovations in less developed nations. Deutschmann's study of the diffusion of innovations in a Colombian village*** stands as a landmark and has led to one focus of communication research upon peasant audiences. Latest and largest of the studies in this tradition is a research project on the diffusion of technological innovations to peasants in Brazil, Nigeria, and India.****

One of the special advantages possessed by the communication research tradition is that it can analyze *any* particular type of innovation. There are no limitations like the education tradition's focus on educational innovations,

*Only one or two studies of news event diffusion, such as the spread of information about Sputnik's launching in Santiago, Chile (Hamuy and others, 1958), were completed outside of a mass media saturated society such as the United States.

**Greenberg's (1964a) analysis of the diffusion of the news of the Dallas assassination is somewhat typical of the approach used in the news event diffusion studies; it is reviewed later in the present chapter. Probably the most noted news event diffusion study, however, is Deutschmann and Danielson (1960).

***The publications from this study are: Deutschmann (1963), Deutschmann and Fals Borda (1962a, b), and Deutschmann and Havens (1965).

****This project, sponsored by the U.S. Agency for International Development, was conducted by the Department of Communication at Michigan State University from 1964 to 1968. A total of 10,000 peasants, village leaders, and change agents were interrogated in the three nations; see Rogers and others (forthcoming) for details.

the rural sociologist's emphasis upon agricultural ideas, or the medical sociologist's concern with family planning methods.* This lack of a *message content orientation* allows the communication researcher to concentrate on the *process* of diffusion, by way of comparing results across a number of different types of innovations. Further, the communication tradition comes equipped with an appropriate toolkit of highly useful concepts and methods (for example, credibility, cognitive dissonance, and the semantic differential) for studying diffusion. In fact, the rapid and enthusiastic way in which some communication scientists have taken to diffusion research makes one wonder why they did not do so sooner.** The multidisciplinary and interdisciplinary backgrounds of the communication field lead one to expect representatives of this tradition to play an important role in the further integration of the various diffusion traditions.

Marketing

Another diffusion tradition that has come on strong in the 1960s is marketing. Marketing managers of firms in the U.S. have long been concerned with how to launch new products most efficiently. Their interest in this topic is sparked by the appearance of large numbers of new consumer products and by the high rate of failure of such products. For instance, it is estimated that only one idea out of every 540 results in a successful new product (Marting, 1964, p. 9). Further, only 8 percent of the approximately 6,000 new consumer items introduced each year have a life expectancy of even one year (Conner, 1964). Commercial companies, therefore, have a vital stake in the diffusion of new products, and a great number of such researches have undoubtedly been completed. However, a large proportion of the research reports are found only in the secret files of the sponsoring companies because of the threat of competitive advantage.

It was not until the early 1960s, when diffusion research began to receive added attention from university faculty members in graduate schools of marketing, that the publicly available literature in the marketing diffusion tradition began to increase. Many of these researches by university professors were conducted with the sponsorship, or at least the cooperation, of the companies manufacturing the new products being studied. Thus, the mar-

---

*In fact, communication researchers have investigated educational, agricultural, and family planning innovations, as well as many others.
**A discussion of the convergence of the diffusion approach with communication research is provided by Rogers (1967b).

keting tradition has an especially strong bias toward producing research results of use to the innovation's source, rather than to the receivers.* But one advantage of this cooperation with the change agency is that marketing researchers often have some control over the diffusion strategies that are used and that will be analyzed. This is a particularly important advantage in the conduct of field experiments.

An illustration of such a field experimental approach in the marketing tradition is provided by Arndt's (1967a) study of a new food product.** He sent a letter about the innovation enclosing a coupon allowing its purchase at one-third price to the 495 housewives living in a married student apartment complex. Personal interviews were then conducted with these consumers after sixteen days of the diffusion campaign. Arndt found that interpersonal communication about the innovation frequently led to its purchase, especially if the interpersonal messages about the new product were favorable. Housewives who perceived the innovation as risky were more likely to seek the advice of opinion leaders about it. Word-of-mouth messages seemed to flow from the low-risk to the high-risk perceivers.

Marketing is an applied field of behavioral science with strong roots in economics. Marketing researchers have been leaders in investigating the effects of economic incentives (as well as advertising campaigns) on the rate of diffusion. The marketing tradition has borrowed heavily from the older diffusion traditions, especially rural sociology. But it has not yet gone international to the extent of the other traditions, perhaps because marketing is not yet a well developed field of study in universities in less developed countries.

Other Traditions

In addition to the seven main diffusion research traditions that we have just discussed, there are a number of minor traditions. Mostly they have developed in recent years. Examples of the minor traditions are:

1. *Agricultural economics* is represented by thirty-nine (3.6 percent) empirical diffusion publications and is similar to the rural sociological approach, although with a primary emphasis on economic variables that explain diffusion and adoption.***

*A similar tendency, although in lesser degree, characterizes most diffusion research, as we show later in this chapter.
**Other publications from this investigation are Arndt (1966a, b, c; 1967a, b).
***It is unfortunate that general economists and agricultural economists have not been

2. *Geography* is represented by seven (0.6 percent) publications, most of them dealing with the simulation of innovation diffusion, with a main emphasis upon spatial variables. This tradition was begun by Torsten Hägerstrand (1952, 1953, 1965a, b, and 1968) in Sweden, and an enthusiastic coterie of geographers is continuing his approach in the U.S. today.

3. *General economics* is represented by sixteen (1.5 percent) empirical diffusion publications and focuses heavily upon the relationships of economic variables to rates of adoption.

4. *Speech* became involved in diffusion as an outgrowth of linguistic interest in tracing the spread of new speech forms. An example is Davis and McDavid's (1949) investigation of the word "shivaree" as it spread in the United States.

5. *General sociology*, a tradition with eighty-seven (8.0 percent) diffusion research publications, is a miscellaneous category for sociologists other than early, rural, and medical sociologists.*

6. *Psychology* is represented by twenty-three (2.1 percent) diffusion publications; many of these studies deal with individual personality variables in diffusion.

We can expect a further proliferation of minor diffusion traditions in the future as the diffusion approach continues to spread to other disciplines with an interest in communication and change. But at the same time we expect a more complete merger of the existing traditions, at least at the conceptual and methodological level. Diffusion researchers in all disciplines will increasingly realize that they are toiling in the same vineyard, even though their grape varieties and their gardening tools differ considerably.

A Typology of Diffusion Research

When showing a large city to a stranger, it is often wise to take him first to the top of a skyscraper so that he may scan the entire landscape prior to being immersed in the details of the city. Likewise, in the present section we hope to

more highly involved in diffusion research, especially in probing the economic characteristics of adopter categories and the role of economic variables in explaining an innovation's rate of adoption. Only 5.1 percent of the empirical diffusion publications were completed by general economists and agricultural economists.

*And for this reason we do not consider such a heterogeneous category as one of our seven main diffusion traditions.

provide the reader with an overall impression of types of diffusion research before moving to a more detailed discussion in later chapters. Our present concern differs from the previous section in that we now shall look at *types* of diffusion research, rather than at the *traditions*.

Table 2-2 shows eight different types of diffusion analysis that are completed or possible and the relative amount of attention paid to each in past inquiry. By far the most popular diffusion research topic has been variables related to individual innovativeness (type 4 in Table 2-2). More than half (58.4 percent) of all the empirical generalizations reported in publications in the Michigan State University Diffusion Documents Center deal with innovativeness.

Perhaps we can make Table 2-2 more intelligible by describing one empirical investigation to illustrate each of the eight types of analysis:

1. *Rate of adoption of an innovation in a social system.* The unit of analysis is the innovation in a study by Fliegel and Kivlin (1966b), which was conducted among 229 Pennsylvania dairy farmers. The investigation utilized farmers' perceptions of fifteen attributes of each of thirty-three dairy innovations to predict the rate of adoption for this sample of Pennsylvania farmers. Innovations perceived as most economically rewarding and least risky were adopted more rapidly. The complexity, observability and trialability of the innovations were less highly related to the rate of adoption, but innovations which were more compatible with farmers' values were adopted more rapidly.

2. *Rate of adoption in different social systems.* Coughenour (1964c) sought to explain the rate of adoption of farm innovations in twelve communities in Kentucky. Farmers in some of these systems had adopted (on the average) many of the innovations, whereas in other communities the members had a much lower level of adoption. Coughenour utilized such variables as the communities' median level of farmers' education, of farmers' mass media exposure, and the degree to which opinion leaders had higher mass media exposure than their followers, to explain the different rates of adoption in the communities.*

3. *Perceived attributes of innovations.* Kivlin and Fliegel (1967a) contrasted two samples of Pennsylvania dairy farmers, one of middle-sized farmers and the other of small-sized farmers, to determine their differences in perceptions of innovations. The small farmers (those with less than sixteen milk cows)

---

*Coughenour's (1964c) inquiry is an illustration of the use of system variables (see Chapter 1), but he used them to predict rate of adoption (also a system variable) rather than to predict individual innovativeness of the system's members, as has been done by researchers in category 4 (Table 2-2).

Table 2-2 Types of Diffusion Research Analysis Completed or Possible

TYPE	DEPENDENT VARIABLE	INDEPENDENT VARIABLES	UNIT OF ANALYSIS	NUMBER AND PERCENTAGES OF GENERALIZATIONS OF THIS TYPE IN THE MSU DIFFUSION DOCUMENTS CENTER	CHAPTER IN THIS BOOK DEALING WITH THIS TYPE OF RESEARCH
1	Rate of adoption of an innovation in a social system	Attributes of innovations (e.g., complexity, compatibility, etc.) as perceived by members of a system	Innovations	82 (1.2%)	Chapter 4—Perceived Attributes of Innovations and Their Rate of Adoption
2	Rate of adoption of innovations in different social systems	System norms; characteristics of the social system (e.g., concentration of opinion leadership); change agent variables (e.g., their strategies of change); types of innovation-decisions	Social systems	159 (2.3%)	Some attention is given in Chapter 7—The Change Agent; and also Chapter 9—Collective Innovation-Decisions
3	Attributes of innovations as perceived by members of a social system (e.g., complexity, compatibility, etc.)	Innovativeness and other characteristics of members of a social system	Members of a social system	0[a]	Some attention is given in Chapter 4—Perceived Attributes of Innovations and Their Rate of Adoption
4	Innovativeness of members of a social system	Characteristics of members (e.g., cosmopoliteness); system norms and other system variables; communication channel usage	Members of a social system	3,974 (58.4%)	Chapter 5—Adopter Categories

[a]There are no generalizations in which the dependent variable is attributes of innovations *per se*, although the Kivlin and Fliegel (1967a) study of larger and smaller-sized farmers is a close approximation.

Table 2-2 Types of Diffusion Research Analysis Completed or Possible—continued

TYPE	DEPENDENT VARIABLE	INDEPENDENT VARIABLES	UNIT OF ANALYSIS	NUMBER AND PERCENTAGE OF GENERALIZATIONS OF THIS TYPE IN THE MSU DIFFUSION DOCUMENTS CENTER	CHAPTER IN THIS BOOK DEALING WITH THIS TYPE OF RESEARCH
5	Earliness of knowing about an innovation by members of a social system	Characteristics of members (e.g., cosmopoliteness); system norms and other system variables; communication usage	Members of a social system	301 (4.5%)	Chapter 3—The Innovation-Decision Process
6	Opinion leadership in diffusing innovations	Characteristics of members (e.g., cosmopoliteness); system norms and other system variables; communication channel usage	Members of a social system	220 (3.2%)	Chapter 6—Opinion Leadership and the Multi-Step Flow of Ideas
7	Communication channel use (e.g., whether mass media or interpersonal)	Innovativeness and other characteristics of members of a social system (e.g., cosmopoliteness); system norms; attributes of the innovations	Members of a system (or the innovation-decision)	458 (6.7%)	Chapter 7—The Change Agent; and Chapter 3—The Innovation-Decision Process
8	Consequences of the innovation	Characteristics of members and the nature of the social system	Members or social systems	16 (0.2%)	Chapter 10—Authority Innovation-Decisions and Organizational Change; and Chapter 11—Consequences of Innovations
Others	—	—	—	1,601 (23.5%)	
			Total	6,811 (100%)	

were slower to adopt new ideas than were the larger farmers (those with six-teen to forty-nine cows). The slower rate of adoption resulted not only from differences in the farmers' perceptions of the innovations but also from their differences in economic resources. The perceived attributes of innovations accounted for 69 percent of the variance in rates of adoption for the small farmers and 51 percent for the larger farmers. The small farmers were quicker to adopt those innovations they perceived as decreasing discomfort, whereas the larger farmers rapidly adopted the new ideas they perceived as economic-ally profitable.

4. *Innovativeness.* A classic diffusion study (Deutschmann and Fals Borda, 1962b) was conducted in a Colombian village to test the cross-cultural validity of correlates of innovativeness derived from U.S. diffusion research. The primary objective of the study was that "taking cultural differences into account the basic pattern of diffusion of information and adoption of new farm practices would be substantially the same in Saucío [the Colombian village] as in the United States." A striking similarity was found between the results obtained from the Colombia study and those reported for Ohio farmers (by Rogers, 1961b): The characteristics of innovators such as greater cosmopolite-ness, higher education, and larger-sized farms were remarkably similar in Saucío and in Ohio. The Deutschmann and Fals Borda (1962b) study in Colombia, along with the Rahim (1961a) investigation in Pakistan, paved the way for a great increase in the number of diffusion investigations among peasants in less developed nations. Until the early 1960s there were few such inquiries in village settings.

5. *Earliness of knowing about innovations.* Greenberg (1964a) attempted to determine what, when, and how people first learned about the news of the assassination of President Kennedy. Data were gathered by telephone interviews with 419 adults in a California city. The respondents were classified as "early knowers" or "late knowers." Most of the early knowers reported that they had heard of the death by radio or TV, whereas most of the late knowers first learned of the assassination by means of interpersonal communication channels.*

6. *Opinion leadership.* The success or failure of programs of directed social change rests in part upon the ability of opinion leaders and their cooperation with change agents. Rogers and van Es (1964) sought (1) to identify opinion leaders in five Colombian villages; (2) to determine their social characteristics,

---

*However, in Chapter 8 we shall show that the relative importance of mass media and interpersonal channels also depend on another important variable, the saliency of the news event to the audience.

communication behavior, and cosmopoliteness; and (3) to determine the differences in these correlates of opinion leadership on the basis of systems with different norms. The data were gathered in personal interviews with 160 peasants in three modern villages and with ninety-five farmers in two traditional communities. Rogers and van Es found that opinion leaders, when compared to their followers in *both* modern and traditional systems, were characterized by: More formal education, higher levels of literacy, larger farms, more agricultural and home innovativeness, higher social status, more mass media exposure, higher empathy, and more political knowledge. In the modern villages, however, the opinion leaders were young and innovative, reflecting the norms, whereas in the traditional systems the leaders were older and not very active in adopting new ideas (Rogers with Svenning, 1969). The leaders tended to reflect the norms of their village, whether modern or traditional.

7. *Communication channel usage.* The Ryan and Gross (1943) investigation of the diffusion of hybrid seed corn in Iowa suggested approaches which have been followed up in a number of subsequent studies. One of these research routes is that dealing with communication channels. The two rural sociologists found that the typical Iowa farmer first heard of hybrid seed from a commercial salesman, but neighbors were the most influential channel in persuading the receivers to adopt the innovation.* Ryan and Gross were the first researchers to suggest that an individual passes through different mental stages (knowledge and persuasion, for example) in adopting a new idea. Different communication channels play different roles at these various stages. Also, salesmen were more important channels about the innovation for earlier adopters, and neighbors were more important for later adopters. This suggests that communication channel behavior is different for the various adopter categories, a proposition widely supported by later diffusion researches.

8. *Consequences of innovation.* The consequences of the use of the steel ax by a tribe of aborigines (Sharp, 1952) illustrates this type of diffusion research. The Yir Yoront were relatively unaffected by modern civilization, owing to their isolation in the Australian bush, until the establishment of a nearby missionary station. The missionaries distributed steel axes among the Yir Yoront as gifts and as pay for work performed. Before the days of the steel ax, the stone ax served the Yir Yoront as their principal tool and as a symbol of masculinity and respect. The men owned the stone axes, but the women and children were the main users of these tools, which they borrowed according

---

*Although later research has generally shown that salesmen are not the most important channel at the knowledge stage.

to a system prescribed by custom. When the missionaries distributed steel axes, they assumed that their axes were much more efficient, and that the use of them would save the Yir Yoront time. However, the innovation contributed little to material progress, as the Yir Yoront used their new-found leisure time for sleep. The impact of the new technological device caused a disruption of status relations among the Yir Yoront and a revolutionary confusion of age and sex roles. Elders, once highly respected, now became dependent upon women and younger men for steel axes. The consequences of the steel ax were mainly unanticipated, far-reaching, and disruptive.

The reader has been provided with a brief glimpse of the diffusion landscape in terms of eight directions (Table 2-2) in which it has been growing. In later chapters of this book, we shall descend the skyscraper and probe these eight types of diffusion research in much greater detail. The generalizations found in these studies will be stated more definitely in the remainder of this book, where further empirical support will be brought to bear on each. Hopefully, the typology of diffusion research just discussed, although brief, will provide the reader with an overall research map of the entire field.

Contributions and Shortcomings of Diffusion Research

Contributions

The status of diffusion research today is impressive. During the 1960s, the results of diffusion research have been incorporated in basic textbooks in social psychology (Secord and Backman, 1964, pp. 213–217), communication (Richardson, 1969; Katz, 1963b and 1964), public relations (Robinson, 1966, pp. 407–421), advertising (Warneryd and Nowak, 1967, pp. 99–111), marketing (Zaltman, 1964), and consumer behavior (Engel and others, 1966, pp. 541–574). Both practitioners (like change agents) and theoreticians have come to regard the diffusion of innovations as a useful field of social science knowledge. Larsen (1964, p. 359) describes diffusion research in laudatory terms, emphasizing its potential for determining communication effects: "Perhaps the most viable area in current communications research . . . is the study of the diffusion of new ideas, products, and practices. Diffusion studies are extensions of traditional research on mass media 'campaigns' that proceed from a more sophisticated frame of reference than does this earlier work."

This generally favorable reputation accorded to the diffusion approach is perhaps a function of the impressive amount of research attention which it has recently received and the conceptual and analytical strength that it gains by incorporating time as an essential element in the analysis of human behavior change. In other words, innovations constitute a type of communication message whose effects are relatively easy to isolate. The case is parallel to the use of radioactive tracers in studying plant growth. One can understand change processes more accurately if he follows the spread of a new idea over time as it courses through the structure of a social system. And because of their salience, innovations often leave deeper scratches on men's minds, thus aiding respondents' recall ability.

Shortcomings

But despite the accolades, the diffusion approach also has its critics; De-Fleur (1966) scores diffusion research for its lack of standardized definitions of concepts, even though he sees it as *potentially* providing "a basis for a quantitative, empirically-based theory of social change."* Two members of the diffusion research fraternity, Fliegel and Kivlin (1966b), complain that their field has not yet received its deserved attention from students of social change: "Diffusion of innovation has the status of a bastard child with respect to the parent interests in social and cultural change: Too big to ignore but unlikely to be given full recognition."**

Therefore, although diffusion research has already made numerous important contributions to our understanding of human behavior change, its potential could be even greater were it not for such shortcomings as the following.

1. *The time dimension.* Diffusion is set off from the broader field of communication by the nature of the *message* that is studied. The message objects are innovations—ideas, practices, and objects perceived as new by the individual. *Time* is therefore an essential element of diffusion but one often not explicitly taken into account in other communication processes such as attitude change

---

*Later in this chapter we shall have much more to say about how this objective might be more effectively reached via the route of middle range analysis.
**Their impression is most directly based upon the writings of LaPiere (1965) and Moore (1963, pp. 85–88).

and persuasion. Carlsson (1965) criticizes most studies of change, including the persuasion tradition, for being inappropriately "timeless" and cites diffusion research as one exception. "Change, whether in political or other attitudes, is usually a time-consuming process, and its time or speed constants are therefore something we ought to study in a much more systematic way than has so far been done."* A process occurs over time and, in a strict sense, has no discernible beginning or end. Although there are blessings that accrue from inclusion of the time variable in diffusion studies (for example, the tracer-like qualities of innovations), there are also two methodological curses.

One weakness of diffusion research is its dependence upon *recall data* from respondents as to their date of awareness or adoption of a new idea. Essentially, the respondent is asked to look back over his shoulder in time and reconstruct his past history of innovation experiences. This hindsight ability is clearly much less than completely accurate (Menzel, 1957; Coughenour, 1965a) for the typical respondent. It probably varies on the basis of the innovation's salience, the length of time over which recall is requested, and on the basis of individual differences in education, mental ability, etc.

Most social science research methods are better suited to obtaining snapshots of behavior, rather than moving pictures, which would be more appropriate for determining the time order of variables. While correlational analysis of survey data is overwhelmingly the favorite methodology of diffusion investigators, diffusion researchers have been especially innovative in pioneering the use of field experiments, prediction, and simulation studies. All these are methodologies which, by their research design, take moving pictures.

2. *Overemphasis on message content.* Since its inception the diffusion field has placed great emphasis upon the *nature of the innovation* as a basis for different diffusion research traditions. Rural sociologists studied agricultural innovations, educational diffusionists investigated new teaching ideas, and so on.** This focus on the message object may be a result of the applied research interest of early diffusion investigators as well as the nature of their research sponsorship, whether by colleges of agriculture or teachers colleges, for example.***

*A similar emphasis on time as an ignored variable in social change was made by Heirich (1964).

**Most rural sociologists studying agricultural diffusion have an undergraduate degree in agriculture, most education diffusion researchers have undergraduate and graduate training in education, etc. Thus, diffusion researchers are generally trained in the technical nature of the innovations they study, which may also be a reason for their emphasis upon the object (the innovation) of the messages in their research studies.

***Another result of these sponsorship and applied influences is the tendency for diffusion researchers to look at the communication process from the *source's* point of view, rather than

The general result of this overemphasis upon the object of the message has not been entirely a blessing to the field. The traditions are thus strengthened and act as partial barriers to the free exchange of diffusion findings and methods, the search for generalizations across the traditions has been retarded, and theoretical potentials have not been fulfilled.*

Most of the innovations that have been studied resulted from physical or biological science research rather than social science research.** Therefore, we know much more about how fertilizers, weed sprays, and antibiotic drugs spread and much less about the diffusion of new ideas in political behavior or human learning. And there are probably important differences in the nature of diffusion of physical and biological "hardware" innovations versus social science "software" items.

Some critics think that the innovations studied have been relatively inconsequential. These "cosmetic" or "band-aid" innovations have usually dealt only with changes in how a social system functions and seldom with its basic social structure. Certainly, we know much less today about the way in which restructuring innovations (e.g., a new organizational form) diffuse than about how functional-technical innovations (e.g., automobile seat-belts or a new food product) spread.

3. *Focus on optional innovation-decisions.* Almost all past diffusion research was concerned with optional innovation-decisions, rather than with decisions of a collective or authority nature. We know more, in comparison, about the psychological process by which individuals adopt or reject new ideas than

the *receiver's*. Thus, we note an assumption in diffusion writings that the rate of adoption should be speeded up, that the innovation should be adopted by receivers, etc. Seldom is it implied in diffusion documents that the source or the channels may be at fault for not providing more adequate information, for promoting inadequate or inappropriate innovations, etc. In fact, "diffusion" research would probably have been called something like "innovation-seeking" if the first studies had been sponsored by farmers, rather than by extension services. The main exception to source-sponsorship of diffusion research is the series of studies under the direction of the late Paul Mort at Columbia University's Teachers College, which were funded by associations of progressive-minded schools.

*Diana Crane, a sociologist at Johns Hopkins University, is currently engaged in an analysis of the sociometrically determined communication patterns among members of the rural sociology diffusion tradition. She is comparing these patterns, obtained by means of questionnaires, with citations of others' work in the footnotes and bibliographies of rural sociology diffusion publications. It would be interesting to utilize her methodology with *all* diffusion researchers, rather than with those in only one tradition. For detail on the method of study and the results of the analysis of the rural sociology diffusion tradition, see Crane (1968c).

**Agricultural innovations have been more widely studied than any other kind of new ideas; they are the focus in 43 percent of all diffusion empirical inquiries, although this proportion has been decreasing in the 1960s.

about the innovation choice-making process in social systems such as communities or bureaucratic organizations.*

4. *Lack of relational analysis.* The individual has been largely used as the unit of analysis in diffusion research, rather than the sociometric dyad, network, or clique, units more appropriate for investigating the *process* aspects of diffusion.

James Coleman (1958) has sagely called for an overhaul of the entire research attack in sociology, which we feel is an even more appropriate reform in diffusion research. He urges us to abandon our primary concern with individuals as units of analysis in favor of relations between individuals as units of analysis. Diffusion processes (and, in a more general sense, all communication processes) are after all a series of transfers of messages from sources to receivers. Therefore, it is entirely appropriate to utilize relationships, transactions, pairings, or chains as our units of analysis in diffusion inquiry, rather than individuals. The use of individuals, Coleman says, produces only a rather poor "aggregate psychology." But very few diffusion researchers have heeded Coleman's admonition.**

Coleman (1958) traces reasons for the overemphasis upon individuals as units of analysis to the neglect of communication relationships. He mainly blames survey research methods, which lead to the neglect of social structure variables and relationships among individuals. "Samples were random, never including (except by accident) two persons who were friends; interviews were with one individual, as an atomistic entity, and responses were coded onto separate IBM cards, one for each person" (Coleman, 1958).

Because the data were gathered from individuals as the units of response, the focus has been upon individual, intrapersonal variables. This largely excludes social structure and interpersonal variables.*** It has been erroneously assumed that because individuals were the units of response, individuals also had to be the units of analysis.

*Chapters 9 and 10 in this book are concerned with summarizing what is known or hypothesized about collective and authority innovation-decisions, respectively, and what should be investigated.
**A content analysis of 1,084 empirical diffusion studies in the Michigan State University Diffusion Documents Center shows only about 136 (or 12 percent) utilized a dyadic approach, the most common type of relational analysis.
***And also, strangely, excludes *personality* variables in diffusion research. This is odd only because personality variables are such an important type of intrapersonal variable. The lack of attention to personality dimensions probably occurred (1) because diffusion researchers were (and are) mostly sociologists, rather than psychologists or social psychologists, and (2) because personality variables are generally difficult to measure in field (as opposed to laboratory) settings.

But recently some social scientists have come to realize that even with the use of survey methods, which are often essential for gathering large-scale amounts of data as a basis for generalization, various techniques of conceptualization,* measurement, data gathering, and data analysis can be utilized to provide a focus on relationships rather than on individuals. In short, the measurement devices center around some type of sociometric question, the data-gathering techniques consist of sampling intact groups (or subsystems) or pairs of individuals (as with so-called "snowball sampling"**), and the data analysis methods use the dyad, chain or network, or subsystem or clique as the unit of analysis.

How might relational analysis be used in diffusion investigations?

(a) *Dyadic* analysis of sociometric data about innovation diffusion entails obtaining information from source-receiver pairs. The communication dyad (or two-person interacting pair) may be identified, for example, by asking a sociometric question of school teachers, such as "Who first told you about modern math?" or "Who convinced you to adopt modern math?" Such questions have been widely utilized in past diffusion studies but not as a basis for forming communication dyads. In one sense the dyad is the most elemental, primitive unit in interpersonal diffusion. It deserves more research attention than the slight bit it has received.***

A variety of important research questions can be answered with dyadic analysis, such as: To what extent does diffusion occur between individuals who are homophilous in their characteristics, behavior, and attitudes? When there is homophily in innovation flows between pairs of individuals, to what extent does the degree of homophily depend upon such variables as the traditionalism of the system's norms, the nature of the variables on which heterophily-homophily is measured (social status and innovativeness, for example), and the nature of the innovation?

---

*One barrier to relational analysis is that we often lack the appropriate concepts to guide our analysis of dyadic or other types of relational data.

**A term used to describe a sampling design in which (1) a random sample of individuals are asked a sociometric question, for example, from whom they obtained information about a new idea; and (2) then the individuals so named are interviewed at a second stage. The snowballing can, of course, be continued to third, fourth, and further stages. An illustration of snowball sampling was encountered earlier in this chapter in our description of the Columbia University drug study.

***In an earlier synthesis of diffusion research, we suggested the metaphor "... students of diffusion have been working where the ground was soft" (Rogers, 1962b, p. viii). The relative softness of the ground, of course, depends on one's digging tools. By using such new methodological approaches as dyadic analysis, for instance, we can dig much deeper into the nature of interpersonal diffusion.

When some heterophily (the opposite of homophily) does occur, do receivers seek sources (pair-wise) who are higher or lower in social status, innovativeness, technical competence, and so on than the sources? In other words, is there a "trickle down" or a "trickle up" of innovations in a social system?

(b) *Network* or chain analysis is essentially similar to dyadic analysis in respect to its dependence upon sociometric data, but it differs in that multiple-person communication chains, rather than dyadic pairs, are the units of analysis. A communication chain or network consists of any number of individuals in a system, starting with a source person and sequentially continuing through all the related individuals who are his direct or indirect receivers. Essentially, chains or networks consist of a number of linked dyads in which the receiver in one dyad is the source in the next.*

(c) *Clique* or subsystem analysis consists of sociometrically determining the communication groupings among the members of a social system.** Once these diffusion cliques are identified, the bases of clique membership can be determined, or the informal cliques in a bureaucratic organization might be compared with the formal organization to see how well the formal structure corresponds to actual diffusion patterns. Also, the liaison individuals*** who link two or more cliques can be located and their characteristics determined.

Diffusion research has suffered by ignoring sociometric relationships as units of analysis. Such an approach offers rich potential for investigating the nature of interpersonal diffusion of innovations.

5. *Concentration in the United States and Western Europe.* Although the number of diffusion researches is impressive and the geographic scope they represent has broadened considerably in the past decade, there are still very important

---

*Statistical methods for the complete analysis of network are not yet well developed, and most diffusion chains are analyzed by means of diagrammatic plotting and visual inspection. Matrix multiplication by computer also provides a statistical means of chain analysis.

**Such clique identification may be accomplished by the visual plotting of sociometric data unless the number of individuals involved is high and/or the interpersonal relationships are too complex. In these cases one can resort to the matrix multiplication procedures suggested by Hubell (1965) or Festinger (1949). Essentially, these techniques consist of reducing the sociometric data about diffusion to a "who-to-whom" matrix in which the source individuals are located on one dimension of the matrix and the receivers on the other. The matrix is squared, then cubed, and so on, usually by computer techniques. Through this procedure, existence of diffusion cliques soon becomes apparent within the total matrix of interpersonal relationships.

***A research focus upon the role of liaison individuals was initiated by Jacobson and Seashore (1951), and followed by Weiss and Jacobson (1955). Recently, Schwartz (1968) utilized their methods of locating liaison individuals in a study of communication patterns in a college of education.

gaps. One of the most striking is the lack of diffusion investigations in Communist countries. One can imagine that the nature of diffusion in these settings would be very different from that in non-Communist nations. Imagine how different hybrid corn diffusion must be in a collective farm in the Ukraine as opposed to Iowa. The lack of an individual profit motive in Russia, Mainland China, and similar countries likely affects the way in which new industrial techniques spread. To date there are only four publications (Galeski, 1965; Makarczyk, 1965, 1967; Szwengrub, 1967) available about agricultural diffusion in Poland. The notion of diffusion research has diffused very little through the Iron Curtain.*

Unfortunately, we possess many more investigations and much better data about modern systems than about traditional systems. This is partly because social science is better financed, larger in scale, and easier to conduct in more developed nations than in less developed countries, and in modern rather than traditional communities. It is difficult to obtain accurate data from an extremely traditional respondent because he is so heterophilous in regard to the investigator. The two seldom share common meanings and beliefs or an effective channel of discourse.** Traditional individuals, whether peasants or urban poor, share a "subculture*** of tradition" whose central elements consist of mutual distrust in interpersonal relations, fatalism, a lack of empathy, and a limited view of the world (Rogers with Svenning, 1969). These attributes not only make it difficult for change agents to diffuse new ideas to the traditional individual;**** they also pose important communication problems for the social scientist seeking to obtain valid and reliable data.

There is a bias in social science research, including diffusion investigations: We know more about modern individuals and modern social systems than

*Although a six-nation, comparative study underway in Europe includes four Communist countries (Czechoslovakia, Hungary, Poland, and Romania). This investigation of agricultural diffusion in modern and traditional villages is directed by Boguslaw Galeski of Poland and Henri Mendras of France, and is sponsored in part by the European Centre for the Coordination of Research and Documentation in Social Sciences, Vienna.

**As was illustrated by Schatzmann and Strauss (1955), who found that Arkansas hillbillies could not effectively communicate their experiences in a tornado to middle-class research interviewers.

***A *subculture* contains many elements of the broader culture of which it is a part but has special characteristics not shared by most members of the larger society.

****A statement which seems to hold true in Communist China, as elsewhere, where innovations are more difficult to introduce to traditional peasants than to urbanites: "We do believe in planned parenthood, but it is not easy to introduce all at once in China and it is more difficult to achieve in rural areas, where most of our people live, than in the cities" (Premier Chou En-lai, as interviewed by Edgar Snow in 1964 at Conakry, Guinea, Africa).

about their traditional counterparts. There is generally a lack of adequate communication between the traditional and modern strata of society. In most nations the power elite make decisions affecting the entire system, including the traditional masses, hopefully for the maximum welfare of all. To do so, the elite must understand the needs, beliefs, and reactions of the masses. However, most communication patterns occur horizontally (homophilously) rather than vertically (heterophilously). That is, the elite interact with each other, and the masses interact with others of the same stratum. Such vertical communication as does occur is mostly downward; the elite do the talking and writing, and the masses are passive recipients of governmental and commercial change campaigns.* Only very rarely do the masses, especially the lower masses, communicate to the elite. So the ruling elite know little about the lower masses, and hence find it particularly difficult to communicate effectively with them. The elite need feedback as to the results of their campaigns of change, launched in order to fill certain needs of the lower masses. Often these programs "scratch where there is no itch" because the elite cannot empathize adequately with the masses whose more traditional style of life is so different from their own.

The inadequate communication between elite and masses may therefore lead to conflict in which the frustrated masses seek to communicate their needs by more violent expression.** Perhaps the high rate of political instability in less developed countries in the 1960s and the summer unrest in the U.S. urban core are manifestations of the communication gap between elites and masses. One reason for the frequent failure of elite-directed programs of change is that we know so little about the traditional audience. Improved communication flows are needed between the traditional and modern sectors of society.***

We must remember throughout this book that *our diffusion generalizations are based more heavily upon data from modern than from traditonal systems.* Fortunately, recent years have seen a great increase in diffusion researches in less developed

---

*"Most of the talking has usually been done by the upper level; the people of the lower [class] sit by quietly, and even sullenly, often without listening" (Gans, 1962, p. x).

**A more detailed discussion of the role of the power elite in diffusion is found in Chapter 11.

***The methods of social science data gathering, although subject to the limitations discussed previously, offer one means, albeit much less than perfect, for the modern elite to gain the perspective of the traditional masses. Illustrations are provided by Gans (1962), Lerner (1958), and Rogers with Svenning (1969). For a detailed discussion of how survey research methods, which originated in more developed countries, ought to be adapted to the sociocultural settings of less developed nations, see Hursh (forthcoming).

countries;* the balance in this particular field of social science is being gradually righted. For instance, in the late 1960s almost as many diffusion studies are being completed in less developed countries as in more developed countries (Figure 2-3).** This recent internationalization of diffusion research provides us an opportunity in this book to draw conclusions concerning the cross-cultural validity of our generalizations, which originated in the United States and Western Europe and have since been tested in nations like Colombia, India, and Nigeria.

We have just reviewed five of the major shortcomings of diffusion research; they led us to conclude that the beginnings of diffusion research left an indelible stamp on the approaches, concepts, methods, and assumptions of the field, some twenty-five years and 1,500 publications later. And often the biases that we inherited from our research ancestors have been quite inappropriate for the important diffusion research tasks of today. It is strange that the study of innovation has itself been so traditional!

Middle Range Analysis

Grand Theory Versus Raw Empiricism

A great deal has been written and said about social change.*** These expressions have come from two quite different schools. On the one hand, we have the abstract speculations of the grand theorists of social change, whose general approach has been to use an entire civilization or society as the unit of analysis and whose product has seldom been translatable into empirically testable hypotheses.**** Although such theorists have undoubtedly sensitized

---

*Of the total of 1,084 empirical diffusion publications in existence as of July, 1968, 69 percent were conducted in more developed countries and 31 percent in less developed countries. Most of the latter were completed after about 1960.

**Especially in India, where 158 (48 percent of the total of 331) of these empirical studies are being conducted; next is Colombia with 20 studies. In comparison, 618 researches were completed in the U.S.

***The main ideas in this section come from Rogers with Svenning (1969) and are used by permission.

****These grand theories seldom pay explicit attention to time, an important variable in any change process. However, the grand theorists implicitly note the importance of consideration of time in their theories when they speak of "dynamics," "processes," and the like.

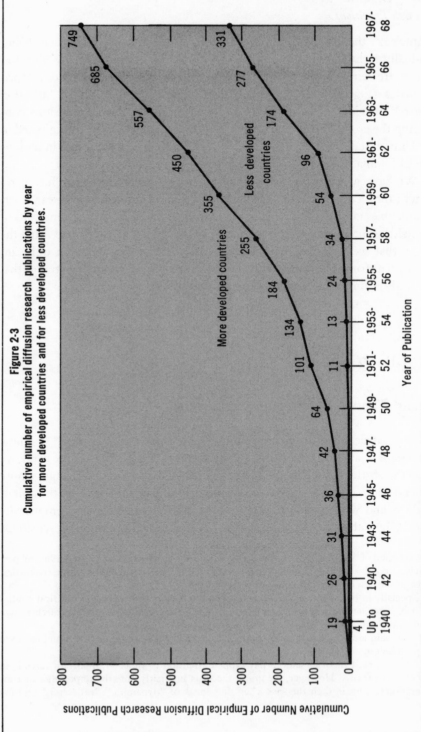

Figure 2-3
Cumulative number of empirical diffusion research publications by year for more developed countries and for less developed countries.

Source: Diffusion Documents Center, Michigan State University, July, 1968.

us to the change process, scientific understanding of change has been advanced but little.

One critic of the grand theorists, C. Wright Mills (1959, pp. 25–31), has taken a lengthy passage from one of the noted works of this school and then succinctly expressed the same propositions in a few words of "plain English," scoring the grand theorists for their verbosity: "Grand theorists are so deeply involved in [unintelligibility] that I fear we really must ask: Is grand theory merely a confused verbiage or is there, after all, also something there?" (Mills, 1959, p. 27). His general conclusion seems to be: Not much. "There is no 'grand theory,' no one universal scheme in terms of which we can understand the unity of social structure. . ." (Mills, 1959, pp. 46–47).

Writings from the grand theory school are indeed difficult to comprehend;* however, we feel there is "something there." That something is a useful set of concepts and typologies and some rather meager propositions relating these concepts. What is missing in the intellectual exercises of the grand theorists is a consistent system of interrelated propositions about human behavior and how it changes, *at a level of generalization that facilitates testing.* The movement away from these classical theories of social change in recent years has occurred because "They turned out to be either untestable, and hence scientifically unacceptable, or only partly true at best" (Etzioni and Etzioni, 1964, p. 7). The major shortcoming of grand theory, then, is neither unintelligibility nor lack of content, but its grand level.

Contrasted to the approach of the grand theorists is the approach of the raw empiricists.** They scurry about the world gathering data concerning the minutiae of change, giving little attention to generalizing their results beyond the particular respondents or social systems that they study. Consider the great number of anthropological and sociological descriptions of peasant villages. There are more than 500 such published accounts of Indian villages alone. And although these ethnographers are to be commended for their energy, they contribute little to our ability to see village life in a general perspective (or to understand how village social structures change). Almost as narrow in scope as the studies of single villages are the multitide of investigations on the diffusion of innovations. Although these researches are less descriptive and somewhat more analytical than the village ethnographies, they

---

*Which is particularly regrettable, as Mills (1959, p. 218) states, because "Such lack of ready intelligibility, I believe, usually has little or nothing to do with the complexity of subject matter, and nothing at all with profundity of thought."
**Mills (1959, p. 55) categorically castigates members of the empirical school: "Abstracted empiricism is not characterized by any substantive propositions or theories."

generally suffer from a similar lack of conceptual orientation.* Whatever its nature, empirical investigation without theoretical basis becomes inevitably bogged down in irrelevant data while ignoring potentially fruitful objectives.

Relating Theory and Research: The Middle Range

We prefer to operate at the middle range,** relating theory to research and research to theory. This means our theoretical basis must be specific enough to be empirically testable, and our data must test theoretical hypotheses. Theory that cannot be tested is useless, and data not related to theoretic hypotheses become irrelevant. The interplay between theoretical concepts and empirical data, although complex in nature, may perhaps be demonstrated by the following illustration of the essential procedural steps in middle range analysis.

1. All concepts must be expressed as variables. A *concept* is a dimension stated in its most basic terms. A conceptual variable utilized throughout this book is innovativeness, defined as the degree to which an individual is relatively earlier in adopting new ideas than other members of his social system. Ideally, a concept should be as general or abstract as possible so that it may be utilized to describe behavior in many different types of social systems. For example, the innovativeness concept has been studied in industry, education, medicine, and among primitive tribes.

2. The postulated relationship between two concepts is called a *general* or *theoretical hypothesis*. An example of a theoretical hypothesis tested in several research studies (that will be cited in future chapters) is: "Innovativeness is positively related to cosmopoliteness." In this example innovativeness and cosmopoliteness are concepts, and the theoretical hypothesis postulates a positive relationship between them. The logic is that individuals who have communication with sources external to their social system are more innova-

---

*One critic of diffusion research described it as "a mile wide and an inch deep," referring to the plethora of empirical research results which are not fully integrated into theoretical meaningfulness.

**This idea comes from Merton (1957, p. 9), who asks for "theories of the middle range," that is, postulated relationships which are testable but that deal with only a rather limited, particular type of behavior. These middle range theories may eventually be consolidated into more abstract general conceptual schemes. We prefer to speak of middle range *analysis*, which is the formulation and testing of theories of the middle range.

tive.* If one has reference groups outside a social system, greater deviation from that system's expectations for one's behavior is likely, and the adoption of new ideas probably results.

Notice that the theoretical hypothesis illustrated here is limited in scope to the diffusion of innovations. This is why our type of analysis is termed "middle range"; application of our hypothesis is explicitly confined to one type of human behavior. This should not prevent, but rather encourage, postulation of similar hypothesized relationships dealing with other types of behavior. Middle range analysis can, therefore, offer one route toward more general theories.

3. A theoretical hypothesis is tested by means of an *empirical hypothesis* (or hypotheses), defined as the postulated relationship between two operational measures of concepts. An *operation* is the empirical referent of a concept; it may be a scale, index, observation, or the answer to a direct question. Whereas concepts exist only at the theoretical level, operations exist only at the empirical level. The degree to which an operation is a valid measure of a concept is

---

*The development of a theoretical hypothesis may also be a result of derivations from other theoretical hypotheses. For example, if concept A is positively related to concept B, and concept B is positively related to concept C, then it may be postulated that concept A is positively related to concept C. Of course, this may not be the case if there are only weak relationships between A and B, and B and C.

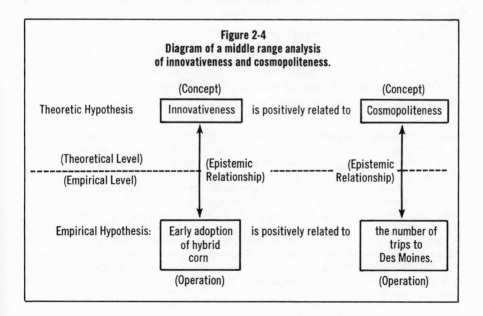

Figure 2-4
Diagram of a middle range analysis
of innovativeness and cosmopoliteness.

called an *epistemic relationship*. Even though it is obviously of great importance, the isomorphism (or "identicalness") of this linkage between concept and operation cannot be tested except by intuitive means.

A middle range analysis of the relationship between innovativeness and cosmopoliteness is illustrated by an example from the Ryan and Gross (1943) hybrid corn study in Iowa (Figure 2-4).

4. An empirical hypothesis may be accepted or rejected on the basis of statistical tests of significance, as well as other criteria such as visual observation of the data. In the hybrid corn study, Ryan and Gross (1943) report a positive, significant relationship between a farmer's time of adoption of hybrid seed and his number of trips to Des Moines.

5. A theoretical hypothesis is supported or rejected on the basis of the tests of corresponding empirical hypotheses. Truth claims may be added to a theoretical hypothesis by similar findings from other analyses of the two conceptual variables in a variety of different social systems. As additional support is added to a general hypothesis, greater confidence may be placed in the relationship between the two concepts, and this relationship may be considered a *generalization* and eventually perhaps a *principle* or even a *law*.*

6. The relationships between each of the two concepts and other concepts may be analyzed, and, as findings of this nature gradually accumulate, a more general body of theory is developed.** Evidence is accumulated in an integrated and consistent manner.

We relate the theoretical and empirical levels by the joint processes of *deduction* (going from theoretical to empirical hypotheses) and *induction* (from empirical results to the conceptual level.) The eventual goal of middle range analysis is the development of an interrelated, integrated series of concepts, linked in a matrix of theories and of established relationships.

---

*Generalizations, principles, and laws represent three points on a continuum which indicates the degree of validity established for a relationship between two or more concepts.
**We are generally taught in graduate research methods courses that the design of an investigation should originate with theory and then move deductively to operational measures. However, inductive processes may be fruitful also, as in the case of serendipity. "Fruitful empirical research not only tests theoretically derived hypotheses, it also originates new hypotheses. This might be termed the 'serendipity' component of research, e.g., the discovery by chance or sagacity of valid results which were not sought for" (Merton, 1957, p. 96). A famous example of serendipity is Sir Alexander Fleming's discovery of penicillin; other examples are Roentgen's detection of X-rays, the discovery of the neutrino, and of the Hawthorne effect. The concept of serendipity comes originally from Horace Walpole's *The Three Princes of Serendip*, a narrative about three adventurers who blundered into fortunate discoveries.

In the present book our objective is to fit a great number of empirical relationships that resulted from the many diffusion investigations into a series of middle range generalizations. These generalizations, which are both the main fruit of this book and its organizational skeleton, are limited in application to the diffusion of new ideas. *These middle range generalizations become stepping stones to more general theories of social change, once they are abstracted to a yet higher level of generality.**

Middle range analysis is a procedure by which theory and research may be related. In the present illustration, the theoretical hypothesis that "innovativeness is positively related to cosmopoliteness" is supported by a finding from the Ryan and Gross (1943) study, which indicates that early adoption of hybrid corn is positively related to the number of trips farmers make to Des Moines. Naturally, a number of other empirical relationships dealing with measures of innovativeness and cosmopoliteness could be related to the same theoretical hypothesis.

Shortcomings of the Present Approach

There are certain cautions which should be expressed about our use of middle range analysis in the chapters that follow.

Cross-Cultural Equivalence

One problem is the cross-cultural comparison (and generalization) of the results of diffusion studies. The investigations reported in this book were conducted in a variety of cultural settings, which is both a blessing in allowing for potential cross-cultural generalization, as well as a possible curse because of the difficulty in abstracting numerous empirical relationships conducted in

---

*Fields of empirical inquiry other than diffusion can yield middle range generalizations which, when combined with those from diffusion studies, will contribute directly toward an emerging general theory of social change. One of these additional, potentially contributory fields is collective behavior. "Generalizations built out of the spread of hybrid corn, clean hog practices, television, and miracle drugs surely would gain by comparison with generalizations from the study of the spread of Negro suffrage in Mississippi, the Ghost Dance, Luddism, and the small family value system" (Ryan, 1965). "Surprisingly, the sociology of diffusion and that of collective behavior, though both are concerned with the appearance of new items in a culture, tend to develop independently of one another" (Pinard, 1968). We feel, however, that codification of generalizations in collective behavior and other fields dealing indirectly with social change lies outside the purview of the present book.

varied locales to one more theoretical generalization. For although the importance of cross-culturally valid generalizations in diffusion research is obvious, we must also be cautious. Kluckhohn and Strodtbeck (1961, p. 92) warn that "anyone who has attempted cross-cultural testing, using the medium of language, is well aware of the deep and as yet bridgeless chasms which separate the linguistically ordered thought-ways of the peoples of varying cultural traditions. . . ." For instance, one can never be certain that what is measured by a test of functional literacy in Colombia (in Spanish) is identical to what one measures with a seemingly similar operation in India (in Hindi). Obviously, one must resort to the level of *conceptual equivalence* in cross-cultural comparative inquiry, rather than pursue the goal of *operational equivalence*,* which is an impossible objective, at least with present data-gathering and analytic tools. But even conceptual equality is difficult to attain, and its assessment must be forged largely on logical rather than empirical grounds.

Arbitrary Categorization of Variables as Dependent and Independent

A further problem with our methodology of theory construction is the rather arbitrary specification of concepts as "dependent" and "independent" in our generalizations. A *dependent variable* is the dimension that we try to predict or explain through its relationship (or covariance) with the *independent variable*. In the illustration in Figure 2-4, innovativeness is the dependent variable and cosmopoliteness is the independent variable. Such categorization is made on the basis of the purposes of the research worker and may not necessarily correspond to the expected time-order in which the variables occur in the real world. In our example cosmopoliteness *may* lead to innovativeness. An individual free from the social control of a social system's norms (due to his external orientation) may adopt new ideas. Perhaps Mrs. B, who was a social stranger in the Peruvian village of Los Molinos (see Chapter 1), is an example of the logic that cosmopoliteness leads to innovativeness. To the contrary, innovativeness *may* lead an individual to be considered a deviant in his social system, and as he seeks social response and friendship outside of the local system, he becomes more cosmopolite. The labeling of variables in our generalizations as dependent and independent does not imply that they are necessarily consequences and antecedents in a time-order sequence.

*In the operational equivalence approach one seeks to develop identical measures of a concept in two or more cultures. This is impossible because of linguistic differences between cultures and because the same concept may have quite different cultural expression in one culture or another.

While the time-order of our concepts does seem more or less logical, this sequence is difficult or impossible to establish empirically. Many of our conceptual relationships are probably interdependent. Such a relationship occurs when "a small increment in one variable results in a small increment in a second variable; then, the increment in the second variable makes possible a further increment in the first variable, which in turn affects the second one, and so this process goes on until no more increments are possible" (Zetterberg, 1965, p. 70). For example, a little increase in cosmopoliteness leads to a small increment in innovativeness, which leads to a little greater cosmopoliteness, and so forth. Who is to say whether cosmopoliteness or innovativeness is the antecedent variable in this relationship? Each is, and yet neither is.

Almost all of the diffusion investigations upon which this book is based are correlational analyses of survey data, rather than before-after analyses of field or laboratory experiments. This means that the time-order nature of our concepts can seldom be definitely determined.* We prefer to speak of our pairs of concepts in theoretical hypotheses as dependent and independent, rather than as necessarily consequent and antecedent. Even the dependent-independent classification is arbitrary, and such specification is ours, not that of the researchers who conducted the studies we are synthesizing.

Theoretical Oversimplification of Two-Concept Generalizations

Another shortcoming of our generalizations in the following chapters is the deceit of their neatness and simplicity. Our generalizations deal almost entirely with pairs of concepts, whereas the real nature of diffusion is certainly a cobweb of interrelationships among numerous variables. For instance, earlier in this chapter we offered the generalization that innovativeness is positively related to cosmopoliteness (Figure 2-4). This two-concept generalization does not indicate that the relationship between innovativeness and cosmopoliteness *may* be due to the relationships of both variables with a third concept such as social status. We know, for example, that more innovative individuals are often of relatively higher socioeconomic status, as are cosmopolites (Chapter 5). Then should not social status also be included in the innovativeness-cosmopoliteness generalization? Unfortunately, it cannot be. Most

---

*Neither can the "forcing quality" of one variable on another be determined, which is the other important aspect of causal relationships in addition to their time-order. Forcing quality, the degree to which one variable is a necessary and sufficient explanation of the other, can seldom be adequately determined by experiments, either, and ultimately rests on logical applications of theoretical insights, rather than directly on solely empirical grounds.

of the empirical diffusion studies reviewed in this book focus upon only two-variable hypotheses, and we cannot summarize findings that do not exist. Further, our ability to understand three-variable, four-variable, and so on generalizations usually suffers in direct proportion to the number of variables included.

Therefore for the sake of clarity and because we lack an empirical basis to do otherwise, the generalizations in this volume, with only a few exceptions, deal with two concepts.* We should not forget that we are artificially and heuristically chopping up reality into conceptual bite-sized pieces. Although such processing may aid digestibility, it also adds an ersatz flavor.

Generalizations which Range Widely in Truth Claims

Some critics of middle range analysis might argue that most scientists intuitively use the essential features of this approach to theory construction without going through the formal mechanics of specifying concepts, operations, and epistemic relationships. This is undoubtedly true, but in the case of diffusion research, there is need to formalize more adequately the wealth of findings available in terms of more general concepts than have been used in most past studies. Middle range analysis is not only useful in synthesizing past research findings, but it also provides useful leads for future diffusion inquiry. In several chapters of this book, especially those dealing with collective innovation-decisions and with authority innovation-decisions in formal organizations (where relatively little past investigation has been done in a diffusion framework), our generalizations possess few claims to truth. In these cases the generalizations more closely resemble a research map for future studies than a summary of past results. The generalizations in this book actually range in degree of existing research support from very little, where the generalizations are not much more than theoretical hypotheses, to a great deal, where the generalizations approach the level of principles.

In the chapters that follow we hope to demonstrate the usefulness of middle range analysis in codifying and synthesizing available diffusion research results. The four perils of (1) the lack of cross-cultural conceptual

*However, where the original research publication provided a basis for doing so, we coded a generalization as "conditional," meaning that the relationship found between two variables depends upon a third variable. An example might be: Innovativeness is positively related to cosmopoliteness, except for those individuals of high social status. However, only 331 (about 5 percent) of the 6,811 empirical generalizations available as of July, 1968, were conditional. All the rest (95 percent) are two-variable generalizations. In Appendix A we consider the conditional relationships as not supporting each diffusion generalization.

equivalence, (2) the arbitrary categorization of concepts in our generalizations as dependent and independent, (3) the theoretical oversimplification of our generalizations owing to their limitation to pairs of concepts, and (4) the range of claims to truth for our generalizations will continue to plague us throughout our work. Therefore, we shall continue to discuss their seriousness and to suggest their possible amelioration throughout.

Summary

A theme of the present chapter is that diffusion research is emerging as a single, integrated body of concepts and generalizations, even though the investigations are conducted by researchers in several scientific disciplines. A *research tradition* is a series of investigations on a similar topic in which successive studies are influenced by preceding inquiries. Diffusion research traditions have acted in the past to partition scholarship on this topic, so that an inadequate flow of diffusion research findings has occurred among diffusion researchers. The consequence has been an unnecessary duplication and unwanted replication of findings. But in the mid-1960s a gradual breakdown of these intellectual boundaries has occurred.

Seven major diffusion traditions, delineated mainly on the basis of the disciplinary affiliation of the researchers, are described: Anthropology, early sociology, rural sociology, education, medical sociology, communication, and marketing.

Eight main types of diffusion research are distinguished in this book, and most are dealt with in detail in future chapters:

1 Rate of adoption of an innovation in a social system.
2 Rate of adoption in different social systems.
3 Perceived attributes of innovations.
4 Innovativeness.
5 Earliness of knowing about innovations.
6 Opinion leadership.
7 Communication channel usage.
8 Consequences of innovation.

There are at least five major shortcomings of diffusion research: (1) its inclusion of the time dimension leads to a dependence upon recall data and to difficulties in determining the time-order of diffusion variables; (2) an over-emphasis upon the nature of the innovation studied leads to separate diffusion research traditions, which in turn impedes the theoretical integration of the field; (3) an overconcern with optional innovation-decisions, largely to the exclusion of collective and authority decisions; (4) use of the individual as the unit of analysis, rather than depending on relational analysis, which is more appropriate for investigating the process aspects of diffusion; and (5) a concentration of diffusion research in the United States and Western Europe, so that cross-cultural testing of generalizations is retarded.

We described our approach to a theory of social change, that of middle range analysis. This approach consists of accumulating and synthesizing middle range generalizations from empirical results on the diffusion of innovations. Such generalizations may become even more general by future incorporation of generalizations from other research fields dealing with changing human behavior.

The problem of much social scientific activity in the past has been the lack of rapprochement between research and theory. We believe that this gap in understanding can be gradually bridged by middle range analysis, which attempts to link empirical data with conceptualization and vice versa.

The first step in this approach is carefully to explicate all essential concepts. A *concept* is a dimension stated in its most basic terms. Next, we postulate a relationship between two concepts in a *general or theoretical hypothesis*. We test this theoretical hypothesis with a corresponding *empirical hypothesis,* which is the postulated relationship between two operational measures of concepts. An *operation* is the empirical referent of a concept. Empirical hypotheses are often accepted or rejected on the basis of statistical tests of significance, but other criteria may be used. Finally, a theoretical hypothesis is supported or rejected by testing corresponding empirical hypotheses, resulting eventually in a series of middle range generalizations. We believe that middle range generalizations are the stepping stones to more general theories of social change, once they are abstracted to a yet higher level of generality.

3 The

Innovation-Decision

One must learn by doing the thing, for though you think you
know it—you have no certainty, until you try.

(SOPHOCLES, 400 B.C.)

Process

— — — — ·   · — — · · — · — — — — · — · · · · · · · ·   — · · · · ·   · · — · — · — — — · · · — ·

T HE *innovation-decision process* is the mental process through which an individual passes from first knowledge of an innovation to a decision to adopt or reject and to confirmation of this decision.* This process should be distinguished from the diffusion process by which new ideas are communicated to the members of a social system. The major difference between the two processes is that diffusion occurs among the units in a social system, whereas innovation decision making takes place within the mind of an individual.

The adoption or rejection of an innovation is a decision by an individual. If he adopts, he begins using a new idea, practice, or object and ceases using the idea that the innovation replaces. The innovation-decision is a special type of decision making; it has certain characteristics not found in other kinds of decision-making situations. In the case of the adoption of an innovation, an individual must choose a *new* alternative over those previously in existence. Therefore, the newness of the alternative is a distinctive aspect of innovation decision making.

The purpose of this chapter is to describe a model of the innovation-decision process, to propose four functions or stages in this process, to

*This was called the "adoption process" in a previous edition of this book (Rogers, 1962b, p. 76), a terminology that implies that all individuals adopt rather than reject new ideas as a result of the process. Further, our original notion of the adoption process did not allow for behavior which takes place after the decision to adopt. We prefer the convention of "innovation-decision process" in the present book, a terminology broader in scope than "adoption process."

present evidence that these functions exist, and to analyze the innovation-decision period. Basically, this chapter blends basic notions of learning, decision making, and dissonance theory into a revised conceptualization of the innovation-decision process. Our main thrust in this chapter deals with optional innovation-decisions (which are made by individuals regardless of the decisions of other members of the social system), although much of what is said contributes a basis for our later discussion of collective and authority decisions in Chapters 9 and 10, respectively.

A Model of the Innovation-Decision Process

Diffusion scholars have long recognized that an individual's decision about an innovation is not an instantaneous act. Rather, it is a *process* that occurs over a period of time and consists of a series of actions. Recent research and conceptualization concentrate on the exact nature of these sequential stages in the process.

The Adoption Process

The traditional view of the innovation-decision process, called the "adoption process," was postulated by a committee of rural sociologists* in 1955 as consisting of five stages:

1 *Awareness stage.* The individual learns of the existence of the new idea but lacks information about it.
2 *Interest stage.* The individual develops interest in the innovation and seeks additional information about it.
3 *Evaluation stage.* The individual makes mental application of the new idea to his present and anticipated future situation and decides whether or not to try it.
4 *Trial stage.* The individual actually applies the new idea on a small scale in order to determine its utility in his own situation.

*The committee was the North Central Rural Sociology Subcommittee for the Study of Diffusion of Farm Practices (1955). Their standardization of the adoption process was influenced heavily by the work of Ryan and Gross (1943) and Wilkening (1952a). The five-stage adoption process was based primarily on theoretical reasoning, rather than empirical evidence, which was not available until the late 1950s (from studies by Beal and others, 1957; Copp and others, 1958).

5 *Adoption stage*. The individual uses the new idea continuously on a full scale.

This conceptualization of the adoption process has been highly favored by diffusion researchers in the past. But recent critics of this model point out that it is too simple. Among its numerous deficiencies are:

1. It implies that the process always ends in adoption decisions, whereas in reality rejection may also be a likely outcome. Therefore, a term more general than "adoption process" is needed that allows for either adoption or rejection.

2. The five stages do not always occur in the specified order, and some of them may be skipped, especially the trial stage. Evaluation actually occurs throughout the process, rather than just at one of the five stages.

3. The process seldom ends with adoption, as further information seeking may occur (Mason, 1964) to confirm or reinforce the decision, or the individual may later switch from adoption to rejection (a discontinuance).

One critic of the adoption process model concludes that only two stages are necessary and sufficient—awareness and adoption—with awareness always occurring before adoption. Past diffusion literature indicates there is little agreement on the number of stages in the process,* although researchers generally recognize that adoption is the result of a *sequence* of events and not random behavior. How can some conceptual order be established as to the nature of stages in the innovation-decision process in the face of the many conflicting theoretic approaches to this process?

Paradigm of the Innovation-Decison Process

Our present model of the innovation-decision process is depicted in Figure 3-1. The present conceptualization consists of four functions or stages:

*For instance, Wilkening (1956) and Emery and Oeser (1958, p. 3) utilized a three-stage process. Wilkening (1953) proposed four stages, as did Ryan and Gross (1943) and Rahim (1961a). Beal and others (1957) and Copp and others (1958) used five stages. Lavidge and Steiner (1961) postulated six stages while Singh and Pareek (1968) proposed an eight-stage model. Not only has there been great diversity in the number of stages used in past inquiries, but there has also been an entertaining variety of terms used for the stages. Among them are: Attention, exposure, initial knowledge, information, application, acceptance, desire, conviction, and deliberation. In the field of marketing, a four-stage model of the innovation-decision process is widely used, the so-called AIDA formula: Awareness, interest, desire, and action (Zaltman, 1964).

Figure 3-1
Paradigm of the innovation-decision process.[a]

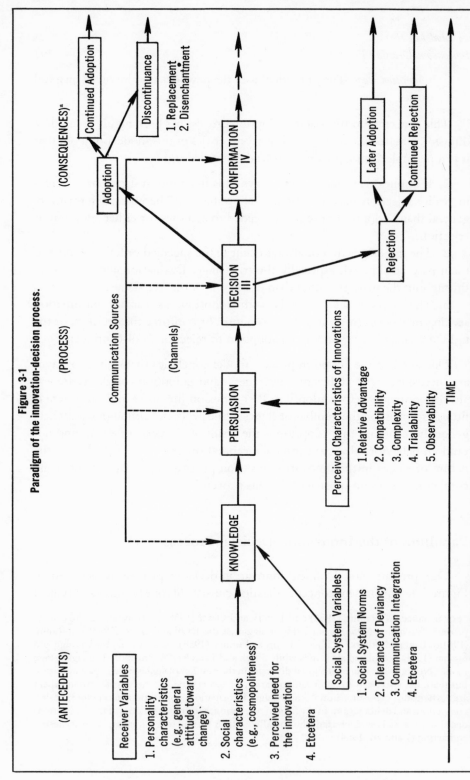

[a]For the sake of simplicity we have not shown the consequences of the innovation in this paradigm but only the consequences of the process.

1 *Knowledge.* The individual is exposed to the innovation's existence and gains some understanding of how it functions.
2 *Persuasion.* The individual forms a favorable or unfavorable attitude toward the innovation.
3 *Decision.* The individual engages in activities which lead to a choice to adopt or reject the innovation.
4 *Confirmation.* The individual seeks reinforcement for the innovation-decision he has made, but he may reverse his previous decision if exposed to conflicting messages about the innovation.

These stages will shortly be described in detail. Our model is designed to account for the major criticisms raised about the five-stage adoption process, to profit from recent researches on the process, and to be consistent with the learning process,* theories of attitude change,** and general ideas about decision making.

The model (Figure 3-1) contains three major divisions: (1) antecedents, (2) process, and (3) consequences. Antecedents are those variables present in the situation prior to the introduction of an innovation. Antecedents consist of: (1) the individual's personality characteristics, such as his general attitude toward change, (2) his social characteristics, such as his cosmopoliteness, and (3) the strength of his perceived need for the innovation. All these variables and others affect the way in which the innovation-decision process occurs for a given individual.

The social system's norms (modern or traditional, for example) serve as incentives or restraints on the individual's decisions. Such system variables as tolerance for deviancy, communication integration, and other characteristics

---

*_Learning_ is a relatively enduring change in a covert or overt response as the result of a perceived stimulus. The way in which new ideas are adopted by an individual is essentially parallel to how any type of learning takes place. In the innovation-decision process, innovation messages reach the individual by various communication channels. The effect of each ensuing message (the stimulus) about the innovation accumulates until the individual interprets it and decides to adopt or reject (the response) the innovation. This is learning.
**The model of the innovation-decision process owes an intellectual debt to the attitude change model proposed by the late Carl Hovland of Yale University, a pioneering social psychologist in the study of attitude formation and change. He proposed three sequences in his attitude change model. (1) arousal of the individual's attention, (2) his comprehension of the message, and (3) his acceptance (Hovland and others, 1953, p. 5). Most of Hovland's research, as well as that of learning psychologists, was conducted in laboratory settings, and we must be cautious in applying his models or findings to innovation decision making in the "real world." However, Hovland (1959) recognized the need for studies using both laboratory and field approaches: "What seems to me quite apparent is that a genuine understanding of the effects of communication on attitudes requires both the survey and the experimental methodologies."

also affect the nature of the innovation-decision processes of the system's members.

Communication sources and channels provide stimuli to the individual during the innovation-decision process. The typical individual gains initial knowledge of the innovation mainly from cosmopolite and mass media channels. At the persuasion function, the individual forms his perceptions of the innovation from more localite and interpersonal channels. An innovation may be adopted at the decision stage in the process and be used continuously or rejected at a later date (a discontinuance). A discontinuance may be due to the innovation's replacement by an improved idea or to disenchantment with the innovation. The new idea may be rejected at the end of the process but adopted at a later date due to changes in how the individual perceives the innovation. Continued information-seeking often occurs throughout the confirmation function, because the individual seeks to reinforce his decision. Sometimes, however, contradictory (to the innovation-decision) messages reach the individual, and this leads to discontinuance or later adoption.

In the following pages we describe in greater detail behaviors which occur at each of the four functions and the purpose which each function fulfills in the innovation-decision process.

Knowledge Function

We conceive of the innovation-decision process as beginning with the knowledge function, which commences when the individual is exposed to the innovation's existence and gains some understanding of how it functions.

Which Comes First, Needs or Awareness ?

Many researchers conceptualize awareness-knowledge as occurring due to random or nonpurposive activities by the individual. He often becomes aware of an innovation quite by accident; he cannot actively seek out an innovation which he does not know exists.* However, Hassinger (1959)

---

*As Tarde (1903, p. 93) points out: "Since the desire for, cannot precede the notion of, an object, no social desire can be prior to the invention." Coleman and others (1966, p. 59) conclude that awareness about a new drug seldom comes from a channel which physicians have to seek out themselves, but that at later stages in the decision process doctors become more active in their quest for information.

criticizes the assumption of nonpurposiveness of awareness-knowledge; he argues that knowledge-seeking must be initiated by the individual and is not a passive activity. The predispositions of individuals influence their behavior toward communication messages and the effects which such messages are likely to have. Generally, individuals tend to expose themselves to those ideas which are in accord with their interests, needs, or existing attitudes. We consciously or unconsciously avoid messages which are in conflict with our predispositions. This tendency is called *selective exposure*.* Hassinger (1959) argues that individuals will seldom expose themselves to messages about an innovation unless they first feel a need for the innovation, and that even if such individuals are exposed to such innovation messages, there will be little effect of such exposure unless the individual perceives the innovation as relevant to his needs and as consistent with his existing attitudes and beliefs.** For example, a farmer can drive past 100 miles of hybrid corn in Iowa and never "see" the innovation. Likewise, we are all exposed daily to hundreds of mass media messages about new products. But few of these register on our minds. Selective exposure and selective perception act as particularly tight shutters on the windows of our minds in the case of innovation messages, because such ideas are new. We can scarcely have consistent and favorable attitudes and beliefs about ideas which we have not previously encountered. There is, then, much in the ideas of selective exposure and selective perception to support Hassinger's viewpoint that need for an innovation must usually precede awareness-knowledge.

But how are needs created? A need is a state of dissatisfaction or frustration that occurs when one's desires outweigh one's actualities, when "wants" outrun "gets."*** An individual may develop a need when he learns that an improved method, an innovation, exists. Therefore, innovations *can* lead to needs, as well as vice versa. Some change agents use this approach to change by creating needs among their clients through pointing out the desirable consequences of new ideas. Thus, knowledge of innovations can create motivation for their adoption.

Selective exposure is the tendency to attend to communication messages that are consistent with one's existing attitudes and beliefs.
**This is *selective perception*, the tendency to interpret communication messages in terms of one's existing attitudes and beliefs. One is reminded of the Philadelphia clothing manufacturer who gained an audience with the Pope while in Rome on business. Upon his return, the businessman's friend asked him what His Holiness looked like. The clothing manufacturer replied: "A forty-one, regular."
***The wants/gets ratio is often seen as a basic ingredient in the modernization of traditional individuals like peasants in less developed nations (Lerner, 1963, p. 349; Rogers with Svenning, 1969, pp. 12–14).

In addition to these specific needs, each dealing with a perceived problem that the individual feels an innovation could help solve, there are also very general needs largely independent of specific innovations. These general needs may result from a desire for a higher level of living, from a favorable attitude toward change, and the like. This type of general need may also lead to awareness of an innovation and to speedy passage through the functions in the innovation-decision process. But it does not cause information seeking about a specific innovation, at least in the same sense that a perceived need for that particular innovation does.

What can we conclude? Does a need precede knowledge of a new idea or does knowledge of an innovation create a need for that new idea? Perhaps this is a chicken-or-egg problem. In any event available research does not provide a clear answer to this question of whether awareness of a need or awareness of an innovation (that creates a need) comes first. The need for some innovations, such as a pesticide to treat a new crop pest, probably comes first. But for other new ideas, the innovation may create the need.

Types of Knowledge

Our discussion up to this point has dealt entirely with one type of knowledge: Awareness that an innovation exists. There are two additional types of knowledge: "How-to"-knowledge and principles-knowledge. These are less likely than awareness-knowledge to occur only at the knowledge function.

"How-to"-knowledge consists of information necessary to use an innovation properly. The adopter must understand what quantity of an innovation to secure, how to utilize it correctly, and so on. In the case of innovations that are relatively more complex, the amount of "how-to"-knowledge needed for proper adoption is much greater than in the case of less complex ideas. And when an adequate level of "how-to"-knowledge is not obtained prior to trial and adoption of the innovation, rejection or discontinuance is likely to result. To date we have few diffusion investigations that deal with "how-to"-knowledge.*

An even more general type of knowledge is that dealing with the functioning principles underlying the innovation. Examples of principles-knowledge are:

*Among the few inquiries of this type are Keith (1968b), who determined the variables correlated with knowledge of fourteen agricultural innovations among 1,347 Nigerian peasants, and White (1967), who studied the correlates of knowledge of innovations among Canadian farmers. Their findings (and those of other researchers) as to the characteristics of early and late knowers about innovations are summarized later in this chapter.

The notion of germ theory, which underlies the functioning of vaccinations and latrines in village sanitation and health campaigns; the fundamentals of human reproduction, which form a basis for all family planning innovations; and the biology of plant growth, which underlies fertilizer innovations. It is usually possible to adopt and use an innovation without possession of principles-knowledge, but the long range competence of individuals to judge future innovations is facilitated by principles know-how.

What is the role of change agents in bringing about the three types of knowledge? Most change agents seem to concentrate their efforts on creating awareness-knowledge, although this goal can be achieved more efficiently in many client systems by mass media channels. Change agents could perhaps play their most distinctive and important role in the innovation-decision process if they concentrated on "how-to"-knowledge, which is probably most essential to clients at the trial and decision function in the process.* Most change agents perceive that creation of principles-knowledge is outside the purview of their responsibilities and is a more appropriate task for formal schooling and general education. It is admittedly difficult for change agents to teach basic understanding of principles. But when such understanding is lacking, the change agent's long-run task remains very difficult. For instance, in India change agents advocate the adoption of new crop varieties to villagers. But because the basic principles of how to evaluate these seed innovations is never developed, the change agents must conduct repeated diffusion campaigns each time a new crop variety becomes available.

Early Versus Late Knowers of Innovations

As pointed out in Chapter 2, we presently have in hand far more diffusion findings regarding variables related to innovativeness than we have studies of knowledge as a dependent variable. In other words we know much more about what innovators and laggards are like than we know about how earlier knowers differ from later knowers.

The following generalizations summarize the results of findings regarding early knowing about an innovation.

Generalization 3-1: *Earlier knowers of an innovation have more education than later knowers.*

*One possible reason for the underemphasis by change agents upon creating how-to knowledge is that change agency policies usually provide greater rewards to change agents for creating awareness-knowledge, perhaps because it is often easier to measure.

Generalization 3-2: *Earlier knowers of an innovation have higher social status than later knowers.*

Generalization 3-3: *Earlier knowers of an innovation have more exposure to mass media channels of communication than later knowers.*

Generalization 3-4: *Earlier knowers of an innovation have more exposure to interpersonal channels of communication than later knowers.*

Generalization 3-5: *Earlier knowers of an innovation have more change agent contact than later knowers.*

Generalization 3-6: *Earlier knowers of an innovation have more social participation than later knowers.*

Generalization 3-7: *Earlier knowers of an innovation are more cosmopolite than later knowers.*

We generally can observe that the characteristics of earlier knowers of an innovation are similar to the characteristics of innovators:* More education, higher social status, and the like. But of course this does not mean that earlier knowers are necessarily the same individuals as innovators.

Knowledge-Behavior Consistency

Knowing about an innovation is often quite a different matter from using the idea.** Most individuals know about many innovations which they have not adopted. Why? One reason is because the individual knows about the new idea but does not regard it as relevant to his situation, as potentially useful. Therefore, attitudes toward an innovation frequently intervene between the knowledge and decision functions.*** In other words the individual's attitudes or beliefs about the innovation have much to say about his passage through the decision process. Consideration of a new idea does not pass beyond the knowledge function if the individual does not define the information as relevant to him or if he does not seek sufficient knowledge to become adequately informed so that persuasion can take place. What do we know about this persuasion function?

*The characteristics of innovators and other adopter categories are summarized in Chapter 5.

**For instance, Keith (1968b) found a correlation of .58 between knowledge and adoption of agricultural innovations by Nigerian peasants; this means about 36 percent of the variance in knowledge and innovativeness occurs together. Several other researchers report a relationship of about the same magnitude between the two variables.

***But there is also a relationship between levels of knowledge and of adoption in a social system, as we shall show in our discussion of the diffusion effect in Chapter 4.

Persuasion Function*

At the persuasion function in the innovation-decision process the individual forms a favorable or unfavorable attitude** toward the innovation. Whereas the mental activity at the knowledge function was mainly cognitive (or knowing), the main type of thinking at the persuasion function is affective (or feeling). Until the individual knows about a new idea, of course, he cannot begin to form an attitude toward it.

At the persuasion stage the individual becomes more psychologically involved with the innovation. Now he actively seeks information about the idea. His personality as well as the norms of his social system may affect *where* he seeks information, *what* messages he receives, and how he *interprets* the information he receives. Thus, selective perception is important in determining the receiver's communication behavior at the attitude formation stage. For it is at the persuasion stage that a general perception of the innovation is developed. Such perceived attributes of an innovation as its relative advantage, compatability, and complexity are especially important at this stage.

In developing a favorable or unfavorable attitude toward the innovation, the individual may mentally apply the new idea to his present or anticipated future situation before deciding whether or not to try it. This might be thought of as a vicarious trial.*** The ability to think hypothetically and counterfactually, to project into the future, is an important mental capacity at the persuasion stage where forward planning is involved.

All innovations carry some degree of subjective risk to the individual. He is unsure of the idea's results and feels a need for reinforcement of his attitudes toward the new idea. He is likely to seek conviction that his thinking is on the right path from peers by means of interpersonal communication channels. Mass media messages are too general to provide the specific kind of reinforcement that the individual needs to confirm his beliefs about the innovation. And

*We do not define persuasion with exactly the same connotation as many communication researchers, who use the term to imply a source's communication with an intent to induce attitude change in a desired direction. Our meaning for persuasion is equivalent to attitude formation and change on the part of a receiver but not necessarily in the direction intended by some particular source, such as a change agent. Our meaning of persuasion is receiver-oriented, rather than source-oriented.

***Attitude* is a relatively enduring organization of an individual's beliefs about an object that predisposes his actions. This definition is based most closely on Rokeach (1966).

***This type of mental trial, however, is distinct from the physical trial of the innovation, which we see as more a part of the decision function.

the peers that he seeks out for such reinforcement are likely to be rather similar to him in their characteristics; such homophily connotes higher credibility of the persuasive innovation messages.

Types of Attitudes

Just as there are at least three levels of knowledge about an innovation, there are at least two levels of attitudes: (1) a specific attitude toward the innovation, and (2) a general attitude toward change. The first, attitude toward the innovation, is our main concern at the persuasion stage in the innovation-decision process. It consists essentially of a favorable or unfavorable belief in the usefulness of the new idea for the individual.

Such a specific attitude, however, has a carryover from one innovation to another. A previous positive experience with the adoption of innovations creates a bank of generally favorable attitudes to change that facilitates the development of a favorable evaluation of the next innovation considered by an individual. On the contrary, a negative experience from an innovation that is perceived as a failure leads to resistance to future new ideas. Change agents should therefore begin their activities in a particular client system with an innovation that possesses a high degree of relative advantage, that is compatible with existing beliefs, and that has a very high likelihood of success. This will help create a general, positive set toward change and grease the ways for later ideas that may be introduced.

One strategy of diffusion for a change agent, then, is to develop a positive general attitude toward change on the part of his clients. Individuals in such a change-oriented system are self-renewing, self-actualizing, open to the new, and active in inquiring about innovations. We just pointed out that one route to developing such a general orientation toward change lies in the proper selection of innovations for sequential introduction. There are also other methods. Proponents of T-group or sensitivity training claim that one result of such an experience is an openness toward new ideas, although this type of training is so costly as to be impractical for a large number of individuals. Another means to create a pro-change orientation is to expose individuals to a bombardment of modernizing messages even though the messages may not deal with specific innovations *per se*. An example of this approach is found among peasants in less developed nations where it is claimed that the mass media of radio, television, film, and newspapers can create a climate for modernization.* The mass media in these settings are filled with pro-change ideas, and one result of peasant exposure to such messages is creation of a favorable attitude toward change, which acts to facilitate the adoption of specific new ideas.

Attitude-Behavior Consistency

The main outcome of the persuasion function in the decision process is either a favorable or an unfavorable attitude toward the innovation. It is assumed that such persuasion will lead to a subsequent change in overt behavior (that is, adoption or rejection) consistent with the attitude held. But there is little evidence that attitude and overt behavior are always consistent, and we know of many cases in which attitudes and actions are quite disparate.

An illustration is provided by a classic social psychological study by LaPiere (1934). A number of U.S. hotel owners and managers were asked if they were willing to house Chinese guests. Most said no. LaPiere and a well dressed Chinese couple then appeared in person at the hotels and requested lodging. Almost all the hotels provided rooms to the guests. These results are cited as evidence that verbally expressed attitudes are not always entirely consistent with actions.**

Recent field investigations conducted under more closely controlled conditions also provide definitive evidence of the distinction between attitudes and actions. For instance Festinger (1964) summarizes the results of three field experiments in which a change in attitude toward an innovation occurred for many respondents, but this attitude change did not result in behavioral change.***

*"The mass media's role in modernizing peasants of a less developed country may be mainly to form a generally favorable attitude toward new ideas, the so-called 'climate for modernization', rather than to provide the specific details needed for adoption of these innovations" (Rogers with Svenning, 1969, p. 110).

**There are numerous possible shortcomings of this study, such as the fact that the experimenter accompanied the Chinese couple, that they were well dressed, and that hotel clerks' actions on one hand are contrasted with the attitudes of hotel owners and managers on the other.

***Rokeach (1967 and 1968) provides some generally confirmatory support for this position. He argues that reasons for attitude-behavior dissonance lie in the personality structure of the individual and that attitudes toward both the innovation and *toward the situation* in which the innovation is introduced need to be considered in order to predict behavioral outcomes. This has seldom been done in past studies.

We should remember that formation of a favorable or unfavorable attitude toward an innovation does not always lead directly or immediately to an adoption or rejection decision. Nevertheless, there is a tendency in this direction, that is, for attitudes and behavior to become more consistent. *Innovation dissonance* is the discrepancy between an individual's attitude toward an innovation and his decision to adopt or reject the innovation. Innovation dissonance is a specific type of cognitive dissonance, and we know from Festinger's (1957) theory that there is pressure in the direction of dissonance reduction. The psychological state of dissonance is uncomfortable, and therefore individuals seek to reduce this tension by bringing their attitudes and their actions into line. Attitude toward an innovation at the persuasion stage in the innovation-decision process is generally (but not perfectly) predictive of a decision to adopt or reject. We shall return to further discussions of innovation dissonance when we deal with the confirmation function, and again in Chapter 10, when we discuss authority decisions.

Decision Function

At the decision function in the innovation-decision process, the individual engages in activities which lead to a choice to adopt or reject the innovation.* Actually the entire innovation-decision process is a series of choices at each function. For instance, in the knowledge function the individual must decide which innovation messages to attend to and which ones to disregard. In the persuasion function he must decide to seek certain messages and to ignore others. But in the decision function the type of choice is different from those previous; it is a decision between two alternatives, to adopt or to reject a new idea. This decision involves an immediate consideration of whether or not to try the innovation, if it is trialable. Most individuals will not adopt an innovation without trying it first on a probationary basis to determine its utility in their own situation. This small-scale trial is often part of the decision to adopt, and is important as a means to decrease the perceived risk of the innovation for the adopter. In some cases, an innovation cannot be divided for trial, and so it must be adopted or rejected in toto.** Innovations which can be divided

---

*As defined in Chapter 1, *adoption* is a decision to make full use of a new idea as the best course of action. *Rejection* is a decision not to adopt an innovation.
**Or it may be given a vicarious trial through the experience of peers.

for trial use are generally adopted more rapidly. Most individuals who try an innovation then move to an adoption decision, if the innovation has at least a certain degree of relative advantage. Methods to facilitate the trial of innovations, such as the distribution to clients of free samples of a new idea, will speed up the rate of adoption. Evidence for this point is provided from a field experiment among Iowa farmers, where it was found that the free trial of a new weed spray speeded the innovation-decision period by about a year.*

We see the decision to adopt or reject and the actual use or non-use of the innovation as somewhat different behaviors. In the case of optional innovation-decisions (the main concern of this chapter), these two behaviors usually occur concurrently and in the same individual. But in the case of authority decisions, for example, the unit of decision and the unit of adoption (or rejection) may be different individuals and the two events (decision and use) may not occur at the same point in time.

It is important to remember that the innovation-decision process can just as logically lead to a rejection decision as to adoption. In fact, each function in the process is a potential rejection point. For instance, it is possible to reject an innovation at the knowledge function by simply forgetting about it after initial awareness. And of course rejection can occur even after a prior decision to adopt. This is discontinuance, which can occur in the confirmation function.

Confirmation Function

Empirical evidence supplied by several researchers** indicates that a decision to adopt or reject is not the terminal stage in the innovation-decision process. For example, Mason (1962a) found that his respondents, who were Oregon farmers, sought information *after* deciding to adopt as well as before. At the confirmation function the individual seeks reinforcement for the innovation-decision he has made, but he may reverse his previous decision if exposed to conflicting messages about the innovation. The confirmation stage continues after the decision to adopt or reject for an indefinite period in time

*Details of this study are provided by Klonglan (1962 and 1963) and Klonglan and others (1960 and 1963).
**Such as Mason (1962b, 1963, 1964, 1966a, b), Francis and Rogers (1962), Ehrlich and others (1957).

(Figure 3-1). Throughout the confirmation function the individual seeks to avoid a state of dissonance or to reduce it if it occurs.

Dissonance

Human behavior change is motivated in part by a state of internal disequilibrium or dissonance, an uncomfortable state of mind that the individual seeks to reduce or eliminate. When an individual feels dissonant, he will ordinarily be motivated to reduce this condition by changing his knowledge, attitudes, or actions. In the case of innovative behavior, this may occur:

1. When the individual becomes aware of a felt need or problem and seeks information about some means such as an innovation to meet this need. Hence, a receiver's knowledge of a need for innovation can motivate information-seeking activity about the innovation. This occurs at the knowledge stage in the innovation-decision process.

2. When he becomes aware of a new idea for which he has a favorable regard. Then the individual is motivated to adopt the innovation by the dissonance between what he believes and what he is doing. This behavior occurs at the decision stage in the innovation-decision process.

3. After the innovation-decision to adopt. The individual may secure further information which persuades him that he should not have adopted. This dissonance may be reduced by discontinuing the innovation. Or if he originally decided to reject the innovation, the individual may become exposed to pro-innovation messages, causing a state of dissonance which can be reduced by adoption. These types of behavior (discontinuance or later adoption) occur during the confirmation function in the innovation-decision process.

These three methods of dissonance reduction consist of changing behavior so that attitudes and actions are more in line. But often it is difficult to change one's prior decision to adopt or reject; activities have been set in motion which tend to stabilize the original decision. Perhaps a considerable cash outlay was involved in adoption of the innovation, for instance. Therefore individuals frequently try to avoid becoming dissonant by seeking only that information which they expect will support or confirm a decision already made. This is an illustration of selective exposure.* During the confirmation stage the in-

---

*Similarly, dissonance can be reduced by selective perception (message distortion) and by the selective forgetting of dissonant information.

dividual wants supportive messages that will prevent dissonance from occurring.

The addition of the confirmation function to the innovation-decision process suggests a new role for the change agent. Whereas change agents have in the past primarily been interested in achieving adoption decisions, the new model gives them the additional responsibility for providing supporting messages to individuals who have previously adopted. Possibly one of the reasons for the relatively high rate of discontinuance of some innovations is that change agents assume that once adoption is secured, it will continue. But without continued effort there is no assurance against discontinuance, because negative messages about an innovation exist in most client systems.

Discontinuance

A *discontinuance* is a decision to cease use of an innovation after previously adopting it. If adoption of an innovation amounts to "unfreezing" some old behavior in order to adopt a new idea, discontinuance happens when "freezing" of the new behavior does not occur.

There have been relatively few researches designed to investigate the nature of discontinuance,* and as a result we know relatively less about this important aspect of diffusion behavior. But one must conclude from these few studies that there is a rather surprisingly high rate of discontinuance for many innovations. In fact, Leuthold (1967, p. 106) concluded from his study among a statewide sample of Wisconsin farmers that the rate of discontinuance was just as important as the rate of adoption in determining the level of adoption of an innovation at any particular time. In other words, for any given year there were about as many discontinuers of an innovation as there were first-time adopters. A similar experience has been reported in India where the rate of discontinuance of birth control loops has been higher than the rate of adopters.** As a result, change agents have had to devote increasing attention to prevent discontinuance of innovations.

*Even from the six or seven studies in which innovation discontinuances were measured, one gains the general impresssion that the discontinuance variable was almost serendipitously chanced upon, rather than planned as a main objective of the study. Most of the discontinuance researches are panel studies in which the adoption and later discontinuance of innovations are measured at two points in time for the same individuals.

**In this case many of the discontinuances are due to a rumor about a husband and wife who could not uncouple because of using the loop.

There are at least two types of discontinuances: Replacement and disenchantment. A *replacement discontinuance* is a decision to cease using an idea in order to adopt a better idea* which supersedes it. In a rapidly changing culture there are constant waves of innovations. And each new idea replaces an existing practice which in its day was an innovation too.

A *disenchantment discontinuance* is a decision to cease using an idea as a result of dissatisfaction with its performance. The dissatisfaction may come about because the innovation is inappropriate for the individual and does not result in a perceived relative advantage over alternative practice.** Or the dissatisfaction may result from misuse of an innovation that could have functioned advantageously for the individual. This later type of disenchantment seems to be more likely among later adopters than among earlier adopters. Laggards have less education and more traditional attitudes and values which might be expected to lead to discontinuance. Unless one possesses a conception of the scientific method, it is difficult to understand how to generalize the results of an innovation's trial to its full-scale use. Laggards also seem to be more submissive to authority in their attitudes toward change agents; they may adopt as the direct result of influence from change agents. When this coercive influence is removed, the innovation is likely to be discontinued. Further, later adopters have fewer resources, which may either prevent adoption or cause discontinuance because the innovations do not fit their limited financial position.

All of this reasoning is consistent with the findings of Johnson and van den Ban (1959) in Wisconsin, Leuthold (1965) in Saskatchewan and (1967) in Wisconsin, Bishop and Coughenour (1964) in Ohio, Silverman and Bailey (1961) in Mississippi, and Deutschmann and Havens (1965) in Colombia. We are led to Generalization 3-8: *Later adopters are more likely to discontinue innovations than are earlier adopters.**** Researchers previously assumed that later adopters are relatively less innovative because they did not adopt or were slower to adopt. But the evidence on discontinuances suggests that many laggards adopt but then discontinue, usually due to disenchantment.

*"Better" in the sense that the individual perceives it as better.
**These disenchantment discontinuances may be caused by overadoption of the innovation, a theory discussed further in Chapter 4.
***For instance, Bishop and Coughenour (1964) reported that the percentage of discontinuance for Ohio farmers ranged from 14 percent for innovators and early adopters, to 27 percent for early majority, to 34 percent for late majority, to 40 percent for laggards. Leuthold (1965) reported comparable figures of 18 percent, 24 percent, 26 percent, and 37 percent, respectively, for Canadian farmers.

Several investigators* have determined the characteristics of those individuals with a high and a low rate of discontinuance. Generally, high discontinuers have less education, low social status, less change agent contact, and the like, which are the opposites of the characteristics of innovators. Discontinuers share the same characteristics as laggards, whom we know to have a higher rate of discontinuance.

The discontinuance of an innovation is one indication that the idea was not integrated** into the practices and way of life of the receivers. Such integration is, of course, less likely (and discontinuance is more frequent) when the innovation is less compatible with the receivers' beliefs and past experiences. An example in the United States occurs in the case of government-sponsored "soil bank" programs in which farmers are paid a subsidy if they retire certain of their fields from cultivation. Some farmers are eager to receive the cash subsidy, but the innovation runs counter to farmers' values on high agricultural production. Even those farmers who adopt are likely to discontinue soil banking once the subsidy is halted. The innovation is simply not integrated into their way of life.

On the other hand, the innovations of hybrid corn and chemical fertilizer are now so firmly integrated into the farming methods in the United States that discontinuance is highly unlikely unless more profitable innovations become available or a penalty were levied against those who continued to use the ideas.***

This discussion suggests that there are innovation-to-innovation differences in rates of discontinuance, just as there are such differences in rates of adoption, and that the perceived attributes of innovations (for example, relative advantage and compatibility) are related inversely to the rate of discontinuance. For instance, we expect an innovation with a low relative advantage to have a slow rate of adoption and a fast rate of discontinuance. Further, innovations that have a high rate of adoption should have a low rate of discontinuance. This theory is supported by the findings of Coughenour (1961) in Kentucky, Silverman and Bailey (1961) in Mississippi, Johnson and

*For example, Leuthold (1967), Leuthold (1965), Deutschmann and Havens (1965), and Wilkening (1952a).
**Integration of an innovation occurs when it has been incorporated into the operations and way of life of members of a social system.
***To protect the health of consumers, the Netherlands government ordered farmers to discontinue the innovation of using hormones in veal production. But this idea was so well integrated in farming practice that use of hormones continued for some years. The case is similar to that of the use of DDT by U.S. farmers, who were ordered by the federal government to discontinue.

van den Ban (1959) in Wisconsin, and Leuthold (1965) in Saskatchewan. Generalization 3-9 is: *Innovations with a high rate of adoption have a low rate of discontinuance.**

The Home Canning Campaign In Georgia**

This case study illustrates how one individual adopted a new idea, home canning of food. Our parenthetical remarks primarily indicate the nature of functions in the innovation-decision process. We see an illustration of the role that interpersonal communication from peers plays in persuading an individual to adopt a new idea.

The Campaign

In Greene County, Georgia, as was the case throughout the cotton plantation area of the South prior to World War II, most tenant families had a diet of fat-back pork, corn meal, and sorghum molasses. In the early 1940s hundreds of low-income farm families in Greene County were convinced to can an average of nearly 500 quarts of food per person each year. Several hundred farm families in the county received loans and supervision from the Farm Security Administration, a government agency, to aid them in adopting the innovation. Most of these families were farm tenants who had canned only an average of twelve quarts per family prior to 1939. (*The members of the social system had little previous experience with the innovation, home canning*). They increased their canning to an average of 225 quarts in 1939, to 350 in 1940, to 386 in 1941, and to 499 in 1942. (*The canning campaign caused major changes in human behavior*). How these families raised their average from a dozen quarts to half a thousand in four years can best be shown by the case of one of the client families.

Lula McCommons' Innovation-Decision

Lula McCommons, a Negro mother of seven children, did not believe in canning. (*The individual has knowledge of the new idea but is not yet persuaded*). She had always gone to the store for something to eat when she could afford to do

*And, one might add, a high degree of integration into the life styles of the receivers.
**Adapted from Raper and Tappan (1943), and used by permission.

so; when she could not, she and her family got along somehow. Lula had never canned more than eight or nine quarts, as "I ain't ever had any more jars, and where can I get the stuff to can?" (*The individual seeks additional "how-to"-knowledge at the persuasion function*). When she was told how many jars she was expected to can, she said, "It ain't any use to tell you I kin can that much. I don't lie. I can't." Every method of conviction was tried. The Farm Security Administration supervisors (*the change agents*) appealed to Lula's husband. Mary, Lula's twelve-year-old daughter, was asked to help. All promised to do what they could, but no canning was accomplished. Lula and Mary were invited down the road to the home of one of the best canners and were left there to talk for more than an hour. (*The individual is exposed to interpersonal communication with peers at the persuasion function*).

A few days later, Lula began to can. She agreed to put up seventy-five quarts. (*The individual decides to try the idea*). She attended church the next Sunday, and between Sunday school and preaching, the women were all talking about how many and what they canned. (*Reinforcement and support for the decision is provided, thus reducing the dissonance of the decision*). Lula said: "I just told myself that if old Mary Rooth and all them others could can all them quarts, I could too. So I just told old Satan to git behind me, for I was goin' to can everything I could git my hands on." And she did (*A decision is made to make full use of the new idea as the best course of action*). She canned 675 quarts that year. The next year she canned nearly 800 quarts. (*Her behavior is further reinforced and strengthened as she continues use of the idea*).

Consequences of the Campaign

One unexpected consequence of the canning campaign in Georgia was the element of prestige that soon came to be associated with canned food. In fact, many families kept their jars on display in the parlor or guest room or on shelves around the kitchen. In their zeal to use all of their jars, some families filled them with sweet potatoes, pumpkins, turnips, and other foods which could be stored without canning. (*The innovation has latent and dysfunctional consequences not intended by the change agents*). Many people were so proud of their canned goods the first year that they would not open their jars. The Farm Security Administration supervisors realized then that they had to teach the families to use the food they had canned. (*The change agents recognize their responsibility to assist their clients in securing the desired consequences of the innovation*). If the change agents had withdrawn when the idea of canning was adopted, there would have been little improvement in the economic or nutritional condition of the individuals because they had not yet adopted the idea that they should also *consume* the food they had canned. (*Clients may perceive a new idea differently from the change agents*).

Are There Functions in the Process?

What empirical evidence is available that the functions posited in our model of the innovation-decision process (Figure 3-1) exist in reality?

Before seeking to answer this question, it should be pointed out that a definitive answer is difficult to provide. Researchers can probe only indirectly the mental processes of individual respondents. Nevertheless, there is tentative evidence from several studies in the United States and in less developed countries that the concept of functions in the innovation-decision process is supported.

Evidence of the Functions

Empirical evidence of the validity of functions in the innovation-decision process comes from an Iowa study (Beal and Rogers, 1960) which shows that most farmer-respondents recognized that they went through a series of stages as they moved from awareness-knowledge to a decision to adopt. Specifically, they realized they had received information from different sources and channels at different stages. Of course, it is possible for an individual to use the same sources or channels, perhaps in a different way, at several functions in the innovation-decision process. However, if respondents report different sources or channels at each function, this tends to indicate some differentiation of the functions. Beal and Rogers (1960) found that all their respondents reported different communication channels for two agricultural innovations at the knowledge and at the decision functions, and there was a good deal of channel differentiation between the knowledge and persuasion stages. There are many other research studies, reviewed elsewhere in this book,* which also indicate a differentiation of sources and channels at different functions in the innovation-decision process.**

*Especially in Chapter 8.
**Herbert Lionberger of the University of Missouri indicated to us that in his research he finds little duplication of the individuals sought by farmers for knowledge versus persuasive messages.

Beal and Rogers (1960) also found that none of their 148 respondents reported adopting immediately after becoming aware of the two new ideas. Instead, 73 percent of the adopters of a new weed spray and 63 percent of the adopters of a new livestock feed, reported different years for knowledge and for the decision to adopt. Most individuals seemed to require a period of time that could be measured in years to pass through the innovation-decision process. This provides some indication that adoption behavior is a *process* that contains various functions and that these stages occur over time.

Yet another type of evidence provided by Beal and Rogers (1960) deals with skipped stages. If most respondents report not having passed through a stage in the innovation-decision process for a given innovation, some question is thus raised as to whether that stage should be included in the model. However, Beal and Rogers found most farmers described their behavior at each of the first three stages in the process: Knowledge, persuasion, and decision.* None reported skipping the knowledge or decision stages, but a few farmers did not seem to pass through the persuasion function, and some did not report a trial prior to adoption.

Farmers in relatively traditional social systems may be more likely to skip the trial, at least in terms of a small-scale initial use of the innovation,** although evidence on this point is rather tentative. Deutschmann and Fals Borda (1962b) found in their analysis of a Colombian peasant community that in only 15 percent of the situations in which innovations could be tried on a partial basis was this done. The villagers tended to go to a full-scale adoption decision without a trial test. However, almost all of the respondents (80 percent) observed the innovation in use on a neighbor's farm before they adopted. Perhaps this observation psychologically fulfilled some of the usual functions of a small-scale trial on their own farm.

Similar results are reported by Rahim (1961a) among Pakistani peasants, many of whom skipped a trial.*** In an investigation of 233 dairy farmers in

---

*Beal and Rogers (1960) used the adoption process model, rather than our present innovation-decision model, and so they did not gather data about the confirmation function.
**A somewhat parallel finding from several studies (Ryan and Gross, 1943; Ryan, 1948; Katz, 1961; Rogers and Pitzer, 1960, p. 28; Haber, 1963) is that earlier adopters (who are generally less traditional in their orientations) try innovations on a smaller scale than later adopters. Not only are more traditional individuals more likely to skip the trial, but they are also less likely to adopt on the installment plan.
***In Pakistan, at least, one possible reason for skipping a trial is that the peasants' fields are simply so small, often only a fraction of an acre, that it is impractical to try an innovation on a partial basis because of the inconvenience of planting, cultivating, and harvesting on such a diminutive scale.

Puerto Rico, Oliver-Padilla (1964) found that more traditional farmers were more likely to skip certain of the stages in the innovation-decision process. In comparison innovators closely followed each of the functions in our model. Perhaps individuals in more traditional social systems and the more tradition-oriented individuals in any system are more likely to skip stages (especially the trial stage) for the following reasons:

1. The norms of the social system have a greater impact upon these traditional individuals, which in essence releases the individual from making his own decision about an innovation on the basis of his own trial. Thus, an optional innovation-decision almost becomes a collective decision.

2. These more traditional individuals are less accustomed to following the elements of the scientific method in their decision making, and hence, are less likely to generalize the results of a trial experiment to wider application. They lack the basic notions of sampling and generalization.

3. The change agent may be a stronger persuasive force to these traditional individuals, who tend to be more submissive to authority than their more modern peers. Therefore, when the change agent introduces a new idea, the traditional individual is more likely to adopt without much further questioning. A seemingly optional innovation-decision thus becomes an authority decision.

On the basis of this reasoning and the rather spotty evidence we have in hand, we suggest Generalization 3-10: *Traditonal individuals are more likely to skip functions in the innovation-decision process than are modern individuals.*

Methodological Problems in Studying Stages

In this section we have reviewed evidence for the existence of functions in the innovation-decision process. We generally conclude that such stages do exist, but the evidence is least clear-cut to support the notion of a confirmation stage. Perhaps this is because few researchers other than Mason (1962b, 1963, 1964, 1966a, b) have conceptualized such a stage in their model. And because investigators like Beal and others (1957), Beal and Rogers (1960), Wilkening (1956), and Copp and others (1958) did not look for the confirmation stage, they did not find it. Certainly, further research is needed on the exact nature of behavior at the confirmation function.

This point raises a more general issue: Is evidence for the presence of stages in the innovation-decision process simply an artifact of a self-fulfilling model in the researchers', rather than the respondents', minds? In other words, is it because researchers set out to look for stages that they find them? The questions asked of respondents are designed to operationalize a multi-stage model. Perhaps the respondents' behavior is thus structured into the researchers' conceptual mold.*

We feel this is not a very serious methodological factor in affecting our judgement of the evidence at hand. It is true that the early investigations of stages, such as Wilkening's (1956) and Beal and others' (1957), utilized direct questions which structured the respondents' replies in the direction of a multi-stage model. But several more recent studies using less structured methods of data gathering still provide evidence that stages exist. For instance, Copp and others (1958) asked their interviewers to describe the sequence of how they adopted certain innovations. Then the data were inspected to determine whether stages existed. A somewhat similar method was used by Singh and Pareek (1968) to determine the presence of stages in the innovation-decision process among Indian farmers. They used depth interviews, which are relatively unstructured methods of data gathering, and found general support for the notion of stages in the process, but there were many exceptions. The same number of stages did not occur for all respondents nor for all innovations. There was some skipping of stages, and the sequence was not always the same. In a few cases the whole process was capsuled into a unit act, an impulse decision. But nevertheless, most of the Indian villagers described different stages in the process.

A further shortcoming of the evidence just reviewed is that all these studies deal with farmer respondents. How do we know that our innovation-decision process model also describes the behaviors of other types of individual and other kinds of innovations? Fortunately, we now have three other studies, one of physicians (Coleman and others, 1966) and two of school personnel (LaMar, 1966; Kohl, 1966). Their results generally support the validity of stages in the innovation-decision process. For instance, Coleman and others (1966) found that most physicians reported different communication channels about a drug innovation at the knowledge function from that of the persuasion function. LaMar (1966, p. 72) studied the innovation-decision process among 262 teachers in twenty California schools. He found that the teachers went

*Further, Campbell (1966) points out that respondents may tend to report a decision making process that is more rational (defined as using the most efficient means to reach a desired goal) than in fact is the case. The seriousness of this type of response bias is difficult to evaluate, but it certainly is a caveat to keep in mind when assessing the evidence presented in the present section. The recall of channels and times at the stages in the process is certainly not completely accurate.

through the stages in the process much as had been found in the studies of farmers. Kohl (1966, p. 68) found that all fifty-eight Oregon school superintendents in his sample reported they passed through all the stages for such innovations as team teaching, language laboratories, instructional television, and flexible scheduling.

In summary, we suggest Generalization 3-11: *There are functions in the innovation-decision process.* We have shown that the evidence is most clear-cut for the knowledge and decision functions and somewhat less so for the persuasion function. There is rather poor data on the nature of the confirmation function. But the boundaries between these stages are indistinct, and it must be admitted that we have rather arbitrarily broken down the innovation-decision process into four functions or stages. This division is (1) consistent with the nature of the phenomena, (2) congruent with existing research findings, and (3) potentially useful for application by change agents. Nevertheless, the number of functions is determined primarily on the basis of ease of conceptualization. The case is somewhat analogous to raising an eyebrow which, on a stopped movie film, looks like several still pictures. But when the film is projected, raising an eyebrow looks like a gesture. Which is it then? The answer would seem to be both, depending on one's point of reference. Raising an eyebrow is a process, but for the heuristic purpose of understanding the stages in this process, we can regard it as a series of acts.

In later chapters we shall present further detail on the importance of various types of communication channels by functions in the innovation-decision process. We shall also present detail on two other than optional types of innovation-decisions where our model must be modified: (1) collective decisions, where a social system, rather than an individual, goes through the process, and (2) authority decisions, where the persuasion function is often of less importance.

Becoming a Marijuana User[*]

Marijuana is a drug usually taken for the pleasureful effects it produces. It does not lead to addiction, as do alcohol and the opiate drugs, and there is no withdrawal sickness. In the United States possession of marijuana is illegal, although its use is becoming fairly widespread, especially among

*Adapted from Becker (1953) and used by permission.

certain segments of society, such as high school and college students. About one-third of all college students have at least some experience with the drug. The mass media carry messages about the apprehension of marijuana users, of course, but do not convey ideas about where to obtain the drug or how to smoke it. Hence, we would expect interpersonal communication channels to play a paticularly important role in the diffusion of marijuana.

There have been a number of investigations by social scientists of marijuana use; we here summarize one of the most discerning of these. It is based on fifty interviews with marijuana users from a variety of social backgrounds and positions in society. The interviews focused upon the innovation-decision process through which the individuals became marijuana users. Typically there are three important stages in this process; although these stages are not exactly identical, they correspond roughly to trial-adoption, knowledge, and persuasion-confirmation. The novel sequence in which these functions occur seems to be partly due to the nature of marijuana use, as well as to the illegal status of the innovation and the absence of mass media messages about details on how to smoke it. In any event the present case materials illustrate the point that the functions in the innovation-decision process do not *always* occur in the order of knowledge, persuasion, decision, and confirmation.

Trial-Adoption

The novice does not ordinarily get "high" the first time he smokes marijuana, and several attempts are usually necessary to induce this state. Unless the drug is smoked properly, in a way that insures sufficient dosage to produce real symptoms of intoxication, one perceives little effect. Therefore, interaction with experienced users is usually a necessary means of obtaining "how-to" information at the trial stage. As pointed out by one respondent, a common mistake by novices is to smoke the drug as if it were tobacco.

> Take in a lot of air, you know, and . . . you don't smoke it like a cigarette, you draw in a lot of air and get it down in your system and then keep it there. Keep it there as long as you can.

Many new users are ashamed to admit ignorance by asking questions and prefer to learn through observation and imitation.

> I came on like I had turned on [smoked marijuana] many times before, you know. I didn't want to seem like a punk to this cat. See, like I didn't know

the first thing about it—how to smoke it, or what was going to happen, or what. I just watched him like a hawk—I didn't take my eyes off him for a second, because I wanted to do everything just as he did it. I watched how he held it, how he smoked it, and everything. Then when he gave it to me I just came on cool, as though I knew exactly what the score was. I held it like he did and took a toke just the way he did.

Knowledge

Even after he learns the proper smoking technique, the novice marijuana user must learn how to get high from others, how to perceive the organic effects of the drug. Thus, interpersonal communication is crucial in structuring the individual's perceptions of the innovation. In this sense one's knowledge of the meaning of marijuana comes *after* trial-adoption.*

I heard little remarks that were made by other people. Somebody said, 'My legs are rubbery. . . .' I was very attentively listening for all these cues for what I was supposed to feel like.

The novice picks up from other users some concrete referents of the term "high" and applies these to his own experience. Then he can locate these symptoms among his own sensations and point out to himself a "something different" in his experience that he connects with drug use. It is only when he can do this that he is high.

How did I know [I was high]? If what happened to me that night would of happened to you, you would've known, believe me. We played the first tune for almost two hours—one tune! Imagine, man! We got on the stand and played this one tune; we started at nine o'clock. When we finished, I looked at my watch, it's a quarter to eleven. . . . And it didn't seem like anything.

For marijuana use to continue, it is necessary not only to use the drug to produce effects but also to learn to *perceive* these effects when they occur. In this way marijuana acquires meaning for the user as an object which can be used for pleasure.

*This is mainly "how-to"-knowledge, rather than awareness-knowledge, which obviously existed prior to trial for the fifty respondents in this study.

Persuasion and Confirmation

One more step is necessary if the user who has now learned to get high is to continue marijuana use. He must learn a favorable attitude toward the innovation, to enjoy the effects he has just learned to experience. Marijuana-produced sensations are not automatically or necessarily pleasurable. The taste for such experience is socially acquired, like a taste for oysters or dry martinis. The marijuana user feels dizzy, thirsty, his scalp tingles, he misjudges time and distance. Are these sensations pleasurable? He is not sure. If he is to continue marijuana use, he must decide that they are enjoyable. Otherwise getting high, although a real enough experience, will be an unpleasant one he would rather avoid.

> A new user had her first experience of the effects of marijuana and became frightened and hysterical. She felt like she was half in and half out of the room and experienced a number of alarming symptoms. One of the more experienced users present said, 'She's dragged because she's high like that. I'd give anything to get that high myself. I haven't been that high in years.'

In short, what was once frightening and distasteful becomes, after a taste for it is built up, pleasant, desired, and sought after. Enjoyment is introduced by the favorable definition of the situation that one acquires from others.

> I was too much, like I only made about four tokes, and I couldn't even get it out of my mouth, I was so high, and I got real flipped. In my basement, you know, I just couldn't stay in there anymore. My heart was pounding real hard, you know, and I thought I was going out of my mind. . . . I walked outside, and it was five below zero, and I thought I was dying, and I had my coat open; I was sweating, I was perspiring. My whole insides were all . . ., and I walked about two blocks away, and I fainted behind a bush. I don't know how long I laid there. I woke up, and I was feeling the worst, I can't describe it at all. . . .

Unless such an individual learns to perceive such effects of marijuana as pleasurable, he will not continue its use. Reinforcement from other users is necessary if adoption is not to be discontinued.

Conclusions

This analysis of the genesis of marijuana use shows that individuals who come in contact with the innovation may respond in a great variety of ways.

If they are to become continued adopters of the drug, they must pass through at least three sequential stages in the innovation-decision process. First, through interpersonal communication with others, the novice must learn the proper techniques of smoking the drug through a *trial* experience. Then, he must gain "how-to" *knowledge* about the effects of marijuana to enable him to get high. Last, he must learn to have a favorable attitude toward the idea, to be *persuaded* by more experienced users that getting high is a pleasurable feeling. Thus, interpersonal communication plays a crucial role at each stage in becoming a marijuana user. The sequence of these functions is quite different from that postulated in our model of the innovation-decision process (Figure 3-1). This suggests that although our model is conceptually useful, we must remember that the exact nature and sequence of the functions can vary by innovation and for different individuals.

Innovation-Decision Period

The *innovation-decision period* is the length of time required to pass through the innovation-decision process.* The time elapsing from awareness-knowledge of an innovation to decision for an individual is measured in days, months, or years. The period is thus a gestation period in which a new idea is fermenting in the individual's mind.

Rate of Awareness-Knowledge and Rate of Adoption

Most change agents wish to speed up the process by which innovations are adopted. One method is to communicate information about new ideas more adequately so that knowledge is created at an earlier date. Another method is to shorten the amount of time required for decision after an individual is once aware of a new idea. Many potential adopters are often aware of an innovation but are not motivated to try it and adopt it. For example, almost all of the Iowa farmers in the hybrid corn study heard about the innovation before

*As pointed out in Chapter 1, the length of the innovation-decision period is usually measured from first knowledge until the decision to adopt (or reject), although in a strict sense it should perhaps be measured to the time of confirmation. This later procedure is often impractical or impossible because the confirmation function may continue over an indefinite time period.

more than a handful were planting it. "It is evident that . . . isolation from knowledge was not a determining factor in late adoption for many operators" (Ryan and Gross, 1950, p. 679). The importance of the innovation-decision period to change agents is thus highlighted; shortening it is one of the main methods of speeding the diffusion of an innovation among their clients. In one sense the main alternative procedure for change agents is to attempt to increase the rate of awareness-knowledge.

Figure 3-2 illustrates these interrelationships between rate of awareness-knowledge, rate of adoption, and the innovation-decision period for one innovation, 2,4-D weed spray. The slope of the curve for rate of awareness-knowledge is steeper than that for the rate of adoption.* These data, plus evidence from supporting studies, suggest Generalization 3-12: *The rate of awareness-knowledge for an innovation is more rapid than its rate of adoption.* For instance, Figure 3-2 shows there are 1.7 years between 10 percent awareness-knowledge and 10 percent adoption, but 3.1 years between 92 percent awareness-knowledge and 92 percent adoption. When looked at in another way, these data (Figure 3-2) indicate that later adopters have longer innovation-decision periods than do earlier adopters, a point to which we shall soon return.

There is a great deal of variation in the average length of the innovation-decision period from innovation to innovation. For instance:

1   9.0 years was the average period for hybrid corn in Iowa (Gross, 1942, p. 57).
2   2.1 years was the average for 2,4-D weed spray in Iowa (Beal and Rogers 1960, p. 10).
3   3.8 years was the average for fertilizer by Pakistani peasants (Rahim, 1961, p. 32).
4   2.0 years was the average for language laboratories in U.S. high schools (Haber, 1961).

How can we explain these differences? Innovations with certain characteristics are generally adopted more quickly; they have a shorter innovation-decision period. For example, innovations that are relatively simple in nature,

---

*In fact, one might wonder why there is such a wide range in the reported dates of awareness-knowledge for this weed spray by Iowa farmers. Mass media channels (such as farm magazines) had carried messages about the innovation to almost all of these respondents since the mid-1940s. Then why is there a seventeen-year range in reported times of knowledge in Figure 3-2? The answer is probably selective exposure and selective perception. Even though a message is presented to a total audience, only certain individuals will receive it and internalize it, depending upon their existing attitudes and beliefs.

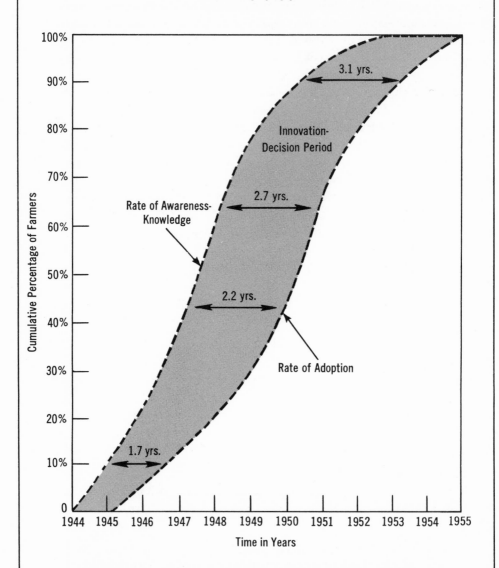

Figure 3-2
Rate of awareness-knowledge, rate of adoption, and length
of the innovation-decision period for Iowa farmers adopting 2,
4-D weed spray by year.

The shaded area in this figure illustrates the aggregate innovation-decision period between awareness-knowledge and adoption of 2,4-D weed spray. Knowledge proceeds at a more rapid rate than does adoption. This suggests that relatively later adopters have a longer average innovation-decision period than do earlier adopters. For example, there are 1.7 years between 10 per cent awareness and 10 per cent adoption, but 3.1 years between 92 per cent awareness and 92 per cent adoption.

Source: A re-analysis of data originally gathered by Beal and Rogers (1960, p. 8), and used by permission.

divisible for trial, and compatible with previous experience usually have a shorter period than innovations without these characteristics. The main dimension of analysis in the following discussion, however, is individual differences in length of the innovation-decision period, rather than differences in periods among various innovations.

Length of the Period by Adopter Category

One of the important individual differences in length of the innovation-decision period is on the basis of adopter category. We pointed out previously that the data in Figure 3-2 show a longer period for later adopters. We show this relationship in greater detail in Figure 3-3, where the average length of the period is shown for the five adopter categories. These data and those from several other studies support Generalization 3-13: *Earlier adopters have a*

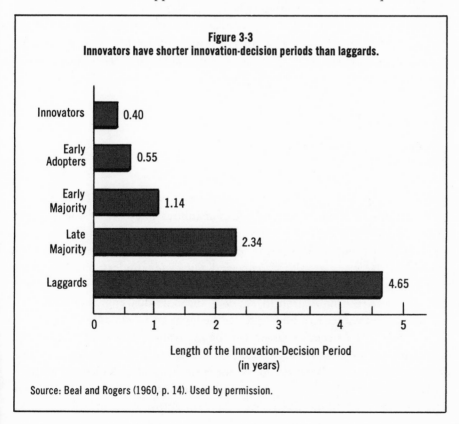

Figure 3-3
Innovators have shorter innovation-decision periods than laggards.

Length of the Innovation-Decision Period
(in years)

Source: Beal and Rogers (1960, p. 14). Used by permission.

shorter innovation-decision period than later adopters. Thus, the first individuals to adopt a new idea (the innovators) do so not only because they become aware of the innovation somewhat sooner than their peers (Figure 3-2), but also because they require fewer months and years to move from knowledge to decision. Innovators perhaps gain part of their innovative position (relative to later adopters) by learning about innovations at an earlier time, but the present data also suggest that innovators are the first to adopt because they require a shorter innovation-decision period.

Why do innovators require a shorter period? Research studies show that innovators have more favorable attitudes toward new ideas and so less resistance to change must be overcome by communication messages about the ideas. Innovators may also have shorter innovation-decision periods because (1) they use more technically accurate sources and channels about innovations, such as direct contact with scientists, and (2) because they place higher credibility in these sources than the average individual. Innovators may also possess a type of mental ability that better enables them to deal with abstractions. An innovator must be able to conceptualize relatively abstract information about innovations and apply this new information to his own situation. Later adopters can observe the results of innovations by earlier adopters and may not require this type of mental ability.

Summary

The *innovation-decision process* is the mental process through which an individual passes from first knowledge of an innovation to a decision to adopt or reject and to confirmation of this decision. Our model of this process consists of four sequential functions or stages: (1) *knowledge*—the individual is exposed to the innovation's existence and gains some understanding of how it functions, (2) *persuasion*—the individual forms a favorable or unfavorable attitude toward the innovation, (3) *decision*—the individual engages in activities which lead to a choice to adopt or reject the innovation, and (4) *confirmation*—the individual seeks reinforcement for the innovation-decision he has made, but he may reverse his previous decision if exposed to conflicting messages about the innovation. We conclude on the basis of research evidence that there are functions in the process. However, this model best fits the case of optional decisions and must be modified somewhat to be appropriate for collective and authority decisions. Further, the model's stages may occur in a different order or in a different way for some individuals and for some innovations. For example, traditional individuals are more likely to skip functions in the innovation-decision process than are modern individuals. We need to know whether the nature of the innovation-decision process for innovators is different from that for laggards.

Earlier knowers of an innovation, when compared to later knowers, are characterized by more education, higher social status, greater exposure to mass media channels of communication, greater exposure to interpersonal channels of communication, greater change agent contact, greater social participation, and more cosmopoliteness.

Innovation dissonance is the discrepancy between an individual's attitude toward an innovation and his decision to adopt or reject the innovation. Throughout the innovation-decision process, the individual seeks to reduce this type of dissonance, especially at the decision and confirmation stages.

A *discontinuance* is a decision to cease use of an innovation after previously adopting it. There are two types of discontinuances: (1) *replacement discontinuances*, in which an innovation is rejected because a better idea supersedes it, and (2) *disenchantment discontinuances*, in which an innovation is rejected as a result of dissatisfaction with its performance. Later adopters are more likely to discontinue innovations than are earlier adopters. Further, innovations with a high rate of adoption have a low rate of discontinuance.

The *innovation-decision period* is the length of time required to pass through the innovation-decision process. Generally, the rate of awareness-knowledge for an innovation is more rapid than its rate of adoption. This means, in other words, that earlier adopters have a shorter innovation-decision period than later adopters.

4 *Perceived Attributes*

of Innovations and

The reception given to a new idea is not so fortuitous and unpredictable as it sometimes appears to be. The character of the idea is itself an important determinant.

HOMER G. BARNETT (1953, p. 313)

It is true that what is settled by custom, though it be not good, at least it is fit. . . . Whereas new things piece not so well; but though they help by their utility, yet they trouble by their inconformity. Besides, they are like strangers, more admired and less favored.

SIR FRANCIS BACON (1597)

Their Rate of Adoption

AMONG the members of a social system, some innovations diffuse from first introduction to widespread use in a few years. For example, in Chapter 2 we showed (Figure 2-2) that one educational innovation, modern math, reached 100 percent adoption by public schools in five years. Yet another innovation, kindergartens, required fifty years. What characteristics of innovations affect the rate at which they diffuse and are adopted?

The purpose of this chapter is to suggest five general characteristics by which any innovation may be described, to show how individuals' perceptions of these characteristics may be utilized in predicting rate of adoption, and to analyze cases of overadoption.

Generally speaking, there have been fewer research studies designed to probe these points than to answer other major questions presented in this book. Out of necessity, the statements here are more hypothetical in nature and have fewer empirical claims to support them. When one peruses the diffusion research literature, he is impressed with how much effort has been expended in studying "people" differences in innovativeness (that is, in determining the characteristics of the different adopter categories) and how little effort has been devoted to analyzing "innovation" differences (that is, in investigating how the properties of the innovation affect its rate of adoption). The latter type of research could be of great value to change agents seeking to base their strategies on diffusion research findings. They could often predict the reactions of their clients to an innovation and perhaps modify certain of these reactions by the way they name and "package" the innovation and by relating the new idea to existing beliefs and attitudes.

But diffusion researchers have often tended to regard all innovations as equivalent units from the viewpoint of study and analysis. This is an oversimplification and a dangerous one. Evidence that all innovations are not equivalent units is that some new products fail and others succeed. The U.S. Department of Commerce estimates that 90 percent of all new products fail within four years of their release.

Many companies try to predict the rate of adoption of an innovation before it is marketed on a full-scale basis. The following example illustrates one of these failures. The case of the pill that failed points out some of the difficulties involved in predicting rate of adoption even after extensive consumer tests.

The Pill That Failed*

The story of Analoze, a combination pain killer and stomach sweetener, provides evidence that even if consumer tests are favorable, a new product still may fail. The business executives who conceived the pill were impressed by the fact that Americans were gulping record quantities of pain killers. Since this was true, they reasoned, would not an analgesic that could be taken without water have a ready market?

The company's laboratories came up with the desired product, a cherry-flavored tablet. Working with a large New York advertising firm, the manufacturer submitted samples of Analoze and competing products to a consumer panel. The verdict: Panel members overwhelmingly preferred Analoze.

The advertising copywriters then developed ads boosting Analoze as the combination analgesic-antacid that "works without water." Tests showed the ads had strong impact. The package was well designed, and the price was right. Backed by heavy advertising outlays, Analoze moved into test markets— Denver, Memphis, Phoenix, and Omaha. Dealers were enthusiastic; prospects appeared bright. Then sales reports began to trickle in. Despite all the careful preparations, the public was buying only small quantities of Analoze. Weeks went by with no improvement, and the manufacturer glumly withdrew Analoze from the market.

*Adopted from Schorr (1961), and used by permission.

After months of post mortem probing, they concluded that the fatal flaw was the "works without water" feature. Headache sufferers unconsciously associate water with a cure, and consequently had no confidence in a tablet that dissolved in the mouth.

This illustration shows that consumers did not perceive the new product as compatible with their existing values on the importance of water as part of a headache cure. Such characteristics of an innovation as its compatibility with values and past experiences have a great deal to do with its rate of adoption. It is the attributes of a new product, not as seen by experts *but as perceived by the potential adopters,* that really matters.*

Attributes of Innovations

We need a standard classification scheme for describing the perceived attributes of innovations in universal terms. Then one would not have to study each innovation in order to predict its rate of adoption. We could say, for example, that team teaching is more like modern math (in the eyes of the adopters) then it is like kindergartens. This general classification system is an eventual objective of diffusion research on innovation attributes. We have not reached this goal, but the present section discusses one approach. Five different attributes of innovations will be described. Each of these five is somewhat empirically interrelated with the other four,** but they are conceptually distinct. Selection of these five characteristics is based on past writings and research as well as on a desire for maximum generality and succinctness. What is needed is a comprehensive set of characteristics of innovations which are as mutually exclusive and as universally relevant as possible. The five attributes of innovations are: (1) relative advantage, (2) compatibility, (3) complexity, (4) trialability, and (5) observability. Each of these attributes is discussed in detail in the following sections of this chapter.

*The crucial importance of perceptions in explaining human behavior was emphasized by an early sociological dictum, "If men perceive situations as real, they are real in their consequences" (Thomas and Znaniecki, 1927, p. 81). This same viewpoint was emphasized by Wasson (1960), who utilized several case examples to show that: "The ease or difficulty of introduction [of ideas] depends basically on the nature of the 'new' in the new product—the new as the customer views the bundle of services he perceives in the newborn."
**But the interrelationships among the five attributes of innovations are quite low, Kivlin (1960) reports.

Throughout this chapter we shall repeatedly emphasize that it is the receivers' perceptions of the attributes of innovations, not the attributes as classified by experts or change agents, which affect their rate of adoption.* Like beauty, innovations exist only in the eye of the beholder. And it is the beholder's perceptions which influence the beholder's behavior.

Relative Advantage

Relative advantage is the degree to which an innovation is perceived as being better than the idea it supersedes. The degree of relative advantage is often expressed in economic profitability, but the relative advantage dimension may be measured in other ways.** For instance, one of the major advantages of 2,4-D spray over previous methods of farm weed control was a reduction in unpleasant labor requirements, rather than a direct financial gain from higher crop yields.

Crisis

The relative advantage of a new idea may be emphasized by a *crisis*. Wilkening (1952, p. 13) investigated the effect of a climatic crisis on the adoption of grass silage by Wisconsin farmers. Adoption of the innovation went from 16 percent in 1950 to 48 percent in 1951. Rain and cool weather in 1951 made the curing of hay difficult, and many farmers turned to grass silage. Had

---

*Tucker (1961) found that twelve agricultural change agents did not perceive the characteristics of thirteen farm innovations differently from twelve farmers residing in the same county. Kivlin (1960) found similar evidence in a Pennsylvania study. There is no other research evidence, however, on whether certain categories (say, innovators) perceive innovations differently from others (say, laggards), but it is reasonable to expect differential perceptions. This was one type of analysis suggested in Chapter 2. There is a long tradition of investigations in social psychology that indicate that an individual's circumstances affect his subjective perception of objects and ideas. For instance, Bruner and Goodman (1947) conducted a study of perceptions of coins by rich and poor children. The perceptions were much different for the two groups, just as laggards and innovators might differentially perceive an innovation. The poor children overestimated the size of the coins more often than the rich children.

**The nature of the innovation may determine what specific type of relative advantage (e.g., economic or social) is important to adopters (Wilkening and Johnson, 1961).

not the relative advantage of the new idea been sufficiently demonstrated before 1951, the weather probably would not have had such an effect that year.

A crisis emphasizes the relative advantage of an innovation and hence, affects its rate of adoption. Mulford (1959) concluded that an economic crisis speeded the rate of adoption of industrial development commissions by Iowa communities. Sutherland (1959) showed that a cotton spinning innovation was adopted more quickly by English firms because of the labor shortage in World War II. Bertrand (1951) found that the crisis of unionized farm laborers and wartime labor shortage aided the rate of adoption of farm mechanization in Louisiana.

Other studies show that a decisive event may retard the rate of adoption of an innovation. However, the members of a social system may make up for lost ground as soon as the crisis is past.* Adler (1955, p. 27) found that depressions and wars retarded the adoption of educational innovations but that the schools he studied accelerated their rate of adoption as soon as the crises were past.

Relative Advantage and Rate of Adoption

Relative advantage, in one sense, indicates the intensity of the reward or punishment resulting from adoption of an innovation. There are undoubtedly a number of subdimensions of relative advantage: The degree of economic profitability, low initial cost, lower perceived risk, a decrease in discomfort, a savings in time and effort, and the immediacy of the reward. This latter factor perhaps explains why *preventive* innovations have an especially low rate of adoption. Such ideas as buying insurance, using auto seat belts, getting innoculations against disease, adopting birth control methods, and using latrines (in peasant villages) are examples. The relative advantage of such preventive innovations is difficult for change agents to demonstrate to their clients, because it occurs at some time in the future.**

A summary of eight different investigations of the relationships between perceived attributes of innovations and their rate of adoption is shown in Table 4-1. Almost every one of these studies reports a positive relationship between relative advantage and rate of adoption. Perhaps this result is so self-evident as to be of little surprise.

*Pemberton (1937) supports this notion.
**This suggests that a lack of observability is also a characteristic of preventive innovations and one that adversely effects their rate of adoption.

Table 4-1 Perceived Attributes of Innovations and Their Rate of Adoption

AUTHOR(S) OF INVESTIGATIONS	TYPE OF RESPONDENTS	NUMBER OF INNOVATIONS STUDIED	NUMBER OF ATTRIBUTES OF INNOVATIONS MEASURED	PERCENTAGE OF VARIANCE IN RATE OF ADOPTION EXPLAINED	ATTRIBUTES OF INNOVATIONS FOUND TO BE SIGNIFICANTLY RELATED TO RATE OF ADOPTION
1 Kivlin (1960); Fliegel and Kivlin (1962a); Fliegel and Kivlin (1962b)	229 Pennsylvania farmers	43	11	51	1 Relative advantage 2 Compatibility 3 Complexity
2 Tucker (1961)	88 Ohio farmers	13	6	—	None
3 Mansfield (1961)	Coal, steel, brewing, and railroad firms	12	2	50	1 Relative advantage (profit-ability) 2 Observability (rate of inter-action about the innovation among the adopters)
4 Fliegel and Kivlin (1966a)	229 Pennsylvania farmers[a]	33	15	51	1 Trialability 2 Relative advantage (initial cost)
5 Petrini (1966a)	1,845 Swedish farmers	14	2	71	1 Relative advantage 2 Complexity
6 Singh (1966a)	130 Canadian farmers	22	10	87	1 Relative advantage (rate of cost recovery, financial return, and low initial cost) 2 Complexity 3 Trialability 4 Observability

[a]The same 229 Pennsylvania farmers are involved in this study as in Kivlin (1960) above, but thirty-three instead of forty-three innovations, and fifteen instead of eleven attributes, were analyzed. Hence, the results are different.

Table 4-1 Perceived Attributes of Innovations and Their Rate of Adoption—continued

AUTHOR(S) OF INVESTIGATIONS	TYPE OF RESPONDENTS	NUMBER OF INNOVATIONS STUDIED	NUMBER OF ATTRIBUTES OF INNOVATIONS MEASURED	PERCENTAGE OF VARIANCE IN RATE OF ADOPTION EXPLAINED		ATTRIBUTES OF INNOVATIONS FOUND TO BE SIGNIFICANTLY RELATED TO RATE OF ADOPTION
7 Kivlin and Fliegel (1967a); Kivlin and Fliegel (1967b)	80 small-scale Pennsylvania farmers (and 229 large-scale farmers)	33	15	51	1	Relative advantage (savings of discomfort)
8 Fliegel and others (1968)	387 Indian peasants	50	12	58	1	Relative advantage (social approval, continuing cost, and time-saving)
					2	Observability (clarity of results)
	80 small-scale Pennsylvania farmers[b]	33	12	62	1	Relative advantage (savings of discomfort, payoff, and time-saving)
	229 large-scale Pennsylvania farmers[c]	33	12	49	1	Trialability
					2	Relative advantage (initial cost)

[b]These are the same eighty small-scale Pennsylvania farmers as in Kivlin and Fliegel (1967a), but twelve instead of fifteen attributes are included in the multiple correlation prediction of rate of adoption; the results obtained are different, therefore.
[c]These are the same 229 large-scale Pennsylvania farmers as in Kivlin (1960) above, but a different number of innovations and attributes are included in the analysis, hence, the results are different.

The respondents in almost all of these eight studies are U.S. commercial farmers, and the motivation for adoption of these innovations is probably centered on pecuniary concerns.* In fact, one study (Kivlin and Fliegel, 1967a) that includes small-scale farmers (who are presumably oriented less to profit maximization) in the U.S. finds that a decrease in discomfort, one subdimension of relative advantage, but not *economic profitability*, is positively related to rate of adoption.

We may summarize this discussion of relative advantage by Generalization 4-1: *The relative advantage of a new idea, as perceived by members of a social system, is positively related to its rate of adoption.***

Peasants and Relative Advantage

Economic profitability as a subdimension of relative advantage is less important as a predictor of rate of adoption for small U.S. farmers than for larger farmers (Kivlin and Fliegel, 1967a). This finding leads one to guess that economic profitability may be even less important for peasant farmers in less developed countries; and indeed, Fliegel and others (1968) find this notion is supported by a cross-cultural comparative study. Punjabi peasants in India seem to behave much like small-scale Pennsylvania dairymen (in fact, even more so) than like large-scale U.S. farmers.*** Economic profitability might be expected to be less crucial in explaining the rate of adoption of innovations among respondents who are peasants oriented largely to subsistence living. For them, other noneconomic dimensions of relative advantage, such as social prestige and social approval, are expected to explain rate of adoption, as would such other attributes of innovations as compatibility with sociocultural values.

---

*As Fliegel and Kivlin (1966a) point out: "Since we are dealing here with innovations having direct economic significance for the acceptor, it is not surprising that innovations perceived as most rewarding and involving least risk and uncertainty should be accepted most rapidly."

**There are at least three possible measures of relative advantage which have been utilized in the investigations which provide evidence for this generalization: (1) as judged by an observer or rater, (2) as perceived in composite by the adopters, and (3) as perceived by each individual adopter separately.

***"Much more than financial incentives will be necessary to obtain widespread and rapid adoption of improved practices. . . . Unlike the Pennsylvania dairy farmers, the Punjabi respondents apparently attach greater importance to social approval and less to financial return" (Fliegel and others, 1968).

The economist Schultz (1964) argued against the prevailing view that sociocultural variables are important determinants of the rate of adoption of innovations among peasants in less developed countries. He says that economic stimuli greatly affect peasants and to such an extent that we do not need to attend to cultural factors. "Studies of the observed lags in the acceptance of particular new agricultural factors [that is, innovations] show that these lags are explained satisfactorily by profitability. . . . Since differences in profitability are a strong explanatory variable, it is not necessary to appeal to differences in personality, education, and social environment" (Schultz, 1964, p. 164). Put in other words, this argument proposes that relative advantage is the *only* attribute of innovations affecting their rate of adoption.

This rather extreme statement has been welcomed by many social scientists, especially economists, in spite of the rather poor evidence that Schultz cites. He ignores a myriad of research studies* showing cultural factors to be of great importance in explaining peasant innovation-decisions, while citing the findings of one of his doctoral students in India (Hopper, 1957) and an anthropological colleague, Tax (1963), who studied a small town in Guatemala.**

But Schultz does have a point, even if it is overstated. Economic considerations are one predictor of the rate of adoption of innovations by peasants, but they certainly are not so strong as to overrule completely all other perceived attributes, such as sociocultural variables. Even were the price of beef to be halved in India, Hindus would not begin eating cows. Nevertheless, there are many innovations which do not run so counter to cultural or social norms, and in these cases the rate of adoption by peasants is likely to be more rapid if the innovation is more economically or socially profitable. Even in these instances, however, the increase in profitability must be rather spectacular to affect the rate of adoption very much. Many students of peasant life feel that the relative economic advantage of a new idea must be at least 25 to 30 percent higher than existing practice for economic factors to affect adoption.*** When an innovation promises only a 5 to 10 percent advantage, a peasant farmer probably cannot even distinguish that it is advantageous.

*Many of these are summarized in books by anthropologists such as Foster (1962), Erasmus (1961), and Arensberg and Niehoff (1964).
**Schultz also cites as evidence of his assertion (that peasants are solely motivated to adopt innovations by economic incentives) the findings from aggregate analyses of price sensitivity among Indian cotton farmers. Unfortunately, reactions to price changes in established crop enterprises tell us little about how peasants adopt new ideas.
***"To induce farmers to change, the potential payoff must be high—not 5 to 10 percent but 50 to 100 percent" (President's Science Advisory Committee, 1967, p. 16).

His limited skill with numbers, his crude accounting schemes, and his lack of finesse with the scientific method of reaching conclusions all act to limit his comparing ability.

We conclude, therefore, that rapid adoption probably depends on such aspects of relative advantage as economic profitability for most individuals,* but this may be less true for peasants.** We need carefully designed research studies to probe the importance of economic profitability, along with the other four attributes of innovations, in explaining rate of adoption among peasants. The definitive investigation on this subject has not yet been done.

Effects of Incentives

Many change agencies award economic incentives or subsidies to their clients in order to speed the rate of adoption of innovations. The function of an incentive is to increase the degree of relative advantage of the new idea.*** Often, however, the effects of such incentives have been rather disappointing, at least in agriculture. Once the subsidy is removed, adoption of the innovation usually stops. The receivers evidently perceive the incentive as separate from the intrinsic relative advantage of the innovation, which is not sufficient to ensure continued adoption of the idea once the incentive stops. Of course,

*A controversy regarding the relative importance of profitability versus compatibility for U.S. farmers can be traced through diffusion literature. Griliches (1957), an economist, explained about 30 percent of the variation in rate of adoption of hybrid corn on the basis of profitability. He used aggregate data from U.S. crop reporting districts and states, and hence, could not claim that similar results would obtain when individual farmers were the units of analysis. Griliches (1957) concluded: "It is my belief that in the long run, and cross-sectionally, [sociological] variables tend to cancel themselves out, leaving the economic variables as the major determinants of the pattern of technological change." Evidence to contradict this assertion has been brought to bear on the controversy: (1) in the case of hybrid sorghum adoption in Kansas (Brandner and Straus, 1959; Brandner, 1960; Brandner and Kearl, 1964), where compatibility was more important than profitability, and (2) for hybrid seed corn adoption in Iowa (Havens and Rogers, 1961b; Griliches, 1962; Rogers and Havens, 1962b), where it was concluded that a combination of an innovation's profitability plus its observability were most important in determining its rate of adoption. For other commentaries in this controversy, see Griliches (1960b) and Babcock (1962).

**Where noneconomic aspects of relative advantage, and compatibility, may be of relatively greater significance in explaining rate of adoption.

***Incentives can also be used to promote interpersonal communication and influence about an innovation. In India an incentive is offered to each individual who brings someone to a health clinic for a vasectomy. This type of incentive increases the observability of an innovation, rather than its relative advantage. Therefore, there are diffusion incentives as well as adoption incentives.

some innovations are very difficult to discontinue, and incentives may be quite effective in achieving initial adoption of these ideas. An example is the family planning innovation of vasectomy (a male sterilization operation), which is almost impossible to discontinue. The government of India pays a small fee to each man who volunteers for a vasectomy.

Incentives can take many forms. Some incentive policies are designed only to encourage a small-scale trial of a new idea; an illustration is the free samples of a new product that are offered by many commercial companies to their customers. The strategy here is that by facilitating trial use, full-scale adoption will follow (if the innovation possesses a potential relative advantage that can be perceived by the receiver). Other incentive policies are designed only to secure adoption of a new idea by earlier adopters; once a level of 20 or 30 percent adoption is reached in a social system, the economic incentive is discontinued by the change agency. In the U.S. many milk companies offered a price incentive to dairy farmers for switching to the bulk tank handling of milk. Once the innovators and early adopters had adopted bulk milk tanks, the incentive price was stopped, and the rate of adoption continued.

In spite of the widespread use of incentive policies by change agencies, especially in such fields as family planning and agriculture, we lack carefully designed and conducted field experiments on the effects of incentives. Such inquiries could provide an empirical basis for change agency decisions on whether to offer economic incentives, on how large such subsidies should ideally be, and whether they should be once only (for initial adoption) or continuing. In a more general sense we need studies of the economics of adoption, a type of investigation that has gone begging because of the relatively minor involvement of economists in diffusion research.

Compatibility

Compatibility is the degree to which an innovation is perceived as consistent with the existing values, past experiences, and needs of the receivers. An idea that is not compatible with the salient characteristics of a social system will not be adopted so rapidly as an idea that is compatible. Compatibility ensures greater security and less risk to the receiver and makes the new idea more meaningful to him. An innovation may be compatible (1) with sociocultural values and beliefs, (2) with previously introduced ideas, or (3) with client needs for innovations.

With Values

The lack of compatibility of beef consumption in India with cultural values prevents the adoption of beef eating, as already pointed out. India has about 520 million people and 200 million sacred cows. No cows can be killed, and the best milkers are not selected for breeding. These facts, plus the poor nutrition of the cattle, results in an average milk yield of only 900 pounds per year. U.S. experts introduced milk goats in 1964 as a substitute for cows, because goats eat only one-fourth as much feed and yield relatively more milk; but the incompatibility of the goats with status and religious factors prevented their adoption. Indian villagers regard goat-raising as an enterprise for "untouchables" only, those at the very bottom of the social structure. Further, the social status of a villager is measured in part by how many cows he possesses. Hence, an innovation that would effectively raise the nutritional level of India's starving millions was rejected because of its incompatibility.

Another example of the importance of culturally learned values in blocking the adoption of an innovation is the case of latrines in Peru. Medical authorities had become discouraged about treating residents of one remote village for intestinal parasites. Within a few weeks after medical treatment, the villagers would be reinfected because of their inadequate sanitation methods. Accordingly, public health officials set about introducing latrines, which the villagers at first seemed to welcome. But the new facilities were seldom used because the villagers were accustomed to defecating whenever and wherever they felt the necessity. Their sphincter muscles, culturally conditioned to a squatting position, were incompatible with use of sit-down latrines. Had the outhouses been designed with the clients' habits in mind, adoption would have been facilitated.

Among the Indians in Bolivia's Andes Mountains, milk is perceived as a type of animal excrement similar to urine. Attempts by change agents to improve the Indians' largely potato diet have failed when it comes to milk, which is locally available from llamas, a goat-like animal. A recent diffusion strategy was to provide free milk flavored with sugar to small schoolchildren. Even the sugar-flavored milk, however, has not been accepted.

In an Asian country steel plows were introduced where a pointed stick had previously served. The peasants accepted the new implements with polite gratitude but used them for ornaments rather than plowing. Why? Because the steel plows required two hands to use; the peasants were accustomed to using only one hand (and driving their bullocks with the other).

In modern urban India today there is a strong norm against eating food with the left hand which is believed to be unclean. This habit began in past centuries when Indian villagers used their left hand for certain functions associated with defecation. At that time there were inadequate washing and sanitary facilities and the left-hand-as-unclean complex was functional. But today it is easy for urban, middle-class Indians to wash their hands before meals. Nevertheless, the unclean-hand habit rigidly persists as a dysfunctional element in urban India. How would you like the role as a change agent responsible for persuading Indians to eat with the left hand? Many change agents face equally difficult assignments in promoting innovations that run counter to strongly held values.*

With Previously Introduced Ideas

An innovation may be compatible not only with deeply imbedded cultural values but also with previously adopted ideas. Compatibility of an innovation with a preceding idea can either speed up or retard its rate of adoption. Old ideas are the main tools with which new ideas are assessed. One cannot deal with an innovation except on the basis of the familiar and the old-fashioned.

Examples of the use of past experience to judge new ideas come from a study conducted in a Colombian peasant community (Fals Borda, 1960). At first the farmers who practiced with a new scythe to cut their grain tended to cut jerkily, as if it were the sickle which they had been using for years and which the scythe replaced. Similarly, some farmers applied chemical fertilizers on top of their potato seed (as they had done with cattle manure), thus damaging their seed and causing a negative evaluation of the innovation. Other peasants excessively sprayed their potatoes with insecticide chemicals, transferring to the new idea their old methods of watering their plants.

The rate of adoption of a new idea is affected by the old idea that it supersedes.** Obviously, however, if a new idea were completely congruent with existing practice, there would be no innovation, at least in the mind of the receiver. Put in other words, the more compatible an innovation is, the less of a change it represents. How useful, then, is the introduction of a very highly compatible innovation? Quite useful, if the compatible innovation is seen as

---

*And often, as this example illustrates, the values reflect a situation as it was some generations earlier. Hence, they are incompatible with the innovation being introduced.
**As we show later in this section in our discussion of innovation negativism.

the first step in a series of innovations which the change agent intends to introduce sequentially. The compatible innovation paves the way for later, less compatible innovations.*

With Needs

One indication of the compatibility of an innovation is the degree to which it meets a need felt by the clients.** One obvious tactic for change agents is to determine the needs of their clients, and then recommend innovations to fulfill these needs. The difficulty often lies in how to become aware of the felt needs; change agents must have a high degree of empathy and rapport with their clients in order to assess their needs accurately. Such techniques as informal probing in interpersonal contacts with individual clients, client advisory committees to change agencies, and surveys are sometimes used to determine needs for innovations.

But often clients do not recognize that they have needs for an innovation because they are not aware of the new idea or of its consequences. In these cases change agents may seek to generate needs among their clients, but this must be done carefully or else the felt needs upon which diffusion campaigns are based may be only a reflection of the change agent's needs, rather than those of his clients. Therefore, one dimension of compatibility is the degree to which an innovation is perceived as meeting the needs of the client system. When felt needs are met, a faster rate of adoption should occur.

Compatibility and Rate of Adoption

A county extension agent in a New Mexico county introduced hybrid seed corn to his clients (Apodaca, 1952). Forty percent of the eighty-four growers in one village planted at least some of the new corn seed in 1946 (Figure 4-1), and the results were spectacular. Their yields were doubled over the yield from the old seed. In the following year over half of the villagers planted hybrid seed, and the extension agent felt his campaign had been successful. But in 1948 half of the adopters discontinued, and by the following year only three growers planted hybrid seed, perhaps because they were close friends of the extension worker.

*This procedure is roughly parallel to Skinner's learning principle of adding sequential elements of material to be learned in very small increments, as in programmed learning.
**The role of needs in the innovation-decision process was discussed in Chapter 3.

Why did the innovation fail after such a rapid increase in rate of adoption? The answer was not found in technical shortcomings of the innovation. Careful observations were made by the extension agent to ensure the success of the innovation; local soils were tested for their applicability in growing the new corn. The change agent set up a demonstration plot near the village in the first year. The hybrid seed yielded three times the normal harvest expected from the old varieties; thus, it had a high degree of relative advantage.

The village farmers discontinued the new idea because their wives did not like the hybrid. Corn was ground to make tortillas, a flat corn bread indispensable to the local diet. The hybrid corn had a strange flavor and did not "hang together well" for tortillas. The norms of the social system favored the old varieties rather than the hybrid. If the change agent had considered local norms as well as local soil conditions, perhaps he could have introduced a hybrid variety that would have resulted in good tortillas as well as high yields. He ignored the incompatibility of the hybrid corn with taste preferences, and the innovation failed.

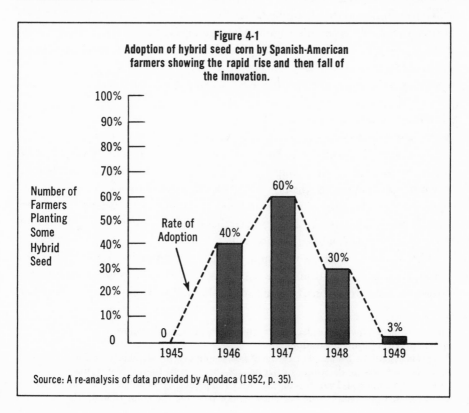

Figure 4-1
Adoption of hybrid seed corn by Spanish-American farmers showing the rapid rise and then fall of the innovation.

Source: A re-analysis of data provided by Apodaca (1952, p. 35).

As a result, the change agent's future promotion of other innovations was also damned. Such innovation negativism* (Arensberg and Niehoff, 1964) is an undesirable aspect of compatibility. When one idea fails, a change agent's clients are conditioned to view all future innovations with apprehension.

Thus, the perceived relationship of an innovation with other ideas can act as a bane, as well as a boon, to adoption. An illustration comes from an investigation in India (Planning Research and Action Institute, 1966, p. 47). Contraceptive devices had been rejected by some Indian villagers because they feared that the family planning change agents were trying to stop births completely. Then a team of public health workers came to the village to give small-pox injections. The vaccinations were widely rejected because the peasants perceived them to be part of the contraceptive campaign, which they already viewed negatively. Thus, an unfavorable association of the vaccination with a previously rejected innovation prevented adoption of the new idea.

Various birth control campaigns have failed among villagers in less developed countries. One of the earlier forms of birth control that was tried in a village in India was the "foaming tablet," which produced contraceptive effects when placed in a woman's vagina. Unfortunately, these tablets were similar in appearance to aspirins and other pills which the village women had received from public health workers. This compatibility of the foaming tablets with a previous innovation led to their eventual rejection, for the women simply ate the tablets.

Foster (1962) cites an illustration of how the functioning of an innovation was changed so as to maximize its compatibility and hence, to facilitate its rate of adoption. Housewives in Sicily enjoy washing their family clothes near wells or streams in the company of other women. When home washing machines were installed, the housewives were quite unhappy at the loss of an opportunity to socialize. Astute change agents moved all the washing machines in each village into a central location so that the housewives could again enjoy their coffee klatches while they washed.

Often the name given to a new product can affect its compatibility and hence, its rate of adoption. This is because words are thought symbols that structure our perceptions.** The Colgate-Palmolive Company committed an

---

Innovation negativism is the degree to which an innovation's failure conditions a client system to reject future innovations.
**This is essentially a statement of the Whorf-Sapir hypothesis, which postulates that an individual's language determines what he sees and what he thinks, that is, his perceptions (Whorf, 1956, pp. 134–159). It is difficult for us to perceive of an object if we do not have a word for it. For example, the Navahos cannot perceive the color orange; they have no

egregious error in introducing its trademarked product "Cue" into French-speaking districts, where the word has an obscene connotation (Martyn, 1964, p. 78).

Hawley (1946) sought to determine why the Roman Catholic religion, as offered by proselytizing Spanish priests, was readily accepted by Eastern Pueblo Indians in Arizona and New Mexico, whereas the Western Pueblos ". . . after a brief taste of Catholicism, rejected it forcefully, killed the priests, burned the missions, and even annihilated the village of Awatobi when its inhabitants showed a tendency to accept the acculturation so ardently proffered." Hawley concluded that the Eastern Pueblos, whose family structue was heavily patrilineal and father-oriented, were attracted by a new religion in which the deity was a male figure. However, Catholicism was incompatible with the mother-centered beliefs of the Western Pueblos. Perhaps if the religious change agents had been able to emphasize the female-image aspect of Catholicism (the Virgin Mary), they would have achieved greater success among the Western Pueblo tribes.

In one section of southern Germany, prior to 1960, the average farm had become fragmented into about seventy-five "postage stamp" plots through land inheritance customs. Government change agencies consolidated these parcels of land in 1961 and redistributed them to the owners so that each farm was in one large plot. Farm mechanization and improved efficiency were expected to result. But farmers' attitudes toward their larger fields had not been changed, and some immediately divided their newly consolidated farms into seventy-five small plots! Their seeding and fertilization rates, they claimed, were on the basis of the tiny plots; so they resisted consolidation efforts. The idea of land consolidation was incompatible with their attitudes and values, and unless these were changed, the new idea could not be forced upon the farmers by government agencies. The case is similar to the problem of urban slum dwellers who are relocated in high-rise apartment buildings without efforts to change their values and to teach them how to use their new facilities. The assumption is that the new environment will change their values and attitudes; what often happens is that the former slum dwellers change their new dwelling to fit their values more closely. What is needed is an educational program for the new residents.

thought symbol for it. On the other hand, one's language may contain many words for objects that are especially important to one's environment. For instance, there are 6,000 Arabic words dealing with camels, including ninety to describe camel pregnancy. The words used to describe an innovation structure our perceptions of the new idea and affect its rate of adoption.

Most of the studies reviewed in this section have dealt with peasant audiences where compatibility may be an especially important predictor of rate of adoption. But studies among more modern individuals also show the importance of perceived compatibility. Graham (1956) found that only 24 percent of his upper-class U.S. respondents adopted television, compared with 72 percent of the lower class. However, the game of canasta was accepted by 72 percent of the upper class but by only 12 percent of the lower class. Upper-class and lower-class recreation values were found to explain these differences in rates of adoption. The compatibility of the innovation with social class values partly determined its speed of adoption.

Carlson (1965) reports an illustration of how an educational innovation's incompatibility with teachers' self-images led to its misuse. The innovation, programmed instruction, allowed students to learn at their own rate. But this main advantage of programmed instruction was not realized because the teachers used various ingenious techniques to equalize the pace for both the slow and the fast learners. "Programmed instruction does not give the teachers much opportunity to perform as they apparently desire; it does not give them sufficient opportunity to teach. In their eyes, because teaching means performing, using programmed instruction is not teaching" (Carlson, 1965, p. 83). The teachers' need to perform was incompatible with proper use of the innovation, its main advantage was not achieved, and its rate of adoption was slowed.

McCorkle (1961) sought to determine why chiropractic, a form of healing by manual adjustment of the spine, was so popular in rural Iowa. He concluded that chiropractic satisfies needs not well covered by orthodox human medicine, and it fits well into the value system of the rural Iowan. Chiropractic offers to put a person in order rapidly so that he may return to work. It is cheap in cost and does not require hospitalization. The 1,000 chiropractors in Iowa are evidence that the idea's compatibility with rural Iowa culture leads to a high level of adoption.

These case study analyses generally support Generalization 4-2: *The compatibility of a new idea, as perceived by members of a social system, is positively related to its rate of adoption.** Statistical analyses of this proposition, which control the effects of other attributes of innovations (Table 4-1), show compatibility to be of relatively less importance in predicting rate of adoption than other attributes, such as relative advantage. This result may be in part an

---

*We previously pointed out that exceptions to this generalization occur when the innovation is compatible with a negative experience, such as a previous idea that failed. This is innovation negativism.

artifact of difficulties in measuring perceived compatibility. In most of the studies shown in Table 4-1, compatibility was found to be positively related to rate of adoption, even though the correlation was often not significant when the effects of other attributes were removed statistically.

Innovation Packages

Innovations often are not viewed singularly by individuals. They may be perceived as an interrelated bundle or complex of new ideas. The adoption of one new idea may trigger the adoption of several others.

One approach that seeks to capitalize on this tendency is the so-called "package programs" of India, Pakistan, and Mexico, which are credited with causing a "Green Revolution" in food production. A bundle of agricultural innovations, usually including improved crop varieties, fertilizer, and other agricultural chemicals, is recommended *in toto* to farmers. The assumption is that the villagers will adopt the package more easily and rapidly than each of the innovations individually. More importantly, by adopting all at once, the farmers get the total yield effects of all the innovations, plus the interaction effects of each practice on the others.*

Unfortunately, the package approach has little empirical basis in diffusion research even though it may intuitively seem to make sense. Naturally, the packaging should be based on the farmers' *perceptions* of the innovations, as well as on their biological yield-increasing interrelationships, but this has not been done. It is a rather simple and direct procedure to factor analyze the intercorrelations among farmers' time of adoption (or their perceptions) of a series of innovations, in order to determine which of the ideas cluster together.**

One of the few investigations of a complex of new ideas is Silverman and Bailey's (1961) analysis of the adoption of three corn-growing innovations by 107 Mississippi farmers. The three ideas (fertilization, hybrid seed, and thick planting) were functionally interrelated in such a way that adoption of the latter innovation without concurrent use of the other two ideas resulted in *lower* corn yields than if none of the ideas were used. Most farmers either adopted all three of the ideas or none of them, but 8 percent used unsuccessful combinations. Silverman and Bailey suggest the need for change agents to show

---

*As an illustration of an interaction effect, fertilizer usually results in a greater yield increase when the farmer applies it to fields planted in an appropriate variety.

**Such a factor analysis was completed by Fliegel (1956) but not as a basis for a package program. He found, as one might expect, that one main factor explained much of the variance in the time of adoption of the individual innovations.

farmers the interrelationships among the three ideas in the corn-growing complex.

Some merchandisers offer tie-in sales, a technique that recognizes the high degree of compatibility among several new products. A new clothes washer may be offered to housewives as a package deal along with a clothes dryer. Some marketing schemes "hook on" an unwanted product to a compatible innovation that possesses a high degree of relative advantage.

There is need to analyze complexes of innovations in future research, to study ideas in an evolutionary sequence, and to determine the degree of compatibility perceived by individuals among interrelated ideas. Then we shall have a sounder basis for the assembling of innovations in easier to adopt packages.

Complexity

Complexity is the degree to which an innovation is perceived as relatively difficult to understand and use. Any new idea may be classified on the complexity-simplicity continuum. Some innovations are clear in their meaning to potential adopters, others are not. Although the research evidence is far from conclusive, we suggest Generalization 4-3: *The complexity of an innovation, as perceived by members of a social system, is negatively related to its rate of adoption.*

Kivlin (1960) found that the complexity of farm innovations was more highly related in a negative direction to their rate of adoption than any other characteristic of the innovations except relative advantage. Similar results were reported by Singh (1966a) in Canada and by Petrini (1966a) in Sweden* (Table 4–1).

There is also case study evidence that perceived complexity is a predictor of rate of adoption. For instance, Graham (1956) sought to determine why canasta and television diffused at different adoption rates in the upper and lower classes. He concluded that one reason (in addition to relative compatibility with recreation values, which were discussed previously, see p. 152) was the difference in complexity of the two ideas. Canasta had to be learned through detailed personal explanation from other card players. Its procedures were com-

---

*In fact, Petrini (1966a) found that complexity, along with relative advantage, explained 71 percent of the variance in the rate of adoption of innovations among Swedish farmers.

plex and difficult to master. Television, however, appeared to be a relatively simple idea that required only the ability to turn a knob.

Trialability

Trialability is the degree to which an innovation may be experimented with on a limited basis.* New ideas that can be tried on the installment plan will generally be adopted more rapidly than innovations that are not divisible. An innovation that is trialable is less risky for the adopter. Some innovations are more difficult than others to divide for trial. Examples of take-it-or-leave-it innovations are bulk milk tanks, home air conditioners, and driver-training education in a school system. In spite of the lack of wide evidence, we suggest Generalization 4-4: *The trialability of an innovation, as perceived by members of a social system, is positively related to its rate of adoption.* Studies by Fliegel and and Kivlin (1966a), Singh (1966a), and Fliegel and others (1968) support this statement (Table 4-1).

There is evidence from several investigations** that relatively earlier adopters may perceive trialability as more important than later adopters. Laggards move from initial trial to full-scale use more rapidly than do innovators and early adopters. The more innovative individuals have no precedent to follow at the time they adopt, while the later adopters are surrounded by peers who have already adopted the innovation. These peers may act as a psychological or vicarious trial for the later adopters; and hence, the actual trial of a new idea is of less significance for them.

Observability

Observability is the degree to which the results of an innovation are visible to others.*** The results of some ideas are easily observed and communicated

---

*We prefer the term "trialability" to "divisibility," our convention in an earlier version of this book (Rogers, 1962b, p. 131), because trialability includes the notion of a psychological trial.

**These studies include Gross (1942, pp. 58–59), Ryan (1948), and Katz (1961). In Chapter 3 we showed that later adopters are more likely to skip the trial than are earlier adopters, a notion that is consistent with the present statement.

***This attribute was less precisely termed "communicability" in an earlier edition (Rogers, 1962b, p. 132).

to others, whereas some innovations are difficult to describe to others. Generalization 4-5 is: *The observability of an innovation, as perceived by members of a social system, is positively related to its rate of adoption.* One illustration of this generalization is the case of preemergent weed killers that are sprayed on a field before the weeds emerge from the soil. The rate of adoption of this idea by Midwestern farmers was very slow in spite of its relative advantage, because there were no dead weeds which the farmers could show their neighbors.

Hruschka and Rheinwald (1965) found that the more observable innovations, which were demonstrated by German "pilot farmers" (demonstrators), diffused more widely than the less visible innovations. For instance, the idea of drying hay on wire racks was known by 76 percent of a demonstration farmer's neighbors, but only 22 percent knew about a new method of feeding calves. The latter practice was less observable, because calf feeding is done inside the farmer's barn whereas the wire racks were easily visible in the demonstration farmer's fields. Demonstrations by change agents are an attempt to increase the observability of an innovation, but the Hruschka and Rheinwald (1965) results imply that some innovations do not lend themselves well to demonstration.

Menzel (1960) hypothesized that attendance at out-of-town medical meetings by doctors would be more highly related to adoption of a drug (an easily communicable idea) than to adoption of modern patient management (a relatively difficult idea to communicate). Menzel's findings did not provide much support for his hypothesis, but his work is suggestive of future research attention. The proposition for testing is that the various types of communication channels play different roles in the innovation-decision process on the basis of the innovation's degree of observability.

Ogburn's (1922) cultural lag theory fits into the present discussion of observability. Ogburn claimed that material innovations diffused and were adopted more readily than nonmaterial ideas.* Linton (1936, pp. 337–338) points out that one reason for this cultural lag (of nonmaterial behind material ideas) is the greater observability of material ideas: "The material techniques and their products are probably the only elements of culture which can be completely communicated, and it is significant that it is usually these elements which are accepted most readily. . . ." The culture lag theory of Ogburn has

---

*The corpus of diffusion research includes focus almost entirely upon material (technological) innovations, but a few studies are concerned with the spread of a nonmaterial idea, as in the news diffusion studies. Our distinction in Chapter 1 between symbolic versus action innovation-decisions has a close parallel to Ogburn's distinction of material versus nonmaterial innovations.

fallen into academic disrepute in recent years. In fact, Boskoff (1957, p. 296) labeled the distinction between material and nonmaterial elements as a theoretical *cul de sac* and recommended a hasty exit. However, we feel that differences in observability from innovation to innovation are important, deserve proper and accurate measurement, and are related to the innovations' rate of adoption. Hence, the distinction between material and nonmaterial innovations is indicative of their observability and has continued relevance in explaining their rate of adoption.

Explaining Rate of Adoption

Rate of adoption is the relative speed with which an innovation is adopted by members of a social system. It is generally measured as the number of receivers who adopt a new idea in a specified time period. For instance, we mentioned earlier that five years were required for modern math to reach nearly 100 percent adoption in U.S. schools, whereas fifty years were necessary for kindergartens. Hence, rate of adoption is a numerical indicant of the steepness of the adoption curve for an innovation (see Figure 2-2, for instance).

We have shown in this chapter that one important type of variable in explaining the rate of adoption of an innovation is its perceived attributes. Table 4–1 indicates that from 49 to 87 percent of the variance in rate of adoption was explained by the five attributes (relative advantage, compatibility, complexity, trialability, and observability).* In addition to these perceived attributes of an innovation, such other variables as (1) the type of innovation-decision, (2) the nature of communication channels used to diffuse the innovation at various functions in the innovation-decision process, (3) the nature of the social system, and (4) the extent of change agents' promotion efforts in diffusing the innovation affect the innovation's rate of adoption (Figure 4-2).**

*Of course, these five attributes are simply a handy and succinct way of classifying the main attributes that have been studied; it is quite possible that other attributes might be included in future studies to predict rate of adoption.

**In addition, the availability of the innovation may affect its rate of adoption in some systems. For instance, in India today, as in many other less developed countries, the limited availability of fertilizer is an important restraint on its adoption by peasants, as Hodgdon and Singh (1963) show. Also, the degree to which potential adopters of an innovation perceive it as risky is related to its rate of adoption. We do not conceptualize risk as one of the five main attributes of innovations, because it overlaps so highly with several of the other five attributes, especially relative advantage, compatibility, and complexity.

The type of innovation-decision is related to an innovation's rate of adoption. We generally expect that innovations requiring only an authority decision will be adopted most rapidly because fewer individuals are involved in the decision-making process. However, if the authority figure is traditional, this may slow down the rate of adoption. Optional innovation-decisions are usually next most rapid to authority decisions. Collective decisions, where a majority of the social system's members must be convinced to favor the innovation, are relatively slower (Chapter 9). Perhaps contingent decisions have a still slower rate of adoption because a sequence of two or more innovation-decisions is involved. Our general theory is that the more persons

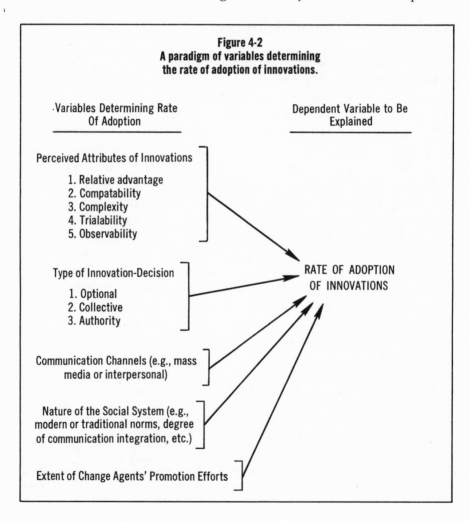

Figure 4-2
A paradigm of variables determining
the rate of adoption of innovations.

·Variables Determining Rate
Of Adoption

Dependent Variable to Be
Explained

Perceived Attributes of Innovations

    1. Relative advantage
    2. Compatability
    3. Complexity
    4. Trialability
    5. Observability

Type of Innovation-Decision

    1. Optional
    2. Collective
    3. Authority

RATE OF ADOPTION
OF INNOVATIONS

Communication Channels (e.g., mass
media or interpersonal)

Nature of the Social System (e.g.,
modern or traditional norms, degree
of communication integration, etc.)

Extent of Change Agents' Promotion Efforts

involved in making an innovation-decision, the slower the rate of adoption. If this assumption is valid, then one route to speeding the rate of adoption is to attempt to alter the unit of decision so that fewer individuals are involved. For instance, it has been found in the U.S. that when the decision to adopt fluoridation of municipal water supplies is made by a mayor or city manager, the rate of adoption is quicker than when the decision is collectively made by a public referendum. We have little research evidence about the relative speed of various types of innovation-decisions, but our assumption about the number of individuals involved seems reasonable.

The communication channels that are utilized to diffuse an innovation also may have an influence on the innovation's rate of adoption. For example, if interpersonal channels must be used to create awareness-knowledge, as frequently occurs among some peasant audiences where mass media channels are unavailable, the rate of knowledge and the rate of adoption will be slowed.

The relationships between communication channels and rate of adoption are even more complicated than this illustration suggests. The attributes of the innovation and the communication channels probably interact to yield a slower or faster rate of adoption. For example, Petrini and others (1968) found differences in communication channel use on the basis of the perceived complexity of innovations among Swedish farmers. Mass media channels, such as agricultural magazines, were satisfactory for less complex innovations, but interpersonal channels with extension change agents were more important for the innovations that were perceived by farmers as more complex. And if an inappropriate channel were used, such as mass media channels for complex ideas, a slower rate of adoption resulted. Similar evidence comes from a field experiment by Spector and others (1963) in Ecuador. For a complex innovation like latrine construction, a combination of mass media channels (radio, posters, and the like) was most effective. But for a relatively simple innovation like vaccination, radio alone was an effective channel.

When discussing the relationship of attributes of innovations and communication channels to rate of adoption,* we cannot forget the functions in the innovation-decision process, for the receiver's perceptions of the properties of the innovation may vary on the basis of these functions.**

---

*One illustration of a study of this type is Krug's (1961) analysis of the relationship of the rate of adoption of ten farm ideas in Wisconsin (1) to attributes of the innovations, and (2) to the extent and type of farm magazine coverage accorded the ideas. However, no strong associations were found among these variables.

**These statements that follow are logical derivations on the basis of the nature of the stages and the innovation characteristics. They are generally not supported by Kohl's (1966)

1   At the knowledge stage, the innovation's complexity and compatibility should be most important.
2   At the persuasion stage, the innovation's relative advantage and observability should be most important.
3   At the decision stage, the innovation's trialability should be most important.

Therefore, in order to maximize an idea's rate of adoption, a change agent's choice of the most appropriate channel depends upon a mix of such considerations as: (1) function in the innovation-decision process, and (2) the perceived attributes of the innovation.

There is also a further consideration (see Figure 4-2): The nature of the social system. Especially important are the norms of the system. In a modern system the rate of adoption is likely to be more rapid because there is less attitudinal resistance to be overcome on the part of the receivers. In a traditional system, however, the rate of adoption is likely to be slower.

Last, as suggested in Figure 4-2, an idea's rate of adoption is affected by the extent of change agents' promotion efforts.* The relationship between rate of adoption and change agents' efforts is not direct and linear, however. There is a greater payoff from a given amount of change agent activity at certain stages in an innovation's diffusion. Stone (1952) and Petrini (1966a) show that the greatest response to change agent effort occurs at about the point when opinion leaders are adopting, which occurs somewhere between 3 and 16 percent adoption in most systems.**

As yet there has been no diffusion research designed to determine the relative contribution of each of the types of variables shown in Figure 4-2 that explain an innovation's rate of adoption. When such inquiry is accomplished, we shall possess a much more adequate basis for planning and allocating the inputs for diffusion campaigns designed to speed the rate of adoption.

study of fifty-eight Oregon school superintendents. Relative advantage and compatibility were most important at the decision stage, while complexity, observability, and trialability were reported to be most important at the knowledge stage. We cannot fully explain the inconsistency of Kohl's results with out hypothesis; further research is needed to settle the matter.

*Very strong empirical evidence to support this statement, which is essentially parallel to Generalization 7–1, is provided by a three-nation diffusion project conducted in Brazil, Nigeria, and India. See Whiting and others (1968), Hursh and others (1969), and Fliegel and others (1967b).

**This relationship is detailed in Chapter 7.

The Diffusion Effect

Not only does change agent effort have a different effect at the different points in the sequence of an innovation's rate of adoption, but the system's self-generated pressures toward adoption also alter as an increasing proportion of the members of the system adopt. We term this informal pressure the "diffusion effect."

The *diffusion effect* is the cumulatively increasing degree of influence upon an individual to adopt or reject an innovation, resulting from the increasing rate of knowledge and adoption or rejection of the innovation in the social system.* For example, when only 5 percent of the individuals in a system are aware of a new idea, the degree of influence upon an individual to adopt or reject the innovation is quite different from what it is when 95 percent have adopted. In other words the norms of the system toward the innovation *change over time* as the diffusion process proceeds, and the new idea is gradually incorporated into the lifestream of the system. The communication environment of the system and its impact upon individuals in the system is dynamic.

There is a complex but important interrelationship between the rate of knowledge about an innovation in a system and its rate of adoption. In one sense the level of knowledge at any given point in time is an indication of the amount of information about the innovation available to the average individual in the system's communication environment. When such a level of information (and accompanying influence) is very low, adoption of the innovation is unlikely for any given individual. As the level of information increases past a certain threshold, adoption is more likely to occur as the self-generated pressures toward adoption increase. This relationship is illustrated for 2,4-D weed spray among Iowa farmers in Figure 4-3, where the rate of adoption is plotted over the rate of awareness-knowledge.** We see that the diffusion

---

*In the previous edition of this book, we preferred to term an essentially similar notion the "interaction effect," which was defined as "the process through which individuals in a social system who have adopted an innovation influence those who have not yet adopted" (Rogers, 1962b, p. 138). Unfortunately, interaction also has a different meaning in a statistical sense, one that is commonly utilized by social scientists. Our present terminology and definition stress the essential and basic implication that the degree of influence on the individual increases as the diffusion process proceeds. It seems appropriate to term it the diffusion effect.

**The plotting of the diffusion effect in this manner should be credited to Dr. V. R. Gaikwad, a staff member at the Indian Institute of Management, Ahmedabad, India, who first became interested in the interrelationship between the rate of adoption of an idea and its rate of awareness-knowledge.

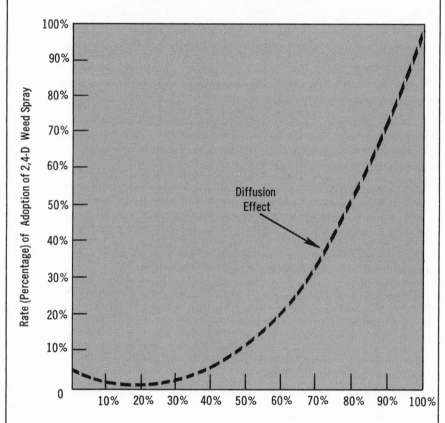

Figure 4-3
The positive (but not direct) relationship between the rate of adoption and the rate of awareness-knowledge for 2,4-D weed spray among Iowa farmers illustrates the nature of the diffusion effect.

Diffusion Effect

Rate (Percentage) of Adoption of 2,4-D Weed Spray

Rate (Percentage) of Awareness-Knowledge of 2,4-D Weed Spray

Source: A re-analysis of the Beal and Rogers (1960, p. 8) data by Dr. V. R. Gaikwad of the (India) National Institute of Community Development. Used by permission.

effect is positive but not linear and direct.* In other words as the rate of awareness-knowledge of the innovation increased up to about 20–30 percent, there was almost no adoption. Then once this threshold point was passed, each additional percentage of awareness-knowledge in the system was associated with several percentage increases in the rate of adoption. This seems reasonable on the basis of our understanding of the diffusion effect. Until an individual has a certain minimum level of information and influence from his system's environment, he is unlikely to adopt. But once this threshold is passed (and the exact threshold point may be different for every innovation and every system),** adoption of the idea is further increased by each additional input of knowledge and influence to the system's communication environment. Further, the threshold seems to occur at about the point where the opinion leaders in a system begin to favor the innovation.***

An investigation of the rate of adoption of five food innovations among 1,028 housewives in five Guatemalan villages provides some further evidence of the importance of the diffusion effect in explaining adoption rates (Mendez, 1968). Faster rates of adoption were found in the more integrated**** villages, where the levels of interpersonal interaction among the villagers were higher and where more of the peasants were reached by the interpersonal communication network. In the less integrated villages the adoption curves approached a straight line with a low degree of upward slope, whereas under conditions where the diffusion effect was greater due to higher integration, the adoption rates followed the more rapid S-shape.

Similarly, Guimarães (1968) reported that innovations had a faster rate of adoption in Brazilian villages with a higher degree of communication integration. Confirming evidence is provided by Yadav (1967) in Indian villages and

*Dr. Gaikwad computed the regression of the rate of adoption on the rate of awareness for 2,4-D spray, using the data shown in Figure 4-3. Linear correlation is .92, which indicates that 85 percent of the variance in awareness and adoption occur together. The curvilinear relationship is, of course, even higher: Correlation is .98, which indicates a covariance of 96 percent. Thus, the relationship is best depicted as curvilinear in Figure 4-3.

**Dr. Gaikwad plotted the diffusion effect for a number of different innovations studied in the United States and in India (in a manner similar to that shown in Figure 4-3), and they bear striking similarity to that shown here for 2,4-D spray in Iowa. One exception is for family planning innovations in India, which may not show much diffusion effect because they are relatively taboo topics for interpersonal discussion among Indian peasants and are not a socially "visible" innovation.

***This suggests that the threshold point may occur relatively earlier in modern systems, where the opinion leaders are more innovative, than in systems with more traditional norms.

*****Communication integration* is the degree to which the units in a social system are interconnected by interpersonal communication channels. This concept is usually measured by an index computed by matrix multiplication of sociometric data.

by Coughenour (1964c) in Kentucky communities, as well as by Coleman and others (1966) in an investigation of the diffusion of a new drug among physicians. In all cases it seems that social systems whose members are in closer interpersonal communication have a greater diffusion effect and a faster rate of adoption of innovations. The logic of our reasoning is thus: Communication integration→diffusion effect→rate of adoption. We conclude this discussion with Generalization 4-6: *The degree of communication integration in a social system is positively related to the rate of adoption of innovations.*

The reader should remember several cautions in his interpretation of this generalization and of the data shown in Figure 4-3. First, our data do not allow us to say definitely that the increasing rate of awareness-knowledge *causes* the increasing rate of adoption. We can only say that they vary together and in a way that we would expect. Further, our data on the rate of awareness-knowledge and on the rate of adoption are dependent upon the accuracy of our respondents' recall, which is certainly much less than perfect.* Nevertheless the type of relationship shown in Figure 4-3 for the diffusion effect is probably generally correct, and it conveys a useful insight that could be used by change agents in conducting a communication campaign for an innovation. Until at least a threshold of information and influence is reached, little adoption occurs. Change agents need to know approximately where this threshold point is and then plan their strategies of change accordingly.

Overadoption

Many past researchers implicitly assumed that adoption of innovations by their respondents is desirable behavior and that rejection of innovations is less desirable. Of course, this is not true in all cases. Few studies are available of *overadoption*,** defined as the adoption of an innovation by an individual when experts feel he should reject. There are several possible reasons for overadoption: Insufficient knowledge about the new idea on the part of the adopter, inability to predict its consequences, a mania for the new. Perhaps this latter type of overadoption is especially common among innovators, who have a particular penchant for the new and thus occasionally appear to be

*It is also probably less accurate for the dates of awareness-knowledge than for the dates of adoption.
**One might speculate why more research has not been completed on overadoption. Perhaps it is because of the greater usefulness of underadoption studies to the research sponsors.

suckers for change. Their venturesomeness may occasionally lead innovators to adopt unsound new ideas. Such neophiles must expect to pay for their suprainnovativeness with an occasional failure.

An example of such enthusiastic overadoption occurred throughout the Midwest in the late 1940s. A chemical weed killer, 2,4-D, was introduced in 1946, and its results were spectacular. Farmers became so enthusiastic about 2,4-D spray, in fact, that it was applied to many cornfields whether the resulting increase in yields justified its use or not. Observers estimate that millions of dollars were lost through overadoption of the weed spray until later years when farmers learned how to use it more wisely. Overadopters sometimes see an innovation as a panacea, as in the case of the home canners in the Georgia diffusion campaign described in Chapter 3.

It is often difficult to determine whether an individual should or should not adopt an innovation. The criterion of rationality* is not easily measurable. The classification can sometimes be made by an expert on the innovation under study.** The classification is often made on the basis of economic criteria. In one sense most individuals perceive their actions to be rational. Through lack of knowledge or through inaccurate perceptions, the individual's evaluation of an innovation does not agree with the expert's. Our main concern is with objective rationality in the present case, rather than with subjective rationality as perceived by the individual.

Most studies of irrational overadoption are concerned with farm innovations of fraudulent or doubtful value. A number of agricultural crazes swept U.S. agriculture in earlier times, such as the Merino sheep, silkworms and Chinese mulberry trees, Rohan potatoes, China tree corn, and silver foxes. In more recent times such noneconomic practices as Canadian oats, fraudulent livestock mineral salts, and one-gallon-to-the-acre liquid fertilizers have been widely purchased by U.S. farmers.

The widespread noneconomic adoption of intrinsically valuable innovations is also often found. Toussaint and Stone (1960) studied a sample of the several thousand North Carolina farmers who had purchased self-propelled tobacco harvesters, which were recommended by economists only if they could be used to full capacity. No farmers were found who used the machine on enough acres for it to be as economical as if they had harvested by hand methods.

Brandner and Straus (1959) found that hybrid sorghums were planted on

---

*We define *rationality* as the use of the most effective means to reach a given goal.
**Homans (1961, p. 80) favors this measure of irrationality: "Behavior is irrational if an outside observer thinks that its reward is not good for a man in the long run."

28 percent of the sorghum acreage in northeastern Kansas the first year they were available, in spite of the fact that this innovation was not recommended by the Kansas Agricultural Experiment Station or Extension Service.

Probably the most adequate analysis of overadoption was a novel study by Goldstein and Eichhorn (1961) of the adoption of four-row corn planters by Indiana farmers. Agricultural economists at Purdue University stated that farmers with sixty acres of corn or less could not economically justify a four-row corn planter. Table 4-2 shows that there were more underadopters than overadopters. The two rational types of individuals outnumbered the two irrational types. A further analysis by Goldstein (1959) indicated that the rational types differed from the irrational types in amount of education and in lack of traditional beliefs. Thus, education is one factor that leads to more rational and discriminating decision-making in adoption decisions (or at least to farmers' decisions that are more like the experts).

Table 4-2   Rationality and Irrationality in the Adoption and Rejection of Four-Row Corn Planters by Indiana Farmers

Individual's Innovation-Decision	EXPERTS' RECOMMENDATION FOR THE INDIVIDUAL	
	Adoption	Rejection
Adoption	Rational adopters 37%	Irrational overadopters 11%
Rejection	Irrational underadopters 19%	Rational rejectors 33%

This paradigm shows the four types of adopters and non-adopters of an innovation, illustrated with data from an investigation of four-row corn planters in Indiana. Rational adopters and rational rejectors (combined) made up 70 percent of the sample. Irrational overadopters, who adopted the innovation under conditions that experts considered more appropriate for rejection, constituted 11 percent. Nineteen percent of the farmers (irrational underadopters) had not purchased the machine, but should have done so in the opinion of experts.
Source: Based on data provided by Goldstein and Eichhorn (1961).

It is often difficult to find, for purposes of a research study, an innovation that is irrational for all receivers. For example, in the Goldstein and Eichhorn (1961) study it is entirely possible that special circumstances might economically justify that an irrational underadopter reject the idea. This sticky problem of securing a universally inappropriate innovation was solved by Francis (1960). He surveyed 88 adopters of grass incubators, an innovation not recommended for any farmer by scientists on the basis of economic and nutritional grounds.

The grass incubator costs several thousand dollars and is manufactured by several U.S. companies. It is a small air-conditioned room in which grain is sprouted and grown to a height of about six inches. The whole plant is then fed to livestock. The grass incubator is not recommended to farmers either by agricultural economists or by animal nutritionists in the U.S. Department of Agriculture, state agricultural experiment stations, or extension services. Thus, all of Francis' (1960) respondents were irrational overadopters. Their characteristics and communication behavior marked them as quite different from the typical U.S. farmer. They were more affluent, more traditional, and more specialized. The grass incubator owners tended to have little communication with government agricultural change agents, and few were innovators of recommended farm ideas. Perhaps extension agents may be more effective in preventing the adoption of nonrecommended innovations than in promoting the adoption of recommended ideas.

It is entirely possible, of course, for a change agent to push his constituents into overadoption in some cases. An example is the fertilizer dealer who wishes to increase sales past the point where farmers are purchasing economical amounts of fertilizer. This may be one reason why U.S. farmers sometimes place little credibility in salesmen.

There is need for further investigation of overadoption* and of the role of change agents in causing or preventing overadoption.

Summary

The purpose of this chapter was to suggest five general attributes of innovations by which any innovation could be described, to see how an individual's perceptions of these characteristics are predictive of rate of adoption, and to analyze cases of overadoption. Throughout this chapter we emphasized that it is the receiver's perceptions of innovations' attributes which affect their rate of adoption.

Relative advantage is the degree to which an innovation is perceived as better than the idea it supersedes. The relative advantage of a new idea, as perceived by members of a social system, is positively related to its rate of adoption.

Compatibility is the degree to which an innovation is perceived as consistent with the existing values, past experiences, and needs of the receivers. The compatibility of a new idea, as perceived by members of a social system, is positively related to its rate of adoption.

*Among such studies, we feel it would be intellectually profitable to determine whether overadopted innovations tend to be discontinued at a later date when the adopter realizes that he has overadopted.

Complexity is the degree to which an innovation is perceived as relatively difficult to understand and use. The complexity of an innovation, as perceived by members of a social system, is negatively related to its rate of adoption.

Trialability is the degree to which an innovation may be experimented with on a limited basis. The trialability of an innovation, as perceived by members of a social system, is positively related to its rate of adoption.

Observability is the degree to which the results of an innovation are visible to others. The observability of an innovation, as perceived by members of a social system, is positively related to its rate of adoption.

Rate of adoption is the relative speed with which an innovation is adopted by members of a social system. In addition to the perceived attributes of an innovation, such other variables affect its rate of adoption as (1) the type of innovation-decision, (2) the nature of communication channels used to diffuse the innovation at various functions in the innovation-decision process, (3) the nature of the social system, and (4) the extent of change agents' promotion efforts in diffusing the innovation.

The *diffusion effect* is the cumulatively increasing degree of influence upon an individual to adopt (or reject) an innovation, resulting from the increasing rate of knowledge and adoption or rejection of the innovation in the social system. It appears that as the rate of awareness-knowledge in a social system increases up to about 20–30 percent, there is almost no adoption; but once this threshold is passed, further increases in awareness-knowledge lead to increases in adoption. The diffusion effect is probably greater in social systems with a higher degree of *communication integration* (the degree to which the units in a social system are interconnected by interpersonal communication channels). The degree of communication integration in a social system is positively related to the rate of adoption of innovations.

Overadoption is the adoption of an innovation by an individual when experts feel he should reject. Overadoption may occur because of insufficient knowledge about the innovation on the part of the adopter, inability to predict its consequences, or a mania for the new.

Needed Research on Attributes of Innovations

We emphasized at the beginning of this chapter that there is only a limited number of diffusion investigations dealing with perceived attributes of innovations. In this closing section we suggest five types of needed research:

(1) measuring perceived attributes at the time of decision, (2) differential perceptions by different groups, (3) improved measurement of perceived attributes, (4) factor analysis of perceived attributes, and (5) studying innovation bundles.

Measuring Perceived Attributes at the Time of Decision

Evidence was presented in this chapter that rates of adoption of innovations depend upon *perceptions* of the characteristics of innovations. Although much further research is needed on this point, there are several difficulties involved in research on perceptions of the attributes of innovations. For one thing, perceptions are always changing and hence they are difficult to measure in retrospect by asking respondents to recall how they perceived an innovation at some previous time, such as before they adopted it.*

As the diffusion process continues, the content of the message often becomes distorted or otherwise changed. Evidence is available from anthropological writings that changes occur in the *meaning* of an innovation and even in the *use* to which it is put as diffusion proceeds (Barnett, 1953, pp. 330–333). Similar evidence of the relatively low fidelity of messages transmitted by interpersonal channels comes from studies of rumor (Allport and Postman, 1947; De Fleur, 1962b; Klonglan, 1963). For this reason we might expect different perceptions of the attributes of innovations by earlier versus later adopters (who often learn of an innovation from earlier adopters).

Likewise, an individual's perceptions of an innovation are likely to change after he adopts it. If his actual experience with the innovation is satisfactory, his perceptions probably will become more favorable. For this reason many of the researches completed to date on perceptions of innovations and their rate of adoption have a very serious weakness. The positive relationship between perceptions and rate of adoption may partly be an artifact of the tendency for individuals who have adopted an innovation to rationalize their decision in terms of relatively positive perceptions.

How might this methodological defect in previous research be overcome? One approach, yet to be tried, is to gather data on perceptions of innovations at the time of an individual's decision to adopt. Certainly this is the point in

*In fact, Festinger's (1957) theory of cognitive dissonance would suggest that many individuals' perceptions of an innovation are determined in part by their adoption or rejection decision. Thus, once a decision is made, perceptions are frequently altered to be consistent with this decision.

time at which perceptions of an innovation are most important in affecting rate of adoption. One can imagine a research design in which data about perceptions of innovations are gathered from adopters at the time of the innovation-decision. An example would be when women come to a clinic to be fitted with a birth control device.

Differential Perceptions by Different Groups

We need research on how perceptions of innovations differ for various groups, such as adopter categories, scientists, change agents, and the like.* The differences in these perceptions could help us predict likely communication "hang-ups" that diffusion campaigns would encounter. For instance, the Apodaca (1952) study of hybrid corn rejection by Mexican-American farmers suggests that whereas the change agent perceived the corn seed mainly in terms of its higher yield, the farmers perceived it as a poor food.

Another illustration of the research approach to differential perceptions of innovations that we are proposing is represented by Kivlin and Fliegel (1967a, b) and Fliegel and others (1968). These researchers, as shown previously (Table 4-1), split their total sample into large- and small-size farmers and found that there were important differences in how each group perceived the same innovations. It is possible (and we feel it would be profitable) to split samples of respondents on variables other than farm size, for example, adopter categories. Innovators undoubtedly perceive the same innovation in different ways than do laggards. And these differences in perception could have important implications for the way change agents might introduce a new idea to each category.

Improved Measurement of Perceived Attributes

The measurement of perceptions of innovations is difficult. In past studies ratings have been made on the basis of respondents' answers to single, direct questions about how they perceive an innovation. One potential improvement in measurement techniques would be to use a scale composed of numerous questions about an innovation. Elliott (1968) utilized the semantic differential, a technique used to measure perceptions of other objects, in order

*Sekhon (1968, p. 120) expected Indian peasants with low change agent contact to perceive innovations less like change agents than did the peasants with high change agent contact, but his hypothesis was not supported.

to operationalize perceptions of the characteristics of innovations. Other measurement techniques should be explored.

Factor Analysis of Perceived Attributes

Factor analysis offers analytical potential for deeper probing of the nature of innovation perceptions. This technique of data reduction could, for example, help to verify whether the five attributes of innovations discussed in this chapter are the most generalized and relevant. Kivlin and Fliegel (1967b) factor analyzed the responses of Pennsylvania farmers to questions measuring perceptions of attributes. These researchers found five factors for large farmers and six for small farmers,* which were roughly translatable to relative advantage, compatibility, complexity, and observability; missing from the five we listed was trialability.

Further factor analyses of adopters' perceptions could eventually lead us to a clearer notion of just how many perceived attributes of innovations we need to conceptualize.** Until then, as pointed out early in this chapter, our postulation of five attributes is primarily a logical, but empirically defenseless, procedure.

Studying Innovation Bundles

A major fault of almost all past diffusion research is its concentration upon single innovations. An implicit and certainly false assumption is that the adoption, and the consequences of an innovation, are completely independent of all other innovations. We know that a receiver perceives any new idea in terms of its degree of compatibility with previously adopted ideas. Further, the consequences that are likely to result from adoption of the new idea are usually highly dependent upon which other innovations the receiver is already using or adopts at the same time. For instance, when a farmer adopts a chemical fertilizer, he is likely to achieve a higher crop yield only if he also adopts an appropriate corn variety and plants his seed more thickly (Silverman and Bailey, 1961). In agriculture and perhaps in most other fields there are impor-

---

*These factors explained 88 percent of the variance for large farmers and 91 percent for small farmers, which is quite high.
**They could lead, as well, to a better understanding of how innovations which are perceived similarly by adopters could be grouped together for promotion by change agents. Elliott's (1968) approach illustrates this method.

tant interactions among the results of innovations.* Therefore, we should study more accurately "bundles" or "packages" of innovations, rather than heuristically conceiving of each innovation as a discreet, separate unit for analysis. In the minds of the adopters, it is not.

Specifically, how could we investigate the adoption of bundles of innovations? A first step is to determine which innovations seem to "bundle together." To do this, a researcher might determine which of a number of possibly interrelated innovations a sample of receivers have adopted. Then, factor analysis of the interrelationships among the adoption of the various innovations would indicate which of the ideas were adopted in unison.** Once such complexes of innovations are determined, change agents would know which innovations to promote as a package. This research strategy is essentially a pragmatic approach in which bundles of innovations are determined in terms of which ones *have been* adopted together.

But of course it is entirely possible that many interrelated innovations, which should have been adopted concurrently simply, were not. In fact, Silverman and Bailey (1961) found this to be the case among Mississippi corn farmers. Another research approach, then, is first to determine the biological interactions among the various innovations (this is the task of the biological or physical scientists who produce the innovations, not the responsibility of diffusion researchers), and then use this knowledge as a basis for innovation bundling.

Perhaps different members of a social system perceive different innovation complexes. Do innovators have the ability to see the bundles more accurately than do laggards? Do they see different bundles? We need to find out.

Further, we need to determine whether the adoption of any single innovation triggers the adoption of other ideas in the complex of innovations. If such trigger innovations exist and can be isolated, the useful implications for change agents are obvious: They can simply promote a trigger innovation and thus secure the adoption of the entire bundle of innovations.

---

*In a statistical sense, *interaction* is defined as the effect of one variable on the joint effects of two other variables. For example, a positive interaction between fertilizer and hybrid seed on corn yields is indicated by the following data.

	No Hybrid Seed	Hybrid Seed
No Fertilizer	40 Bushels	45 Bushels
Fertilizer	50 Bushels	60 Bushels

The yield increase due to hybrid seed is 5 bushels, and to fertilizer alone is 10 bushels. When both hybrid seed and fertilizer are used, the yield increase is 20 bushels, indicating an interaction effect of the seed and fertilizer of 5 bushels $(20 - 5 - 10 = 5)$.

**As Fliegel (1955) and Elliott (1968) have done.

5 *Adopter*

The innovator makes enemies of all those who prospered under the old order, and only lukewarm support is forthcoming from those who would prosper under the new . . . because men are generally incredulous, never really trusting new things unless they have tested them by experience.

NICCOLO MACHIAVELLI (1961, p. 51)

A slow advance in the beginning, followed by rapid and uniformly accelerated progress, followed again by progress that continues to slacken until it finally stops: These are the three ages of real social beings which I call inventions. . . . If taken as a guide by the statistician and by the sociologist, [they] would save many illusions. GABRIEL TARDE (1903, p. 127)

Categories

▪— ▪—▪▪ ▪▪▪ ▪▪▪▪▪ ▪— ▪ —▪   ▪—▪▪— ▪— ▪— ▪ ▪▪—   ▪▪▪ ▪—▪▪ ▪▪ ▪ ▪▪▪   ▪— ▪

ALL individuals in a social system do not adopt an innovation at the same time. Rather, they adopt in an ordered time sequence, and they may be classified into adopter categories on the basis of when they first begin using a new idea. We could describe each individual adopter in a social system in terms of his time of adoption, but this would be tedious work. It is much easier and more meaningful to describe adopter categories,* each containing individuals with a similar degree of innovativeness. There is much practical usefulness for change agents if they can identify potential innovators and laggards in their client audience and utilize different change strategies with each such subaudience.

We know more about innovativeness, the degree to which an individual is relatively earlier in adopting new ideas than other members of his social system, than any other concept in diffusion research (Chapter 2). The expressed, short-term goal of most change agencies is to facilitate the adoption of innovations by their clients. Because increased innovativeness is the objective of change agencies, it has become the main dependent variable in the diffusion research these change agencies sponsor. A further reason for the prime focus on innovativeness in diffusion research, especially in less developed countries, is that innovativeness is the best single indicator of

*Adopter categories are the classifications of the members of a social system on the basis of innovativeness.

175

modernization* (Rogers with Svenning, 1969, p. 292). Innovativeness indicates behavioral change, the ultimate goal of modernization programs, rather than cognitive or attitudinal change.

This chapter suggests one method of categorizing adopters and demonstrates the usefulness of this technique with research findings from both more developed and less developed nations. We shall discuss the normality of adopter distributions, the method of classifying adopters, characteristics of adopter categories, and predicting innovativeness.

Innovativeness and Adopter Categories

The Need to Standardize Categories

Titles of adopter categories are about as numerous as diffusion researchers themselves. The inability of diffusion researchers to agree on common semantic ground in assigning terminology has led to a plethora of adopter descriptions. The most innovative individuals have been termed progressists, high-triers, experimentals, lighthouses, advance scouts, and ultradopters. Least innovative individuals have been called drones, parochials, and diehards. The fertile disarray of adopter categories and methods of categorization, illustrated by the adopter categories, emphasizes the need for standardization. How can a reader compare research findings about adopter categories until there is standardization of both the nomenclature and the classification system? Fortunately, one method of adopter categorization has gained a dominant position in recent years, one based upon the S-shaped curve of adoption.

S-Curve of Adoption Over Time

It is the time variable which allows researchers to classify adopter categories and to plot diffusion curves.** Research has generally shown that the

Modernization is defined as the process by which individuals change from a traditional way of life to a more complex, technologically advanced, and rapidly changing style of life. This is best indicated by an individual's actual use of new ideas in agriculture, health, family living, and other fields.
**As such, an adopter distribution is one type of diffusion curve, which represents the number of knowers or adopters of an innovation per unit of time. In the present book we utilize

adoption of an innovation follows a normal, bell-shaped curve when plotted over time on a frequency basis. If the cumulative number of adopters is plotted, the result is an S-shaped curve. Figure 5-1 shows that the same adoption data can be represented by either a bell-shaped (frequency) or an S-shaped (cumulative) curve.

the term "adopter distribution," rather than "diffusion curve," for the sake of greater precision.

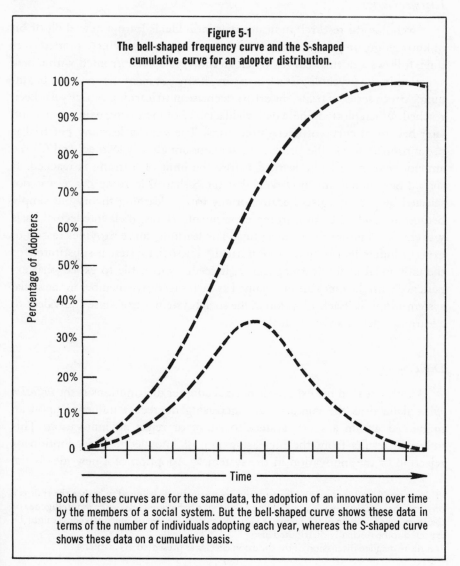

Figure 5-1
The bell-shaped frequency curve and the S-shaped
cumulative curve for an adopter distribution.

Both of these curves are for the same data, the adoption of an innovation over time by the members of a social system. But the bell-shaped curve shows these data in terms of the number of individuals adopting each year, whereas the S-shaped curve shows these data on a cumulative basis.

The S-shaped adopter distribution rises slowly at first when there are few adopters in a time period. Then it accelerates to a maximum when half of the individuals in the system have adopted. It then increases at a gradually slower rate as the few remaining individuals finally adopt. This S-shaped curve is normal. Why?

Learning Curves

Psychological research indicates that individuals learn a new skill, or bit of knowledge, or set of facts through a learning process which, plotted over time, follows a normal curve. When an individual is confronted with a new situation in the psychologist's laboratory, he makes many errors at the beginning. After a series of trials, the errors decrease until learning capacity has been reached. When plotted, these data yield a curve of increasing gains at first and later become a curve of decreasing gains. The gain in learning per trial is proportionate to (1) the product of the amount already learned, and (2) the amount remaining to be learned before the limit of learning is reached. It should be pointed out, however, that the S-shaped learning curve was not selected by psychologists because of any formal learning theory but simply because it resulted from learning experiments. Thus, their intellectual basis was empirical rather than conceptual. The learning curve provides reason to expect adopter distributions to be normal.* If a social system is substituted for the individual in the learning curve, it seems reasonable to expect that experience with the innovation is gained as each successive member in the social system adopts it. Each adoption in the social system is in a sense equivalent to a learning trial by an individual.

Diffusion Effect

Another reason for expecting normal adopter distributions is the *diffusion effect*, defined as the cumulatively increasing degree of influence upon an individual within a social system to adopt or reject an innovation. This influence results from the increasing rate of knowledge and adoption or rejection of the innovation in the system.** Adoption of a new idea is the

---

*It has been found that many human traits are normally distributed, whether the trait is a physical characteristic, such as weight or height, or a behavioral trait, such as intelligence or the learning of information. Hence, a variable such as degree of innovativeness might be expected to be normally distributed also.

**A more detailed discussion of the diffusion effect was presented in Chapter 4.

result of human interaction. If an innovation is introduced in a social system, there are theoretical grounds for expecting the number of adoptions over time to be normally distributed. If the first adopter of the innovation discusses it with two other members of the social system, and these two adopters each pass the new idea along to two peers, the resulting distribution follows a binomial expansion.* This mathematical function follows a normal shape when plotted.

Of course, several of the assumptions underlying this hypothetical example are seldom found in reality. For instance, members of a social system do not have completely free access to interact with one another. Status barriers, geographical location, and other variables affect diffusion patterns. The diffusion effect begins to level off after half of the individuals in a social system have adopted, because each new adopter finds it increasingly difficult to tell the new idea to a peer who has not yet adopted, for such nonknowers become increasingly scarce.

Testing Adopter Distributions for Normality

It has generally been found that *adopter distributions follow a bell-shaped curve over time and approach normality.*** There are useful implications to be found for a standard method of adopter categorization, as we shall soon see.

Eight adopter distributions for single innovations tested by Rogers (1958b) were bell-shaped and all approached normality, although half of those tested were found to deviate significantly from normality (Table 5-1). Similar evidence is provided by Bose (1964b), who found the adoption of crop chemicals was normally distributed in each of seven Indian villages, even though these villages differed widely in their norms, size, and social make-up. Further, Andrus (1965) found that innovativeness scores (based on the adoption of new household products by a national sample of 11,000 U.S. families) formed a bell-shaped, but not exactly a normal, curve. Table 5-1 shows that the distribution of innovativeness scores for samples of Iowa and Ohio farmers are normal.

*A pattern that is similar to that of an unchecked infectious epidemic. Bailey (1957, pp. 29–37, 155–159) provides epidemiological models for this biological phenomenon that could usefully be applied to the diffusion of innovations.
**DeFleur (1966) suggests that the curve of dicontinuance may also approach normality, although there is yet no research evidence on this point.

Table 5-1 Normality of Adopter Distributions

INNOVATION (OR INNOVATIVENESS SCORES)	NORMALITY OF ADOPTER DISTRIBUTION	RESEARCH STUDY
1 2,4-D weed spray in Iowa (all adopters)	Normal	Beal and Rogers (1960)
2 2,4-D weed spray in Iowa (beginning farmers excluded)	Normal	Beal and Rogers (1960)
3 Antibiotics in Iowa (all adopters)	Not normal[b]	Beal and Rogers (1960)
4 Antibiotics in Iowa (beginning farmers excluded)	Not normal[a]	Beal and Rogers (1960)
5 Hybrid corn (Iowa)	Not normal[b]	Ryan (1948)
6 Hybrid corn (Virginia)	Not normal[b]	Dimit (1954)
7 2,4-D weed spray (Ohio)	Normal	Rogers (unpublished data)
8 Warfarin rat poison (Ohio)	Normal	Rogers (unpublished data)
9 Adoption of farm innovations scores (Iowa)	Normal	Rogers (1958b)
10 Adoption of farm innovations scores (Ohio)	Normal	Rogers (1958b)

[a]Deviation from normality is significant at the 5 percent level.
[b]Deviation from normality is significant at the 1 percent level.
Source: Rogers, 1958b; used by permission.

A Method of Adopter Categorization

A researcher seeking standardization of adopter categories faces three problems: (1) determining the number of adopter categories to conceptualize, (2) deciding on the portion of the members of a system to include in each category, and (3) determining the method, statistical or otherwise, of defining the adopter categories.

There is no question, however, about the criterion for adopter categorization. It is innovativeness, the degree to which an individual is relatively earlier in adopting new ideas than other members of his social system.* Innovative-

---

*One of the major problems in measuring innovativeness is the inaccuracy of recall by some individuals and for some innovations. In one study which collected data both by recall and from records, it was found that doctors' statements tended to make them appear more up-to-date than did druggists' prescription records (Menzel, 1957). On the other hand, Havens (1962b) found that Ohio dairy farmers were quite accurate in reporting from memory their date of adoption of bulk milk tanks. Havens checked the adoption dates with records of the farmers' milk purchasers. The different results from the two studies may be due to the type of innovation, the latter requiring a substantial cash outlay as well as a change in behavior.

ness is a "relative" dimension, in that one either has more or less of it than others in a social system. Innovativeness is a continuous variable, and partitioning it into discrete categories is only a conceptual device, much like dividing the continuum of social status into upper, middle, and lower classes.

Before describing a method of adopter categorization, it is important to specify the characteristics which a set of categories should possess. Ideally, categories should: (1) be *exhaustive,* or include all the respondents of the sample, (2) be *mutually exclusive,* or exclude from any other category a respondent who appears in one category, and (3) be derived from *one classificatory principle* (Jahoda and others, 1951, p. 264).

We have previously demonstrated that adopter distributions closely approach normality. This is important because the normal frequency distribution has several characteristics which may be used in classifying adopters. One of these characteristics or parameters is the mean $(\bar{x})$, or average, of the sample. Another parameter of a distribution is the standard deviation (sd), a measure of dispersion about the mean. The standard deviation explains the average amount of variance on either side of the mean for a sample.

These two statistics, the mean $(\bar{x})$ and the standard deviation (sd), can be used to divide a normal adopter distribution into categories. If vertical lines are drawn to mark off the standard deviations on either side of the mean, the curve is divided into categories in a way that results in a standardized percentage of respondents in each category. Figure 5-2 shows the normal frequency distribution divided into five adopter categories: (1) innovators, (2) early adopters, (3) early majority, (4) late majority, and (5) laggards. These five adopter categories and the approximate percentage of individuals included in each are located on the adopter distribution in Figure 5-2.

The area lying to the left of the mean time of adoption minus two standard deviations includes the first 2.5 percent of the individuals to adopt an innovation—the *innovators*. The next 13.5 percent to adopt the new idea are included in the area between the mean minus one standard deviation and the mean minus two standard deviations; they are labeled *early adopters*. The next 34 percent of the adopters, called *early majority*, are included in the area between the mean date of adoption and minus one standard deviation. Between the mean and one standard deviation to the right of the mean are located the next 34 percent to adopt the new idea, the *late majority*. The last 16 percent are called *laggards*.

This method of adopter classification is probably the most widely used in current diffusion studies. However, as can be observed, it is not a symmetrical classification in that there are three adopter categories to the left of the mean and only two to the right. One solution would be to break laggards into two categories, such as early and late laggards, but laggards seem to be a fairly homogeneous category. Similarly, innovators and early adopters could be combined into a single class to achieve symmetry, but their quite different characteristics mark them as two distinct categories.

Another difficulty in our method of adopter classification is incomplete adoption, which occurs for innovations that have not reached 100 percent use at the time of their study. This means that our five-fold classification scheme is not completely exhaustive. But the problem of incomplete adoption or nonadoption is eliminated when a series of innovations is combined into a composite innovativeness scale.

Three principles of categorization were suggested earlier in this section. Innovativeness as a criterion fulfills each of these requirements. The five adopter categories are exhaustive (except for nonadopters), mutually exclusive, and are derived from one classification principle. The method of adopter categorization just described is the most widely used in diffusion research today.

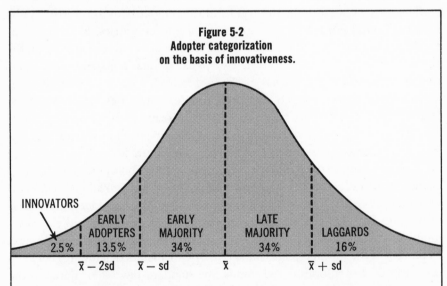

Figure 5-2
Adopter categorization
on the basis of innovativeness.

INNOVATORS

| EARLY ADOPTERS | EARLY MAJORITY | LATE MAJORITY | LAGGARDS |

2.5% | 13.5% | 34% | 34% | 16%

$\bar{x} - 2sd$ $\bar{x} - sd$ $\bar{x}$ $\bar{x} + sd$

The innovativeness dimension, as measured by the time at which an individual adopts an innovation or innovations, is continuous. However, this variable may be partitioned into five adopter categories by laying off standard deviations from the average time of adoption.

Adopter Categories as Ideal Types

The five adopter categories set forth in this chapter are ideal types. *Ideal types* are conceptualizations that are based on observations of reality and designed to make possible comparisons. The traditional and modern norms described in Chapter 1 are examples of ideal types. The function of ideal types is to guide research efforts and serve as a framework for the synthesis of research findings.

Actually, there are no pronounced breaks in the innovativeness continuum between each of the five categories. Ideal types are not simply an average of all the observations about an adopter category. Exceptions to the ideal types must be found. If no exceptions or deviations are located, ideal types would not be necessary. Ideal types are based on abstractions from empirical cases and are intended as a guide for theoretical formulations and empirical investigations. However, they are not a substitute for these investigations.

We now present a thumbnail sketch of the dominant characteristics and subcultural values of each adopter category, which will be followed by more detailed generalizations. Actually, there are few adequate investigations completed of the values of each adopter category; hence, the following section is an abstraction from a variety of studies not aimed specifically at determining value differences among adopter categories.

Innovators: Venturesome

Observers have noted that venturesomeness is almost an obsession with innovators. They are eager to try new ideas. This interest leads them out of a local circle of peers and into more cosmopolite social relationships. Communication patterns and friendships among a clique of innovators are common, even though the geographical distance between the innovators may be great. Being an innovator has several prerequisites. These include control of substantial financial resources to absorb the possible loss due to an unprofitable innovation and the ability to understand and apply complex technical knowledge.

The salient value of the innovator is venturesomeness. He desires the hazardous, the rash, the daring, and the risky. The innovator also must be willing to accept an occasional setback when one of the new ideas he adopts proves unsuccessful.

Early Adopters: Respectable

Early adopters are a more integrated part of the local social system than are innovators. Whereas innovators are cosmopolites, early adopters are localites. This adopter category, more than any other, has the greatest degree of opinion leadership in most social systems. Potential adopters look to early adoptors for advice and information about the innovation. The early adopter is considered by many as "the man to check with" before using a new idea. This adopter category is generally sought by change agents to be a local missionary for speeding the diffusion process. Because early adopters are not too far ahead of the average individual in innovativeness, they serve as a role model for many other members of a social system.* The early adopter is respected by his peers. He is the embodiment of successful and discrete use of new ideas. And the early adopter knows that he must continue to earn this esteem of his colleagues if his position in the social structure is to be maintained.

Early Majority: Deliberate

The early majority adopt new ideas just before the average member of a social system. The early majority interact frequently with their peers, but leadership positions are rarely held by them. The early majority's unique position between the very early and the relatively late to adopt makes them an important link in the diffusion process.

The early majority may deliberate for some time before completely adopting a new idea. Their innovation-decision is relatively longer than that of the innovator and the early adopter. "Be not the last to lay the old aside, nor the first by which the new is tried," might be the motto of the early majority. They follow with deliberate willingness in adopting innovations, but seldom lead.

Late Majority: Skeptical

The late majority adopt new ideas just after the average member of a social system. Adoption may be both an economic necessity and the answer to increasing social pressures. Innovations are approached with a skeptical and cautious air, and the late majority do not adopt until most others in their

*It will be pointed out in Chapter 6 that the degree of opinion leadership possessed by each adopter category depends in part on the social system's norms.

social system have done so. The weight of system norms must definitely favor the innovation before the late majority are convinced. They can be persuaded of the utility of new ideas, but the pressure of peers is necessary to motivate adoption.

Laggards: Traditional

Laggards are the last to adopt an innovation. They possess almost no opinion leadership. They are the most localite in their outlook of all adopter categories; many are near isolates. The point of reference for the laggard is the past. Decisions are usually made in terms of what has been done in previous generations. This individual interacts primarily with others who have traditional values. When laggards finally adopt an innovation, it may already have been superseded by another more recent idea which the innovators are already using. Laggards tend to be frankly suspicious of innovations, innovators, and change agents. Their tradition direction slows the innovation-decision process to a crawl. Adoption lags far behind knowledge of the idea. Alienation from a too-fast-moving world is apparent in much of the laggard's outlook. While most individuals in a social system are looking to the road of change ahead, the laggard has his attention fixed on the rear-view mirror.

Characteristics of Adopter Categories

From content analyses of research publications in the Diffusion Documents Center at Michigan State University, we have gleaned over 3,000 findings relating various independent variables to innovativeness.* Research findings on the characteristics of adopter categories are summarized as generalizations under the following headings: (1) socioeconomic status, (2) personality variables, and (3) communication behavior.

Socioeconomic Characteristics

Generalization 5-1: *Earlier adopters are no different from later adopters in age.* There is inconsistent evidence about the relationship of age and innovativeness; about half of the 228 studies on this subject show no relationship,

---

*Nearly 60 percent of the relationships produced by the content analysis have innovativeness as the dependent variable.

20 percent show that earlier adopters are younger, and 30 percent indicate they are older.*

Generalization 5-2: *Earlier adopters have more years of education than do later adopters.*

Generalization 5-3: *Earlier adopters are more likely to be literate than are later adopters.*

Generalization 5-4: *Earlier adopters have higher social status than later adopters.*** Status is indicated by such variables as income, level of living, possession of wealth, occupational prestige, self-perceived identification with a social class, and the like. But however measured, about two-thirds of such inquiries find a positive relationship of status with innovativeness.

Generalization 5-5: *Earlier adopters have a greater degree of upward social mobility than later adopters.* Although definitive empirical support is lacking, our evidence suggests that earlier adopters are not only of higher status but are on the move in the direction of still higher levels of social status. In fact, they may be using the adoption of innovations as one means of getting there.

Generalization 5-6: *Earlier adopters have larger sized units (farms, and so on) than later adopters.*

Generalization 5-7: *Earlier adopters are more likely to have a commercial (rather than a subsistence) economic orientation than are later adopters.* A subsistence orientation is typified by a traditional peasant who produces only for his own consumption and not for sale. Greater innovativeness comes with the advent of a commercial orientation in which farm products are raised for market.

Generalization 5-8: *Earlier adopters have a more favorable attitude toward credit (borrowing money) than later adopters.*

Generalization 5-9: *Earlier adopters have more specialized operations than later adopters.*

The social characteristics of earlier adopters thus mark them as better educated, of higher social status, and the like. They are wealthier, more

---

*Rogers (1962b, pp. 173–174) reanalyzed Gross' (1942) original data to demonstrate there are wider differences in age between adopter categories when age at the time of *adoption* of hybrid seed was used, rather than age at the time of *interview*.

**Not only do earlier adopters have higher social status, but they may also have a higher degree of "status inconsistency," the degree to which an individual's various status dimensions, such as income, education, and occupation, are interrelated. Thus, an individual with high status inconsistency might be relatively high in education and in occupational prestige but low in income. Wells and MacLean (1962) found that more innovative Michigan farmers had higher status consistency, contrary to our expectations. This study was tentative in nature but suggests a lead for further research on status inconsistency and innovativeness.

specialized, and have larger sized units. Wealth and innovativeness appear to go hand-in-hand.* Do innovators innovate because they are rich, or are they rich because they innovate? The answer to this cause-and-effect question cannot be answered on the basis of available correlational data. However, there is adequate reason why wealth and innovativeness vary together. Greatest profits go to the first to adopt; therefore, the innovator gains a financial advantage through his innovations. Some new ideas are costly to adopt and require large initial outlays of capital. Only the wealthy units in a social system may be able to adopt these innovations. The innovators become richer and the laggards become poorer through this process. Because the innovator is the first to adopt, he must take risks that can be avoided by later adopters. Certain of the innovator's new ideas are likely to fail. He must be wealthy enough to absorb the loss from these occasional failures. It should be pointed out that although wealth and innovativeness are highly related, economic factors do not offer a complete explanation of innovative behavior (or even approach doing so). For example, although agricultural innovators tend to be wealthy, there are many rich farmers who are not innovators.

Personality Variables

Personality variables associated with innovativeness have not yet received their share of research attention, perhaps because of difficulties of measuring these dimensions in field interviews.**

Generalization 5-10: *Earlier adopters have greater empathy than later adopters.* Empathy is the ability of an individual to project himself into the role of another person. This ability is an important quality for the innovator, who must be able to think counterfactually, be imaginative, and take the roles of heterophilous others in order to communicate effectively with them.

Generalization 5-11: *Earlier adopters are less dogmatic than later adopters.* Dogmatism is a variable representing a relatively closed belief system, a set of beliefs that are strongly held. The highly dogmatic person would not welcome new ideas; he prefers to hew to the past in a closed manner.

*Cancian (1967) argues that the relationship of wealth and innovativeness may be curvilinear because of the intervening variable of perceived risk.
**Harp (1960) feels that the inclusion of personality variables in analyses of innovativeness will contribute little. He states that if other sociological variables are included in investigations of innovativeness, the effect of ". . . personality may disappear." This is, of course, an empirical question, yet to be fully answered.

Generalization 5-12: *Earlier adopters have a greater ability to deal with abstractions than later adopters.* Innovators must be able to adopt a new idea largely on the basis of abstract stimuli, such as are received from the mass media. But later adopters can observe the innovation in the here-and-now of a peer's operation. Therefore, they need less ability to deal with abstractions.

Generalization 5-13: *Earlier adopters have greater rationality than later adopters.* Rationality is use of the most effective means to reach a given end.

Generalization 5-14: *Earlier adopters have greater intelligence than later adopters.*

Generalization 5-15: *Earlier adopters have a more favorable attitude toward change than later adopters.*

Generalization 5-16: *Earlier adopters have a more favorable attitude toward risk than later adopters.*

Generalization 5-17: *Earlier adopters have a more favorable attitude toward education than later adopters.*

Generalization 5-18: *Earlier adopters have a more favorable attitude toward science than later adopters.* Because most innovations are the products of scientific research, it is logical than innovators should be more favorably inclined toward science.

Generalization 5-19: *Earlier adopters are less fatalistic than later adopters.* Fatalism is the degree to which an individual perceives a lack of ability to control his future. How can a change agent convince a client to adopt innovations that will control the size of his family and give him better health and a higher level of living when the client believes that his future is determined by fate?

Generalization 5-20: *Earlier adopters have higher levels of achievement motivation than later adopters.* Achievement motivation is a social value which emphasizes a desire for excellence in order for an individual to attain a sense of personal accomplishment.

Generalization 5-21: *Earlier adopters have higher aspirations (for education, occupations, and so on) than later adopters.*

Communication Behavior

Generalization 5-22: *Earlier adopters have more social participation than later adopters.*

Generalization 5-23: *Earlier adopters are more highly integrated with the social system than later adopters.* Communication integration is the degree to which the units in a social system are interconnected by interpersonal communication channels.

Generalization 5-24: *Earlier adopters are more cosmopolite than later adopters.* The innovators' reference groups are more likely to be outside rather than within their social system. They travel widely and are involved in matters beyond the boundary of their local system. For instance, as shown in Chapter 2, Iowa hybrid corn innovators traveled more often to urban centers like Des Moines than did the average farmer (Ryan and Gross, 1943). Medical doctors who innovated a new drug attended more out-of-town professional meetings than noninnovators (Coleman and others, 1966). Industrial and educational organizations that are markedly innovative are more likely to hire consultants, a cosmopolite influence. Thai peasant innovators were found to visit Bangkok more frequently (Goldsen and Rallis, 1957, pp. 25–28). Chaparro (1955) concluded that Costa Rican innovators were amazingly cosmopolite. Over 84 percent had visited the U.S., 62 percent had traveled to Europe, and 67 percent had visited Mexico.

Generalization 5-25: *Earlier adopters have more change agent contact than later adopters.*

Generalization 5-26: *Earlier adopters have greater exposure to mass media communication channels than later adopters.*

Generalization 5-27: *Earlier adopters have greater exposure to interpersonal communication channels than later adopters.*

Generalization 5-28: *Earlier adopters seek information about innovations more than later adopters.*

Generalization 5-29 *Earlier adopters have greater knowledge of innovations than later adopters.*

Generalization 5-30: *Earlier adopters have a higher degree of opinion leadership than later adopters.* Although we find that innovativeness and opinion leadership are positively related, we know that the degree to which these two variables are related depends in part on the norms of the system. In a modern system opinion leaders are more likely to be innovators than in traditional systems (Chapter 6).

Generalization 5-31: *Earlier adopters are more likely to belong to systems with modern rather than traditional norms, than are later adopters.*

Generalization 5-32: *Earlier adopters are more likely to belong to well integrated systems than are later adopters.* The internal "trickle-down" of new ideas in a well integrated system is faster, enabling the members of such systems to learn about new ideas more rapidly.

In summary, we see that most of these independent variables in the thirty-two generalizations are positively related to innovativeness (Figure 5-3). This means that innovators will score higher on these variables than laggards. For instance, Rogers with Svenning (1969, p. 300) found that in traditional Colombian villages the innovators averaged thirty trips a year to cities whereas the laggards averaged only 0.3 trips.

A few variables, such as dogmatism and fatalism, are negatively related (Figure 5-3), and opinion leadership is greatest for early adopters, at least in most systems.

Thus, a set of characteristics of each adopter category emerges from past diffusion research. The important differences among these categories suggest that change agents might utilize somewhat different strategies of change with

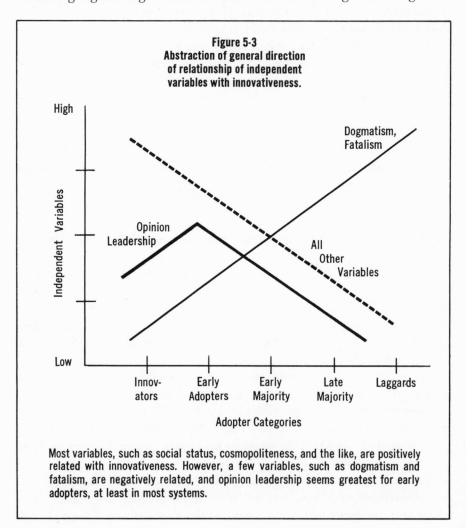

Figure 5-3
Abstraction of general direction of relationship of independent variables with innovativeness.

Most variables, such as social status, cosmopoliteness, and the like, are positively related with innovativeness. However, a few variables, such as dogmatism and fatalism, are negatively related, and opinion leadership seems greatest for early adopters, at least in most systems.

each. Thus, one might appeal to innovators to adopt an innovation because it was soundly tested and developed by credible scientists, but this approach would not be effective with laggards.

Predicting Innovativeness

One of the goals of social science is to provide an empirical basis for predicting human behavior. The empirical prediction of behavior is not meaningful unless it is theoretically based and logically consistent. Although such prediction is of clear import to the scientist, it is perhaps even more useful and relevant for those such as change agents, development planners, and administrators, whose immediate concern is action.

Approaches to Prediction

The two general approaches to predicting human behavior are *clinical* and *statistical*. The clinical method of prediction is used primarily in medicine, psychiatry, and clinical psychology where each case is viewed as somewhat unique, requiring an assessment of the total complexity of antecedent symptoms before a prediction can be intuitively offered. In their attempt to predict the success or failure of change agents in introducing innovations in less developed countries, Niehoff and Anderson (1964a) and Niehoff (1966a) represent one of the few uses of clinical prediction methods in diffusion research.

Statistical methods of prediction are more commonly used to predict innovativeness. Probability (or stochastic) models are used to forecast the likelihood of individuals behaving in a particular manner. Multiple correlation techniques and the configurational method* are perhaps most useful for predicting innovative behavior.

*The configurational method of prediction consists of dividing a sample of respondents into relatively homogeneous subsamples on the basis of the independent variables. Each subsample is regarded as a separate unit for analysis, since it has a unique configuration of independent variables. After these successive breakdowns on the basis of the independent variables, the probability of a desired outcome (e.g., adoption) is calculated. For illustrations of the configurational approach to predicting innovativeness, see Findley (1968), Rogers and Havens (1962a), Bonilla (1964), Rogers and Ramos (1965), Ross and Bang (1965), Keith (1968a), Keith (1968b), Herzog and others (1968).

Multiple Correlation Approach

Multiple correlation is a statistical procedure designed to analyze and explain the variance in a dependent variable in components due to the effects of various independent variables. The goal of the multiple correlation approach is to predict a maximum of the variance in the dependent variable, which in the present case is innovativeness. Table 5-2 shows a summary of multiple correlation analyses of innovativeness. We notice a general trend for the percentage of explained variance in innovativeness to increase with the passing of years until in the mid-1960s up to about 80 percent of the variance in innovativeness was explained. This may be partly attributable to the advent of computer data analysis which allows the inclusion of a greater number of independent variables in these analyses. Further, a greater variety of independent variables were included in these studies; economic and social psychological dimensions are evidenced in the recent studies, along with variables indicating social structural aspects.

One advantage of the multiple correlation approach is that it discloses the degree to which each independent variable is related to innovativeness, while controlling the effects of all other independent variables. This yields an indicant of the novel contribution of each independent variable in explaining innovativeness. For instance, we have just reviewed evidence in this chapter that social status and cosmopoliteness are both related to innovativeness. But we also generally find that status and cosmopoliteness are correlated. The relationship of cosmopoliteness to innovativeness may be due to the relationship of both with social status. Multiple correlation methods help us to untangle the complex webs of interrelationships among our independent variables as they relate to innovativeness.

Despite the fact that some multiple correlation analyses have explained up to 80 percent of the variance in innovativeness, most attempts are much less successful. Further research is needed to raise the level of predictive accuracy. Further, most of the studies reviewed in Table 5-2 are actually "postdiction" rather than prediction in that they did not validate their predictions on a future sample of similar respondents.

Simulation of Innovation Diffusion

Computer Simulation

The investigations of diffusion researchers have traditionally been bound by their research tools to examinations of slices or cross-sections of the

Table 5-2 Summary of Multiple Correlation Analyses of Innovativeness

	INVESTIGATOR	RESPONDENTS	PERCENTAGE OF VARIANCE IN INNOVATIVENESS EXPLAINED (%)	NUMBER OF INDEPEN-DENT VARIABLES UTILIZED
1	Copp (1956)	Kansas farmers	50.0	5
2	Fleigel (1956)	Wisconsin farmers	32.0	6
3	Copp (1958)	Wisconsin farmers	52.0	4
4	Rogers (1957a)	Iowa farmers	17.0	5
5	Armstrong (1959)	Kentucky farmers	42.1	3
6	Ramsey and others (1959)	New York dairy farmers	9.6	4
7	Hobbs (1960)	Iowa farmers	29.7	7
8	Sizer and Porter (1960)	West Virginia farmers	25.9	4
9	Straus (1960)	Wisconsin farmers	33.6	3
10	Kimball (1960)	Michigan farm families	25.0	6
11	McMillion (1960)	Large farmers in New Zealand	39.9	5
12	Rogers and Havens (1961b)	Ohio farmers	56.4	5
13	Flinn (1961)	Truck growers in 7 Ohio communities	56.6	4
14	Cohen (1962)	New Jersey families	54.8	3
15	Rogers and Havens (1962a)	Ohio farmers	64.1	5
16	Deutschmann and Fals Borda (1962b)	Colombian farmers	56.3 (and 68.9 when using 27 variables)	8
17	Junghare (1962)	Farmers in India	23.8	7
18	Madigan (1962a)	Heads of households and other males in the Philippines	17.1	3
19	Neill (1963)	Ohio farmers	40.5	6
20	Havens (1963a)	Colombian farmers	47.3	3
21	Flinn (1963)	Truck growers in Ohio	64.1	5
22	Jain (1965)	Farmers in Canada	50.3	7
23	Haring (1965)	Wisconsin farmers	50.2	34
24	Andrus (1965)	U.S. consumers	41.0	21
25	Rogers (1966a)	Colombian farmers in five communities	From 24.1 to 39.0	6
26	Morgan and others (1966)	U.S. household heads	16.0	5
27	Beal and Sibley (1966)	Guatemalan Indian farmers	78.0 (42.0 when using 6 variables)	51
28	Moulik and others (1966)	Farmers in India	81.0	4
29	Whittenbarger and Maffei (1966)	Colombian farmers	44.4	5
30	Ramos (1966a)	Colombian farmers	12.9	9
31	Singh (1966b)	Indian farmers	63.5	6
32	Wish (1967)	Retail food stores in Puerto Rico	87.5	35
33	Chattopadhyay and Pareek (1967)	Indian peasants	59.0	3
34	Herzog and others (1968a)	Brazilian peasants	43.0	13
35	Roy and others (1968)	Indian farmers	50.0	15
36	Ascroft and others (1969)	Nigerian peasants	42.0	13

process at one point in time. Methodological limits have necessitated slow-motion analyses which hold a slice of the process stationary while observing the dynamics of diffusion. Now with the flexible time considerations provided by the computer, it is possible to fuse the stationary analysis with the continuing process and capture the important variables in action. This can be done with the technique of computer simulation.

The result of computer simulation is the reproduction of the social process that one seeks to mimic.* If the simulated process does not correspond to reality data, then one knows that he must adjust his model or set of rules governing the simulated process.

Torsten Hägerstrand, a quantitative geographer at the Royal University of Lund, Sweden, is the father of diffusion simulation research. His work on computer simulation began in the early 1950s but was published only in Swedish. For many years the "paper curtain" of language barriers prevented the diffusion of his work to U.S. researchers. From the mid-1960s, however, his work has been carried forward in a series of investigations by quantitative geographers. Examples of simulations are the diffusion of deep well drilling in Colorado (Bowden, 1965a) and agricultural innovations in Colombia (Hanneman, 1969). These studies and others like them suggest that computer simulation of diffusion holds promise as a means to explore the complexities of the process as it unfolds over time. This potential, however, has not been very fully realized.

Simulation Training Games

Not only does diffusion simulation have an important research potential, but it may also be useful as a training device. A simulation game called "Change Agent" has been developed in the Department of Communication at Michigan State University. This game, which requires about one hour to play, asks the participant to take the role of a change agent in a peasant village of 100 families. His objective is to choose optimally among various strategies of change so as to obtain as close as possible to 100 percent adoption of an innovation by the villagers. For example, the change agent can choose to locate the important opinion leaders in the village (at a "cost" of twenty days of effort each) or he can broadcast a radio program about the innovation (which requires ten days to prepare). The game's scoring system converts each of the

*Probably one of the best known computer simulations was the prediction of voting behavior in the 1960 national presidential election (Pool and Abelson, 1961).

player's decisions into the percentage of clients who adopt the innovation as a result. The player can even choose to obtain feedback as to the percentage adoption of the innovation that he has obtained at any given point during the playing of "Change Agent" (at a cost of thirty days for hypothetically surveying his peasants).

The advantage of training games like "Change Agent" is that they teach future change agents the realistic nature of their job before they are actually in it. Games seem especially important in creating interest and involvement in learning more about the diffusion of innovations.

Summary

Adopter categories are classifications of individuals within a social system on the basis of innovativeness, the degree to which an individual is relatively earlier in adopting new ideas than other members of his system. A variety of categorization systems and titles for adopters have been used in past research studies. This chapter suggests a possible conceptual standardization of these categories.

The research of sociologists, learning psychologists, and students of the diffusion effect provide theoretical reasons for expecting adopter categories to be normal. The *diffusion effect* is the cumulatively increasing degree of influence upon an individual within a social system to adopt or reject an innovation. This influence results from the increasing rate of knowledge and adoption or rejection of the innovation in the system. Adopter distributions tend to follow an S-shaped curve over time and approach normality.

The continuum of innovativeness can be partitioned into five adopter categories (innovators, early adopters, early majority, late majority, and laggards) on the basis of two characteristics of a normal distribution, the mean and the standard deviation. These five categories are an arbitrary classification system.

The suggested adopter categories are ideal types, conceptualizations based on observations and designed to institute comparisons. Dominant values of each category are: innovators—venturesome; early adopters—respectable; early majority—deliberate; late majority—skeptical; and laggards—traditional. The relatively earlier adopters in a social system tend to have more education, a higher social status, more upward social mobility, larger units, a commercial rather than a subsistence orientation, a favorable attitude toward credit, and more specialized operations. Earlier adopters also have greater empathy, less dogmatism, greater ability to deal with abstractions, greater rationality, and more favorable attitudes toward change, risk, education, and science. They are less fatalistic and have higher achievement motivation scores and higher aspirations for their children. Earlier adopters have more social participation, are more highly integrated with the system, are more cosmopolite, have more change agent contact, have more exposure to both mass media and interpersonal channels, seek information more, have higher knowledge of innovations, and have more opinion leadership. They usually belong to systems with modern norms and to well integrated systems.

Social scientists have used two approaches to predicting innovative human behavior: Clinical and statistical analysis. Multiple correlation attempts to explain the maximum variation in the dependent variable resulting from several independent variables.

A relatively new research tool is the computer, which has been used to simulate diffusion processes. But diffusion simulation is more than a predictive device; it has great value as a training tool for change agents.

6 *Opinion*

Leadership and the

▄ ▄▄   ▄▄▄ ▪ ▄▄   ▪▪ ▄ ▪ ▪▪ ▄▄▄ ▄▪   ▪▄▪▪ ▪ ▪▄   ▄▪▪ ▪ ▪▄▪ ▪▪▪ ▪▪▪▪ ▪▪ ▪▄

Every herd of wild cattle has its leaders, its influential heads.

GABRIEL TARDE (1903, p.4)

What we shall call opinion leadership, if we may call it leader-ship at all, is leadership at its simplest: It is casually exercised, sometimes unwittingly and unbeknown. . . . It is the almost invisible, certainly inconspicuous, form of leadership at the person-to-person level of ordinary, intimate, informal, every-day contact. ELIHU KATZ AND PAUL L. LAZARSFELD (1955, p. 138)

Multi-Step Flow of Ideas

COMMON observation and empirical research show that certain individuals have the ability to influence other peoples' behavior in a desired way.* This is leadership. Such individuals are often sought for information and advice on specific topics about which others feel they are expert. These individuals may be in formal leadership positions, but their influence is exerted informally through interpersonal communication networks. The term "opinion leader" is often applied to these individuals who lead in influencing others' opinions in informal ways; it contrasts with formal leadership, which is exercised by virtue of the formal office an individual holds. We define *opinion leadership* as the degree to which an individual is able to influence informally other individuals' attitudes or overt behavior in a desired way with relative frequency.

The concept of opinion leadership was first developed by Lazarsfeld and others (1944, p. 152) in a study of political behavior in the 1940 Presidential election. Various researchers have used different terms for a more or less similar concept, as illustrated by the following list.

Fashion leaders
Gatekeepers
Influencers
Information leaders

*Certain ideas in this chapter are drawn from Rogers (forthcoming).

> Key communicators
> Sparkplugs
> Style setters
> Tastemakers

These terms all refer to the same basic dimension: Opinion leadership. There is an obvious need to standardize both the terminology and the criteria for selecting opinion leaders. In the present chapter we use the term opinion leader to refer to those individuals who play a role in diffusing innovations, whether the new ideas are in fashion, politics, sports, or other fields.

We should not forget that opinion leaders can be either hot or cold toward innovations; they can either speed the diffusion of a new idea or fight it. The respondents in most of the research studies reviewed in this book are probably pro-change opinion leaders, and as Klapper (1960, p. 35) points out, practically no research attention has been paid to the role of anti-change leaders in discouraging diffusion. Near the end of this chapter we encounter a case of just such leaders in traditional Colombian villages, and we see how they lost their followings as the village norms became more modern.

In order to understand the nature of opinion leadership, we shall discuss (1) the various models of mass communication flow, such as the two-step flow and its latter-day revisions, (2) how homophily-heterophily affects the flow of communication, (3) measures of opinion leadership, (4) characteristics of opinion leaders, and (5) monomorphic and polymorphic opinion leadership. First, however, we look at an illustration of the role of opinion leaders in diffusing modern math among schools.

Opinion Leadership and the Diffusion of Modern Math*

Insight into the nature of opinion leadership is provided by a study of the spread of an educational innovation, modern math, among the thirty-eight school superintendents in Allegheny County, Pennsylvania.

The innovation began in the early 1950s when top mathematicians in the U.S. assembled to overhaul the nature of math training being offered in public

*Adapted from Carlson (1965, pp. 17–21) and used by permission.

schools. Out of their efforts came "modern math," a new approach which was packaged to include new textbooks, audio-visual aids designed for teaching the new concepts, and summer training institutes to retrain school teachers in the new subject matter. The innovation spread relatively quickly (Figure 2-2) because of powerful sponsorship at the national level by the National Science Foundation and the U.S. Office of Education.

Modern math entered the local scene of Allegheny County by means of one school superintendent, shown in Figure 6-1 as "I", who adopted in 1958. This innovator was a sociometric isolate in that he had no interpersonal communication with any of the other school superintendents in the County. We know from Chapter 5 that innovators are frequently disdained by their fellow members in a local system and that they interact primarily with cosmopolite friends.

Figure 6-1 is a sociogram, a communication map on which we trace diffusion of the innovation. The arrows show the patterns of friendship among the superintendents. The dotted line circles six friends who constitute a clique or informal friendship group. All the superintendents in this clique received at least two friendship choices from within the circle while no one outside received two such choices. In fact, fifteen of the seventeen friendship choices made by these six men were directed within the circle.

This clique played a central role in the diffusion of modern math. Once the clique members (and especially the three main opinion leaders who decided to use the innovation in 1959 and 1960) adopted, the rate of adoption began to climb rapidly in the system. As Figure 6-1 shows, there was only one adopter in 1958 (the innovator), five by the end of 1959, fifteen by 1960, twenty-seven by 1961, thirty-five by 1962, and thirty-eight by the end of 1963. The rapid spurt in 1959, 1960, and 1961 appears as a direct result of the opinion leaders' communication behavior.

Later in this chapter we review evidence that opinion leaders are highly conforming to the norms of a system, and we see support for this generalization in the present case. The cosmopolite innovator was *too* innovative to serve as an appropriate role-model for the other superintendents; they waited to adopt until the opinion leaders in the six-member clique favored the innovation. If one were a change agent with a responsibility to diffuse modern math in Allegheny County, he would be wise to concentrate his efforts on these opinion leaders.

Further, we see evidence in Figure 6-1 of a rather high degree of homophily in time of adoption. Many of the friendship arrows are between superintendents who adopted in the same year or within one year of each other. And

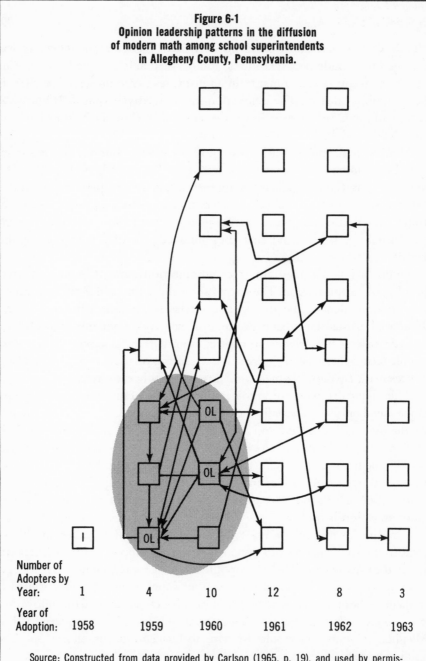

Figure 6-1
Opinion leadership patterns in the diffusion
of modern math among school superintendents
in Allegheny County, Pennsylvania.

Number of
Adopters by
Year: 1   4   10   12   8   3

Year of
Adoption: 1958   1959   1960   1961   1962   1963

Source: Constructed from data provided by Carlson (1965, p. 19), and used by permission. Notice that for the sake of simplicity, only thirty-two of the thirty-eight superintendents who adopted modern math are shown.

when two superintendents in a dyadic relationship adopt in different years, the difference tends to be slight, suggesting the notion of "tolerable hetero-phily," where the source is different enough from a receiver to be perceived as competent but not so much different as to be an inappropriate role-model.

We must point out that the sociometric data shown in Figure 6-1 are but one case study of diffusion and may not be entirely typical. For instance, not *all* innovators are sociometric isolates from the rest of their system, especially if the system's norms are very modern. And the opinion leaders often do not constitute a clique, as they did among these school superintendents.

Models of Mass Communication Flows

In order to understand better the nature of opinion leadership in diffusion, we shall now examine several models of mass communication flows, roughly in the temporal sequence of their entrance on the communication research scene.

Hypodermic Needle Model

The *hypodermic needle model* postulated that the mass media had direct, immediate, and powerful effects on a mass audience.* The effects attributed to the mass media bore a close parallel to the stimulus-response ideas current in psychological research in the 1930s and 1940s. Based on the S-R principle, the hypodermic needle model came prior to recognition by communication researchers of the many intervening variables operating between the initial communication stimulus (source) and the ultimate response by an audience (receivers). The model pictured the mass media as a giant hypodermic needle, pecking and plunging at a passive audience (Figure 6-2).**

The hypodermic needle model also drew support from the development of a mass society*** in the U.S. during this era. Observers noted a trend to

---

*The hypodermic needle model was essentially a one-step flow model in which the mass media carried messages to an audience with direct effects.

**"The model in the minds of early researchers seems to have consisted of: (1) the all-powerful media, able to impress ideas on defenseless minds; and (2) the atomized mass audience, connected to the mass media but not to each other" (Katz, 1963b, p. 80).

***A *mass society* consists of a mass audience of standardized and atomized individuals bound only loosely by interpersonal relationships.

homogeneity in dress, speech patterns, and values that seemed to result from mass media exposure and mass production, and leading toward a mass culture. The mass media were conceived as an all-powerful influence on human behavior. The omnipotent media were pictured as sending forth messages to atomized masses waiting to receive them, with nothing intervening (Katz and Lazarsfeld, 1955, p. 16). Evidence of the great manipulative power of the mass media came from such historical events as: (1) the role of U.S. newspapers in arousing positive public opinion toward the Spanish-American War with such shibboleths as "Remember the Maine," (2) the apparent power of Goebbel's propaganda machine in World War II, and (3) fear of Madison Avenue's influence on consumer and voting behavior.

Eventually, the more sophisticated research methods that came to be utilized in communication inquiry cast considerable doubt on the hypodermic needle model. It was based primarily on intuitive theorizing about historical events and was too simple, too mechanistic, too gross to account accurately for mass media effects.

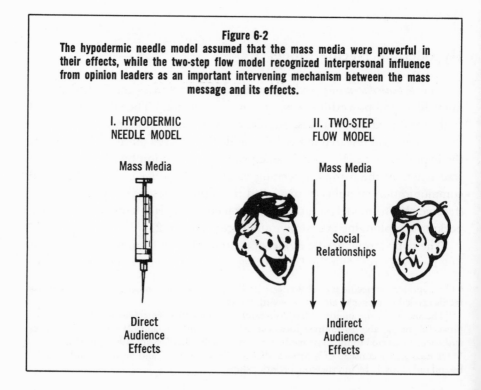

Figure 6-2
The hypodermic needle model assumed that the mass media were powerful in their effects, while the two-step flow model recognized interpersonal influence from opinion leaders as an important intervening mechanism between the mass message and its effects.

I. HYPODERMIC
NEEDLE MODEL

II. TWO-STEP
FLOW MODEL

Mass Media

Mass Media

Social
Relationships

Direct
Audience
Effects

Indirect
Audience
Effects

Two-Step Flow Model

The decisive discarding of the hypodermic needle model resulted serendipitously from a classic study of the 1940 Presidential election (Lazarsfeld and
others, 1944).* This inquiry was designed with the hypodermic needle model
in mind and was aimed at analyzing the role of mass media in clinching political
decisions. To the researchers' surprise, the evidence indicated that almost no
voting choices were directly influenced by the mass media.** Rather, the data
seemed to indicate " . . . that ideas often *flow* from radio and print *to* opinion
leaders and *from* these to the less active sections of the population" (Lazarsfeld
and others, 1944, p. 151). The first step, from sources to opinion leaders, is
mainly a transfer of *information*, whereas the second step, from opinion
leaders to their followers, involves also the spread of *influence*. Because the
conceptualization was not part of the original design, this so-called *two-step
flow hypothesis* was not really well documented by the data. Later investigations
tested the two-step flow model in a great variety of communication situations
and found generally that this model provided a usable conceptual framework
for examining mass communication phenomena.***

The two-step flow model helped focus attention upon the role of mass
media-interpersonal interfaces. Instead of assuming, as did the hypodermic
needle model, that the masses were a large body of disconnected individuals
hooked up to the media but not to each other, the two-step flow model viewed
the masses as interacting individuals. It put people back into mass communication. It implied that the mass media were not so powerful nor so direct as once

*A parallel rejection of the hypodermic needle model by researchers investigating the
diffusion of agricultural innovations occurred at about the same time. They rejected the
notion of all-powerful mass media, because "the rural sociologists never assumed, as
students of mass communication had, that their respondents did not talk to each other"
(Katz, 1960).

**As Lazarsfeld and Menzel (1963, p. 96) admit: "This study went to great lengths to deter-
mine how the mass media brought about such changes. To our surprise we found the effect
to be rather small. . . . People appeared to be much more influenced in their political
decisions by face-to-face contact with other people . . . than by the mass media
directly."

***This model served as a basis for a series of investigations conducted by communication
researchers of Columbia University's Bureau of Applied Social Research: The Rovere
study of opinion leadership in a suburban community (Merton, 1949, pp. 180–219), the
Decatur study of consumer decisions (Katz and Lazarsfeld, 1955), the Elmira study of
election behavior (Berelson and others, 1954), and the medical drug diffusion study
(Coleman and others, 1966). These studies utilized the two-step flow model in their
design and amplified our understanding of mass communication flows and of opinion
leadership.

thought. One may be exposed to a new idea either through mass media or interpersonal channels; he then engages in communication exchanges about the message with his peers. In most mass communication flows (whatever one takes as his point of reference in the process), there is likely to be a flow of messages from a source by mass media channels to a receiver, who in turn reacts to the message and/or passes it on to those individuals with whom he interacts (Rogers, 1962b, p. 313).

Communication research of the last twenty-five years has profited greatly from use of the two-step flow model; with modification it is probably the most popular framework, explicitly or implicitly, utilized in diffusion research. But this research has also demonstrated several shortcomings inherent in the two-step flow model. Basically, the model does not tell us enough. Research on the diffusion of innovations brought many of these shortcomings to the fore because these investigations necessarily include *time* as a variable, which the original 1940 study did not.* Six limitations of the two-step flow model are:

1. The two-step model implied that individuals active in information seeking were opinion leaders and that the remainder of the mass audience were passive.** The activity of the opinion leaders was thought to provide the main thrust to initiate the communication flow. A more accurate reflection of reality would probably be a model which indicates that opinion leaders can be either active or passive, that they seek receivers, are actively sought by them, and that opinion leaders often play *both* active and passive roles in most communication situations (Rogers with Svenning, 1969, p. 222).***

2. The view that the mass communication process consists essentially of two-steps limits analysis of the process. The mass communication process may involve more or fewer than two steps. In some instances there may be only one step; that is, the mass media may have direct impact on a receiver. In other instances the impetus of the mass media may lead to a multi-stage communi-

---

*Diffusion inquiries are helpful in further analyzing the flow of communication messages because they deal primarily with innovations. It is relatively easier to isolate the effects of messages whose main theme is a new idea because respondent recall is facilitated. New ideas seem to leave deeper impressions on men's minds than do more routine messages, as we pointed out in Chapter 2.
**The two-step flow theory was proposed at a period when the conception of a passive audience was widely accepted in communication research (Bauer, 1963).
***This idea recognizes that either opinion leaders or their receivers are capable of active and/or passive roles in the communication flow, an idea which comes from Sicnski (1963) and Wright and Cantor (1967).

cation process. By focusing only on the two-step aspects of the process, research that mirrors reality is severely limited.*

3. The two-step flow model implies a reliance by opinion leaders on mass media channels and most past research on the two-step flow assumed the primacy of mass media channels for opinion leaders. But recently there have been some indications that opinion leaders obtain messages from other channels than mass media. For village leaders in less developed countries, where little or no mass media are available, other channels like personal trips to cities, conversations with change agents, and the like can be the initiating force (Rogers with Svenning, 1969). The specific channels utilized by opinion leaders depend on such considerations as the nature of the message, its origin, and the location of the opinion leaders in the social structure. Opinion leaders' channels are those *most relevant* for initiating the mass communication flow, and they are not mass media in all cases.

4. The original 1940 inquiry did not take into consideration the different channel behavior by receivers on the basis of their time of knowing about a new idea. Diffusion research results show that relatively earlier knowers and adopters of innovations utilize mass media channels much more than later knowers and adopters.** The opinion leaders may simply be the earlier knowers of new ideas, and their dependence upon mass media channels may be a function of their early knowing, rather than their opinion leadership *per se*. Earlier knowers must necessarily depend upon mass media channels because at the time of their awareness of new ideas, few of their peers in the system are yet knowledgeable about the innovation. Interpersonal channels could, therefore, hardly function as very important creators of knowledge for the earlier knowers.

5. Different communication channels function at different stages in the receiver's innovation-decision process. The original two-step flow model did not recognize the role of different communication channels at the varying stages of innovation-decision (van den Ban, 1964b). We know from Chapter 3 that individuals pass from (1) *awareness-knowledge* of an innovation, (2) to *persuasion* of a favorable or unfavorable attitude toward the innovation, (3) to *decision* to adopt or reject, and (4) to *confirmation* of this decision. Mass media channels are primarily knowledge creators, whereas interpersonal channels

---

*As Menzel and Katz (1955) concluded from their study of the diffusion of a new medical drug: "We have found it necessary to propose amendments to the two-step flow of communications: by considering the possibility of multi-step rather than two-step flow."
**As we showed in Chapter 3.

are more important at persuading, that is, forming and changing attitudes.*
This notion was masked in the original statement of the two-step model
because the time sequence involved in decision-making was explicitly ignored.
Such channel differences at the knowledge versus persuasion stage exist for
both opinion leaders and followers. Thus, it is not only the opinion leaders who
use mass media channels, as the original statement of the two-step flow model
seemed to suggest.

6. An audience dichotomy of opinion leaders versus followers was implied
by the two-step flow model (Katz, 1957). First, we know that in fact opinion
leadership is a continuous variable and should be conceptualized (even though
it can not always be measured) as such. Second, many "non-leaders" are not
followers of the leaders, at least in any direct sense.**

The overall criticism of the two-step flow model, as originally postulated,
is mainly that *it does not tell us enough*. The flow of communication in a mass
audience is far more complicated than two steps. What is known about the mass
communication process is too detailed to be expressed in one sentence or in
two steps. Nevertheless, two intellectual benefits from the two-step flow
hypothesis are evident in communication research: (1) a focus upon opinion
leadership in mass communication flows, and (2) several revisions of the
two-step flow, such as the one-step and multi-step flow.

One-Step Flow Model

The *one-step flow model* states that mass media channels communicate
directly to the mass audience, without the message passing through opinion
leaders; the message does not equally reach all receivers, nor does it have the
same effect on each (Troldahl, 1967). The one-step flow model probably
results from a refinement of the hypodermic needle model discussed earlier.
But the one-step model recognizes: (1) the media are not all-powerful;

*Troldahl (1967) proposes on the basis of dissonance theory that followers who are exposed
to mass media messages that are inconsistent with their predispositions will initiate interpersonal communication with opinion leaders to reduce their dissonance. Troldahl found
tentative support for his hypothesis.
**In order to identify the followers from the non-followers among the category of non-
leaders, researchers should use leader-follower sociometric *dyads* as units of analysis, rather
than *individuals*, as has generally been the case in past inquiry on opinion leadership (and
in research on diffusion, as shown in Chapter 2). Often in this present chapter we speak of
"followers," really meaning "non-leaders." This is only for shorthand convenience,
because in a strict sense, not all non-leaders are followers of opinion leaders, as Troldahl and
van Dam (1965) have demonstrated.

(2) the screening aspects of selective exposure, perception, and retention affect message impact; and (3) differing effects occur for various members of the receiving audience. Further, it allows for direct effects of communication emanating from mass media channels. The one-step flow model most accurately describes the flow of messages to a mass audience when the saliency of the message is extremely high or perhaps very low.* For medium-salient news events, awareness-knowledge of the ideas is created almost entirely by the mass media, indicating a one-step flow. This notion is supported by Deutschmann and Danielson's (1960) conclusion from a study of the diffusion of six news events: "Initial mass media information on important events goes directly to people on the whole and is not relayed to any great extent."

Multi-Step Flow Model

The multi-step flow model incorporates all of the other models previously discussed. The multi-step flow model is based on a sequential relaying function that seems to occur in most communication situations. It does not call for any particular number of steps nor does it specify that the message must emanate from a source by mass media channels. This model suggests that there are a variable number of relays in the communication flow from a source to a large audience. Some members will obtain the message directly through channels from the source, while others may be several times removed from the message origin. The exact number of steps in this process depends on the intent of the source, the availability of mass media and the extent of audience exposure, the nature of the message, and salience of the message to the receiving audience.

Today most communication researchers place credence in the multi-step conceptualization of the mass communication process and at least intuitively subscribe to it, as evidenced in their research designs. Perhaps the model permits a more accurate analysis of the mass communication process, because it allows the researcher to account for different variables in different communication situations. The multi-step model is the least specific or restrictive of the models we have considered.

*Greenberg (1964b) argues that the amount of interpersonal relaying that occurs in the case of news event diffusion is greatest when the event is of medium saliency. When a news event is either of very low or very high saliency (for example, in the case of a routine local government decision versus President Kennedy's assassination), mass media channels are of greatest importance.

Homophily-Heterophily and the Flow of Communication

The importance of the interpersonal relaying process in mass communication was mentioned previously in our discussion of the multi-step flow model. Understanding the nature of communication flows through interpersonal channels can be enhanced by an examination of the concepts of homophily and heterophily. The nature of *who* relays to *whom* is brought out in such research.

Homophily-Heterophily

One of the most obvious and fundamental principles of human communication is that the transfer of ideas most frequently occurs between a source and a receiver who are alike, similar, homophilous.* *Homophily* is the degree to which pairs of individuals who interact are similar in certain attributes, such as beliefs, values, education, social status, and the like (Figure 6-3).** Although a conceptual label—homophily—has been assigned to this phenomenon only in recent years by Lazarsfeld and Merton (1964, p. 23), the existence of homophilic behavior was noted a half century ago by Tarde (1903, p. 64): "Social relations, I repeat, are much closer between individuals who resemble each other in occupation and education."

Why does homophily occur? Because *better communication occurs when source and receiver are homophilous* and this effective communication is rewarding to those involved in it.*** When source and receiver share common meanings, attitudes and beliefs, and a mutual language, communication between them is likely to be effective. Most individuals enjoy the comfort of interacting with others who are quite similar. Interaction with those quite different from our-

---

*Homans (1950, p. 184) noted: "The more nearly equal in social rank a number of men are, the more frequently they will interact with one another."

**Heterophily* is the degree to which pairs of individuals who interact are different in certain attributes.

***Support for this proposition comes from numerous small group laboratory studies, which show that heterophilous communication leads to message distortion, one type of ineffective communication (Barnlund and Harland, 1963).

selves involves more effort to make communication effective. Heterophilic interaction is likely to cause cognitive dissonance, because the receiver is exposed to messages that may be inconsistent with his existing beliefs, an uncomfortable psychological state. Homophily and effective communication breed each other. The more communication there is between members of a dyad, the more likely they are to become homophilous;* the more

*Although similarities in age and other demographic characteristics could hardly be explained as the result of previous communication leading to increased homogeneity.

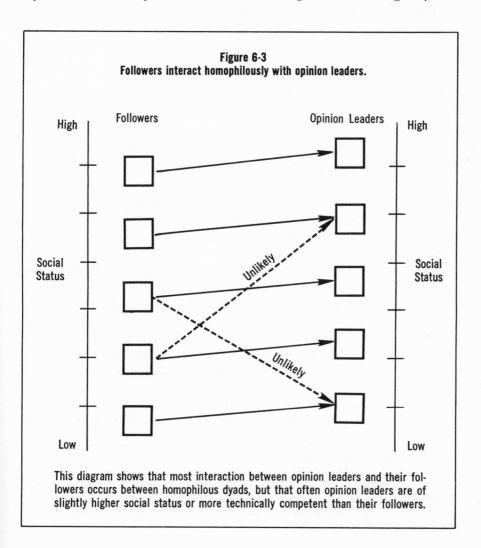

Figure 6-3
Followers interact homophilously with opinion leaders.

This diagram shows that most interaction between opinion leaders and their followers occurs between homophilous dyads, but that often opinion leaders are of slightly higher social status or more technically competent than their followers.

homophilous they are, the more likely it is that the communication will be effective.* Individuals who break the homophily boundary and attempt to communicate with others quite different from themselves are beset with the frustrations of ineffective communication. Differences in technical competence, social status, attitudes, and beliefs all contribute to heterophily in language and meaning, thereby leading to messages that go unheeded.

Homophily as a Barrier to Diffusion

Homophily can act as an invisible barrier to the rapid flow of innovations within a system. New ideas usually enter by means of higher status and more innovative members of the system. When a high degree of homophily is present, these elite individuals interact mainly with each other, and there is little "trickle down" of the innovations to nonelites. Homophilic diffusion patterns cause new ideas to spread horizontally, rather than vertically, within a system. Homophily, therefore, acts to slow down the rate of diffusion. One implication of homophily as a barrier to diffusion is that change agents should work with different opinion leaders throughout the social structure. If a system were characterized by extreme heterophily, a change agent would need to concentrate his efforts on only one or a few opinion leaders near the top in social status or in innovativeness (Figure 6-3).

We need to know (1) how much homophily is involved in interpersonal diffusion, and (2) on what variables or characteristics of source-receiver pairs there is more homophily or heterophily. Researches conducted both in more and in less developed countries suggest seven generalizations, which follow:

Generalization 6-1 : *Interpersonal diffusion is mostly homophilous.* Although the exact degree of homophily** in interpersonal diffusion varies with the system's norms, the nature of the innovation, and the like, a general pattern of homophily has emerged from studies in Iowa, Missouri, the Netherlands, India, and

---

*Therefore, homophily may be the *result* of interaction or the basis of *choice* of those with whom one interacts, as Lazarsfeld and Merton (1964, p. 34) note. In the 1940 election study, Lazarsfeld and others (1944, p. 139) concluded: "The changes in vote intention increase group homogeneity. . . . The majority of voters who change at all change in the direction of the prevailing vote of their social group."

**The measures presently used by communication researchers to measure homophily for continuous variables are: (1) Pearsonian zero-order correlation at the dyadic level, and (2) indicants derived from computer matrix multiplication procedures at the system level (Ho, 1969). For dichotomous variables some version of Coleman's (1958) index may be utilized (Signorile and O'Shea, 1965).

Colombia. For instance, seldom do those of highest status in a system interact directly with those of lowest status. Likewise, innovators seldom converse with laggards. Although this homophily pattern in interpersonal diffusion acts to slow the diffusion of innovations within a system, it also has some benefits. For example, a high status opinion leader may be an inappropriate role-model for someone of lower status; any interaction between them may not be beneficial to the latter. An illustration of this point comes from an investigation by van den Ban (1963b) in a Netherlands agricultural community. He found that only 3 percent of the opinion leaders had farms under fifty acres in size, but 38 percent of all farms in the community were smaller than fifty acres. The wisest farm management decision for the large farmers was to purchase mechanized farm equipment, such as tractors and milking machines, as a substitute for hired labor. However, the best economic choice for those on farms under fifty acres was to ignore power equipment and begin intensive horticultural farming. However, as might be expected, the small farmers were following the example of the opinion leaders on the large farms, even though the example was inappropriate for their situation. In this case a high degree of homophily, in which small farmers would interact mainly with opinion leaders who were small farmers also, would probably be beneficial.

Now we look at generalizations which specify six characteristics on which homophily-heterophily has been studied:

Generalization 6-2: *When interpersonal diffusion is heterophilous, followers seek opinion leaders of higher social status.*

Generalization 6-3: *When interpersonal diffusion is heterophilous, followers seek opinion leaders with more education.*

Generalization 6-4: *When interpersonal diffusion is heterophilous, followers seek opinion leaders with greater mass media exposure.*

Generalization 6-5: *When interpersonal diffusion is heterophilous, followers seek opinion leaders who are more cosmopolite.*

Generalization 6-6: *When interpersonal diffusion is heterophilous, followers seek opinion leaders with greater change agent contact.*

Generalization 6-7: *When interpersonal diffusion is heterophilous, followers seek opinion leaders who are more innovative.*

Each of these six generalizations indicates a tendency for followers to seek information and advice from opinion leaders who are perceived as more competent than themselves. This perceived competency may be expressed as higher status, greater innovativeness, or more exposure to mass media communication channels. An illustration of this tendency for followers to look up the competency scale is provided by Rogers with Svenning (1969,

p. 237). When opinion leader-follower dyads were classified into adopter categories, it was found that:

Only 19 percent of the dyadic choices were to a less innovative adopter category.

30 percent were to an equally innovative category.

51 percent were to a more innovative category.

Thus, when heterophily occurs, it is usually in the direction of greater competency but not *too* much greater.* We should not forget that the general pattern is one of homophily in interpersonal diffusion.

Homophily-Heterophily in Traditional and Modern Systems

What differences exist between modern and traditional systems in the degree of homophily in interpersonal diffusion? Rogers with Svenning (1969, pp. 237–238) found that more traditional Colombian villages were characterized by a greater degree of homophily in interpersonal diffusion.** Only when the norms of a village became more modern did diffusion become more heterophilous. And this breakdown of homophilous diffusion patterns acted to make the system even more modern, by facilitating the trickle down of innovative messages within the village. Bose (1967) found a very high degree of homophily among the residents of an Indian village on the basis of caste ranking, education, and farm size. In the nearby city of Calcutta, however, caste was unimportant in structuring interaction patterns, but income was very important. The exact attributes on which homophily occurs seem to vary with the nature of the system as well as the nature of the innovation.

We conclude, although definitive evidence is still lacking, with Generalization 6-8: *Interpersonal diffusion is characterized by a greater degree of homophily in traditional than in modern systems.*

*Perhaps there is a "tolerable" degree of heterophily between a source-receiver pair in which the source is different enough from the receiver to be perceived as competent but not so much different as to lack comparison as a role model. This tolerable range of heterophily may provide energy to the communication exchange and hence, facilitate effective communication. There is no research to date designed to determine whether such tolerable heterophily exists, or the exact range in tolerable heterophily, but the notion is suggested by Lionberger's (1959) data. The idea of tolerable heterophily was implied by Tarde (1903, p. 224), writing over sixty years ago. He stated that diffusion is a "kind of social *water-tower*, whence a continuous water-fall of imitation may descend. . . . The influence of the model's example is efficacious inversely to its *distance* as well as directly to its superiority."

**This finding is supported by the work of Yadav (1967) in India but not by Ho (1969) in Brazil.

Further, the Rogers with Svenning (1969, p. 238) investigation in Colombia suggests that the *nature* of heterophilous diffusion, when it occurs, is different in modern from that in traditional systems. In modern villages farmers sought opinion leaders who were more competent (in innovativeness, social status, and so on) than themselves, as our Generalizations 6-2 through 6-7 imply. But in the traditional villages, the heterophilous sociometric arrows pointed to *less* technically competent opinion leaders. The basis of attraction to these traditional leaders was their gregariousness, sociability, and age, rather than their technical competence with new ideas.

Thus, we suggest Generalization 6-9: *In traditional systems followers interact with opinion leaders less (or no more) technically competent than themselves, whereas in modern systems opinion leaders are sought who are more technically competent than their followers.* This hypothesis should be tested in future research.

Measuring Opinion Leadership

Three main methods of measuring opinion leadership have been utilized in communication research: (1) sociometric, (2) informants' ratings, and (3) self-designating techniques. These methods are compared in Table 6-1.

The sociometric method consists of asking questions of respondents as to whom they sought (or "hypothetically" *might* seek) for information or advice about a given topic, such as a technological innovation or a news event. Opinion leaders are those members of a system who receive the greatest number of sociometric choices. Undoubtedly, the sociometric technique is the most valid measure of opinion leadership as it is measured through the eyes of the followers. However, the sociometric approach necessitates interrogating a large number of respondents in order to locate a small number of opinion leaders. And the sociometric method is most applicable to a sampling design where all members of a social system are interviewed, rather than where a small sample within a large population is contacted.*

An alternative to sociometry is the selection of judges or key informants who are especially knowledgeable about the patterns of influence in a system. For example, in a Latin American village, the priest may be able to identify the

---

*It is possible, however, to locate sociometric opinion leaders with survey sampling by means of such designs as "snowball sampling," in which an original sample of respondents in a system are interrogated. Then the individuals sociometrically designated by this sample are interviewed as a second sample, and so on.

Table 6-1   Advantages and Limitations of Three Methods of Measuring Opinion Leadership

MEASUREMENT METHOD	DESCRIPTION	QUESTIONS ASKED	ADVANTAGES	LIMITATIONS
1 Sociometric method	Ask system members to whom they go for advice and information about an idea	Who is your leader?	Sociometric questions are easy to administer and are adaptable to different types of settings and issues. Highest validity	Analysis of sociometric data is often complex. Requires a large number of respondents to locate a small number of opinion leaders. Not applicable to sample designs where only a portion of the social system is interviewed
2 Informants' ratings	Subjectively selected key informants in a social system are asked to designate opinion leaders	Who are the leaders in this social system?	A cost-saving and time-saving method as compared to the sociometric method	Each informant must be thoroughly familiar with the system
3 Self-designating method	Ask each respondent a series of questions to determine the degree to which he perceives himself to be an opinion leader	Are you a leader in this social system?	Measures the individual's perceptions of his opinion leadership, which influence his behavior	Dependent upon the accuracy with which respondents can identify and report their self-images

persons of greatest local influence. The judges or key informants are asked to identify the opinion leaders for a given topic or topics.

The self-designating technique asks respondents to indicate the tendency for others to regard them as influential. A typical self-designating question is: "Do you think people come to you for information or advice more often than to others?" Instead of asking respondents, "Who is your leader?" as in the case of the sociometric method, the respondents are asked, "Are you a leader?" The self-designating method depends upon the accuracy with which respondents can identify and report their self-images. This measure of opinion leadership is especially appropriate when interrogating a random sample of respondents in a system, a sampling design that often precludes use of sociometric methods. An advantage of the self-designating technique is that it measures the individual's perceptions of his opinion leadership, which is actually what affects his behavior.

When two or three types of opinion leadership operations have been utilized with the same respondents, positive correlations among the three measures have been obtained, although these relationships are much less than perfect.* This finding suggests that the choice of any one of the three methods can be based on convenience; all three are about equally valid.

Characteristics of Opinion Leaders

How do opinion leaders differ from their followers? The following generalizations summarize a wealth of empirical studies designed to answer this question.** In each we refer to "opinion leaders" and "followers" as if

*Among these studies are Rogers and Burdge (1962) in Ohio, Rogers with Svenning (1969, pp. 224–225) in Colombia, Sollie (1966) in Mississippi, and Abelson and Rugg (1958) in the U.S.

**All the generalizations which follow state two-variable relationships, dealing with the dependent variable of opinion leadership and some other independent variable. There have been several recent multi-variate attempts to explain the variance in opinion leadership scores with a multiple correlation approach. Generally, the results have been rather desultory. Rogers and Burdge (1962) explained only 26 percent of the variance in opinion leadership with three independent variables: Innovativeness, conformity to system norms, and social status. Rogers with Svenning (1969, p. 232) explained from 33 to 59 percent of the variance in opinion leadership in five Colombian villages. Three variables were consistently the best predictors in all five analyses: Innovativeness, social status, and political knowledge. The correlates of opinion leadership (innovativeness, status, and so on) are themselves highly interrelated. The multiple correlation approach is potentially useful in determining the amount of variance in opinion leadership explained by each of these variables when the effect of the others is controlled statistically.

opinion leadership were a dichotomy and as if nonleaders were all followers. These oversimplifications are for the sake of clarity.

External Communication

Generalization 6-10: *Opinion leaders have greater exposure to mass media than their followers.* As stated in the original conception of the two-step flow, opinion leaders attend more to mass media channels. One of the ways in which opinion leaders gain their competency is by serving as an avenue for the entrance of new ideas into their social system. The linkage may be provided by mass media channels, by the leader's cosmopoliteness, or by the leader's greater change agent contact. "It is the opinion leader's function to bring the group into touch with this relevant part of its environment through whatever media are appropriate" (Katz, 1957).

Generalization 6-11: *Opinion leaders are more cosmopolite than their followers.*

Generalization 6-12: *Opinion leaders have greater change agent contact than their followers.*

Accessibility

In order for opinion leaders to relay their personal messages about innovations, they must have direct dialogue with their followers.* That is, opinion leaders must be accessible. One indicant of such accessibility is social participation; face-to-face communication about new ideas occurs at meetings of formal organizations and through informal discussions.

Generalization 6-13: *Opinion leaders have greater social participation than their followers.* An illustration of this point is provided by the key opinion leader in a Colombian village once studied by the senior author. The leader lived at the fork of a Y-shaped foot trail over which the villagers traveled daily. His social accessibility undoubtedly contributed to his high degree of influence.

Social Status

We expect that followers dyadically seek opinion leaders of somewhat higher status than their own, as suggested in a previous generalization (6-2).

*Becker (1968b) provides some tentative evidence in the case of public health officials that their centrality as opinion leaders may result *from* innovativeness, rather than vice versa. The officials in state departments of health who were first to adopt new public health programs were sought by their peers for information about these innovations.

Opinion leaders, on the average, should be of higher status. This was stated by Tarde (1903, p. 221): "Invention can start from the lowest ranks of the people, but its extension depends upon the existence of some lofty social elevation." Generalization 6-14: *Opinion leaders have higher social status than their followers.*

Innovativeness

If opinion leaders are to be recognized by their peers as competent experts on innovations, it is likely that they adopt new ideas before their followers. There is strong empirical support for this supposition. Generalization 6-15: *Opinion leaders are more innovative than their followers.* The research findings do not indicate, however, that opinion leaders are necessarily innovators. At first glance, there appears to be contradictory evidence on whether opinion leaders are innovators. For example, Menzel and Katz (1955) reported that in one study of new drugs, the leaders were innovators. In another drug diffusion investigation, however, there was little overlap between innovator-physicians and opinion leaders (Coleman and others, 1966). Wilkening (1952b) found only three of fifteen farm innovators to be opinion leaders, but Lionberger (1953) and Rogers and Burdge (1962) found considerable overlap.

What explains this seeming paradox? A partial solution to this inconsistency is provided by considering the effect of system norms on the innovativeness of opinion leaders.

Innovativeness, Opinion Leadership, and Norms

Homans (1961) stated that leaders obtain their position of influence by rendering valuable and rare services to their system. Leader conformity to norms is a valuable service to the system in that the leader thus provides a living model of the norms for his followers. "A man of high status [that is, leadership] will conform to the most valued norms of his group as a minimum condition of maintaining his status" (Homans, 1961, p. 339).

How can opinion leaders be most conforming to system norms and also lead in the adoption of new ideas? The answer is expressed as Generalization 6-16: *When the system's norms favor change, opinion leaders are more innovative, but when the norms are traditional, opinion leaders are not especially innovative.** In

*Obviously, opinion leaders can either favor or oppose change and influence their followers to do the same. Arndt (1968g) concludes: "Those receiving favorable word-of-mouth communications (from opinion leaders) were three times as likely to buy the new product as were those receiving unfavorable word-of-mouth."

traditional systems the opinion leaders are usually separate individuals from the innovators (the first to adopt new ideas); the innovators are perceived with suspicion and often with disrespect by the members of traditional systems. For instance, in a study of Colombian peasants Rogers with Svenning (1969, pp. 230–231) found that opinion leaders in the relatively more modern villages were more innovative than their followers, but in the traditional villages the opinion leaders were only slightly more innovative than their followers and were older and less cosmopolite.

Data from various other inquiries in India, Poland, and the U.S. generally support the notion of opinion leaders as highly conforming to system norms. For instance, Herzog and others (1968, p. 72) conclude from their study of Brazilian villages: "In most traditional communities, neither the leaders nor their followers are innovative, and as a result, the community remains traditional. In the most modern communities, community norms favor innovativeness and both the leaders and followers are innovative. In the middle range communities, where modernization is just getting underway, divisions occur and the community opinion leaders lead the way toward modernization, by trying new ideas before the other farmers in the community."

There is an important implication for change agents in the present generalization about opinion leader conformity to norms on change. A common error made by change agents is that they select opinion leaders who are too innovative. We point out in this book (Chapter 7) that change agents work through opinion leaders in order to close the heterophily gap with their clients. But if opinion leaders are too much more innovative than the average clients, the heterophily (and accompanying ineffective communication) that formerly existed between the agent and his clients now exists between the leaders and their fellow clients. This is why innovators are poor opinion leaders in traditional systems; they are too elite and too change-oriented. They serve as an unrealistic model for the average client, and he knows it.

An investigation of German demonstration farmers by Hruschka and Rheinwald (1965) provides insight into the nature of opinion leadership and conformity to norms. The demonstrators ranged in their degree of sociometric opinion leadership, although all were selected by extension agents on the basis of their influence with their fellow villagers. In the villages where the demonstrators had the highest degree of opinion leadership, they were most effective in diffusing farm innovations to their peers, as might be expected. The demonstrators who were highly innovative but not high in opinion leadership were ineffective in diffusing the innovations that they demonstrated to their fellow villagers. These demonstrators reported that they suffered from their neighbors' sarcastic remarks, that they now felt they were outsiders in the village, and that even their children were teased. Further, they said they would never agree to serve as a demonstration farmer again. On the other hand, the demonstrators who were high in opinion leadership, while agreeing that being a demonstrator was a very "tricky task," expressed the view that being a demonstrator fit into the role they already held in the village.

A parallel case to that of leader-follower homophily among farmers is found in the case of the so-called "laboratory schools" in the U.S. These schools were usually affiliated with a college of education, located on a university campus, and utilized for the introduction and trial of new teaching methods. The typical lab school had almost unlimited funds, and its student body was composed of bright faculty children. Supposedly, the lab school was an attempt to demonstrate educational innovations which would then spread to other schools. But the lab schools, with their enriched environments and talented students, were perceived as too heterophilous by the average school. Visiting teachers and administrators would come to the lab schools, impelled by a curiosity motive, but would go away unconvinced by the innovations they had observed. As a result, laboratory schools throughout the U.S. have fallen into disrepute as a means of diffusion, and most of them have been terminated in recent years.

Sometimes change agents identify potentially effective opinion leaders among their clients, but they concentrate their change efforts too much on the leaders, who soon become innovators and lose their former following.* An illustration of this problem comes from East Pakistan, where village leaders travel to a training center for weekly sessions about agricultural and other innovations. The opinion leaders then return to their villages, where they tell their peers about the new ideas. When working in their muddy rice paddies, these villagers ordinarily wear the bottoms of their *lungis* (a sort of shirtlike work clothes) high up on their thighs. Nonpeasants, such as the change agents at the training center, wear their *lungis* at ankle length as a status symbol. As time went on, the village leaders in Pakistan began to wear their *lungis* lower and lower, mimicking the apparel of the change agents and thus symbolizing their own modernization. But when the leaders' *lungis* crept below their knees, they found they had lost their followers. The villagers felt their former leaders had become too much like the change agents, and their allegiance swung to a new corps of leaders. The caution for change agents is to guard against making opinion leaders too innovative, else they will become ex-leaders, who will

*This possibility is implied in Hardin's (1951) analysis of soil conservation opinion leaders in the U.S.

lose their following by deviating from the system's traditional norms (Rogers with Svenning, 1969, p. 231). The modernized leaders move outside of their followers' range of tolerable heterophily.

Changing the Guard in a Colombian Peasant Village[*]

Not only can opinion leaders be stripped of their influence if they become too innovative in a traditional system; deviation from the norms can occur in the opposite direction as well. If the norms of a system change and the leaders do not, the leaders lose their leadership. Such an instance can happen in a rapidly modernizing village, as Lazarsfeld and Menzel (1963, p. 100) hypothesize: "New kinds of opinion leaders seem to come to the fore as traditional folk communities make the transition to the modern industrial world. When this happens, the community's traditional elders tend to lose much of their preeminence to individuals whose positions allow them to act as pipelines to the great world outside."

The senior author encountered an instance of just such a turnover of leadership in the Colombian village of Pueblo Viejo that he once studied. This social system, prior to the initiation of intensive change agent activities in 1959, was "ruled" by a small corps of traditional opinion leaders. When extension workers began to spend one day each week in the village, they were opposed by these leaders, and so they concentrated their attention on a new group of younger, emerging leaders like Miguel Gomez. Miguel served as chairman of the village cooperative, which was organized by the extension workers, and as president of the Village Development Committee. Miguel was well liked by most of his fellow villagers (our data show that he received a high percentage of the sociometric opinion choices), and he served as an effective liaison between the change agents and the villagers. His farm was located at the mid-point of the foot trail over which the peasants traveled, and this physical and social accessibility contributed to his position of influence. Through the efforts of Miguel and the extension workers, a new road was constructed, a piped water system was installed, and the cooperative store flourished. Agricultural and health innovations were adopted.

[*]This case illustration is adapted from Rogers with Svenning (1969, pp. 219, 338–340), and is used by permission.

The relative power of the older, traditional leaders gradually shrank in the face of Miguel's development successes until at the time of our data gathering in 1963, the traditional leaders had only a handful of followers. The rise and decline of these leaders demonstrates that opinion leaders must conform closely to a system's norms. When the norms changed in Pueblo Viejo, so did the leadership. It might have been possible for the leaders to modernize along with the system's norms, and this probably happens when the metamorphosis occurring in the community is not so rapid.

Monomorphic and Polymorphic Opinion Leadership

Is opinion leadership universal? That is, is there one set of all-purpose opinion leaders in a system, or are there different opinion leaders for each issue? *Polymorphism* is the degree to which an individual acts as an opinion leader for a variety of topics (Merton, 1949). Its opposite, *monomorphism*, is the tendency for an individual to act as an opinion leader for only one topic. The degree of polymorphic opinion leadership in a given social system seems to vary with such factors as the diversity of the topics on which opinion leadership is measured, whether system norms are modern or traditional, and so on.

Katz and Lazarsfeld (1955, p. 334) concluded their analysis of opinion leadership among housewives in Decatur, Illinois, for four different topics: "The hypothesis of a generalized leader receives little support in this study. There is no overlap in any of the pairs of activities. Each arena, it seems, has a corps of leaders of its own." Their data seem less definitive than their conclusions, however, because they found that one-third of the opinion leaders exerted their influence in more than one of the four areas (fashions, movies, public affairs, and consumer products).*

System norms are thought to affect the degree of polymorphism among opinion leaders in a social system. Generalization 6-17 states: *When the norms*

---

*Marcus and Bauer (1964) reanalyzed the Katz and Lazarsfeld data, and their recalculations showed the actual frequency of polymorphic opinion leadership was greater than chance. Marcus and Bauer (1964) concluded, contrary to Katz and Lazarsfeld, that there was some degree of polymorphism among the Decatur housewives. These different conclusions drawn from the same data point to the difficulty in constructing a standardized measure of polymorphism that is not open to subjective interpretation.

of a system are more modern, opinion leadership is more monomorphic.* As the technological base of a system becomes more complex, a division of labor and specialization of roles results, which in turn lead to different sets of opinion leaders for different issues. In more traditional systems there is less role differentiation on the basis of occupation. The leaders in such systems are more likely to serve as opinion leaders for all issues in the system. This point is illustrated by Lerner's (1958, p. 26) interview with the chief of a Turkish village who, when asked on what topics he was influential, replied: "About all that you or I could imagine, even about their [the villagers'] wives and how to handle them, and how to cure their sick cow."

Summary

Opinion leadership is the degree to which an individual is able to influence informally other individuals' attitudes or overt behavior in a desired way with relative frequency. Opinion leaders play an important role in the diffusion of innovations, whether the new ideas are in sports, fashion, politics, or other fields. The concept of opinion leadership was originally labeled by Lazarsfeld in 1940 as part of the two-step flow model, which hypothesized that communication messages flow from a source, via mass media channels, to opinion leaders, who in turn pass them on to followers.

The two-step flow model challenged the popular *hypodermic needle* model, which hypothesized that the masses were a large body of disconnected individuals connected to the media but not to each other. Research has since expanded our knowledge of the variable number of steps in the relay process. The predominant model for mass communication flows today is a multi-step flow in which receivers are a variable number of times removed from the message origin. The exact number of steps in the process depends on the intent of the source, the availability of mass media and the extent of audience exposure, the nature of the message, and the message's salience to the receiving audience.

*The evidence is not completely clear-cut on this hypothesis. Rogers with Svenning (1969, p. 227) found that opinion leadership in modern Colombian villages was no more monomorphic than in the traditional villages. This may have been due to a rather restricted range in the degree of modernization of the village norms. On the other hand the investigations of Yadav (1967), Sen (1969), and Sengupta (1968) in India, and Attah (1968) in Nigeria, seem to support the notion of greater monomorphism in opinion leadership in more modern villages.

Homophily is the degree to which pairs of individuals who interact are similar in certain attributes, like beliefs, values, education, and social status. *Heterophily* is the degree to which pairs of individuals who interact are different in certain attributes. Interpersonal diffusion is mostly homophilous, the research evidence indicates. Such homophily can act as an invisible barrier to the rapid flow of innovations within a social system.

Compared to followers, opinion leaders have greater mass media exposure, more cosmopoliteness, greater change agent contact, greater social participation, higher social status, and more innovativeness. Opinion leaders conform more closely to a system's norms than do their followers. When the system's norms favor change, opinion leaders are more innovative; but when the norms are more traditional, leaders are not especially innovative.

Polymorphism is the degree to which an individual acts as an opinion leader for a variety of topics, while *monomorphism* is the tendency to act as an opinion leader for only one topic. When the norms of a system are more modern, opinion leadership is more monomorphic.

7 *The*

— .... .   — .—. .... .— — .— —. .   .— —. . — . — . — — — ....

One of the greatest pains to human nature is the pain of a new idea. It is, as common people say, so "upsetting"; it makes you think that after all, your favorite notions may be wrong, your firmest beliefs ill-founded. . . . Naturally, therefore, common men hate a new idea, and are disposed more or less to ill-treat the original man who brings it.   WALTER BAGEHOT (1873, p. 169)

Change Agent[*]

$$\cdots \cdot - \; -\cdot \; -- \cdot \; \cdot \; \cdot - \; --\cdot \; \cdot \; - \; -- \cdot \; - \; -\cdots \cdot \; -\; -\cdot \; \cdots \cdot - \; - \; -\cdot \; \cdot \; \cdot \; -\; --$$

MOST change is not a haphazard phenomenon but the result of planned, premeditated actions by change agents. Such architects of change include extension agents, Peace Corps volunteers, commercial salesmen, political precinct workers, teachers, VISTA volunteers, and many others.

This chapter is about the role of the change agent, his relationships with clients, and various strategies of change he may employ to increase his success in securing desired changes in his clients' behavior. A *change agent* is a professional who influences innovation-decisions in a direction deemed desirable by a change agency.[**] In most cases he seeks to secure the adoption of new ideas, but he may also attempt to slow the diffusion and prevent the adoption of certain innovations.

---

[*]This chapter is written with John A. Winterton, Research Assistant in Communication, Michigan State University. Several of our central views in this chapter come from Rogers with Svenning (1969, pp. 169–194).
[**]Our definition of change agent is generally consistent with others' conceptions. "Change agent refers to the helper, the person or group who is attempting to effect change" (Bennis and others, 1962, p. 5). Other authors (Lippitt and others, 1958) feel that the change agent must be exogenous to the system. We maintain that he is set off from his clients by nature of his professional status (that is, employment by a change agency), rather than whether he lives in or out (or considers himself a member) of a particular system.

The Role of the Change Agent

The change agent functions as a communication link between two or more social systems (Figure 7-1). For example, a company saleman provides linkage between his change agency, a drug manufacturer and the client system of medical doctors whom he contacts. Similarly, a technical assistance worker provides linkage between a more developed nation and clients in the less developed country in which he is introducing innovations.

There is often a social chasm between the system represented by agents of

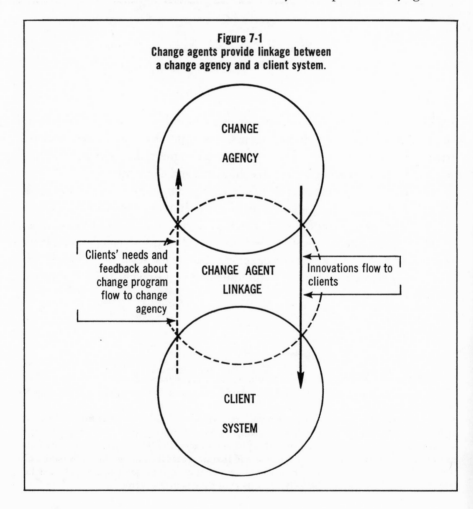

Figure 7-1
Change agents provide linkage between
a change agency and a client system.

CHANGE

AGENCY

Clients' needs and
feedback about
change program
flow to change
agency

CHANGE AGENT

LINKAGE

Innovations flow to
clients

CLIENT

SYSTEM

change and the clients' system. Typical disparities between such systems include subcultural language differences (even though both may ostensibly share a common tongue), socioeconomic status, technical competence, and beliefs and attitudes. Change agents, even though they link the two social systems, may be quite heterophilous in relation to their clients or to their superiors in the change agency. This heterophily gap on both sides of the change agent creates role conflict and problems in communication, among other things, as we see later in this chapter. As a bridge between two differing systems, the change agent is necessarily a marginal man with one foot in each of two worlds. His success in linking the change agency with his client system lies at the heart of the process of planned change.

The Sequence of Change Agent Roles

Seven roles are isolated in the process by which a change agent introduces an innovation to his clients.*

1. *Develops need for change.* A change agent is often initially required to help his clients become aware of the need to alter their behavior. This is especially true in less developed countries. The short planning horizons, low achievement motivation, high fatalism, and low aspirations characteristic of most peasants mean that the change agent must serve a catalytic function for client needs. In order to initiate the change process the change agent points out new alternatives to existing problems, dramatizes these problems, and convinces clients that they are capable of confronting these problems. He not only assesses clients' needs at this stage but also helps to create these needs in a consultive and persuasive manner.

2. *Establishes a change relationship.* Once a need for change is created, a change agent must develop rapport with his clients. He may enhance his relationship with his clients by creating an impression of credibility, trustworthiness, and empathy with their needs and problems. Clients must accept the change agent before they will accept the innovations he promotes.

3. *Diagnoses the problem.* The change agent is responsible for analyzing his clients' problem situation in order to determine why existing alternatives do

*Adapted from Lippitt and others (1958), and from Rogers with Svenning (1969, pp. 171–173). There is obviously a close parallel between these seven roles and the four functions that we distinguished in the innovation-decision process (Chapter 3).

not meet their needs. In arriving at his diagnostic conclusions, the change agent must view the situation empathically from his clients' perspective, not his own. He must psychologically zip himself into their skins, put himself in their shoes, see their situation through their eyes. This empathic transferral is difficult.

4. *Creates intent to change in the client.* After a change agent explores various avenues of action that his clients might take to achieve their goals, he should encourage an intent to change, a motive to innovate. But the change must be client-centered, rather than a change for change's sake. Here the change agent's role is to motivate.

5. *Translates intent into action.* A change agent seeks to influence his clients' behavior in accordance with his recommendations which are based on the clients' needs. In essence the agent works to promote compliance with the program he advocates. This means more than simply agreement or intent; it means action or behavioral change.

6. *Stabilizes change and prevents discontinuances.* Change agents may effectively stabilize new behavior by directing reinforcing messages to those clients who have adopted, thus "freezing" the new behavior. This assistance frequently is given when the client is at the trial-decision or confirmation function in the innovation-decision process.

7. *Achieves a terminal relationship.* The end goal for any change agent is development of self-renewing behavior on the part of his clients. The change agent should seek to put himself out of business by developing his clients' ability to be their own change agents. In other words the change agent must seek to shift the clients from a position of reliance on the change agent to reliance on themselves.

Thrown Onto the Edge of Asia*

Active involvement in the process of directed change is a difficult and sometimes frustrating experience for change agents. An autobiographical report from a less-than-successful Peace Corps volunteer illuminates the pressures on change agents who bridge the gap between widely divergent social systems, where heterophily poses great problems to effective communication. This Peace Corpsman was assigned to introduce innovations to

*Adapted from Anonymous Peace Corps Volunteer (1967, pp. 54–63) and used by permission.

peasants in an Asian nation. Following is an account of his cross-cultural problems in client relationships, written in the form of a letter home.

Life is very slow here and plenty real. It will be sporting for you at first—to be constantly adored as the strangest and most god-like individual ever to have turned his eye onto an isolated village, otherwise unknown. Very soon, I think, the game goes out of it. Then you must know what you're there for, what you are going to do, and why. I think you should have the answers ready because you're going to need them.

Let me describe a somewhat typical day in my life as a change agent. A chicken wakes me up; he is crowing about something indiscernible to me: Not even the chicken speaks English. Language is essential. . . . I brush my teeth for a while, not that they really need brushing, but this is a habit I've formed. The lady next door is watching from a screenless window. I feel a sense of pride that, having been awake for only ten minutes, I am already functioning as an agent of change. Trouble is, my neighbor's teeth are hopelessly blackened. She's about as likely to adopt my technique as I am to begin chewing betel nut.

Later, at a restaurant in the village marketplace, run by a retired prostitute, I order a bottle of orange soda. I've been ordering this every morning partly because I don't trust the water . . . and partly because it is the nearest equivalent to orange juice. She looks at me, in disgust, as if to say, "What the hell kind of man would drink orange soda pop for breakfast?"

A crowd begins to gather. Eating breakfast is my first big performance of the day. I sit as tall as I can and eat with vigor. I invite the nearest villagers to sit down with me—the ultimate of democratic gestures—but they refuse, as expected, because it's not their place to be seen eating with a superior. They're not sure whether I'm a superior or not, but they know I was placed there by some distant and very important machinery of government; they are, consequently, playing it safe.

Small talk concerning hunger, food customs, and health soon leads to a question beyond my language capabilities. "Eh?" This is the word which I use most often during the day. The villager repeats his question, leaving me even more confused. One of the other peasants explains to the questioner that he must speak very slowly because the foreigner does not know "how to listen to our language." A couple of children giggle and repeat: "Foreigner." In their language the word also means the guava fruit, and, with a short modifier, bird manure.

The crowd thins. This curious new thing in town, a full-grown man who does not know how to speak, can sustain interest for only so long. I walk around the market, and stop to thump a few pineapples. I don't know how a good pineapple should thump, but this seems to be the proper thing to do, like kicking tires in a car lot.

Later, with a government change agent who is my counterpart, I visit the local barber shop. There are a half-dozen men seated in the barber shop and my colleague says a few brisk words to them in his language. The man in the barber's chair gets up and takes a place on a bench. I try to indicate that I want to wait my turn, but my language fails me. The barber, smiling, waits with cloth in hand. All eyes watch me. My counterpart says, "You sit down there." Everyone seems happy about it.

We retire to a tavern for rice whisky. There are five glasses on the table: one for me, one for my colleague, three for some soldiers. There are a dozen more persons around the table, but they are only peasants, watching the soldiers drink.

The sergeant at my right seems to be in charge; he is a counter-insurgency soldier. He explains that they have come to check over the village. He wears a pistol. He is drunk. He hates Communists.

By means of his G. I. English and my meager knowledge of his language, we reach a few understandings. He questions me about who I am and what I am doing. I have written my English name, to which I add "Peace Corps, U.S.A." He assumes that we are in the same line of work. He is anxious to prove to me his knowledge of propaganda and other counter-insurgency methods.

"Allies," he says in English.

"Allies," I agree.

"The American country and our countries—allies. You and me. We know how to kill the Communists together." He eyes the intent crowd keenly.

I hold up my hands. I announce, for the second time, as clearly and publicly as my vocabulary will allow, "I am not a soldier. I am a volunteer come from America to help the peasant, because you and the American are a pair of friends together."

The sergeant has another drink. "Allies," he says, and then: "Help the rotten villager. Look at this man." He orders him to stand up. "We help this man. This man is always lazy. This man is always dirty. Look."

The best way to handle those soldiers is to stay away from them. I decide to leave. I look over at the peasants who sit with their hands folded soberly. I had not even remembered to invite them to drink with us.

By way of a parting gesture, I order two bottles of whisky. The soldiers protest; this is their party, they pay. Then I order six glasses, which I pass out to the villagers. We drink. I shake hands with each person, including the soldiers and my counterpart. This is not the customary manner of taking leave, but they know that Americans always shake hands and say "okay," or "hello," or "goodby."

As I leave, the sergeant winks at me and says one English word: "Psychology." I wink back. "Psychology."

This Peace Corps volunteer, like many other agents of change, faces difficult problems in successfully communicating with his heterophilous clients. Linguistic and cultural differences prevent fulfilling even the first two roles for directed change which were discussed previously: First, to develop a need for change, and second, to establish rapport and a change relationship with the clients. A final note: The change agent who wrote this report shortly thereafter resigned from the Peace Corps and returned to the U.S.

Change Agent Effort

Why are some change agents relatively more successful in introducing innovations? The answer seems to lie in a number of reasons, which will be summarized in the remainder of this chapter.

One of the factors in change agent success is the extent of effort he expends in change activities with his clients. Numerous researches suggest Generalization 7-1: *Change agent success is positively related to the extent of change agent effort.** The strongest support for this proposition comes from a three-nation comparative investigation of the relative success of planned change programs in sixty-nine Brazilian communities, seventy-one Nigerian villages, and 108 Indian villages. Similar concepts and equivalent research procedures were utilized in each country, so that a general picture of variables related to change

*The degree of success of change agents is usually measured (in the studies synthesized in this chapter) in terms of the adoption of innovations by members of the client system (similar to the rate of adoption dimension used as the dependent variable in Chapter 4). This measure is frequently used because the main objective of most change agents is to secure adoption of new ideas by their clients. In some respects (as will be discussed in Chapter 11), an improved measure of a change agent's success is the degree to which desired consequences of innovation adoption occur among his clients, consequences such as improved levels of living, higher incomes, and the like.

agent success would emerge. The most important predictor of the success of village programs of agricultural change is the extent of change agent effort.* "Success" villages, as contrasted with "failure" villages, were characterized by change agents who contacted more clients, who spent fewer days in their offices and more in the villages, and who generally played an active rather than a passive role in the change process. Increased interpersonal communication with clients, then, is crucial to change agent success.

Further evidence of the relationship between adoption of innovations and the extent of change agents' efforts comes from an investigation by Deutschmann and Fals Borda (1962b) in a Colombian peasant community. They found that two farm innovations promoted by a change agent were adopted much more quickly than were two other farm ideas which the change agent had not emphasized as part of his program of directed social change.

Another investigation dealing with change agent success and utilizing yet another research approach arrived at similar conclusions: Change agent effort leads to success in introducing innovations to clients. Niehoff** concluded from his analysis of several hundred case studies, each dealing with a change agent's attempt to transfer an innovation cross-culturally, that one of the most fundamental factors in success is the extent of change agent and client contact. This communication interface lies at the heart of the diffusion process, particularly when the clients are peasants in less developed countries.

Our generalization is by no means limited to this setting. There is confirmatory evidence from a variety of diffusion studies in the U.S. and Sweden. For example, Hoffer (1944) correlated various promotional activities by Michigan extension agents (for example, their number of visits with clients, newspaper articles published, and circular letters written) with the adoption of innovations by farmers. Ross (1952) found that the rate of adoption of driver training programs by high schools was much more rapid than for other educational innovations (see Figure 2-2). He attributed this rapid rate of adoption to the promotional efforts of car dealers, insurance companies, and other commercial change agents.

Stone (1952, p. 16) analyzed the amount of effort by eighteen Michigan county extension agents from 1943 to 1950 in promoting the adoption of a

---

*The detailed findings from this three-nation diffusion project are presented in Whiting and others (1968), Hursh and others (1969), and Fliegel and others (1967b). An overall synthesis of the investigation is Rogers and others (forthcoming). In each country the village is the unit of analysis, rather than the individual respondent as in most past diffusion studies. Of course, massive research resources are required for an inquiry of this sort, where data are gathered from a sample of client systems.

**A detailed discussion of this research approach is found in Niehoff (1964a and 1966a).

new idea, the artificial breeding of dairy cattle. In the first four years of the diffusion campaign the adoption of the innovation roughly paralleled the amount of change agents' efforts, as measured by the number of agent days a year devoted to the innovation. However, after about 30 percent adoption was reached, the extension agents' efforts decreased, whereas the farmers continued to adopt the new idea at almost a constant rate.

Figure 7-2 illustrates the general relationship between change agent effort and the diffusion of an innovation, as loosely based on data reported by Stone (1952) and Petrini (1966b and 1967a). We see that at Stage I, in the very early phase of an innovation's adoption, change agent activity has little effect on rate of adoption. Then, when the adoption curve starts to climb (from perhaps 5 to 20 percent adoption), increased inputs of change agent activity result in direct gains in rate of adoption (at Stage II). But after about 15 to 20 percent

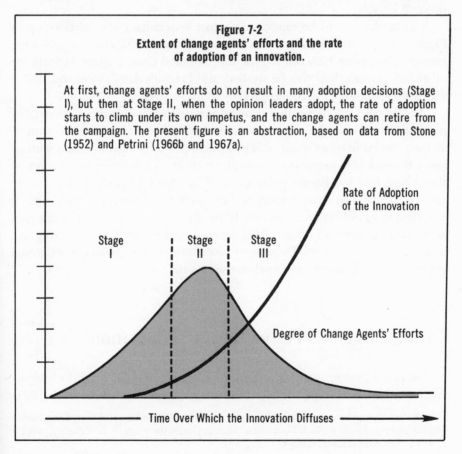

Figure 7-2
Extent of change agents' efforts and the rate
of adoption of an innovation.

At first, change agents' efforts do not result in many adoption decisions (Stage I), but then at Stage II, when the opinion leaders adopt, the rate of adoption starts to climb under its own impetus, and the change agents can retire from the campaign. The present figure is an abstraction, based on data from Stone (1952) and Petrini (1966b and 1967a).

Rate of Adoption
of the Innovation

Stage
I

Stage
II

Stage
III

Degree of Change Agents' Efforts

Time Over Which the Innovation Diffuses

adoption, further change agent inputs seem to have no direct effects on the rate of adoption.

Figure 7-2 shows that the relationship (stated in Generalization 7-1) about change agent effort and success is not linear. Why isn't it? Probably because of the influence potentials of the particular clients who adopt at Stages I, II, and III.* The innovators who adopt at Stage I are seldom active in diffusing the innovation by word-of-mouth channels to their peers. But as we saw in Chapter 5, the early adopters (who are mainly adopting during Stage II) possess a high degree of opinion leadership in most systems. These individuals influence their peers to adopt the innovation. This causes the adoption curve to shoot upward in a self-generating fashion, and the change agent can begin to retire from the scene. In Stage III, the adoption curve continues to climb, independent of change agents' efforts, under further impetus from the opinion leaders.

The reader should be cautioned against generalizing the relationship in Figure 7-2 to all diffusion situations. However, Stone (1952) claims a somewhat similar relationship between rate of adoption and change agents' efforts for other farm innovations that he studied, and Petrini's data (1966b and 1967a) are based upon numerous diffusion campaigns in Sweden.

Another point to remember is that it is entirely possible that the extent of promotional efforts by change agents results *from* a rapid rate of adoption of an innovation! In other words it is possible that change agent success with an innovation leads to increased change agent effort. For instance, let us suppose that a new idea has been rapidly adopted by about 10 percent of a change agent's clients. The change agent is then likely to interpret the rapid rate of adoption as an indication of his clients' need for the new idea and decide that he should concentrate more of his promotional efforts upon securing yet wider adoption of the idea. Thus, rapid adoption may cause greater promotional efforts by change agents, rather than the reverse.

Change Agency Versus Client Orientation

Because a change agent's position is located midway between the bureaucracy to which he is responsible and the client system in which he works, he is

---

*There is an obvious parallel between the present discussion and that presented in Chapter 4 to explain the diffusion effect. Both discussions deal with the point in the diffusion process at which the opinion leaders adopt.

necessarily subject to role conflicts. The change agent is often expected to engage in certain behaviors by the change system, and at the same time he is expected by his client system to carry on quite different actions.

Studies by Wilkening (1957) in Wisconsin and by Bible and Nolan (1960) in Pennsylvania indicate there is considerable disagreement between the role expectations by local clients for their county extension agent and the agent's self-definition of his role. For example, the change agents perceive their role as one of basic education about agriculture and other topics, but their clients expect them to provide services such as finding speakers for farm organization meetings, culling chickens, and the like.

Gans' (1962) investigation of Italian-American "urban villagers" in Boston describes the breakdown in communication between social workers and their clients. He attributed this block primarily to the bureaucratic orientation of change agents, as well as to the agents' lack of empathy with their clients. The social workers perceived themselves as "caretakers" of foolish, inept clients to whom they often referred possessively as "my people." Meanwhile, clients sought to "put-on" the change agents by manipulating them to provide needed services, while immunizing themselves from the agents' attempts at change. The change agents' lack of client orientation led to their failure in changing client behaviors.

Preiss (1954) studied role conflicts among Michigan extension agents, and concluded that their success was associated with a disregard for the expectations of the extension service bureaucracy in favor of their local clients' expectations. This suggests Generalization 7-2: *Change agent success is positively related to his client orientation, rather than to change agency orientation.** Three reasons for this relationship are that client-oriented change agents are more likely to be feedback minded, to have close rapport and high credibility in the eyes of their clients, and to base their programs of change on clients' needs.

Compatibility with Clients' Needs

One of the most important and difficult roles for the change agent is diagnosing clients' needs.** Diffusion campaigns often fail because change

---

*This generalization (as well as Generalization 7–3) essentially restates a basic proposition about communication: That effective communication is positively related to being receiver-oriented.

**The diagnosis of needs is facilitated by client participation in planning change programs. This strategy also (1) increases client commitment to decisions which are made, as a result

agents are more innovation-minded that they are client-oriented. They "scratch where their clients do not itch." We suggest Generalization 7-3: *Change agent success is positively related to the degree to which his program is compatible with clients' needs.** There is much evidence for this proposition from descriptive case studies.

Change projects not based on the clients' felt needs often go awry or produce unexpected consequences. For example, one Indian village was provided with development funds to construct irrigation wells which could approximately double their crop yields. But the villagers wanted wells for drinking, because they had to carry their water about two miles from a river. The peasants built the wells in the village center, rather than in their fields, and drank the water, rather than irrigating their crops. If the change agent had based his program upon the felt needs of the villagers, he would have agreed to provide at least one well for drinking purposes, or else he could have tried to develop a felt need for irrigation by pointing out the financial advantages of this innovation.

In one extremely traditional community in Colombia crop yields were severely depressed by high soil acidity. Local change agents were unsuccessful in motivating farmers to apply lime to the soil to correct the acidity because the peasants were ignorant of the principles of soil chemistry. However, when a severe outbreak of insects occurred on their crops, the farmers were easily convinced by the change agents to apply lime. The farmers felt a need to control the insects, which they could see, whereas they had not seen nor understood the problems of soil acidity (Rogers, 1966a).

Attempts to eradicate the habitat of the tsetse fly in northern Nigeria, in hope of reducing the incidence of sleeping sickness, were largely unsuccessful because villagers did not believe there was a relationship between the insect and the illness. Likewise, an irrigation engineer from an Asian country who had received training in the U.S. returned home convinced of the value of building wells in order to irrigate rice. Over 100 wells were constructed in

of their participation in the decision-making process, and (2) helps legitimize collective innovation-decisions, because the system's power holders are thus involved in the planning process. There is some reason to believe that change agents are more authoritarian in their dealings with more laggardly and lower status clients (Rogers with Svenning, 1969, p. 181), and hence these clients participate less in planning change programs. As a result, the change programs are less appropriate for the non-elite clients and are less likely to be adopted.

*This generalization has an obvious similarity to Generalization 4-2, which we encountered in an earlier chapter: *The compatibility of a new idea, as perceived by members of a social system, is positively related to its rate of adoption.*

isolated villages before the engineer realized they were not being used. Villagers regarded the irrigation water from wells as "artificial," not as natural as rainfall. Since they feared it would harm their crops, they refused to adopt the use of irrigation water and all 100 wells passed into complete disuse.

Many change programs fail because they seek to swim against the tide of clients' cultural values, without steering toward clients' perceived needs. Change agents must have knowledge of their clients' needs, attitudes, and beliefs, their social norms and leadership structure, if programs of change are to be tailored to fit the clients. Mead (1955, p. 258) stated: "Experience has taught us that change can best be introduced not through centralized planning, but after a study of local needs."

It is possible to allow clients to pursue the solution to their needs so completely that they commit errors. Niehoff (1964b) recounts a case of an unsupervised self-help program in Laos which led to unexpected results. Leaders in each village were allowed to decide on their own development projects; then the U.S. Agency for International Development provided construction materials, such as cement, hardware, and roofing materials. Hundreds of village projects were carried out, including building schools, roads, markets, irrigation canals, and dams. But it soon became apparent that half of the construction projects were Buddhist temples, a result hardly expected or desired by the change agency.

Change agents must be aware of their clients' felt needs and adapt their programs to them. However, they should not relinquish their role in developing and shaping these needs, so as to benefit the clients' welfare in the long run.

If change agent-client heterophily were not present, clients' needs would be identical with those of the change agents. To assess his clients' needs, a change agent must be able to empathize with his clients, to see their problems through their eyes.

Change Agent Empathy

Change agent empathy with clients is especially difficult when the clients are very different from the change agents, as in the earlier case of the Peace Corps volunteer in Asia. We expect change agents to be more successful if they can empathize with their clients. Although there is little empirical support for this expectation, we tentatively suggest Generalization 7-4: *Change agent success is positively related to his empathy with clients.*

Although this statement holds for most situations it is possible that an exception occurs when the change agent is so empathic that he completely takes the role of his clients and does not wish to change them. Such over-empathy probably rarely occurs, but it is possible. An example is provided by the U.S. anthropologist who empathized so highly with his Pueblo Indian respondents that he joined the tribe.

If empathy is so important in change agent effectiveness, how can it be increased? One method lies in the selection of change agents; those who have once been in the client's role are probably better able to empathize with it.* For example, agricultural change agencies often seek to employ change agents who come from farm backgrounds.

Sometimes novice change agents are given empathy training by living with a client family for some weeks or a month, so that they are able to see the world through the eyes of their clients. Likewise, role-playing (in which the change agent is asked to act hypothetically in the role of the client) is some-times utilized as a training technique to teach change agent empathy with clients. For example, the police department in one California city instructs its officers to play the role of criminal offenders and to subject themselves to arrest from brother officers, in order to develop empathy with real lawbreakers whose behavior they hope to change. This kind of initial empathy with clients is most effectively continued by being feedback-minded and receiver-oriented. In turn the change agent's capacity to obtain feedback from his clients depends in part upon the closeness of his rapport with them.

Homophily and Change Agent-Client Contact

As previously defined, homophily is the degree to which pairs of individuals who interact are similar in certain attributes, and heterophily is the degree to which they differ. Change agents usually differ from their clients in most respects and tend to interact with those clients who are most like themselves. This general statement leads to a series of generalizations for which there

---

*There are other selection techniques. One method is to test the empathic capacity of applicants for change agent positions in an interview situation. The applicant might be asked to play the role of a change agent or a client in order to demonstrate his level of empathy.

is rather strong empirical support, both in more and in less developed countries:*

Generalization 7-5: *Change agent contact is positively related to higher social status among clients.*

Generalization 7-6: *Change agent contact is positively related to greater social participation among clients.*

Generalization 7-7: *Change agent contact is positively related to higher education and literacy among clients.*

Generalization 7-8: *Change agent contact is positively related to cosmopoliteness among clients.*

The logic behind all of these generalizations is that more effective communication with clients (expressed specifically as greater change agent success in this particular case) occurs when source and receiver are homophilous. A homophilous dyad share common meanings and interests; they are better able to empathize with each other because the other's role is similar.** An illustration of a change agent's difficulties in communicating effectively with heterophilous clients is provided by one experienced change agent: "The people that I work with seemed very eager for help and guidance, but their responses to my suggestions were sullen and sometimes even resentful. . . . The people and I seemed to be living on two different levels of thinking. . . . We spoke the same language, but we didn't communicate" (Weller, 1965, p. 1).

One implication of the effective communication-homophily proposition

---

*In addition, three generalizations about change agent contact were encountered in previous chapters. Generalization 3-5 states: *Earlier knowers of an innovation have greater change agent contact than later knowers.* Generalization 5-25 is: *Earlier adopters of innovations have more change agent contact than later adopters.* Generalization 6-12 is: *Opinion leaders have greater change agent contact than their followers.*

**But the homophily-seeking nature of change agent-client contact raises an important ethical problem for change agents: They fail to interact with those clients who most need their help. In effect change agents concentrate their activities upon those clients who have higher social status, more education, and larger incomes; this leads to the relatively earlier adoption of innovations by these elite clients than by less advantaged clients. The net result is often that the rich get richer, an occurrence which change agencies seldom officially recognize in their statement of objectives.

The greater effectiveness of communication is not the only explanation for change agent-client homophily, however. There are organizational pressures on a change agent to produce results, usually in the form of client adoption of innovations that he is promoting. This forces the change agent to interact most frequently with those clients who are most responsive, that is, those of higher social status and greater innovativeness. Further, if the change agent plays a passive role in which he does nothing more than simply make himself available to his clients, it is *they,* not he, who determine their degree of contact. Those clients with the most favorable attitudes toward change will most frequently seek the change agent.

for change agencies is that they should select change agents who are as much like their clients as possible.* If most clients possess only a few years of formal education, a university-trained change agent will likely face greater communication difficulties than if he had less education. Evidence supporting this statement comes from a study by the Allahabad Agricultural Institute (1957) in India. Village-level change agents with only an elementary education were more effective in reaching illiterate Indian villagers than were change agents with high school or university education. Similar findings are reported by Rahudkar (1960), also in India, which leads to Generalization 7-9: *Change agent success is positively related to his homophily with clients.***

In addition to selecting change agents who are homophilous with their clients, another approach to effective communication is to raise the technical competence level of the clients. In effect this decreases change agent-client heterophily and facilitates more effective communication. In some helping professions, however, it appears that change agents work hard to maintain a "safe" distance from their clients in technical competence and in social status. This may be designed to maintain the change agent's image as a credible source of ideas, even though it interferes with his effective transmission of messages.

One problem for the change agent, especially when he is dealing with heterophilous clients, is that they may perceive his role quite differently from the way he perceives it. For instance, change agents often see themselves as primarily disseminators of technical expertise. This self-image may contrast with clients' perceptions of the change agent's role; they may see him in terms of his ethnic background, age, education, marital status, or other personal characteristics, as well as his technical ability. The social workers and their

---

*However, this approach to lessening change agent-client heterophily leads to widening of the heterophily gap between the change agent and his superiors in the change agency. A similar occurrence is reported by Rogers and others (forthcoming) in the case of research interviewers who were selected so as to be homophilous with their peasant respondents, thus causing greater heterophily with their scientist supervisors.

**This generalization implies that effective communication (that is, change agent success) results from change agent-client homophily. Actually, the reverse may be the case. When a dyad interacts effectively over an extended period of time, they are likely to become more homophilous in attitudes, beliefs, and certain behaviors. Of course, homophily in such static characteristics as age, race, education, and so on cannot change as a result of effective communication. Not only can the client become more like the change agent through intensive interaction, but the change agent can also become more like ("corrupted" by) his client. A literary example is provided by the film *Never on Sunday*, in which a well meaning American tourist attempts to teach a Greek prostitute to appreciate art, fine music, and the like. In the process of the conversion, however, the tourist comes to see that the happy prostitute has an enviable style of life.

welfare clients in Gans' (1962) study failed to "read" correctly each other's roles. Obviously, it is how the clients perceive the change agent that matters most in explaining his success or failure in reaching them.

Almost every analysis shows that change agents have more communication with higher status than with lower status members of a social system, as we indicated in a previous generalization. The way in which a change agent handles his status relationships with clients obviously has much to do with his relative effectiveness in introducing new ideas. Erasmus (1961) points out that change agents in some less developed countries avoid the stigma of working with their hands to demonstrate new ideas, because manual labor is symbolic of lower status. These change agents are more likely to *tell* farmers what to do rather than to *show* them. "In Colombia, I have seen highland agricultural extension agents go into the field in a black double-breasted suit with tie, scarf, and black homburg. They were far more eager to demonstrate their social distance from the farmer than to demonstrate improved agricultural practices" (Erasmus, 1961, p. 84).

Yet a further method remains which change agents frequently employ to bridge the heterophily gap: They work through opinion leaders in the system to halve the social distance between themselves and the majority of their clients.* Then the original wide heterophily gap is shortened to two smaller gaps. The use of leaders may also gain credibility for the change agents' innovations by gaining the tacit endorsement of the opinion leaders.

Opinion Leaders

Opinion leadership is the degree to which an individual is able to influence informally other individuals' attitudes or overt behavior in a desired way with relative frequency. Diffusion campaigns are more likely to be successful if change agents identify and mobilize opinion leaders. Therefore, Generalization 7-10 is: *Change agent success is positively related to the extent that he works through opinion leaders.*

The time and energy of the change agent are scarce resources. By focusing his communication activities upon opinion leaders in a social system, he can

---

*Ramos (1966a) found that the more social distance Colombian peasants perceived between themselves and extension agents, the less favorable attitudes they held toward the change agents, the less credibility they placed in them, and the less interpersonal communication they had with the change agents.

hasten the rate of diffusion. Economy of effort is achieved because the time and resources involved in meeting with opinion leaders is far less than if each member of the client system were to be consulted. Essentially, the leader approach magnifies the change agent's efforts. He can communicate the innovation to a few opinion leaders and then let word-of-mouth communication channels spread the new idea from there.

Furthermore, by enlisting the aid of leaders, the change agent provides the aegis of local sponsorship and sanction for his ideas. Directed change takes on the guise of spontaneous change. Working through leaders improves the credibility of the innovation, thereby increasing its probability of adoption. In fact, after the opinion leaders in a social system have adopted an innovation, it may be impossible to stop its further spread.

Change agents sometimes mistake innovators for opinion leaders. They may be the same individuals, especially in systems with very modern norms, but often they are not. Opinion leaders possess a following, whereas innovators excel at being the first to adopt new ideas. When the change agent concentrates his communication efforts on innovators, rather than on opinion leaders, the results may help to increase awareness-knowledge of the innovations, but few clients will be persuaded to adopt. The innovators' behavior is not likely to convince the average client to follow suit. A related difficulty occurs when a change agent correctly identifies the opinion leaders in a system but then proceeds to concentrate his attention so much on these few leaders that they may become *too* innovative in the eyes of their followers, or become perceived as too friendly and overly identified with the change agent.*

Change Agent Credibility

Credibility is the degree to which a communication source or channel is perceived as trustworthy and competent by the receiver. A basic proposition from laboratory experimental studies in communication is that an individual's attitude change is positively related to the credibility with which he perceives the source (or channel**) of persuasive messages. If a client perceives that a

*As was described in Chapter 6.
**It is often difficult to distinguish between source and channel credibility in most non-laboratory communication situations, although the convention is to speak of "source credibility."

change agent possesses relatively higher credibility than various other sources and channels, the client will be more receptive to messages from that change agent. Although there is little evidence from empirical diffusion studies, it seems logical to suggest Generalization 7-11: *Change agent success is positively related to his credibility in the eyes of his clients.*

Rogers with Svenning (1969, pp. 184–186) determined the relative credibility that Colombian peasants placed in extension change agents in comparison with five other sources of information about agricultural innovations. The respondents were sequentially presented with these six sources in all possible pairs and asked which source in each combination they felt was more credible.* Figure 7-3 shows highest credibility was attributed to the

*This method of determining source credibility is called the paired comparison technique. Its advantage is that it simplifies the stimuli alternatives presented to the respondent in each question; its disadvantage is that it can require considerable interview time if the number of alternatives is large.

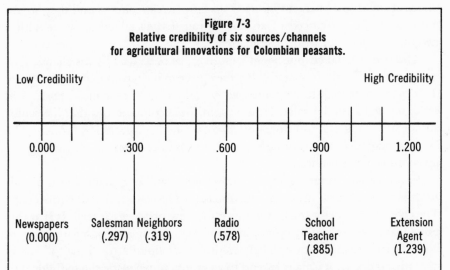

Figure 7-3
Relative credibility of six sources/channels
for agricultural innovations for Colombian peasants.

Low Credibility   High Credibility

0.000   .300   .600   .900   1.200

Newspapers   Salesman Neighbors   Radio   School   Extension
(0.000)   (.297)   (.319)   (.578)   Teacher   Agent
                                                           (.885)   (1.239)

The paired comparison technique was used to obtain these data, which show the relative credibility of sources (or channels) of agricultural innovations range from extension change agents (highest credibility) to newspapers (lowest credibility). The reader is cautioned that these particular findings may be somewhat idiosyncratic to the Colombian villages of study, due to their modern norms and the technical competence of the change agent, although they are supported by Herzog's (1967b) data from Brazil.

Source: Rogers with Svenning (1969, p. 186) and used by permission.

extension agent, followed by the school teacher, radio, neighbors, commercial farm salesman, and newspapers. Almost identical credibility ratings were provided by 1,307 Brazilian peasants (Herzog, 1967b).

Figure 7-3 masks individual differences in the way different peasants perceived the credibility of the extension change agent. Obviously, some peasants viewed the change agent as having more credibility than others. The clients with highest confidence in the credibility of the change agent were those with whom he worked most closely; they had a higher degree of change agent contact, more social status, and larger farms.

We see that the Colombian peasants placed relatively low credibility in the commercial salesman, a finding supported in the U.S.* The adoption of a new idea almost always entails the sale of a new product. For some innovations and under some conditions commercial change agents are of great significance in the diffusion of innovations. But commercial change agents often carry the albatross of low credibility in the eyes of their clients. The senior author found in one study that 97 percent of his Ohio farmer respondents said they would more likely be convinced of an innovation if they talked about it with a neighbor rather than with a salesman.

The commercial change agent's motives, as perceived by his clients, may be one reason for the low credibility they place in his recommendations. They feel that he may seek to promote the overadoption of new ideas, perhaps in order to secure higher sales. However, in some communities that the senior author studied, local farm store dealers were widely respected by farmers. The farmers regarded these dealers as friends, rather than as change agents promoting new products.

Commercial change agents are more important as a channel at the trial-adoption stage than at any other function in the innovation-decision process.** The client purchases a small amount of the new product for trial. It is at this point that he relies heavily upon commercial change agents for information on how to use the innovation. Their credibility is limited to "how-to" information and does not usually extend to an ability to persuade the individual to a favorable attitude toward the innovation. Such persuasive credibility is accorded to peers, noncommercial change agents, and other sources who do not have anything to gain, at least to the extent the commercial agent does.

*Only newspapers received a lower credibility rating by the Colombian peasants, a fact explained by Rogers with Svenning (1969, p. 185) as due to the lack of content about agricultural innovations in these media.

**Support for this statement is provided by the findings of Ryan and Gross (1943) for hybrid corn, by Beal and Rogers (1957) for a new clothing fabric, and by Copp and others (1958) for three farm innovations.

Clients' Evaluative Ability

One of the change agent's unique contributions to the change process is his technical competence, which allows him to provide his clients with expertise in making innovation-decisions. But if the change agent takes a long-range approach to change, he should seek to raise his clients' technical competence and ability to evaluate potential innovations. Then, eventually, the clients could become their own change agents. This suggests Generalization 7-12: *A change agent's success is positively related to his efforts in increasing his clients' ability to evaluate innovations.* The limited empirical support for this statement comes largely from descriptive case studies.

An example of the ill results that may occur if adopters are not provided with the background competence to evaluate innovations comes from India (Rogers, 1966a). Peasants in India commonly refer to superphosphate fertilizer as "sugar" because of its white crystalline form. A popular saying is: "The more sugar added, the sweeter the yield." However, these Indian villagers had no idea of the biochemistry of fertilizer effects on crop yields nor of the importance of soil testing as a basis for wise fertilizer application. Therefore, they applied overdoses of superphosphate and damaged their crops. This illustration emphasizes the importance of educating clients as to the basic scientific reasons why an innovation has its desired effects. Otherwise, the clients are likely to misuse the innovation, even when the entire diffusion campaign is repeated. If this happens, the change agent has sacrificed long-term progress for the sake of short-run gain. An illustration comes from an agricultural change agent in India who persuaded his clients to adopt nitrogen fertilizer as the result of an energetic communication campaign but did not in the process teach them anything about the way fertilizers stimulate plant growth. When superphosphate fertilizer became available the next year, the change agent had to repeat his campaign approach, because his clients still had not gained an ability to evaluate innovations by themselves.

Alers-Montalvo (1957) studied a Costa Rican community in which many local families had adopted vegetable gardens as the result of a diffusion campaign, but the innovation was later discontinued. The major reason for discontinuance was that the families lacked a basic knowledge of the role of vitamins in adequate nutrition. They never really understood *why* they had adopted vegetable gardening.

Similarly, in the previous example of insects and soil acidity in a Colombian community, one wonders if adoption of lime will continue in the future after the insects are gone, when the peasants do not correctly understand why higher crop yields result from applying lime.

The present generalization about improving clients' competence in evaluating innovations implies a long-range rather than a short-range approach to introducing new ideas in a social system. It might mean, for example, that a change agent in a traditional social system would encourage innovativeness through youth organizations. There are few research studies on the effectiveness of youth organizations in introducing innovations. Straus and Estep (1959) and Olson (1959b) found that adult U.S. farmers who had been in 4-H clubs as youths were more innovative. Research is needed in other settings to determine the effectiveness of introducing innovations through youth clubs.

Clients' evaluative capacities should not be neglected by change agents, who are often more concerned with short-range goals of escalating the rate of adoption of innovations. *Self-reliance and self-renewing client behaviors should be the goal of planned change programs,* leading to termination of client dependence upon the change agent.

Summary

A *change agent* is a professional who influences innovation-decisions in a direction deemed desirable by a change agency. The change agent often fills seven roles in the change process: (1) he develops a need for change on the part of his clients, (2) establishes a change relationship with them, (3) diagnoses their problems, (4) creates intent to change in his clients, (5) translates this intent into action, (6) stabilizes change and prevents discontinuances, and (7) achieves a terminal relationship with his clients.

Propositions were suggested regarding a change agent's relative success in securing the adoption of innovations by his clients. Change agent success is positively related to: (1) the extent of change agent effort, (2) his client-orientation, rather than change agency-orientation, (3) the degree to which his program is compatible with clients' needs, (4) the change agent's empathy with clients, (5) his homophily with clients, (6) the extent he works through opinion leaders, (7) his credibility in the eyes of his clients, and (8) his efforts in increasing his clients' ability to evaluate innovations.

Further, we proposed that change agent contact is positively related to: (1) higher social status among clients, (2) greater social participation, (3) higher education and literacy, and (4) cosmopoliteness.

8   *Communication*

It was the pressure of communications which brought about the downfall of traditional societies.   LUCIAN W. PYE (1963, p. 3)

Innovative educational techniques, such as television, transistor radios, and movies should be employed on a massive scale to decrease illiteracy, improve extension education, and increase opportunities for communication of information on family planning, nutrition, food production, food preservation, and for generally increasing the educational level of the population.

President's Science Advisory Committee (1967, p. 35)

Channels

.— —. —. . .—.. ...   —.—. ——— —— —— .·— —. .. —.—. .— — .. ——

In the introductory chapter of this book we compared the diffusion process to the S–M–C–R communication model (Berlo, 1960, p. 72) and concluded that the two were essentially similar. Our main focus in this chapter is upon the channels by which messages flow from source to receiver. We know from past research that certain channels are more effective than others for certain sources, with certain messages, and for certain receivers. This chapter attempts to summarize what is known about the role of communication channels in the diffusion of innovations.

Types of Communication Channels

It is often difficult for individuals to distinguish between the source of the message and the channel which carries that message. A *source* is an individual or an institution that originates a message. A *channel* is the means by which a message gets from a source to a receiver. There is an analogy to the channel that a barge takes as it carries its cargo (messages) across a harbor from ship to shore (source to receiver).

Schramm (1964, p. 123) points out that there are certain tasks which one channel can do that others cannot do. For these reasons communication

251

channels often can be combined to advantage.* Despite their importance, relatively little research has focused on communication channels in the diffusion process.

Researchers categorize communication channels as either (1) interpersonal or mass media in nature, and (2) as originating from either localite or cosmopolite sources. Past research studies show that these channels play different roles in creating knowledge or in persuading individuals to change their attitudes toward innovations. They also are different for earlier adopters of new ideas than for later adopters. Studies reviewed in this chapter suggest that the role of interpersonal and mass media channels (and also localite and cosmopolite channels) is somewhat different in the diffusion of new ideas in less developed and in more developed countries.

Categorizing Channels

Mass media channels are all those means of transmitting messages that involve a mass medium, such as radio, television, film, newspapers, magazines, and the like, which enable a source of one or a few individuals to reach an audience of many. Mass media communication is sometimes labeled "mediated" or "interposed" because of the print or electronic channel which links the source to the receiver. Mass media can:

1 Reach a large audience rapidly.
2 Create knowledge and spread information.
3 Lead to changes in weakly held attitudes.

However, the formation and change of strongly held attitudes is best accomplished by interpersonal channels.

Interpersonal channels are those that involve a face-to-face exchange between two or more individuals. These channels have greater effectiveness in the face of resistance or apathy on the part of the communicatee. What can interpersonal channels do?

1 Allow a two-way exchange of ideas. The receiver may secure clarification or additional information about the innovation from the source individual. This characteristic of interpersonal channels sometimes

*As we see later, mass media and interpersonal channels play complementary roles rather than competing roles in the transmission of messages, and hence, they may be combined in media forums to yield maximum results.

allows them to overcome the social and psychological barriers of selective exposure, perception, and retention.

2   Persuade receiving individuals to form or change strongly held attitudes.

Some of the important distinguishing characteristics of interpersonal and mass media channels are:

Characteristics	Interpersonal Channels	Mass Media Channels
Message flow	Tends to be two-way	Tends to be one-way
Communication context	Face-to-face	Interposed
Amount of feedback readily available	High	Low
Ability to overcome selective processes (primarily selective exposure)	High	Low
Speed to large audiences	Relatively slow	Relatively rapid
Possible effect	Attitude formation and change	Knowledge change

The most appropriate channel should be selected in terms of the goal of the source and the context of the message if it is to affect a given set of receivers.

We now present a case illustration that provides insight into the way in which interpersonal channels affect the rate of an innovation's diffusion.

Interpersonal Channels in the Diffusion of Home Air Conditioners[*]

In order to assess the importance of interpersonal communication channels in the diffusion of a new idea, we began counting air-conditioning boxes sticking out of Philadelphia row houses. Air conditioners were chosen because they are a prime example of a new product climbing the adoption curve from the luxury to the necessity category. They also happen to be highly visible, and thus provide a convenient index of how individuals are influenced by one another.

Obviously, consumers buy new products to keep up with the Joneses; they also buy air conditioners to keep cool. However, it became clear, as our

*Adapted from Whyte (1954) and used by permission.

experiment proceeded, that other, less "logical" forces had also been at work. As the location of conditioned houses was plotted on a map, a curious distribution pattern began to show up. It could be explained only by the presence of a vast and powerful interpersonal network among the adopters.

What proved to be most significant, however, was the way the conditioners were located within white-collar neighborhoods. While the percentage of conditioners in the whole area usually ran around 20 percent, this figure varied widely from block to block. Despite the fact that most of the residents were of the same age, and had similar backgrounds and income, one block of 52 homes might show only three conditioners, while the very next block might show 18.

No "logical" factors could explain it. It was just as hot in one block as in the other, and there had been no local selling campaigns by vigorous dealers to explain the difference. It became apparent that the clusters were the symbols of a powerful communication network.

These clusters were based on two recurrent factors. The first is what could be called social traffic. We noted that the main clusterings of air conditioners went up and down the sides of the block, rather than across the street. We also found that where there was a row of conditioners along one side, there were likely to be more conditioners on the other side of the alley. We came to realize that the pattern of communication within the block was the explanation. The "block," socially, is not the row of houses on either side of the street, but on either side of the alleyway. Children play in the cemented back alleys, and there the mothers gather to keep an eye on them, gossip, and in many cases, attend to the slowly vanishing clothesline. The children often play out in front also, and on Saturday afternoons and Sundays fathers work there at their lawns and chat with the people next door. It is only a stone's throw to the other side of the street, but young children cross only rarely, and thus the street is a formidable barrier.

But this pattern of interpersonal communication is common to every block, so it does not explain the block-to-block differences in adoption rates for air conditioners. One more factor is needed before a group coalesces—the catalytic influence of a leader. The leaders are not uniformly distributed; some blocks have three or four leaders, others none. Nor are the leaders all of a type. The housewife who is a hard worker on civic affairs, for example, is not necessarily the person around whom neighbors make a social grouping. Couples with the most money are not necessarily the leaders either—indeed, they may be so out-of-tune with their neighbors as to be a standing reproach that is resented.

If any one member of a group is influenced, the initial stimulus is immensely multiplied. First, there is the plain, garden variety of word-of-mouth. As the children play together in the backyard and the girls sit around chatting, Ethel Ostermeyer says that since they got the conditioner she only has to vacuum once a week. On a Sunday afternoon, while Al and Dick work on their lawns, they talk about other aspects of new products.

The indirect word-of-mouth can be potent. Once a group is established, the members are highly sensitized to what the others don't say—or what you think they would say if you weren't around. The couple who held out on getting a television set, for example: When the Wednesday night card group starts talking about some network special, the TV-less couple can become rather uneasy because they don't know what the others are talking about. Or [are they] thinking that people without TV are trying to be highbrow? That at least they should think of their children? The imagined word-of-mouth, even when it is the result of a tortured imagination, can be as much of a reality as the spoken word. Thus the great importance of the remarks that filter up from the junior network. "Mommy, why don't we have an automatic washer too?" The question, many housewives feel, comes from the group as much as from the child.

Amid these uncertainties, word-of-mouth is tremendously important, for it above all is the best guide in the delicate job of keeping in tune with the life style of the moment. And our investigation of the diffusion of home air conditioners suggests that these interpersonal channels may be at least as important in clinching innovation-decisions as company advertising and dealer promotion.

Channels by Functions in the Innovation-Decision Process

Mass Media Versus Interpersonal Channels

Generalization 8-1 is: *Mass media channels are relatively more important at the knowledge function, and interpersonal channels are relatively more important at the persuasion function in the innovation-decision process.* The importance of interpersonal and mass media channels in the innovation-decision process is illustrated by data obtained from 175 Pennsylvania farmers (Copp and others,

1958). The results indicate a greater importance of interpersonal channels at the persuasion function than at the knowledge function.

Sill (1958) found that if the probability of adoption is to be maximized, communication channels must be utilized in an ideal time sequence, progressing from mass media to interpersonal channels. Copp and others (1958, p. 70) found: "A temporal sequence is involved in agricultural communication in that messages are sent out through media directed to awareness, then to groups and finally to individuals. A farmer upsetting this sequence in any way prejudices progress at some point in the adoption process." The greatest thrust out from the knowledge function was provided by mass media, while interpersonal channels were salient in moving individuals out of the persuasion function. Using a communication channel that was inappropriate to a given function in the innovation-decision process was associated with later adoption of the new idea, because such use delayed progress through the process.

Data on the relative importance of interpersonal and mass media channels at each function in the adoption of 2,4-D weed spray were obtained by Beal and Rogers (1960, p. 6) from 148 Iowa farmers. The percentage of respondents mentioning an interpersonal channel increased from 37 percent at the knowledge function to 63 percent at the persuasion function. But mass media channels, such as farm magazines, bulletins, and container labels, were more important than interpersonal channels at the knowledge function for this innovation.

The evidence just presented in support of Generalization 8-1 was drawn from research done in the United States where all of the mass media are widely available. However, studies done in peasant settings indicate that a first condition for mass media effects—availability of the media—may not be met in less developed countries. For example, Deutschmann and Fals Borda (1962b, p. 33) found that interpersonal channels were heavily used even at the knowledge function by Colombian villagers. In an agricultural diffusion study in an East Pakistan village, Rahim (1961a and 1965) found that mass media channels were seldom mentioned as channels of information about agricultural innovations, whereas cosmopolite interpersonal channels were very important, and *in some ways seemed to perform a similar role to that played by mass media channels in more developed countries.*

Figure 8-1 is based on a composite of data from the U.S., Canada, India, Pakistan, and Colombia.* Although mass media channels seem to be of

---

*In cases where a study included more than one innovation, each innovation was considered as a unit of analysis. The data in Figure 8-1 represent twenty-three different innovations, mostly of an agricultural variety.

Figure 8-1
Importance of (1) mass media and (2) cosmopolite interpersonal channels by functions in the innovation-decision process for more developed and less developed countries.

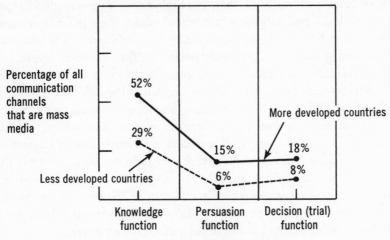

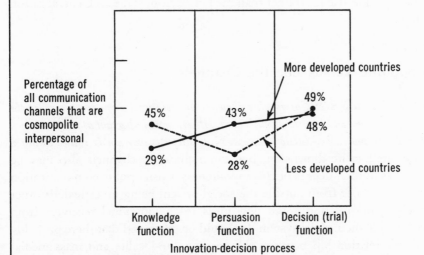

These data indicate that mass media channels (1) play their most important role at the knowledge function and (2) are of greater importance in more developed than in less developed countries. Cosmopolitan interpersonal channels are more important in less developed than in more developed countries at the knowledge function. This suggests that a portion of the role played by mass media channels in creating awareness-knowledge in more developed countries is filled by cosmopolitan interpersonal channels in less developed countries.

Source: A composite of data from the following studies: Beal and Rogers (1957), Beal and Rogers (1960), Copp and others (1958), Jain (1965), Rahim (1961a), Rogers and Meynen (1965), Ryan and Gross (1943), Sawhney (1966), Singh and Jha (1965), Wilkening (1956).

relatively greater import at the knowledge function in both less developed and in more developed countries, there is a higher *level* of mass media channel usage in more developed nations, as we would expect. Less reliance on mass media channels in less developed nations may be attributed to: (1) less mass media exposure by audiences, especially peasants, (2) low literacy levels, and (3) a lack of relevancy of messages in the mass media channels that do exist. Myren (1962) concluded from his study of channels in the diffusion of innovations in Mexico: "The hypothesis about the impact of the mass media can be applied only in areas where media circulate widely, and where, equally important, they command attention and deal with questions of interest to farmers in comprehensible terms." There is very little message content in the mass media devoted to agricultural innovations in less developed countries, even though farmers often constitute a majority of the population. So the *potential* role for mass media channels in diffusing innovations is high, even though it is not being reached in less developed countries today.

We find, then, lower levels of mass media channel use in less developed countries, but the few times that these channels are cited by respondents are mainly at the knowledge (rather than the persuasion) stage, as Generalization 8-1 implies.

Cosmopolite Versus Localite Channels

Generalization 8-2: *Cosmopolite channels are relatively more important at the knowledge function, and localite channels are relatively more important at the persuasion function in the innovation-decision process.* Not only do *individuals* range along a cosmopolite-localite dimension, but communication channels also may be classified as to their degree of cosmopoliteness. Cosmopolite communication channels are those from outside the social system being investigated; other channels of information about new ideas reach individual receivers from sources inside their social system. It should be cautioned that there probably is a direct relationship between the cosmopolite-localite and mass media-interpersonal channel categorizations. Nevertheless, the two classifications are conceptually distinct; clearly interpersonal channels may be either local or cosmopolite. Mass media channels are almost entirely cosmopolite.

Figure 8-1 shows general support from studies in both less developed and more developed countries for Generalization 8-2, if cosmopolite interpersonal and mass media channels are combined to form the composite category of cosmopolite channels. In more developed nations the percentage

of all channels that are cosmopolite is 81 percent at the knowledge function
and 58 percent at the persuasion function. In less developed nations the
percentages are 74 percent at the knowledge function and 34 percent at the
persuasion function. These data (Figure 8-1) hint that perhaps the role played
by mass media channels in more developed countries (creating awareness-
knowledge) is partly replaced by cosmopolite-interpersonal channels in less
developed countries. These channels include change agents, visits outside of
the local community, and visitors to the local system from the city.

Communication Channels by Adopter Categories

The preceding discussion of communication channels by functions in the
innovation-decision process ignored the effects of the respondents' adopter
category. Now we probe channel usage by different adopter categories.

Generalization 8-3 is: *Mass media channels are relatively more important than
interpersonal channels for earlier adopters than for later adopters.* This generalization
seems logical, since at the time that innovators adopt a new idea there is no one
else in the system who has experience with the innovation. Later adopters need
not rely so much on mass media channels because there is a "bank" of interpersonal, local experience that has accumulated in the system by the time they
decide to adopt. Perhaps interpersonal influence is not so necessary to motivate
earlier adopters to decide favorably on an innovation. They possess a need for
venturesomeness, and the mass media message stimulus is enough to move them
over the mental threshold to adoption. But the less change-oriented later adopters require stronger influence, like that resulting from interpersonal channels.

There is strong support for Generalization 8-3 from researches both in
more developed and in less developed nations. Data illustrating the proposition
are shown in Figure 8-2 for the adoption of a weed spray by Iowa farmers.

Similar reasoning to that just presented leads to Generalization 8-4:
*Cosmopolite channels are relatively more important than localite channels for earlier
adopters than for later adopters.** Innovations enter a system from external
sources; those who adopt first are more likely to depend upon cosmopolite
channels. Then these earlier adopters, in turn, act as interpersonal and localite
channels for their more laggardly peers.

*This proposition bears close resemblance to Generalization 5-24 which states that earlier
adopters are more cosmopolite than later adopters. However, Generalization 8-4 refers to
cosmopolite *channel* usage, rather than to cosmopolite behavior in general.

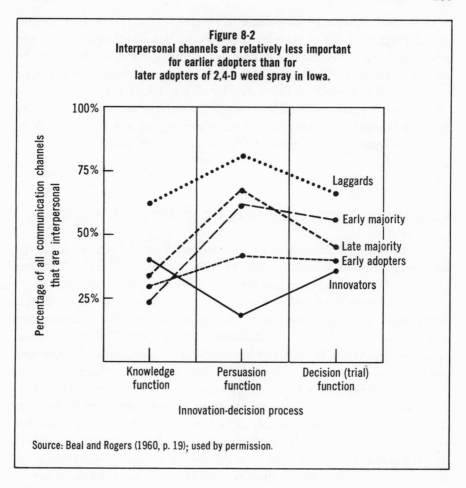

Figure 8-2
Interpersonal channels are relatively less important
for earlier adopters than for
later adopters of 2,4-D weed spray in Iowa.

Source: Beal and Rogers (1960, p. 19); used by permission.

Combining Mass Media
with Interpersonal Channels*

A combination of mass media and interpersonal communication channels is the most effective way of reaching people with new ideas and persuading them to utilize these innovations. Media forums developed originally in

*This section is drawn from Rogers (forthcoming), and Rogers with Svenning (1969, pp. 134–145) and used by permission.

Canada among farm families and later spread to such less developed countries as India, Nigeria, Ghana, Malawi, Costa Rica, and Brazil.* *Media forums* are organized small groups of individuals who meet regularly to receive a mass media program and discuss its contents. The mass media linked to the forum may be a radio, as in the India forums or the *radiophonics* schools of Latin America, printed fare as is usually the case in Communist Chinese study groups, or television as in the Italian *telescuola*. In all cases media forums represent a combination of mass media with interpersonal channels.

Types of Media Forums

Radio Forums

Undoubtedly the largest, most thoroughly-researched media forum program today is India's, representing "a degree of experience with the radio rural forum unequalled in the world" (Schramm and others, 1967, p. 107). There are currently about 12,000 forums enrolling a quarter of a million Indian peasants in twice-weekly meetings. Radio forums help make farmers aware of agricultural and health innovations and encourage them to try these new ideas. Regularly scheduled radio programs beamed at meetings of forum members gathered in homes or public places to hear the broadcast serve as impetus for the group discussion that follows. The forums usually provide regular feedback reports of decisions and questions of clarification to the broadcaster. Using the same format, but exchanging the radio for television, UNESCO has sponsored experimental television listening groups in France and Italy, and the government of India has established teleclubs in some villages.

Mass Media "Schools"

Media schools attempt to provide a basic education, including literacy training, for peoples living in remote rural areas. The Italian *telescuola* and the *radiophonics* programs in Latin America are examples of such schools, as well as Cruzada ABC in Brazil, which uses printed materials. The radiophonic broadcasters intersperse their "lessons" with news, agricultural programs, religious training, and music. Each school group is led by a trained monitor who helps the students learn and encourages them to listen regularly.

*However, the Danish folk school, which is somewhat like a media forum, has existed for centuries.

Chinese Communist Study Groups

The Chinese Communist Party has employed magazine and newspaper discussion groups as a means of indoctrination and learning among their own party cadres and recruits for 35 years. Approximately 60 percent of the adult Chinese population regularly participates in study groups where printed material is read and discussed (Hiniker, 1968). Strict control of discussion is maintained by the cadre leader who forces each member to take a position on each issue and voice his opinion to the group. Study groups are considered essential elements in the special communication campaigns launched to achieve such varied goals as fly killing, river swimming, anti-spitting, family planning, farm production, and "Mao learning."

In all these various types of media forums currently in operation, some form of mass media communication is combined with interpersonal communication in small groups. The forum seems to be an important element in moving the individuals toward acceptance of the messages being transmitted through the mass media. The media forums are used primarily in less developed countries, chiefly to introduce new ideas to vast audiences. With proper adaptation, they could be utilized in educational or political campaigns in more developed nations.

Effects of Media Forums

Although there are important country-to-country and program-to-program differences in the types of media forum systems just reviewed, they possess certain common elements. All utilize a mass medium (radio, television, or print) to carry the major load of disseminating messages about technical innovations to the discussion forums. All feature small-sized groups (usually with fifteen to twenty members) who are exposed to the mass media channel and who then participate in discussion of the message. All of the media forum programs seem to be generally effective in creating knowledge, forming and changing attitudes, and in catalyzing behavioral change. But adequate scientific evidence of these media forum effects is rare; one exception is Neurath's (1960 and 1962) field experiment with India's radio forums.

Neurath designed his field experiment so that comparisons could be made in knowledge increase among peasants who lived in three types of villages: (1) those in which radio forums were established; (2) those in which radios were already present, but no forums were organized; and (3) those with neither a radio nor a forum. Forum villages had much greater gain in knowledge of innovations,* than did the control villages. In fact, the nonforum villages with a radio showed only very slight gains in knowledge level. These results suggest Generalization 8-5: *The effects of mass media channels, especially among peasants in less developed countries, are greater when these media are coupled with interpersonal communication channels in media forums.***

Why do the media have greater effects on individuals when they are members of media forums? Some of the reasons follow:

1 Interest in attendance and participation is encouraged by group pressure and social expectations.
2 Attitude change appears to be more readily achieved when individuals are in groups. Further, group decisions are more likely to be accepted by the individual if he participates in making the decision, as usually occurs in the media forums.
3 The novelty effect of new channels of information and the subsequent high credibility that may be attached to these media (both electronic and interpersonal) may account for some of the success of the media forums.***

Implications for Change Agents

In the foregoing section we have seen that media forums offer several advantages over purely interpersonal or mass media communication campaigns. By incorporating the advantages of each type of channel into a single propelling force, the change agents can reach a greater percentage of their clients and better persuade them to utilize new ideas.

We should not forget that even very traditional societies had mass-like interpersonal communication systems. An example comes from the state of Orissa, one of the most traditional and least developed parts of India. For

---

*To measure knowledge, each respondent was queried about six topics that were included in the forum broadcasts, both before and after twenty forum broadcasts.
**Evidence for this generalization is also provided by Menefee and Menefee (1967), who found much greater effects (on political knowledge) of a community weekly newspaper in Indian villages when the newspapers were read and discussed in forums.
***For instance, Rhoads and Piper (1963) found that new literates in *radiophonics* schools in Honduras and El Salvador said they believed everything they read.

centuries Orissa has been the site of wandering story singers called *cakulia pandas,* who play an important role in introducing news and ideas to villagers. These storytellers walk from village to village with bells tied to their thighs to announce their coming. Then they promenade the main street of a village, singing recent news to guitar or accordion accompaniment. The story-singers are literate and generally well read. In addition their cosmopolite travel brings them into contact with ideas external to the village. Further, the story-singers meet in occasional assemblies to exchange ideas and to agree on the regions that they will cover.

The story-singers are highly credible channels for villagers; the content of their messages represents a combination of religion, traditional stories, and modern technology. One storyteller was heard to sing about a new fertilizer, which he advocated to villagers, and also about the number of tons of wheat being imported to India. Nevertheless, the story-singer's messages are largely limited to relatively simple facts. As Schramm (1963, p. 38) noted, "The mass media are necessary . . . because the traveler and ballad singer come too seldom and know too little." In fact, in Orissa today the number and importance of story-singers is decreasing, because their role is being usurped by the transistor radio and the tabloid.

Another example of the use of traditional mass media, capitalizing on its connections with interpersonal channels is in the family planning campaign in India. An elephant is being used to create awareness-knowledge of family planning. He walks from village to village with family planning placards on his side, causing considerable attention. Further examples of traditional mass media are the amateur dramatic and song-and-dance troupes which travel within many less developed countries.

A point which is clear from the examples we have just described is that combining mass media and interpersonal channels reduces the effect of selective exposure. Interpersonal communication functions to multiply and increase the effect of the mass media messages, and the media forums serve to heighten the impact of change-oriented messages by reducing the possibility of selective exposure and selective perception (Hiniker, 1968).

The idea behind the use of elephants or storytellers in disseminating new ideas is similar to the radio forums, except that the latter are more formal, more frequent, and generally have more information to disseminate. The unique feature of study groups, forums, and radio or television schools is that they involve a greater amount of intervention in the normal diffusion process than change agents have undertaken in the past. The limited research evidence we have suggests this is an efficient strategy for change agents.

Needed Research

Most of the channel studies reviewed in the present chapter are based on the recall ability of the respondents.* Methods of gathering data about channel behavior of respondents is much in need of improvement.

Considering the channel variable alone is insufficient to explain behavior of a receiver in the innovation-decision process. Various message strategies may increase or decrease the effectiveness of a channel throughout the innovation-decision process. For example, a message which discusses both the pros and cons of an issue is more effective in the long run than a one-sided presentation (1) when the audience initially disagrees with the source's position, or (2) when the receivers are exposed to subsequent counterpropaganda. To date, the message content carried on diffusion channels has been totally ignored by researchers.

No attempt has been made in this chapter to describe which communication channels to use on the basis of characteristics of the innovation. We suspect that some channels (for example, mass media) may prove more effective for simple innovations than for complex innovations, but further evidence is necessary to support this proposition.

Little work has yet been done on the effect of channel credibility in the diffusion process. An understanding of which channels have relatively higher credibility could be extremely useful to change agents in selecting diffusion channels.

Research is needed regarding media forums to ascertain (1) whether the knowledge gains found by Neurath (1960) in the India radio forums were partly a function of the short-term novelty effect of the forums or whether they are indeed widely reproducible, (2) how important the make-up of the forum membership is in terms of its homogeneity or heterogeneity of age, literacy, social status, and so on, (3) how important feedback from the forums to the mass media institution that produces the messages is, and (4) whether or not knowledge gains that occur in media forums lead to actual adoption.

*Katz and others (1963) are highly critical of past research on communication channels: "If there is any single thing wrong with contemporary studies of diffusion . . . [it] is that there is too much emphasis on channels. The typical design for research . . . has been based, almost exclusively, on the assumption that people can be asked to recall the channels of information and influence that went into the making of their decisions to adopt an innovation."

Summary

Communication channels are the means by which messages travel from a source to a receiver. Channels provide the vehicle for getting communication from one person or institution to another. Researchers categorize communication channels as either interpersonal or mass media in nature, and as originating from either localite or cosmopolite sources.

Mass media channels are all those means of transmitting messages that involve a mass medium, such as radio, television, film, newspapers, and the like, which enable a source of one or a few individuals to reach an audience of many. *Interpersonal channels* are those that involve a face-to-face exchange between two or more individuals.

Mass media channels are relatively more important at the knowledge function, and interpersonal channels are relatively more important at the persuasion function in the innovation-decision process. Cosmopolite channels are relatively more important at the knowledge function, and localite channels are relatively more important at the persuasion function in the innovation-decision process.

Mass media channels are relatively more important than interpersonal channels for earlier adopters than for later adopters. Cosmopolite channels are relatively more important for earlier adopters than for later adopters.

Media forums are organized, small groups of individuals who meet regularly to receive a mass media program and to discuss its contents. They are a method of combining mass media and interpersonal channels to maximize communication effects. Media forums have been used in India, Nigeria, Colombia, Brazil, France, China, and other countries. The effects of mass media channels, especially among peasants in less developed countries, are greater when these media are coupled with interpersonal communication channels, as in the media forums. Media forums appear to have a greater effect because they exert social pressure on attendance and participation and on attitude change in small groups, because of a high-credibility "novelty" effect from the media, and because feedback to the broadcaster is comparatively immediate.

9   *Collective*

Ideas confine a man to certain social groups and social groups confine a man to certain ideas. Many ideas are more easily changed by aiming at a group than by aiming at an individual.

JOSEPHINE KLEIN (1961, p. 119)

Innovation-Decisions[*]

$-\ \cdot\cdot\ \text{—}\text{—}\text{—}\ \text{—}\cdot\quad\text{—}\cdot\cdot\ \cdot\ \text{—}\cdot\text{—}\cdot\ \cdot\cdot\ \cdot\cdot\cdot\ \cdot\cdot\ \text{—}\text{—}\text{—}\ \text{—}\cdot\ \cdot\cdot\cdot\qquad\text{—}\cdot\text{—}\cdot\ \text{—}\text{—}\text{—}\ \cdot\text{—}\cdot\cdot\ \cdot\text{—}$

ALMOST all past diffusion research has been based upon the implicit assumption of optional decision-making by individuals, rather than collective decision-making *within* (as opposed to *between*) social systems. What is different about the diffusion and adoption of innovations when the decision is made by a collectivity rather than by an individual? We shall try to find out in this chapter.

Types of Innovation-Decisions and Rate of Adoption

Let us briefly review our earlier discussion (in Chapter 1) of types of innovation-decisions.[**] They are:

I Authority decisions, which are forced upon an individual by someone in a superordinate power position.

---

[*]This chapter was written with the assistance of Graham Kerr, Assistant Professor of Sociology, State University of New York at Buffalo.
[**]These types of innovation-decisions are presented as discrete types, but in reality they undoubtedly fall along a continuum.

II Individual decisions, in which the individual has influence.
1 Optional decisions, made by an individual regardless of the decisions of other members of the social system.
2 Collective decisions, made by the individuals in a social system by consensus.

In addition, there are *contingent innovation-decisions*, choices to adopt or reject which can be made only after a prior innovation-decision. Thus, a decision to adopt a new teaching method can be made by a faculty member only after its prior sanction by the entire school faculty. This illustrates an optional decision that follows a collective decision, but other sequential combinations of two or more of the three types of innovation-decisions can also constitute a contingent decision.

Collective decisions are the main focus of the present chapter. We shall look specifically at past research on community power structure and decision-making studies and at small group investigations of member participation, to arrive at a series of generalizations and hypotheses about the way in which collective innovation-decisions are made.

By ignoring within-system decision-making, past inquiry distorted the reality of how many innovation-choices are actually made. And we have not studied change in a structural context that could likely contribute to theoretical understandings. The present essay, as well as the following chapter which concentrates on authority innovation-decisions in organizations, are designed to help correct this shortcoming.

It is well to remember that none of the community power or small group participation investigations were conducted, at least explicitly, in a diffusion research framework,* and hence, we must apply them on a *post hoc* basis to suit our concepts and propositions. A priority step for diffusion researchers is to design future analyses of collective innovation-decisions more adequately to test the tentative generalizations suggested in this chapter.**

*However, many of these investigations, especially the community power studies, were designed to study decisions about new ideas as they were initiated and carried out in communities. Nevertheless, these studies did not place much emphasis upon the time dimension through which these ideas were accepted.
**Since this chapter was first drafted, several of its central principles have been tested by Rahim (1968) in his investigation of collective and optional innovation-decisions in eighty villages in East Pakistan. The collective innovations he studied included credit and mechanization cooperatives. Rahim (1968, p. 116) found: "Collective innovativeness is something different than just the sum of individual innovativeness of the members of the collectivity." A measure of village (collective) innovativeness was not related to the average of individual innovativeness scores in each village, nor were these two innovativeness variables related to the same independent variables.

The more individuals involved in an innovation-decision, the slower it will proceed. When information about a new idea must be communicated to a larger number of individuals, there is greater opportunity for message distortion, more room for differential perceptions of an identical stimulus, and a greater likelihood that consensus will be reached more slowly. Each individual brings to a joint discussion his own storehouse of opinions and beliefs, and these color his attitudes toward the innovation in a way different from that of his peers.

We expect that authority innovation-decisions made by a superior within a social system will proceed more quickly than if a multitude of optional innovation-decisions were made by many individuals within that system. Likewise, an optional innovation-decision should occur more rapidly than a contingent decision which rests upon prior approval of the innovation by authority or by the consensus of members of the social system. A collective innovation-decision, in contrast, requires the involvement of all members of the social system (or at least the leaders) and approval of the innovation by a majority of the system's members. The collective decision is thus likely to be slower than even the contingent decision.

Evidence of the relative slowness of collective decisions is provided by the following experience: The Pakistan Ministry of Health feared that husbands' antagonism to family planning innovations would have to be won over before adoption rates could approach widespread use. Accordingly, the Ministry instituted a form indicating desire for an intrauterine device, the loop, which had to be signed by *both* husband and wife. In essence this changed an optional innovation-decision by the wife alone to a collective innovation-decision involving both family members. However, the rate of adoption of the loop dropped to almost zero once the form was initiated, and it had to be discontinued. Health officials thus learned that optional adoption usually proceeds more rapidly than collective decisions. However, one consequence of collective decision-making is increased stability of change, once effected, for a decision made by a group can often be changed (discontinued) only by that group.

The collective decision involves the greatest number of participants in the social system, and the authority decision involves the least. In some respects the collective decision is the most "sociological" type of innovation-decision in that it largely rests on the nature of social relationships and on the structure of the social system. We now look at what is known about collective innovation-decisions by synthesizing past research on community power and decision-making.

Community Power and Decision-Making

Social scientists have concentrated on the concept of power since Plato's speculations about philosopher-kings in his utopian state. But the work that set off a flurry of empirical investigations into the nature of community power structure was undoubtedly Floyd Hunter's (1953) study of "Regional City." Hunter, a sociologist, utilized the "reputational method" to identify a small, closely knit coterie of business executives who dominated the important community decisions of the city, such as the decision to construct a Negro swimming pool. Hunter's approach consisted of first asking knowledgeable citizens to indicate the names of the main power holders in the city. Then, he interviewed the power leaders to determine their characteristics, the web of social relationships among them, and their roles in various community decisions. He found that occasionally the power holders played behind-the-scenes roles in the collective decision-making process by operating through lieutenants or by informally guiding figurehead decision-makers.

Hunter's investigation immediately set off a barrage of criticism of his methods and results, largely from political scientists led by Robert Dahl of Yale University.* The opponents of Hunter's approach utilized a decision-making method of investigation in which collective decisions were analyzed retrospectively by means of a case study approach. Dahl's (1961) inquiry into the nature of power in New Haven, Connecticut, illustrates this decision-making approach. In this community Dahl found a pluralistic power structure in which different leaders played different roles for different community decisions. Leadership was highly specialized; only 3 percent of New Haven's 1,029 leaders were engaged in more than one of the three major community issues (urban redevelopment, political nominations, and public education) studied by Dahl (1961, p. 175). "A leader in one issue-area is not likely to be influential in another. If he is, he is probably a public official and most likely the mayor" (Dahl, 1961, p. 183). This point also illustrates another finding by Dahl and his disciples which contrasts with the results by Hunter and his students. The former report the importance of elected or appointed political

---

*Dahl (1961, p. 185) said about Hunter's concept of a monolithic power structure: "Although careful analysis has shown that the conclusions . . . [of Hunter and his followers] often rest upon dubious evidence and even that some of the data found in the works themselves actually run counter to the conclusions, some communities do seem to have conformed to this pattern in the past and some may today."

leaders in community power analyses, whereas the latter stress the dominance of captains of industry, a type of economic elite.

We see, then, that two different methods have been utilized by sociologists and political scientists in investigating community power structure. The *reputational* method asks informants to identify the most influential people in the community. In this type of survey-sociometric approach, the power-holders may be identified directly by the respondents, or a two-step nomination procedure may be utilized in which respondents name a panel of leaders who in turn are interviewed to determine the top power-holders. The *decision-making* method consists of historical reconstructions of community decisions with data obtained from records or by interviews with active participants in the decisions. The power-holders are those who played important roles in the community decisions.

Thus, the second method focuses primarily upon the process of community decision-making and somewhat secondarily upon the different roles played by leaders in this process. It is a type of historical, reconstructed case study approach that assumes leadership is situational. In contrast, the reputational method explicitly ignores the process of community decision-making, and focuses instead upon the leaders, assuming that community power-holding is a generalized ability somewhat independent of specific issues or decisions made.

After reviewing thirty-three studies of community power dealing with fifty-five communities, Walton (1966) concluded that sociologists more frequently employ the reputational technique whereas political scientists prefer the decision-making method. Thus, the discipline of the investigator has largely determined the method of investigation used. The method of study in turn affects the nature of results from the investigations. Use of the reputational method (mainly by sociologists) tends to result in findings of a pyramidal* power structure; use of the decision-making method is associated with finding a factional, coalitional, or amorphous power structure.** A comparison of the reputational versus the decision-making method of investigating community power structure is depicted in Table 9-1.

The polemical argument between the reputational and the decision-making

---

*A *pyramidal* power structure consists of a monopolistic, monolithic, or single cohesive leadership group in the community (Walton, 1966).

**A *factional* power structure is composed of two or more competing leadership groups in the community. A *coalitional* power structure consists of fluid leadership groups in the community that vary with issues. An *amorphous* power structure is the absence of any persistent leadership group in the community (Walton, 1966).

**Table 9-1   Paradigm Summarizing the Two Primary Methods of
Investigating Community Power Structure**

METHOD OF INVESTIGATION	INDIVIDUAL CHIEFLY CREDITED WITH FOUNDING THE METHOD	PREDOMINANT DISCIPLINE OF INVESTIGATOR	TYPE OF POWER STRUCTURE USUALLY FOUND	ASSUMPTIONS ABOUT COMMUNITY POWER-HOLDING
1 Reputational (survey sociometry)	Floyd Hunter	Sociology	Pyramidal (monolithic)	Community power is a generalized ability
2 Decision-making (reconstructed case study)	Robert Dahl	Political Science	Factional, Coalitional, or Amorphous (pluralistic)	Community power is situational to the decision at issue

schools has partially abated in recent years with the advent of community power studies that utilize both methods of investigation (Kellstedt, 1965). An example is Presthus' (1964) analysis of two small communities in New York State. He demonstrated convincing advantages of using both the reputational and the decision-making methods, and he concluded that their joint use was superior to either alone. Similar conclusions are evident from a series of power studies using both approaches in four U.S. cities (Agger and others, 1964) and an investigation in two Missouri communities (Watson and Lionberger, 1967).

In testing the generalizations about collective innovation-decisions that follow in the present chapter, the decision-making approach of community power studies will be more useful than the reputational method. This is because our central interest from a diffusion research viewpoint lies in the analysis of the decision-making process and the roles played by different individuals as this process unfolds. We are less concerned with whom the general power holders are: Our focus is more upon the innovation and the way it is collectively adopted and less upon the leaders who make the decisions.

Although these two methods of study have been utilized to research only community power and decision-making, it is possible that they can be broadened to any particular type of social system in which collective decisions are made. Similarly it seems reasonable to extend certain of the findings from community studies as hypotheses or tentative generalizations about collective innovation-decisions in almost any type of social system, such as a bureaucracy, small group, or family. The issues that have been analyzed in the community studies are all actually innovations: New swimming pools, recreation centers, or hospitals; creation of new formal organizations; urban redevelopment programs; fund-raising campaigns such as Community Chest, and the like. All are ideas perceived as new by individuals. All are collective decisions in that the individuals in the social system agree by some sort of consensus to adopt or reject the innovation.

Stages and Roles in Collective Decision-Making

It is obvious that collective innovation-decisions are considerably more complex than optional decisions. One reason is because the collective decision process is really composed of a multitude of individual decisions: To initiate a new idea in a social system, to adapt the new proposal to local conditions, to sanction the idea, to support the innovation, and so on. Each of these different behaviors may be carried out by different individuals in the collectivity. In the case of optional innovation-decisions all these different activities occur within the mind of a single individual and culminate in the adoption or rejection of the innovation.

Figure 9-1 presents a simplified paradigm of the collective innovation decision-making process and is a synthesis of several conceptions by different researchers.* This process is generally viewed as a series of steps, stages, or subprocesses. The steps are not necessarily mutually exclusive, nor do they always occur in the exact chronological order depicted in Figure 9-1. Nevertheless, this oversimplified paradigm is useful because it provides a general framework for analyzing collective innovation-decisions.

One can see a close parallel between the steps in collective innovation-decision-making and the functions that were described in Chapter 3 in the individual innovation-decision process.**

*The most direct basis for our present model is Beal (1957, pp. 17–22) and Beal and others (1964). Agger and others (1964, p. 40) conceive of a series of six stages in the collective decision-making process; their stages bear striking resemblance to ours, especially to stimulation, initiation, and legitimation, as do the three stages that Bales (1950) described in the group problem-solving process.

**However, individual and group decision-making are not the same, as Kelley and Thibaut (1954, p. 738) point out: "Group problem-solving processes are not the same, *but much more*. . . . Thinking done by a group member occurs in a different context from that done by the 'isolated thinker.'"

Collective Innovation-Decision Steps	*Individual Innovation-Decision Functions*
1 Stimulation	Knowledge (especially awareness-knowledge about the innovation)
2 Initiation	Knowledge
3 Legitimation	Persuasion
4 Decision	Decision
5 Action	—

Of course, there is an important difference between the two types of innovation-decisions. In the collective decision process the *social system* is the

---

Figure 9-1
Paradigm of the collective innovation decision-making process.

1. STIMULATION of interest in the need for the new idea (by stimulators)

2. INITIATION of the new idea in the social system (by initiators)

3. LEGITIMATION of the idea (by power-holders or legitimizers)

4. DECISION to act (by members of the social system)

5. ACTION or execution of the new idea

The collective innovation decision-making process is usually conceived as five or more steps or subprocesses from original realization of a need for the new idea (stimulation), to final action or carrying out the new idea in the social system. This conception has mainly evolved from research on community decision-making (see Beal and others, 1964, p. 7), but it should be generally applicable to most other types of social systems, such as bureaucracies, committees, and families.

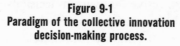
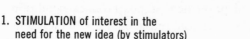
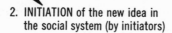
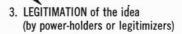

decision-making unit, whereas the *individual* is the unit in the case of individual innovation-decisions.

Stimulation

Stimulation is the subprocess in collective innovation decision-making at which someone becomes aware that a need exists for a certain innovation within a social system. Up to this point in time neither the innovation nor the need that the innovation might fulfill are perceived to be important to members of the social system. This lack of awareness may result because no one of the system's members knows about the innovation, because no individual recognizes the existing problem, or because no one has linked the existing problem to the innovation. The stimulator(s) very often is an outsider to the social system or else is a system member who is oriented externally through social relationships with members of other systems. For example, a new school superintendent enters a community and realizes that the old school building is a fire trap. He plays the role of stimulator if he calls the inadequate school plant to public attention. Stimulators play an important role in relating the social system to the outside world. In this sense they act as one type of gatekeeper* who controls the communication flow by which new ideas of a collective nature enter a social system.

Generalization 9-1: *Stimulators of collective innovation-decisions are more cosmopolite than other members of the social system.* The cosmopoliteness of stimulators may be expressed by wide travel, readership of nonlocal publications, affiliation with national or international organizations, or by membership in professional occupations associated with occasional migration, such as teacher, minister, or salesman. Perhaps the stimulator may not even be a regular member of the social system but only have some external contact with it.

The cosmopoliteness of stimulators provides them with relatively early access to innovations and with a comparative frame of mind which allows them acutely to perceive needs and problems in the social system which are not seen by the system's more localite members. Cosmopoliteness may also lead the stimulator to have a generally favorable attitude toward change, relative to

*Lewin (1943) originally coined the term "gatekeeper" in his studies during World War II to describe the individuals who introduced new foods to their families. The term has since been broadened by communication researchers to include any individual who is located in a communication structure so as to control which messages flow through a communication channel (Katz and Lazarsfeld, 1955, p. 119).

others in the local system.* This desire for new approaches is not balanced by a vested interest in the status quo. Longevity in the system acts to inhibit stimulation activity. For instance, Hage (1963) found that the longer a physician served on a hospital staff, the less likely he was to recognize a need for change.

Must the perceived need for a collective innovation originate inside a social system, or may such perceptions start with a state or national decision (such as for school desegregation), and then result in the innovation-decision being thrust upon the local community? As shown by the decision-making paradigm, it does not matter whether the collective innovation-decision is directed by external sources or whether it arises spontaneously within the system. In most past inquiry stimulation has come directly or indirectly from sources external to the system, but this may simply be due to the fact that all the systems studied were U.S. communities. But wherever perception of the need originates, the collective idea must enter the system through a stimulator before the decision-making process can begin. "Whether someone 'borrows' a policy formulation from outside of the community or whether a policy is of local origin, it must at some time become a preference of a person within the polity in order to become part of a community decision-making process" (Agger and others, 1964, p. 41).

Initiation

Initiation is the subprocess in collective innovation decision-making by which the new idea receives increased attention by members of the social system and is further adapted to the needs of the system. Whereas stimulators perceive a need or problem in the system and suggest a new idea that might help solve the problem, initiators incorporate the innovation into a specific plan of action that is adapted to the conditions of the social system. This role involves intimate knowledge of the social system, including the ability to predict certain consequences of the new idea, once it is adopted. Thus, whereas the stimulator is an "outside man" with far-ranging contacts, the initiator is an insider, a localite, whose forte is that he knows the system. The stimulator's expertise lies in knowing the innovation; he is message-oriented. The initiator, in contrast, is receiver-oriented.

Initiation may result from the activities of more than one individual, and so we sometimes speak of an "initiating set." In an investigation of local

*But the cosmopolite may be relatively more traditional than members of other systems to which he is oriented.

decisions to construct new hospitals in 218 U.S. communities, one person initiated the action in only about one-third (32 percent) of the cases (Miller, 1953, p. 185). About 90 percent of the vocational school bond issues in Iowa were initiated by sets of school officials (Beal and others, 1967c).

There must, of course, be at least some communication between the stimulators and initiators. A community study in Costa Rica (Edwards, 1963) showed that such influential persons as stimulators and initiators talked with other influentials more than they talked with non-influential persons. The top influentials in San José, Costa Rica, were the persons who had the highest degree of communication with other leaders (Edwards, 1963, p. 113). In some cases the initiators in San José acted as stimulators at the previous stage in the collective innovation-decision process, and this may be a frequent occurrence elsewhere.

A nice illustration of the intricate relationship between stimulator and initiator is provided by Dahl's (1961, p. 128) reconstruction of how urban redevelopment ideas occurred in New Haven, Connecticut. The cosmopolite stimulator was "an imaginative Belgian, a professional city planner, who spun off ideas as a pin wheel throws off sparks. And, like sparks, his ideas often vanished into darkness. But his presence in New Haven, where he headed a firm of city planning consultants with a world-wide clientele, insured that his ideas would be heard. In a few places, the sparks fell on tinder, smoldered, and finally burst into flame." The man who helped this stimulator's innovations progress from idea into reality was the administrator of New Haven's urban redevelopment program. This initiator assessed the costs and potential gains of each proposal, and after judging the few that seemed worthwhile, passed them along to the city's mayor for approval. We see that in New Haven, the chief stimulator was not an initiator, although the two were in close contact.

A division of labor is even more likely between the initiators and the legitimizers, those who sanction the idea for the social system. In the collective decision process the functions performed by initiators and by legitimizers are quite different, as are their social characteristics. For instance, legitimizers usually are found to have high social status in the system, whereas the initiators are more often noted for their highly favorable attitude toward change and for their intimate knowledge of the system.*

Generalization 9-2: *Initiators of collective innovation-decisions in a social system are unlikely to be the same individuals as the legitimizers.* To return again to the case of New Haven, Dahl (1961, p. 128) found that the initiator, who was the chief

---
*Obviously, the initiators must also possess social accessibility to the legitimizers.

administrator of the urban redevelopment program, passed the innovations he judged to be worthwhile along to the mayor, who was the city's chief legitimizer.* Agger and others (1964, p. 47) postulated that "participants in one stage [of collective decision-making] may or may not be the same as participants in other stages." Their findings from four communities supported this notion.

Most of what we know about collective innovation decision-making comes from investigations in communities, rather than in complex organizations. However, Chandler (1962) studied the diffusion of the idea of decentralized organization within major U.S. industrial firms. "At DuPont, General Motors, and Jersey Standard, the initial awareness of the structural inadequacies caused by the new complexity [that is, the need for decentralization] came from executives close to top management, but who were not themselves in a position to make organizational changes [that is, they were *initiators,* not *legitimizers*]. In all three, the president gave no encouragement to the proposers of change. . . . In all three companies, it took a sizable crisis to bring action. Yet all three presidents had received proposals for reorganization before that crisis made their usefulness apparent."

Thus, we see that the initiators are usually different individuals from the legitimizers.

Legitimation

Legitimation is the subprocess in collective innovation decision-making at which a collective innovation is approved or sanctioned by those who informally represent the social system in its norms and values and in the social power they possess. Although the role of the legitimizer is mainly that of screening new ideas for approval, he may often alter or modify the proposals put to him by the initiators. However, seldom will legitimizers actively promote an idea for collective approval after giving their own approval. They generally play a more passive role in the collective decision-making process. Legitimizers thus give sanction, justification, the license to act.

*Throughout the present chapter we use the terms legitimizer and power-holder almost synonymously, although perhaps there is a shade of difference (largely resulting from the selection of units of analysis and methodological stances by the reputational and the decision-making schools). The legitimizer functions in a sanctioning role in specific collective decisions, whereas the power holder supposedly exerts his power across a broader range of collective ideas.

Generalization 9-3: *Rate of adoption of a collective innovation is positively related to the degree to which the social system's legitimizers are involved in the decision-making process.* It may be possible for the initiators to proceed successfully without consulting the legitimizers in a social system, but this decreases the chances of securing adoption of the collective innovation. Usually the legitimizers can kill an idea if they are not consulted. Initiators may circumvent the legitimization stage in the process for three reasons: (1) because they do not know who the legitimizers are for the idea they are initiating; or (2) perhaps because the initiators know who the legitimizers are but lack social access to these power leaders;* or (3) because the initiators wish to save time. For whatever reason, when legitimizers are ignored, they are more likely than not to scuttle the collective innovation.

Rosenthal and Crain (1965) analyzed the role of mayors, one type of formal legitimizers, in the fluoridation decisions of 362 U.S. cities. They found mayors to be of great significance. Adoption occurred in 61 percent of the cases where the mayor favored fluoridation. When he was opposed or neutral, adoption resulted in only 6 percent of the cities.

Pluralism of Legitimizers

To what extent is legitimizing ability a general versus a specialized ability? In other words are the legitimizers for a new school building in a community the same individuals who sanction an urban renewal program? The answer seems to depend in part on the researcher making the study. We pointed out earlier that Dahl (1961, p. 175) found little overlap among the New Haven power-holders for politics, education, and urban redevelopment. However, such a pluralistic power structure was not reported by the sociological followers of Floyd Hunter, who generally found one tight-knit clique of all-purpose power-holders in the communities they studied. Undoubtedly, the pluralism** of legitimizers varies with such variables as (1) their breadth of experience, personality, and interests; (2) the size (smaller communities, for example, should be less pluralistic***) and social structure of the system; and (3) the nature of the collective innovations studied (for instance, pluralistic legitimizers are more likely to be found in connection with two closely

*Or else the initiators know that the legitimizers are opposed to the innovation.

**Pluralism* of legitimizers is the tendency to exert influence only on one rather narrowly defined issue or topic; thus there is a set of legitimizers for each issue or topic in a system. One can see a parallel to the case of monomorphic opinion leadership (Chapter 6).

***Evidence that legitimizers are more pluralistic in larger communities is provided by Clark (1968) in his study of fifty-one U.S. communities.

related innovations, such as new schools and new swimming pools, than for a pair of quite different collective ideas).

Social Status

Generalization 9-4: *Legitimizers of collective innovation-decisions possess higher social status than other members of the social system.* Dahl (1961, p. 169) found that power leaders in New Haven were typified by higher income, education, and other indicators of social status than their fellow citizens. A national investigation of community decisions to adopt new hospitals indicated that those in professional occupations were ninety times more likely to be legitimizers than were manual laborers (Miller, 1953, p. 22). In the four communities they analyzed, Agger and others (1964, p. 279) found that the highly educated and the middle- and upper-income citizens were over-represented among the legitimizers. In San José, Costa Rica, the power holders were older and better educated (Edwards, 1963, p. 112).

Along with their high social status, legitimizers usually possess considerable social resources such as wealth, formal position, influence over others, and knowledge. They frequently hold high, informal positions in powerful friendship cliques. Their high reputation in the system lends credibility to their decisions, an essential quality if their choices are to be accepted by the members of the social system. However measured or expressed, the legitimizer has a virtual monopoly on two of the scarce resources of any social system: Status and power.

Concentration of Power

A general finding from past investigations on community power is that power is often highly concentrated in the hands of a few persons. *Power* is defined as the degree to which an individual has the capacity to influence the beliefs, decisions, and actions of others. Following are some reported results from inquiries into the concentration of community power.*

1. In a New York rural village, "Springdale," important collective decisions were made by a political machine of four power leaders who worked

Power concentration is the degree to which one or more units of a social system possess power in greater amounts than other units in that system.

behind the scenes to control the affairs of the community (Vidich and Bensman, 1958). Yet the residents of Springdale claimed that one advantage of their village life over city life was equality and neighborliness.

2. In Atlanta, a city of half a million population at the time of the study, forty key power-holders were influential in important community decisions, such as expanding the city limits or starting a community fund (Hunter, 1953). These first-rate power holders were supported by second-rate lieutenants who helped carry out the legitimizers' decisions.

3. Dahl (1961, p. 115), who studied the community decision for urban redevelopment in New Haven, concluded: "In origins, conception, and execution, it is not too much to say that urban redevelopment has been the direct product of a small handful of leaders." Furthermore, he found that only three to six leaders (about one-twentieth of 1 percent of the city's registered voters) controlled the political nominations in each of the two major political parties of the city (Dahl, 1961, p. 106).

Public Visibility of Power

Legitimizers must possess informal power in the system; they *may* also have formal positions of high authority, but this is not necessary. It was pointed out earlier that the studies by Hunter's reputational school generally find that legitimizers are informal, behind-the-scenes manipulators of community power. In contrast Dahl (1961) found that the key power holder in New Haven was the chief executive of the city, the mayor. Whether the legitimizers are formal as well as informal power wielders or not, their decision-making activities are usually private rather than public affairs. In fact, stimulation and initiation, as well as legitimation, commonly occur in smoke-filled rooms. The public often becomes informed about the collective innovation (and involved in it) only at the fourth and fifth stages in the decision-making model, when the decision to act is made by the members of the system and when the decision is carried into action. This does not mean that the public is simply a rubber stamp for the legitimizers; the polity can countervail against the power-holders. However, this is unlikely because the legitimizers hold their influential position in the system because they represent the system's norms and values. Their decisions must be made largely for the system's benefit, rather than for their own private gain. A legitimizer holds his leadership position only so long as he is responsive to the wishes of his followers, or so long as his followers perceive him to be responsive.

The legitimizers, then, act as a type of normative screen for the social system, keeping out collective ideas that they feel would not benefit the system. Innovations are also selectively screened out at the stimulation and initiation stages, as implied earlier. The stages in the decision-making process are like a series of sieves; at each stage many ideas are discarded. Only a small portion of the hopeful candidates for approval at the stimulation stage survive the decision process and are put into action.

Power Concentration and Rate of Adoption

We concluded previously that power is usually concentrated in a few hands, at least in the communities that have been investigated by social scientists analyzing power and decision-making. But it is important to remember that the exact degree to which power is concentrated varies widely from system to system. Where power is more concentrated (and the power leaders are not opposed to change), we expect that collective innovations will be adopted more rapidly because fewer individuals are involved in the decision-making process.* When power is more widely distributed, a longer period of time is necessary for adoption because more individuals must be informed, persuaded, and convinced of the innovation's merits.

Generalization 9-5: *The rate of adoption of collective innovations is positively related to the degree of power concentration*** in a system.**** Past research provides some support for this proposition. Hawley (1962) tested the general notion that "the greater the concentration of power in a community, the greater the probability of success in any collective action affecting the welfare of the whole," with data from ninety-five U.S. cities. He found that a city's likelihood of adopting urban renewal programs was greater if power were concentrated in fewer hands. Similarly, Gamson (1968) and Rosenthal and Crain (1968) provide evidence from analyses of collective adoption of fluoridation by U.S.

*In fact when power is extremely concentrated, as in the hands of only one person, a collective innovation-decision becomes an authority decision.
**An appropriate index of power concentration is the Gini ratio, which expresses the degree of inequality of posesssion of any specific characteristic by members of a social system. A description of the Gini ratio and a formula for its computation are provided by Wunderlich (1958). Crude approximations of the Gini ratio of power were used by Hawley (1962) and Clark (1968).
***This proposition might be stated in other words: "The rate of adoption of collective innovations is negatively related to the degree of pluralism of legitimizers in the social system."

cities that adoption is related to power concentration. However, Clark (1968) reports contradictory data in the case of the adoption of urban renewal programs.*

All these studies were conducted in the U.S. Does our proposition hold when the systems analyzed are peasant villages in less developed nations? The answer seems to be "yes, but weakly." Low but positive correlations between village power concentration and adoption of innovations are reported by Hursh and others (1969) in Nigeria, Fliegel and others (1967b) in India, and Whiting and others (1968) in Brazil.** We generally find support for Generalization 9-5, although further inquiries with large samples of systems are needed to provide more definitive evidence.

A rather considerable amount of time appears to be required for the collective decision-making process to occur in communities, but the exact amount of time undoubtedly varies with the degree of power concentration. For example, one investigation reported that most decisions to secure new hospital facilities required from two to ten years between stimulation and construction of the hospital building (Miller, 1953, p. 20). Obviously, the length of the collective innovation-decision process also depends on the nature of the innovation. But in any event the process is often time-consuming, and it may be difficult for change agents to speed the process.

Participation in Collective Decisions

At the fourth stage in the collective decision-making process, the focus is upon the decision to act by members of the social system. The public's preferences may be expressed in a variety of ways: A survey may be conducted, a referendum may be held on the issue, petitions may be circulated, or a public meeting or hearing may provide the means for expression. Regardless of the mode of expression, it is usually thought to be advantageous to have widespread participation in the choice process by members of the system.

*He explains the discrepancy between his results and those of Hawley (1962), who also studied the adoption of urban renewal programs in the 1950s, by arguing that such collective ideas were no longer innovative in nature by the time his data were gathered in the 1960s.

**However, the dependent variable in the studies in Brazil, Nigeria, and India was the adoption of a mixture of optional and collective innovations in agriculture.

Participation is the degree to which members of a social system are involved in the decision-making process. This concept plays a major role in contemporary organizational research and theory.* In the present section we shall cite evidence from participation studies in industrial organizations, in small groups, and in community decision-making to support our generalizations.

Satisfaction and Participation

Generalization 9-6: *Satisfaction with a collective innovation-decision is positively related to the degree of participation of members of the social system in the decision.*** There is extensive empirical support for this generalization, although recent research suggests certain qualifications. Researchers reporting a direct relationship between satisfaction and participation in collective decision-making include Morse and Reimer (1956) and French and others (1958). Confirmation is also provided by Seashore and Bowers (1963) in their experimental study of an industrial organization. They increased the participation of employees in work decisions, the amount of supportive behavior on the part of supervisors, and the amount of interaction among workers. As a result, employees' satisfaction with their work increased, and there was a rise in productivity of the organization.

Among the supportive-but-with-qualifications studies is that of French and others (1960). They found that participation led to satisfaction only when the system's members felt that their participation in decision-making was legitimate, rather than superficial. Hamblin and others (1961) found satisfaction related to participation only if the members perceived that their leaders were not very competent. Furthermore, we should remember that the individual's *perceived* participation is more important in explaining his satisfaction with collective decisions than is his *objective* participation (that is, his degree of participation as judged by others).***

Why should members of a social system be more satisfied with a collective decision if they feel they are involved in making that decision?

*Especially in the work of Likert (1961), Likert (1967), and Galbraith (1967).
**This generalization could be tested with future data of two sorts: (1) with the individual member as the unit of analysis, and (2) with the social system as the unit of analysis. In the former case our dependent variable would be within-system individual differences in participation; in the latter case we would measure system-to-system differences in levels of member participation.
***This distinction between psychological or perceived participation and objective participation is suggested by Vroom (1960).

1. Through participation in the decision-making process, individual members learn that most others in the system are also willing to go along with the decision. Participation is a means of revealing to the individual the extent of group consensus and commitment. If the individual member knows of group support for the decision, he is more likely to be satisfied with it himself.

2. The decision to accept or to reject is more appropriate to the needs of the system's members if they take part in reaching such a decision. In most cases we would expect a system's members to know their own needs more accurately than would their change agents or their administrators.

3. Widespread participation allows the opinion leaders in the system to assume a major part in making the decision. Hence, the opinion leader's position is reinforced, and the members are induced to abide by the decision and be more satisfied with it.

This same general logic for the relationship of satisfaction to participation should also explain why acceptance of collective decisions is related to participation in decision-making.

Member Acceptance of Collective Decisions

We should not forget that the average member of a social system has an important role to play in collective decisions. After the system's legitimizers have ruled on the innovation, it is up to the members to accept or reject that decision. But this acceptance-rejection decision by members is not strictly analogous to a series of optional decisions because the members' relationships with the social system and its legitimizers and their degree of participation in the decision-making process have much to say about their eventual reaction to the collective idea.

Generalization 9-7: *Member acceptance of collective innovation-decisions is positively related to the degree of participation in the decision by members of the social system.** One of the most oft-cited evidences of this proposition is the

*Somewhat related to this generalization is the finding that higher productivity (usually in an industrial work-output sense) is positively related to member participation. Examples of these studies on productivity and participation are Coch and French (1948), French and others (1958), French and others (1960), Lawrence and Smith (1955), and Seashore and Bowers (1963). More closely related to Generalization 9-7 is the theory that changes in attitudes and behavior are easier to bring about through a group approach rather than by working with individuals separately. Studies by Lewin (1943), Jacques (1948), Coch and French (1948), Levine and Butler (1952), and Giffin and Ehrlich (1963) support this proposition.

investigation by Lewin (1943) on lecture versus group discussion methods in convincing housewives to adopt new food products. In the groups where the respondents were asked to discuss the advantages and disadvantages of the new meats, the women were ten times more likely to prepare the innovations for their families during the following month (Figure 9-2). Of course, other variables besides degree of member participation were involved in the experiment, such as differences in message content, length of the meeting, whether the women were asked to make a public commitment (by raising their hands if they intended to serve the new meats), and the like. Nevertheless, Lewin's

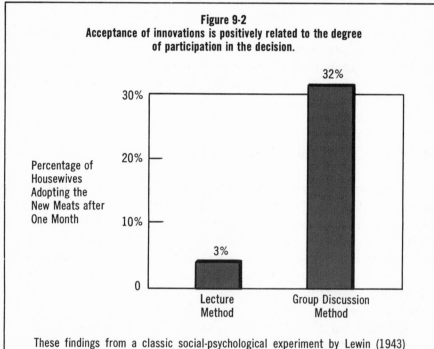

Figure 9-2
Acceptance of innovations is positively related to the degree of participation in the decision.

These findings from a classic social-psychological experiment by Lewin (1943) illustrate that a much higher degree of acceptance of an innovation results when members of an audience are more highly involved in the decision-making process. The respondents were housewives in an Iowa city, and the new idea was serving sweetbreads (a new kind of meat) to their families. This message was presented via two methods: lecture and group discussion (where members were involved in reaching a decision). Unfortunately for our present purposes, the innovation-decisions were optional rather than strictly collective in nature, but Lewin's findings are nonetheless suggestive of the generalization that member acceptance of innovations is positively related to the degree of participation in the decision-making process.

results are suggestive, and his leads have been followed up in later investigations which generally suggest that acceptance of a collective decision is more likely when there is higher member participation in the choice process.

An illustration from this later research* is the Levine and Butler (1952) study of factory foremen who were urged by means of lecture and group discussion to adopt a new method of rating their employees' performance. An attempt was made to control or remove the effects of message content and time differences between the lecture and discussion methods. The discussion groups featuring greater member participation were again much more effective in securing acceptance of the new rating procedures. In fact, the work of Bennett (1952) indicates that it is member participation, resulting in the individual's perceptions of group consensus, that leads to acceptance of collective innovation-decisions, rather than such other variables involved in the original Lewin experiment as group discussion *per se* or public identification of members' intentions.

Finally, a follow-up investigation by two of Lewin's proteges (Radke and Klisurich, 1947) showed that not only is there a greater number of attitude and behavioral changes as a result of member participation in decision-making, but also that these changes are more permanent over time. It seems that there is not only more widespread acceptance of collective decisions as a result of member participation, but also that the degree of commitment to such acceptance is higher.**

Unfortunately for our present purposes, the Lewin (1943), Levine and Butler (1952), Giffin and Ehrlich (1963), and Bennett (1952) inquiries were all more concerned with optional decisions than with collective innovation choices.*** Nevertheless, their results have a high degree of relevance to our generalization. And fortunately, Davis (1965) provides us with supportive evidence about member participation in collective decisions. He studied the adoption of education innovations in two liberal arts colleges—one innovative and one noninnovative. The progressive organization featured a high degree of involvement of faculty in collective innovation decision-making. Such participation built commitment to the total institution and encouraged acceptance of collective decisions. It allowed innovations to be initiated by individuals

*Other examples, with largely confirmatory results, are Giffin and Ehrlich (1963) and Coch and French (1948).
**Jain (1969b) also found commitment to adoption related to participation among Indian peasants in radio forums.
***As were the investigations of Wallach and Kogan (1965) and Wallach and others (1965), which show that participation in decision-making leads particularly to the acceptance of risky ideas.

at all hierarchical levels, rather than only by administrators for top-down adoption by the faculty.

In a somewhat similar study, Queeley and Street (1965) investigated two elementary schools in Chicago—one innovative and one laggardly. They found wider teacher participation in innovation-decisions in the more progressive school system.*

One exception to our generalization about participation and adoption is provided by Crain and others (1969, p. 228), who studied the diffusion of fluoridation among U.S. cities. They claim this exception to the participation principle is because "fluoridation is a technical issue, the advantages of which are rather small from the citizen's point of view, . . . and because the opposition can easily implant doubt. Doubt takes root and blossoms the more the issue is discussed." Nevertheless, the overwhelming weight of evidence is in support of the principle that member participation in collective decisions leads to acceptance of the decisions.

Member Acceptance and Cohesion

Not only does member acceptance of collective decisions vary with participation but also with the degree to which the individual perceives himself to be attached to the group. Generalization 9-8 is that: *Member acceptance of collective innovation-decisions is positively related to member cohesion with the social system.*** *Cohesion* is defined as the degree to which members perceive themselves to be strongly tied to the social system. Group pressures to change beliefs or behavior will be more strongly felt by those most attached to the group.

One study demonstrating this generalization is Kelley and Volkhart's (1952) experiment with Boy Scout troops. The researchers first determined which scouts had highest cohesion with scouting by means of a questionnaire. Then a speaker attacked the value of camping and woodcraft, two central elements in Boy Scout ideology. The scouts who changed their beliefs the

---

*Lin and others (1966, p. 2) somewhat similarly found that public school teachers in Michigan with more favorable attitudes toward educational innovations were more likely to feel they played a meaningful part in school decision-making. Further, Beal and others (1967c, p. 51) found that most Iowa communities adopting vocational education bond issues had a Parent Teachers Association (providing for participation), while none of the rejecting communities had a PTA.

**This generalization, like the one preceding it, could be tested with either the individual member or the social system as the unit of analysis.

most were those with the lowest cohesion or attraction to scouting. The group members who changed most, in a direction away from the group's beliefs, were those who least valued the Boy Scout troop as a reference group.

Eibler (1965) studied five of the most innovative and five of the most laggardly high schools in the Detroit metropolitan area. One characteristic of the innovative schools was a high degree of staff cohesion. Again, unfortunately for our purposes, the innovations were more of an optional type than of a collective variety, but we see some suggestive evidence for our present proposition.

The Role of Change Agents in Collective Decisions

Change agents can perform the role of stimulator and perhaps initiator in the collective decision-making process, but seldom are they legitimizers. As we pointed out previously, the legitimizer is usually characterized by seniority, high status, and established respect in a social system. Rarely do change agents possess these characteristics. They usually are perceived as social strangers temporarily alight in the system, of possibly high technical competence but of low general status and social power, and of relatively low credibility regarding decisions for members of the social system. They simply are not members of the Establishment.

The change agent possesses excellent qualifications, however, to function in stimulating and initiating collective innovations. His widespread social relationships and technical competence in his speciality provide a firm basis for calling new ideas to the attention of the system's leaders. The change agent can offer helpful advice on the nature of the decision-making process itself. For instance, he can help identify legitimizers for a given issue and urge the initiators to utilize them. He can also help his clients more carefully evaluate the social costs of the collective decisions they are considering. Sometimes, for example, communities underestimate the finances, time, energy, and other social resources required to put a collective innovation into action. As a result, members bite off more than they can chew, and the collective idea fails.

Insight into the role of change agents in collective decisions is provided by Rahim's (1961b) investigation of the adoption of diesel water pumps by

Pakistani villagers. These irrigation pumps would enable the villagers to raise an extra crop of rice each year but cost so much that individual farmers could not afford to purchase them. As a result, an entire village typically founded a cooperative which bought the pump and rented the water to its members. Rahim (1961b) gathered data through personal interviews with villagers and their leaders in each of five East Pakistani villages that had adopted the diesel pumps. Essentially, he was able to reconstruct the collective innovation-decision process by which the cooperative was formed and the pump was purchased. He found a close correspondence between the model of the collective decision stages postulated in this chapter and the actual process in the five villages.

The stimulators in each case were extension change agents who called the idea of pump irrigation to the attention of village leaders. In four of the five cases this stimulation took place in informal conversations in the village market place. The change agent was invariably supported by a local village initiator with whom the extension worker had friendly relations. The legitimation stage seemed to be somewhat less clear-cut, but the change agent was not directly involved in this sanctioning decision. The idea of an irrigation pump cooperative was presented to the total village in informal meetings organized by the initiators (Rahim, 1961b, p. 24). "Here each individual was given a chance to express his views in the complex process of achieving a group decision." Following group acceptance of the innovation, the leaders proceeded to purchase the pumping equipment, establish priorities and rates for its use, and so on. At this execution stage the change agent again entered the process to provide technical advice and organizational support to the cooperative leaders.

The Battle at Stevens Point:
*Fluoridation's Waterloo**

Counter-campaigns to new ideas are often launched by subgroups within a social system who oppose a collective innovation. An illustration comes from collective decisions by U.S. cities to adopt fluoridation of water supplies. One might think that introducing a trace of fluoride into the municipal water supply, a measure of proven effectiveness in protecting children's teeth from

*This section is largely an adaptation of McNeill (1957) and is used by permission.

decay, would not be controversial. Yet in Seattle opponents claimed that fluoridation is "mass medication, socialized medicine, rat poison, and Nazi; that it ruins car batteries, radiators, and lawns; that it causes hardening of the arteries and veins, premature aging, loss of memory, and nymphomania" (Davis, 1959). The opponents of fluoridation are well organized and vocal; they have been successful in defeating fluoridation in about half of the approximately 3,000 U.S. cities where public referendums have been held (Crain and others, 1969, p. 4).* We turn now to a detailed discussion of one of the first cities in which a counterattack on fluoridation was launched: Stevens Point, Wisconsin. The so-called "Battle at Stevens Point" signified a turning point in the success of fluoridation campaigns. After 1950 U.S. citizens no longer perceived fluoridation as a matter of public health but as a flaming political issue. The violent conflict which erupted at Stevens Point sounded an ominous note for the future rate of adoption of an innovation which already had received the endorsement of the American Dental Association, representing 80,000 dentists, and the U.S. Public Health Service, a powerful federal change agency.

The fight for fluoridation at Stevens Point began in May, 1949, when the local American Legion Post sponsored a forum on the subject. A Madison dentist, Dr. John Frisch, and an assistant director of the state hygiene laboratory, Starr Nichols, outlined results of their research on fluoridation and urged the community to join many other Wisconsin cities in adopting the idea. An "unofficial" public spokesman rose from the audience to question the two speakers. He asked why they advocated "putting poison" in the water. The critic was Alexander Y. Wallace, age 67, long the "watchdog of the public treasury" at Stevens Point. Townspeople knew him for his strenuous crusades, vivid prose, and recurrent letters to the editors of local, Madison, and Milwaukee newspapers.

Two months after the American Legion forum, the city council voted "no" on a joint request of the county dental society and the city board of health for fluoridation of the municipal water supply. Council members said the idea was still too experimental. Surprised by this action, fluoridation's proponents decided to appeal to the women's clubs for support of their idea. Two housewives, Mrs. Nelson Bell and Mrs. Armin Manske, led the campaign. With the Stevens Point Chapter of the American Association of University Women issuing a call to arms, fifty-four women's groups met in November, 1949, to discuss methods of pressuring the city council into an adoption decision. The

---

*Another example of such counter-campaigns in the U.S. are those conducted by white citizens' councils to oppose school desegregation.

women presented a petition to council members that was signed by 1,000 voters. It urged immediate action on fluoridation.

Wallace appeared at the next council meeting to scoff at the lengthy (thirty-seven feet) petition and hinted that he also might launch a writing campaign. At the end of his arguments the city council rejected Wallace's demand for a referendum on the issue and voted to purchase fluoridation equipment. The next day Wallace wrote a letter to the local newspaper editor acknowledging that he had taken a "bad beating." He termed the fight a "clash of opinions," thanked the women for a "nice clean campaign," and concluded by bemoaning the fact that he was "getting too old to organize for a fight."

This letter motivated two local men, Ben LaHaye, railroad repairman, and Albert Skalski, a retailer of sewer pipe, to rally to Wallace's side. Their strategy was to get the issue on the ballot of the next city election. This required the signatures of 1,000 voters; by May, 1950, the trio had collected 1,318. The city council then had two choices: Accept the petition and withdraw its approval of fluoridation, or place the question on the ballot. The council chose the referendum but continued plans to put fluoride in the city water supply.

Battle lines were drawn more sharply when on May 25, 1950, Stevens Point began fluoridation of all drinking water. Officials made no public announcement of the event for nearly a week. Later in the campaign Wallace used this effectively as evidence that council members had tried to fool the public.

At this juncture the three opponents of fluoridation began a massive letter-writing campaign answering all arguments of the "fluoriders" and accumulating material to discredit the claims of proponents. Wallace sent a form letter to the dean of every U.S. dental school asking if fluoride was safe to drink and if the treatment was out of the experimental stage. The opinions of certain cautious deans had a powerful influence in deciding the referendum. The *Stevens Point Journal* attempted to point out that Wallace refused to use the replies of deans who favored fluoride, but this was a weak plea in the face of the opposing "evidence."

While the champions of fluoride made charts showing statistical trends, Wallace and his lieutenants had a broadside printed with inch-high letters reading "Poison! Poison! Posion!" serving as a border for the phrase: "Get the Poison Out of Our Drinking Water." They also staged mass meetings in the public square broadcasting anti-fluoride parodies on "Good Night, Irene" and "The Old Oaken Bucket."

This burst of activity convinced proponents of fluoride that they should ignore the trio in the belief that Wallace and his group would become the laughing stock of the community with their oversimplified and low brow appeal.

On September 19, 1950, Stevens Point voters went to the polls to decide one of the most bitter struggles in the city's history. They voiced a resounding rebuttal, 3,705 to 2,166, to the city council's decision to fluoridate the municipal water supply.

Using the Stevens Point campaign as a guide, dissenters across the United States began rallying an opposition network to the spread of fluoridation. The U.S. Public Health Service estimates that since 1950, 80 to 90 percent of all fluoridation referenda have failed in U.S. communities. And meanwhile, the teeth of thousands of U.S. children needlessly decay.

What does this case illustration of fluoridation's rejection at Stevens Point tell us about the nature of collective decision-making? It illustrates the cosmopolite nature of the idea's *stimulators*, who traveled to the community from the state capital. The two housewives played the role of *initiators* by calling the innovation to the attention of the *legitimizing* city council. In this particular example there arose an anti-innovation set of power leaders who were successful in securing rejection of the idea.

Needed Research

Generalizations presented in the present chapter are based on very limited evidence. Many of them are little more than untested hypotheses. As yet, there has been almost no research explicitly addressed to exploring theoretical hypotheses about collective innovation-decisions. The data we have used were not gathered in accordance with the framework spelled out in this chapter. To this extent the entire chapter is an essay on needed research on collective decision-making.

Our methodology, especially our research design, certainly needs major overhauling. We know a great deal about power and decision-making in each of thirty-five to fifty U.S. communities that have been studied in depth.* In

*In one of the very few investigations conducted outside of the U.S., Edwards (1963) found largely confirmatory support for conclusions about community decision-making in the U.S.

fact, we know too much about the idiosyncrasies of each of these case studies and almost nothing about what holds true across communities. We need to know more about less (but a more relevant "less") in *each* system and from an adequate sample of such systems. In the case of community studies our primary problem is one of the limited generalizability of results.

The same shortcoming faces us at another level. The sparse research to date on collective decisions has been conducted mostly in communities. To what extent are these tentative results generalizable across different types of social systems, such as industrial plants, government bureaucracies, and small groups such as the family? We need to find out. The handful of extant studies about small groups, such as those of the Lewin tradition, suffers (for our purposes) from the fact that the decisions were mainly optional choices, rather than collective choices.

Summary

This chapter focused upon collective innovative-decisions where the members of a social system agree to adopt or reject a new idea by consensus. Unfortunately, diffusion researchers have largely neglected this realm of inquiry, and therefore we necessarily drew upon community power studies in sociology and political science and upon small group investigations in social psychology to support eight propositions which we formulated.

Sociologists in the Hunter tradition, using the reputational method of determining community power-holders, found a small, monolithic group of decision-makers within the community. Political scientists, following the lead of Dahl (1961) and Polsby (1963), employed a decision-making model in their community power studies which resulted in findings of a factional, coalitional, or amorphous power structure. This latter approach is more in harmony with the diffusion tradition, which attempts to analyze the innovation decision-making process and to determine the roles played by various individuals as the process unfolds. We tend to concentrate more upon the new idea and the way it is adopted within a social system and less upon the power-holders or influential persons who make the decisions.

Collective innovation decision-making can be thought of as a series of subprocesses from (1) *stimulation* of interest in the new idea, (2) to *initiation* of the new idea in the social system, (3) to *legitimation* of the idea by power holders, (4) to a *decision* to act, (5) to *action* or execution of the new idea.

A collective decision is actually composed of a multitude of individual decisions, but each of these individual decisions or behaviors may be carried out by different individuals within the collectivity. In the case of optional innovation-decisions all five subprocesses occur within the mind of a single individual.

Stimulators of collective innovation-decisions are more cosmopolite than other members of the social system. This characteristic provides them with easier access to innovations and the ability to perceive needs and problems of the social system. The initiators of collective innovation-decisions are unlikely to be the same individuals in a social system as the legitimizers. Initiators are noted for their favorable attitudes toward change and for their intimate knowledge of the system. Legitimizers are the high status power-holders of the system who sanction the change. The rate of adoption of a collective innovation is positively related to the degree to which the legitimizers are involved in the decision-making process and to the degree of power concentration in the social system.

Participation is the degree to which members of a social system are involved in the decision-making process. Member satisfaction with, and acceptance of, collective innovation-decisions is positively related to the degree of participation in the decision by members of the social system. Finally, member acceptance of collective innovation-decisions is positively related to member cohesion with the social system.

Change agents possess qualifications which allow them to act as stimulators and initiators of collective innovation-decisions. But seldom can they be legitimizers of collective decisions because they lack the seniority, high status, social power, and established credibility within the social system that a power holder must have to sanction new ideas.

10   Authority
Innovation-Decisions

Rome falling to the barbarians, an old family firm going into bankruptcy, and a government agency quietly strangling in its own red tape have more in common than one might suppose.   JOHN W. GARDNER (1963, p. 3)

The behavior of people in organizations is still the behavior of individuals, but it has a different set of determinants than behavior outside organizational roles. Modifications in organizational behavior must be brought about in a different manner.

DANIEL KATZ AND ROBERT L. KAHN (1966, p. 391)

*and Organizational Change**

THREE types of innovation-decisions were described in Chapter 1: (1) optional decisions, (2) collective decisions, and (3) authority decisions. Earlier chapters of this book deal mostly with the diffusion and adoption of innovations involving optional decisions, and Chapter 9 focused on collective innovation decisions. But we have paid very little attention so far in this book to authority innovation-decisions.

We intend to deal here with authority decisions and changes in formal organizations, such as government bureaucracies, factories, and schools (rather than informal social systems such as a peasant village or a community of doctors) that result from authority decisions. Why do we focus on authority and change in formal organizations? First, authority innovation-decisions are much more common in formal organizations than in any other type of social system. Second, most of the theoretical and empirical evidence relevant to authority decisions comes from literature dealing with organization theory and organizational change. There is very little diffusion research completed on the process of authority decisions. We intend to forge a convergence between diffusion researches and organizational change; the two have much in common but have not been wedded.

*This chapter was written in part by Nemi C. Jain, Assistant Professor of Communication, University of Wisconsin, Milwaukee.

Adoption of Visual Aids in Adams High School[*]

John Quincy Adams High School was built in an Eastern city during the 1930s to provide technical training skills for a major industry in the area. Mr. A became principal of this school in 1960 when his predecessor was transferred to another school because of discipline problems at Adams High which resulted in lower quality instruction. Mr. A's previous school pioneered a pilot program in the intensive use of visual aids in teaching. His experience with this innovation convinced him it could improve the quality of teaching at Adams High School. As a result of his personal observations and informal discussions with the department chairmen at Adams High, Mr. A recognized that the teachers in his school were not correctly or adequately using visual aids in the classroom. He saw this as one reason for the poor quality of teaching. Mr. A decided to initiate a visual aids program.

The principal began by discussing the importance of visual aids with his assistant principal, his department chairmen, and his teachers. He had a general outline of the visual aids program in his mind (based on his previous experience), and he wanted to initiate a program suited to the needs of Adams High.

Within two years Mr. A appointed a committee consisting of the assistant principal and two department chairmen. The committee was delegated the responsibility of assessing the situation, working out the details of a visual aids program, and implementing the program in the school.

The visual aids committee made detailed plans for popularizing the innovation among the teachers, who were the functional units of adoption. The committee's report included details about how the program should be carried out, what was expected of each teacher in terms of utilization, how teachers might be rewarded for making innovative use of the aids, and the like.

Principal A approved the plan of action. However, he included some modifications to make it appear more "democratic."

After approval of the program by the principal, projectors, tape recorders, and a host of other audio-visual equipment started pouring into the school. The committee issued memos and directives to the teachers explaining proper

---

[*]This is a hypothetical case designed specifically for the present chapter but based on ideas taken from Atwood (1964), who discussed resistance to the introduction of a high school guidance program.

use of the equipment. The department chairmen met with the teachers in their departments to demonstrate the new ideas. The principal congratulated the committee for its fine work and prepared to reap the rewards of audio-visual instruction in his school.

But the teachers were not simply a passive audience, and many of their decisions were antagonistic. In the first few months of the program the teachers flooded the assistant principal's office with requests, questions, and complaints, some of which were irrelevant to the program. When they did not receive satisfactory responses, the teachers began to ignore the program. Soon they were ridiculing the use of audio-visual materials to their department chairman. Some teachers liked visual aids and made effective use of them. Most teachers, however, were only "going through the motions" of using the aids to keep their jobs. Soon there was an active resistance movement among the teachers. And it grew.

In response to the growing resistance and improper use of the audio-visual aids in the classroom, Mr. A was forced to suspend the visual aids program. The expensive audio-visual equipment was stored in the basement of Adams High School. There it remains today.

What went wrong? The full answer lies in the nature of authority innovation-decisions.

What Is an Authority Innovation-Decision?

Authority innovation-decisions are those forced upon an individual by someone in a superordinate power position. The individual (or any other type of adoption unit) is ordered by someone in a position of higher authority to adopt or reject an innovation. The individual is not free to exercise his choice in the innovation-decision process. He is forced by someone with more authority in the social system to adopt or reject the innovation. Thus, the authority structure of the social system (in plainer language "the boss") influences the individual to conform to the decision.

We see this point in our case study of Adams High School. The decision about the visual aids program was made by the principal (Mr. A) and the visual aids committee, who were in a position of higher authority than the teachers. The teachers were "forced" to conform to this decision to adopt the innovation.

There are at least two kinds of units involved in authority innovation-decisions:

1. The *adoption unit,* which is an individual, group, or other unit that adopts the innovation. Teachers were the adoption unit for visual aids in the case study.

2. The *decision unit,* which is an individual, group, or other unit that has a position of higher authority than the adoption unit, and which makes the final decision as to whether the adoption unit will adopt or reject the innovation. The principal and the visual aids committee in Adams High represent the decision unit.

How do authority decisions differ from optional and collective decisions? The main difference lies in the nature of social system influence on an individual member's decision to adopt an innovation. In optional decisions there is very little influence from the social system on an individual's decision. In the authority decision there is much influence from the social system through its authority structure on the individual's decision. Such decisions represent two extremes on the continuum shown in Figure 10-1, with collective decisions falling somewhere between them.

The following characteristics distinguish authority innovation-decisions:

1 The individual is not free to exercise his choice in adopting or rejecting an innovation.

2 Decision-making and adopting are activities of two separate individuals or units.

3 The decision unit occupies a position of higher authority in the social system than the adoption unit.

4 Because of this hierarchical relationship between the decision unit and

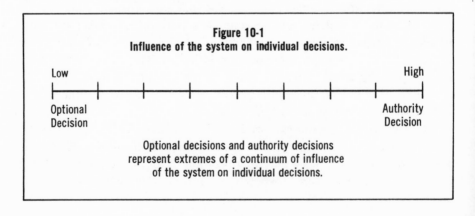

Figure 10-1
Influence of the system on individual decisions.

Low   High

Optional   Authority
Decision   Decision

Optional decisions and authority decisions
represent extremes of a continuum of influence
of the system on individual decisions.

the adoption unit, the decision unit can force the adoption unit to conform to its decision.

5   Authority innovation-decisions occur more frequently in formal organizations than in informal social systems.

Characteristics of Formal Organizations

In order to understand the process of authority decisions and the factors which influence this process, we shall examine the characteristics of formal organizations, which are the social systems in which these decisions generally occur.

It is easier to give examples of formal organizations than to define the term. Adams High School is a formal organization; so is the Red Cross, General Motors, a city hospital, Stanford University, and the corner drug store. Like a family, village, or a community of doctors (which are informal social organizations), these groups are bound by a common goal and share similar social norms, beliefs, and values. But what distinguishes formal organizations from informal systems?

The main distinguishing feature of formal organizations is that they are *deliberately established for accomplishing certain predetermined goals*. In contrast to a social system where men are living together by the fiat of nature, formal organizations are consciously created to achieve defined goals and objectives (Blau and Scott, 1962).

If accomplishing an objective requires collective effort, men set up an organization designed to coordinate the activities of many people to accomplish that objective. For example, industrial organizations are established to produce goods that can be sold for a profit. Universities are organized to provide higher education to the youth of society. In these cases the goals and functions to be achieved are predetermined. The rules that the members of the organization are expected to follow are formulated to guide the members of the organization. Since these organizations are *formal*, established for the explicit purpose of achieving specified goals, the term "formal organization" is used.

A *formal organization* is a social system that has been deliberately established for achieving certain predetermined goals; it is characterized by prescribed roles, an authority structure, and a formally established system of rules and regulations to govern the behavior of its members. Characteristics which set a formal organization off from other types of social systems include:

1. *Predetermined goals.* Organizations are formally established for the explicit purpose of achieving certain predetermined goals. The goals for which the organization is established determine to a large extent the structure and function of the organization. For example, Adams High School was established to provide technical training for students. This goal has much to say about the organization of the school staff.

2. *Prescribed roles.* Organization tasks are distributed among the various positions as prescribed roles or duties. A role is a set of activities to be performed by an individual occupying a given position. Positions are the "boxes" on an organizational chart; for each position there is a prescribed role. Individuals may come and go in an organization, but the positions continue.

3. *Authority structure.* In a formal organization all positions do not have equal authority. The principal in Adams High has more authority than his department chairmen, who in turn have more authority than teachers. Positions are organized in a hierarchical authority structure which specifies who is responsible to whom.

4. *Rules and regulations.* A formal, established system of rules and regulations governs decision-making among organizational members. There are prescribed rules and regulations for hiring new members, for promotion, for discharging unsatisfactory employees, and for coordinating the control of various activities to insure uniform operations.

5. *Informal patterns.* Every continuing formal organization is characterized by various kinds of informal practices, norms, and social relationships among its members. These informal practices emerge over a period of time and represent an important part of any organization. In the present chapter we recognize the informal aspects of formal organizations, since it is impossible to understand the organizations otherwise. Nevertheless, the intent of bureaucratic organization is to depersonalize human relationships by standardizing and formalizing them.

The Process of Authority Innovation-Decisions

It should be recognized at the outset that authority innovation-decisions are considerably more complex than optional decisions. One reason is that the authority innovation-decision process involves two kinds of units, as previously pointed out. Because the decision unit has more authority than the adoption unit and hence, can order the latter to conform to its decision, the decision unit has to know about the new idea, evaluate the idea, and decide if it should be adopted in the system. Once the decision unit has accepted the idea, this decision is communicated to the adoption unit, which must then take action. Each of these different activities may be carried out by different individuals in the organization. In the case of optional decisions all these different activities occur within the mind of a single individual.

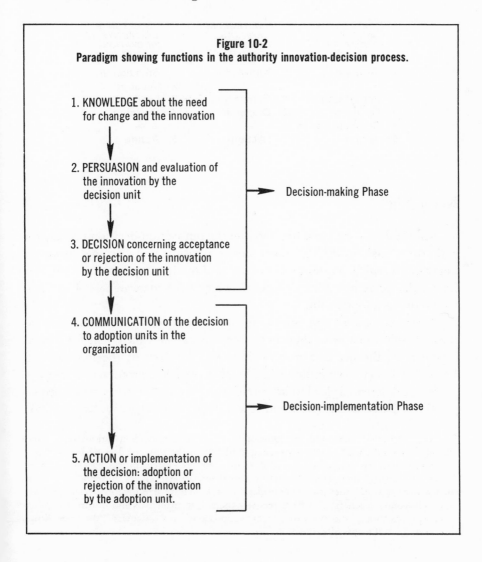

Figure 10-2
Paradigm showing functions in the authority innovation-decision process.

1. KNOWLEDGE about the need
   for change and the innovation

2. PERSUASION and evaluation of
   the innovation by the
   decision unit   ───▶ Decision-making Phase

3. DECISION concerning acceptance
   or rejection of the innovation
   by the decision unit

4. COMMUNICATION of the decision
   to adoption units in the
   organization

                                          ───▶ Decision-implementation Phase

5. ACTION or implementation of
   the decision: adoption or
   rejection of the innovation
   by the adoption unit.

Figure 10-2 presents a simplified paradigm of the functions involved in the authority innovation-decision process. These functions are not necessarily mutually exclusive, nor do they always occur in the exact chronological order depicted in Figure 10-2.

One can see a close parallel between these functions in the authority innovation-decision process and (1) the functions (described in Chapter 3) in the individual innovation-decision process and (2) the steps (Chapter 9) in the collective innovation-decision process.

FUNCTIONS IN AUTHORITY INNOVATION-DECISIONS	FUNCTIONS IN INDIVIDUAL INNOVATION-DECISIONS	STEPS IN COLLECTIVE INNOVATION-DECISIONS
1 Knowledge	1 Knowledge	1 Stimulation
		2 Initiation
2 Persuasion	2 Persuasion	3 Legitimation
3 Decision	3 Decision	4 Decision
4 Communication	—	—
5 Action	4 (Action)	5 Action

Knowledge

Knowledge is the most basic step in the innovation-decision process for authority decisions. At this function the decision unit becomes aware of innovations which are destined for later trial by the adoption unit. And it is knowledge *on the part of the decision unit* which leads to awareness of an innovation by the adoption unit.

Awareness-knowledge about an innovation may be communicated through subordinates to their boss; the subordinates then wait for formal approval by the decision unit. This is upward flow of the innovation.* Empirical observations indicate that this pattern often is prone to problems.

Status differentials lead to inaccuracies in message flows between unequal positions. Three studies (Back and others, 1950; Thibaut, 1950; Kelley, 1951)

---

*Rogers and others (1968), studying diffusion of certain educational innovations in Thailand secondary schools, found some surprising evidence of the upward flow of innovations. The researchers expected a downward flow from top executives in the Thai Ministry of Education, through hierarchical levels of regional educational officers, to school principals and teachers. In reality they found a considerable upward flow of innovations from teachers, especially young teachers who were recently trained, to principals and in turn to higher officials. Even though the innovations were supposed to flow downward, the actual flows were heavily in the opposite direction.

hint that selective screening of information from low to high status members may be used as a psychological substitute for actual achievement by low status members. Further, Festinger (1950) observes that hierarchical structures often incorporate restraints to contain subordinates' criticisms or perceived threats from reaching higher authorities. Subordinates try to protect their position in the hierarchy by screening facts so that they are in accord with the expectations of their superiors (Katz and Kahn, 1966). This screening of upward communication takes place to some extent in all organizations but especially in more autocratic systems. Message accuracy is also affected by the interpersonal trust of the subordinate for his superior (Read, 1962; Likert, 1961; Likert, 1967). When the relationship between a superior and a subordinate is characterized by mutual trust and positive regard, there is greater potential for effective communication. We conclude with Generalization 10-1: *A supportive relationship between the adoption unit (a subordinate) and the decision unit (a superior) leads to more upward communication about the innovation.*

The decision unit can also become aware of the innovation from external sources. Some researchers (Griffiths, 1964; Miles, 1964) suggest that external sources are the most important catalysts of change for administrators of formal organizations. Studies dealing with large-scale organizational change show that external sources, such as consultants, play a crucial role in developing awareness of needed changes in formal systems (Rice, 1958; Seashore and Bowers, 1963).

In Adams High School the principal's personal interest and commitment to audio-visual aids led to introduction of the innovation when he was hired. Changes in personnel, then, as well as employee training by external institutions, are also important avenues for the entry of awareness-knowledge into formal organizations.

Persuasion

No matter what the source of input, the decision unit evaluates an innovation in light of the organization's needs. The persuasion function is characterized by detailed information seeking and the evaluation of costs, feasibility, and possible contingencies. In effect the formal organization is considering a hypothetical trial.

At Adams High the principal involved only the visual aids committee in the persuasion stage. Mr. A himself evaluated the feasibility of the innovation in terms of his experience at a previous school.

An important difficulty in evaluating organizational innovations is their low visibility. One reason for the laggardliness of large organizations, particularly regarding changes in the formal structure of the organization itself, is that economic or psychological advantages cannot be as readily perceived as can the advantages of innovations such as hybrid seed or penicillin.

Furthermore, hard data on the outcomes of adoption are often difficult to gather; therefore, the criteria for costly and far-reaching adoption or rejection decisions often are subjective judgments or intuitions of decision-makers. If we could more accurately assess the consequences of innovations in large organizations, we would be better able to decide which ones to embrace and which ones to spurn. However, innovations presently can be evaluated only about as accurately as fads and fashions, and consequently a high rate of discontinuance occurs as each new idea is tandemly replaced by the next fad-like innovation. Kerr (1964, pp. 106–107) commented on this difficulty of evaluating innovations within one type of bureaucratic organization when he wrote of the university: "The external origin of most change raises very grave problems: How to identify the 'good' and the 'bad,' and how to embrace the good and resist the bad. . . . These obligations fall primarily on the reluctant shoulders of the administrator."

Decision

Thus far the decision unit has gathered awareness-knowledge about an innovation and has evaluated it in terms of its perceived advantages, its feasibility, and its expected consequences. *At the decision stage a formal choice is made by the decision unit to adopt or reject the innovation.*

Perhaps the most important element in the decision function is the degree to which *the adoption unit participates in decision-making.* In Adams High there was negligible involvement of the adoption unit in decision-making. The teachers did not participate in the early stages when the audio-visual committee developed a course of action to implement use of the innovation by the teachers. The teachers' participation was allowed only at a point where they had already developed a strong psychological resistance to the change.

In Chapter 9 we discussed the relationship of an individual's participation in collective decisions to his acceptance of and satisfaction with the collective decisions. Although Generalizations 9-7 and 9-8 were examined mostly in the light of evidence based on small groups and organizational studies (Morse and

Reimer, 1956; Coch and French, 1948; Lewin, 1943; Jacques, 1948), these findings seem equally relevant to the authority decision process.

Extrapolating the evidence discussed earlier, we can speculate about the effect of the adoption unit's participation in decision-making on its subsequent behavioral responses to the innovation. We should remember that although the adoption unit may conform to the executive decision overtly, it may reject the decision attitudinally. This may lead to consequent disruptions in organizational procedures or eventual discontinuance of the innovation. Thus, *attitude toward* an innovation and *satisfaction with* the decision are two important dependent variables; the adoption unit's participation in the decision stages is a predictor of both acceptance and satisfaction.

Generalization 10-2: *An individual's acceptance of an authority innovation-decision is positively related to his participation in innovation decision-making.**

Generalization 10-3: *An individual's satisfaction with an authority innovation-decision is positively related to his participation in innovation decision-making.***

Communication

When the decision unit has chosen the innovation alternative it wishes to adopt, messages must be transmitted in a *downward flow* from superiors to subordinates, following the authority pattern of hierarchical positions, to the adoption unit.*** In the individual innovation-decision process there is no need to include a communication stage, since both the decision-making and decision-implementing functions occur in the mind of an individual. In authority decisions, however, communication is a crucial function, because adoption or rejection cannot be implemented until a transfer of meaning from the decision unit to the adoption unit has taken place. The flow of downward communication from the visual aids committee at Adams High consisted of directives, memos, and verbal explanations about the innovations to the teachers.

*This proposition is similar to Generalization 9-7 (for collective innovation-decisions): *Member acceptance of collective innovation-decisions is positively related to the degree of participation in the decision by members of the social system.*

**This statement is parallel to Generalization 9-6 (dealing with collective innovation-decisions): *Satisfaction with a collective innovation-decision is positively related to the degree of participation by members of the social system in the decision.*

***Key figures in such vertical communication flows may be *liaison individuals*, who connect two or more subgroups within the system. A research focus upon liaison individuals was initiated by Jacobson and Seashore (1951) and followed by Weiss and Jacobson (1955) and Schwartz (1968).

It is generally recognized that when there is contact between individuals of different status, communication from superior to subordinate takes place more easily than from subordinate to superior. But it is also recognized that the wider the status differential (greater heterophily), the more restricted are the channels of communication and the more likelihood there is that the same channels will be used repeatedly (Barnlund and Harland, 1963, p. 468). An experimental laboratory study of a small organizational unit (Gerard, 1957) indicated that designating one member of a four-member group as the boss increased the frequency of communication from that individual to other members of the team (downward communication). Existence of a hierarchical structure affects communication patterns in a system.

But what of the accuracy of such communication? Research shows that message content is transformed in organizations through the use of omitting devices, by reducing content through special code categories, by by-passing segments of the communication network, and by neglecting input. If the subordinate in fact insulates his superior from clear knowledge of work-related problems, then "good" decisions at the top of the hierarchy are effectively blocked by the lower levels (March and Simon, 1958). The messages flowing downward may become dysfunctional to the extent that the superior lacks awareness of the problems affecting his subordinates. Cyert and March (1963, p. 110) noted that "one of the ways in which the organization adapts to the unreliability of information is by devising procedures for making decisions without attending to apparently relevant information," that is, developing special coding categories.

Another important aspect of downward communication is its acceptance by subordinates. Likert (1961) suggests that in "authoritative" organizations, downward messages are viewed with great suspicion, creating problems of misunderstanding and attitudinal rejection. In "participative" organizations downward communication is more often accepted.

Action

Action refers to the adoption or actual use of the innovation by the adoption unit. In a sense this is the final step in the authority innovation-decision process. Often the behavioral consequences of an innovation become most visible, whether rewarding or disappointing, at the action stage.

In Adams High School the basement filled with the expensive audio-visual equipment symbolizes the attitudinal discrepancy of the teachers. They were exposed to a dissonant situation between their negative attitudes toward the audio-visual materials and the principal's order to adopt. This condition, which we term "innovation dissonance," is illustrated in Table 10-1.

Innovation dissonance in a formal organization is the discrepancy between an individual's attitude toward an innovation and the overt behavior (adoption or rejection) demanded by the decision unit. Types I and IV in Table 10-1 are consonant because the member's attitudes and beliefs are in accord with his behavior. Types II and III both lead to cognitive dissonance (Festinger, 1957) because the member's attitude toward the innovation is discrepant with his behavior in the organization. The balance theory of tension reduction suggests a tendency over time for Type IIs and Type IIIs to: (1) change their attitudes to make them consonant with the behavior demanded by the organization, or (2) discontinue the innovation, misuse the innovation, or circumvent the adoption edict to make their organizational behavior consonant with their attitudes. This strain toward balance will cause Type IIs and Type IIIs to become either Type Is or Type IVs.

Table 10-1   Four Dissonant-Consonant Types on the Basis of Individual Attitudes Toward an Innovation and the Overt Behavior Demanded by the Organization

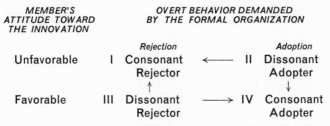

MEMBER'S ATTITUDE TOWARD THE INNOVATION		OVERT BEHAVIOR DEMANDED BY THE FORMAL ORGANIZATION		
		Rejection		*Adoption*
Unfavorable	I	Consonant Rejector	⟵ II	Dissonant Adopter
Favorable	III	Dissonant Rejector	⟶ IV	Consonant Adopter

Source: Adapted from Knowlton (1965, p. 53), and used by permission. Note that the arrows in the table indicate pressures toward consonance.

Generalization 10-4: *When an individual's attitudes are dissonant with the overt behavior demanded by the organization, the individual will attempt to reduce the dissonance by changing either his attitudes or his behavior.*

To test this dissonance hypothesis, we need data on individual members' attitudes toward innovations and on authority decisions over time. Balance theory suggests that an individual will attempt to resolve imbalance in his cognitive system by taking that alternative which is easiest for him to achieve. In formal organizations this usually means that the individual will change his attitudes over time to conform to his public behavior; he may have little opportunity to change his organization behavior. On the other hand the individual may find ways of circumventing the authority decision of the organization and of resolving the dissonance by making his behavior conform to his private attitudes. Of course, there are other ways of resolving dissonance, such as quitting the organization, living with the imbalance, and ignoring the dissonance, but these are less likely than the actions suggested by Generalization 10-4.

In the case of authority innovation-decisions the action by the adoption unit is based on the supervisors' directives. The situation is similar to what Kelman calls *compliance*, which occurs when an individual is influenced by another person or group because he hopes to earn a favorable reaction from the other (Kelman, 1961). The adoption unit complies with the decision of his supervisor so as to maintain a favorable relationship with him. In compliance situations the change in behavior is usually temporary and needs surveillance for it to continue.

When studying authority innovation-decisions, we find that *attitude toward the innovation* is often a more suitable dependent variable for study than overt behavior or actual use of the innovation (Lin, 1966). This is because overt behavior can be manipulated by the organization, at least in the short range, but the member's attitudes affect continued adoption versus discontinuance.

Approaches to Organizational Change

Thus far we have examined the interrelated functions involved in the authority innovation-decision process. These functions can be handled in several ways. For instance, the decision function can be unilaterally made by the higher authority (as by the principal in Adams High School), or the adoption units can be involved in decision-making. There are, therefore, different approaches to authority decisions in order to achieve organizational change.*

One convenient way to look at the various approaches to organizational change is in terms of the power distribution style, the manner in which decision-making power is distributed among those affected by the change (Leavitt, 1965). One can think of a style in which the power for making change

*For a detailed discussion of various approaches to organizational change, see Greiner (1965), Leavitt (1965), Katz and Kahn (1966), and Barnes (1967).

resides in a single authority. The other extreme is a style in which power is widely shared by the individuals affected by the change.

There are two basic approaches: (1) the *authoritative approach,* in which there is a very unequal distribution of power, and decisions about change are made by a centralized power position, others being required to obey the decision, and (2) the *participative approach,* in which there is wide sharing of power, and decisions about change are made in consultation with those affected by the change.

Authoritative Approach

The authoritative approach to organizational change involves a one-way announcement originating with a person of high formal authority and passed on to those in lower positions. This "decree approach" (Taylor, 1911; Gouldner, 1954; Greiner, 1965) represents unilateral decisions by the authority figures. Those who are affected do not have any say in how or when to change. Without question, the authoritative approach is more prevalent in industry, as compared to the participative approach (Barnes, 1967). In terms of the authority innovation-decision process, this approach involves minimum participation of the adoption unit in the decision process.

Participative Approach

This approach to organizational change involves a two-way interaction between executives who initiate change and those who are affected by the change. Decision-making power is shared by all concerned at different stages in the organizational change.* In authority decisions the participative approach involves wide involvement of the adoption units, especially at the decision function. The adoption unit may be involved in identifying relevant innovations, in evaluating them, and deciding to accept or reject them.

Decisions are made by the top executive in the authoritative approach. Organizational change is thus faster than the participative approach. Therefore, when speed of change is required, authoritative approaches can bring about quick results. But the change may not be permanent, and sometimes organizational members resist the innovation, leading to discontinuances (as we saw in Adams High).

*Examples of organizational change using the participative approach include Coch and French (1948), Seashore and Bowers (1963), and Sofer (1961).

Generalization 10-5: *The rate of adoption of authority innovation-decisions is faster by the authoritative approach than by the participative approach.*

Generalization 10-6: *Changes brought about by the authoritative approach are more likely to be discontinued than those brought about by the participative approach.*

Need for an Adaptive Unit

The high degree of social structure in a formal organization often acts to impede communication flows. The hierarchy is often blamed for its effects on message distortion, lack of feedback, and even for causing information overload. For instance, in their investigation of large public agencies, Janowitz and Delany (1957) found that the higher the position of an individual in a bureaucracy, the less knowledge he had about the clients of his organization. Gardner (1963, pp. 78–79) points out the dangers of "filtered experience" as a cause of bureaucratic rigidity. His argument is: As organizations become larger and more complex, the men at the top must depend less on firsthand experience and more and more on information which is processed up through lower levels of the hierarchy. But this information-processing filters out emotions, sentiments, and other sensory impressions not easily expressed in words or numbers, the organizational code. The picture of reality that reaches the top of the bureaucracy is often a dangerous mismatch with the real world.

The solution, Gardner (1963, p. 79) suggests, is for the top administrator to "periodically emerge from his world of abstractions and take a long unflinching look at unprocessed reality." There are also other methods of obtaining feedback from the lower levels of the hierarchy: The organizational structure can be decentralized to minimize the number of hierarchical layers between top and bottom, between client and president. The organization can also create an atmosphere which gives the staff an opportunity to pass along informally the types of feedback which cannot be expressed in organizational code.

The social structure of a formal organization can be made to work *for* effective communication, rather than to impede it. One method is to create an adaptive unit as part of the organization's structure. It may be called a research and development unit or some other such euphemistic title. The purpose of the adaptive unit is to sense the changes in the environment, to determine the need for changes in the organization, to identify suitable innovations, and to evaluate innovations.

In a sense adaptive units perform the role of change agents. "Perhaps what every corporation (and every other organization) needs is a department of continuous renewal that would view the whole organization as a system in need of continuing innovation" (Gardner, 1963, p. 76). This self-renewal unit should be located near the top of the hierarchy power structure. If it is to operate as an effective change agent, it must have skilled personnel and adequate funds to influence potential adopters, wherever they may be in the organization.

Summary

This chapter focused on authority innovation-decisions and change in formal organizations. *Authority innovation-decisions* are forced upon an individual by someone in a superordinate power position. The individual (or any other adoption unit) is ordered by someone in a higher authority position to adopt or reject an innovation. Optional and authority decisions represent two extremes on a continuum representing the influence of the social system on individuals' decisions.

Authority decisions are more common in formal organizations than in other types of systems. A *formal organization* is a social system that has been deliberately established for achieving certain predetermined goals and is characterized by prescribed roles, an authority structure, and a formally established system of rules and regulations to govern the behavior of its members. Besides these formal aspects, every enduring formal organization has various kinds of informal practices, norms, and social relationships among its members.

Diffusion researchers have largely neglected the study of diffusion and adoption of innovations involving authority innovation-decisions; we know very little about the process by which authority decisions are made and organizational changes implemented. We draw upon the literature on organization theory and studies dealing with organizational change to analyze the process of authority innovation-decisions.

The authority innovation-decision process can be conceived as a process involving five interrelated functions: (1) *knowledge* about the need for change and the innovation on the part of the decision unit, (2) *persuasion and evaluation* of the innovation by the decision unit, (3) *decision* as to acceptance or rejection of the innovation by the decision unit, (4) *communication* of the decision to the adoption units in the organization, and (5) *action* or implementation of the decision by the adoption units.

Knowledge about the need for change and the innovation can come from sources internal or external to the organization. Upward communication of ideas from subordinates to their boss is often problem-prone and depends largely on the nature of relationships between the adoption unit and the decision unit. A supportive relationship between the adoption unit (subordinate) and the decision unit (superordinate) leads to more upward communication about the innovation.

The adoption unit's participation in decision-making is highly related to its attitude toward and satisfaction with the authoritative innovation-decision. An individual's acceptance of and satisfaction with an authoritative innovation decision is positively related to his participation in the decision-making.

In authority innovation-decision situations, an individual's attitude toward an innovation may not coincide with the overt behavior demanded by the organization, leading to innovation dissonance. *Innovation dissonance* in a formal organization is the discrepancy between an individual's attitude toward an innovation and the overt behavior (adoption or rejection) demanded by the decision unit. When an individual's attitudes are dissonant with the overt behavior demanded by the organization, the individual will attempt to reduce the dissonance by changing either his attitudes or his behavior.

The two basic approaches to organizational change are (1) the *authoritative approach,* in which there is an unequal distribution of power, decisions about change being made by a centralized power position, with others obeying the decision, and (2) the *participative approach*, in which there is wide sharing of power, decisions about change being made in consultation with those affected by the change. The rate of adoption of authority innovation-decisions is faster by the authoritative approach than by the participative approach, but changes brought about by the authoritative approach are more likely to be discontinued than those brought about by the participative approach. In either case there seems to be a need for an adaptive unit in formal organizations to facilitate the process of authority decisions and organizational change.

11 *Consequences*

Changing people's customs is an even more delicate responsibility than surgery. EDWARD H. SPICER (1952, p. 13)

I have never learned to accustom myself to innovations, and
I fear that above everything else, for I know full well that in
making innovations, safety can in no way be preserved.

Proclus, to the Roman Emperor Anastasius

of *Innovations*[*]

—. ——— ...—.— .— — —.. ——— —. ...   —.—. ——— —. .... . ——.

T HREE subprocesses of social change were outlined in Chapter 1: Invention, diffusion, and consequences. *Consequences* were defined as the changes that occur within a social system as a result of the adoption or rejection of an innovation. An innovation is of little use until it is distributed to others and put to use by them. Thus, invention and diffusion are but immediate means to an ultimate end: The consequences from adoption of an innovation.

In spite of the importance of consequences, they have received very little study by diffusion researchers. The data we have about consequences are rather "soft" in nature; most of them are based on case studies only. Lack of research attention and the nature of the data make it difficult to generalize about consequences. We can only describe the subprocess of consequences in social change and establish categories for classifying consequences.

Researchers have given little attention to consequences; so have change agents. They often assume that adoption of a given innovation will produce only beneficial results for its adopters. Change agents should recognize their responsibility for the consequences of innovations they introduce. They should be able to predict the advantages and disadvantages of an innovation before introducing it to their clients, but this is seldom done.

The introduction of wagons to an Indian tribe illustrates this neglect of responsibility. Every program of planned change produces social and econo-

*This chapter was written with B. E. Davis, Associate Professor of Communication, Abilene Christian College.

mic reactions that run throughout the social structure of the client system, as
we see here.

In the Wake of the Wheel*

The Papago Indians of southern Arizona knew nothing of the wheel until
its introduction by white men. Because of their relative isolation in the desert,
they were not aware of wagons until after 1900. Papago acceptance of wheeled
vehicles came about as the result of a deliberate program by change agents
from the U.S. Bureau of Indian Affairs. No one anticipated the far-reaching
consequences of the wagon upon the life style of the Papagoes.

The present case study centers on the first wagon to be introduced in a
remote Papago village. Three brothers of the village headman saw wagons
used off the reservation in railroad construction. When they returned to their
village, they persuaded the headman to request a wagon from the Indian
agent, who was eager to provide the innovation. When the wagon arrived about
a year later, it began immediately to cause changes in the Papago culture. The
wagon was used intensively from the beginning and soon replaced pack
horses which had been the Papagoes' primary means of transportation. The
vehicle was used to carry water from springs to the village households.
Pottery jars had previously been used for fetching water, but these were easily
broken by the wagon. The Papagoes turned to unbreakable metal barrels, and
the women stopped making pottery for use as water jars.

Changes in other aspects of Papago culture soon resulted from the wagon.
A road was constructed to facilitate wagon transportation. Long trading
journeys into Mexico by pack horses were replaced by short wagon trips into
nearby towns. Crops and firewood were marketed through use of the wagon.
The gathering of firewood soon developed as an occupation for men instead
of piecemeal gathering by women and children. The shift from subsistence to
market economy and the need for manufactured harness and metal barrels
linked the Papagoes with the national society in new ways.

In the Papago case little conscious control of the innovation was exercised
by the change agent. If you were the Indian agent, would you have expected
the many consequences that ensued? The need for roads and the disuse of pack
horse transportation might have been expected by anyone who planned
ahead; but could you have predicted that the wagon would produce shifts in

*Adapted from Bliss (1952, pp. 23–33) and used by permission.

the division of labor, increase dependence on the national economy, and influence the relations of Papagoes with surrounding peoples? Thus, we see the difficulty of predicting the consequences of an innovation.

A Model for Studying Consequences

We propose two specific illustrations of a general model for studying consequences; both examples stress a new approach in diffusion research in that the main focus is upon consequences. Our model is first illustrated with the consequences of educational change.

Consequences of Educational Innovations

There is an almost total concentration by educational diffusion researchers on the dependent variable of innovativeness, with no attention given to the subsequent consequences. Very little evidence is now available regarding the desirable and undesirable results actually achieved by educational innovations diffused throughout public school systems. As a result, new education ideas must be adopted largely on faith, rather than on a more rational basis in which expected consequences are considered.

Carlson (1965, pp. 74–84) found that programmed instruction, an innovation that supposedly allows different individual rates of student achievement, was *misused* by a Pennsylvania school system so that differences in individual learning rates were actually held down. The consequences were not what change agents expected, and the potential desirable effects of the innovation were not achieved.

Consider the Michigan school teacher forced to adopt modern math. She had twenty-three years' teaching experience, little understanding of the innovation, and a strong dislike for it. Her school system established a deadline for adoption one September, when all of the staff were to be teaching modern math. Miss "Laggard," our unwilling dissenter, was sent to a modern math teachers' institute during the previous summer and returned to her school in the fall with a little more understanding and much greater aversion to the innovation. By October she was aware of her inability to handle the new material; in November she threatened to resign; and in December she had a nervous breakdown. Certainly this is a highly atypical case, but it does illustrate the undesirable consequences that can accrue from an innovation generally regarded as high in relative advantage. Anticipation of consequences should be one main focus of future research in educational change.

It has been said that this is a time of great innovation but little change in U.S. education. Many new ideas are promoted and adopted, but change in education is almost nil. Many of the innovations are merely educational fads, and after widespread adoption of them, it is difficult to measure an increase in achievement. Many educational innovations with a low degree of relative advantage have been adopted and then discontinued after a short time.

There is need for a new model of change in education, one that stresses consequences. Past researchers asked: "What are the correlates [that is, antecedents] to educational innovativeness?" Such research was concerned with the *wrong dependent variable*. Our investigations should try to explain the consequences of innovation in education, especially educational quality, rather than innovativeness *per se*. Figure 11-1 shows a proposed new model for studying consequences in education. Innovativeness, the main dependent variable in most past research, now becomes only a predictor of a more ultimate dependent variable, consequences of innovation. The new model seeks to explain consequences, a research goal that is closer to the objectives of most change agencies. They aim to bring about desirable consequences among their clients, not simply the adoption of innovations *per se*.

Consequences among Peasants

There is a similar need for analyzing innovation consequences in studies of peasants in less developed nations, where emphasis has been largely on determining attributes of innovators, early adopters, and so on. These analyses would be extended in utility by concentrating on explaining certain consequent variables, such as increased farm production, higher income, and the like (Figure 11-1).

An illustration of the use of the new model of consequences is provided by a recent study. Mason and Halter (1968) first determine variables related to innovativeness among Oregon farmers. Then they include innovativeness with many other variables to explain farm production levels, one type of desired consequence of the adoption of agricultural innovations. They predict about 50 percent of the variance in farm production and find that innovativeness makes a unique contribution in raising yields. Inquiries such as this demonstrate an approach that can potentially provide quantifiable and predictive generalizations about consequences.

Figure 11-1

A new model for studying change in education and among peasants.

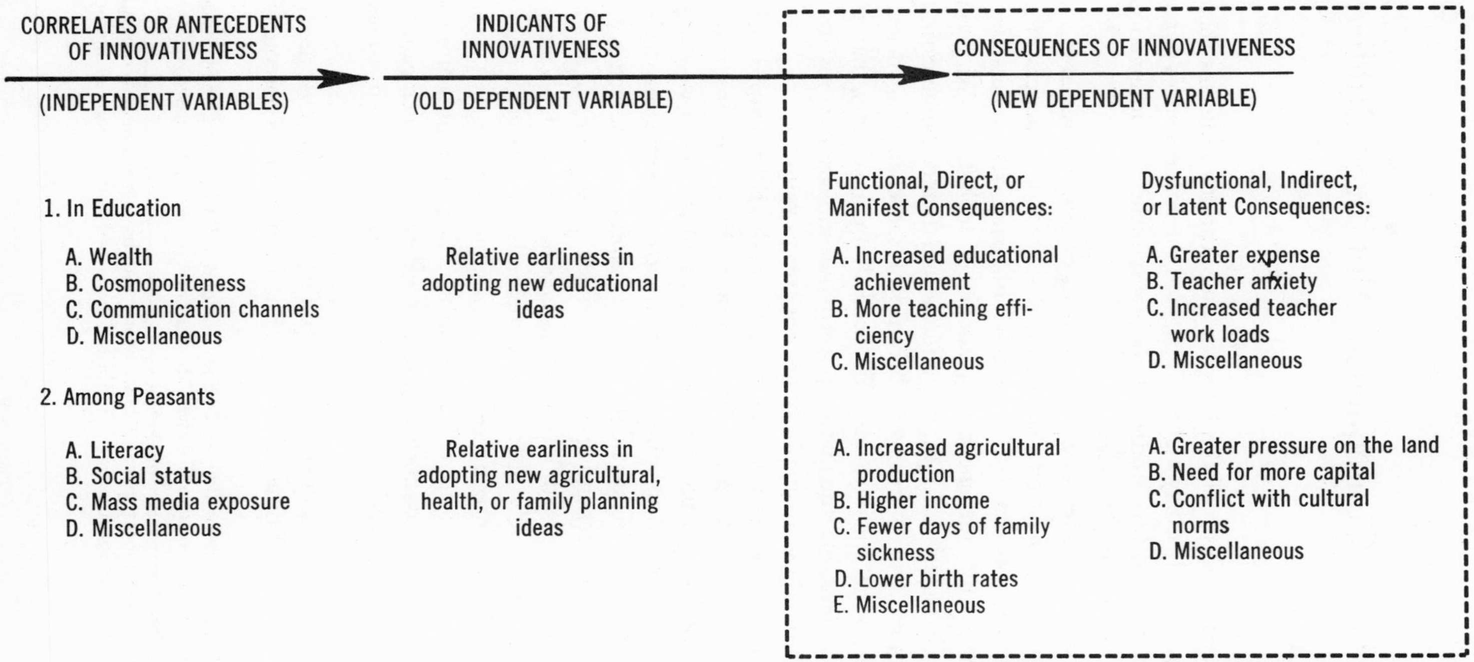

CORRELATES OR ANTECEDENTS
OF INNOVATIVENESS

(INDEPENDENT VARIABLES)

1. In Education

A. Wealth
B. Cosmopoliteness
C. Communication channels
D. Miscellaneous

2. Among Peasants

A. Literacy
B. Social status
C. Mass media exposure
D. Miscellaneous

INDICANTS OF
INNOVATIVENESS

(OLD DEPENDENT VARIABLE)

Relative earliness in
adopting new educational
ideas

Relative earliness in
adopting new agricultural,
health, or family planning
ideas

CONSEQUENCES OF INNOVATIVENESS

(NEW DEPENDENT VARIABLE)

Functional, Direct, or
Manifest Consequences:

A. Increased educational
achievement
B. More teaching effi-
ciency
C. Miscellaneous

A. Increased agricultural
production
B. Higher income
C. Fewer days of family
sickness
D. Lower birth rates
E. Miscellaneous

Dysfunctional, Indirect,
or Latent Consequences:

A. Greater expense
B. Teacher anxiety
C. Increased teacher
work loads
D. Miscellaneous

A. Greater pressure on the land
B. Need for more capital
C. Conflict with cultural
norms
D. Miscellaneous

Note: The area outlined in dotted lines represents the additional element of
consequences that should be considered in diffusion research.

Why Have Consequences Not Been Studied More?

A recent analysis of the diffusion publications indicates that of nearly 1,500 studies, only thirty-eight investigated the consequences of innovations. Why are there so few studies of the consequences of innovation? Several possible reasons may be offered.

1. Change agencies, often the sponsors of diffusion research, overemphasize adoption *per se*, tacitly assuming that the consequences of innovation decisions will be positive. Typically, diffusion researchers devote much attention to the antecendents of adoption, including socioeconomic and personal characteristics of the respondents and their communication behavior, for example. This is primarily because of the change agencies' *ipse dixit* assumptions that the innovation is "needed" by the clients, that its introduction will be desirable, and that adoption of the innovation represents "success." But we know these assumptions are not always valid.*

2. Perhaps the usual survey research methods are inappropriate for the investigation of innovation consequences. Extended observation over a period of time might prove more useful, or an in-depth case-study approach might produce more insights. The participant observation technique used widely by anthropologists might be more helpful than the survey approach,** in that it does not depend upon the receivers' perceptions of an innovation's consequences. Because diffusion researchers have been highly stereotypical in relying almost entirely upon survey methods of data-gathering, they have ignored studying consequences, a type of inquiry where the one-shot survey method is not very effective. But the anthropological approaches suffer in that they largely yield idiosyncratic, descriptive data from which generalization to other innovations and other systems is difficult or impossible, as Ryan (1965) points out.

The study of consequences is complicated by the fact that they occur over extended periods of time. The study of an innovation's consequences cannot

---

*Although they are more likely to be considered as valid by change agents and by clients in a change-oriented system like the U.S. than in a more traditional system such as a less developed country.

**Perhaps it is significant that anthropologists, who have investigated consequences more than any other diffusion research tradition, have not utilized survey methods in their inquiries.

be accomplished simply by adding an additional question to a survey instrument, another 100 respondents to the sample population, or another few days of data-gathering in the field. Instead, a long-range research approach must be taken in which consequences are analyzed as they unfold over time. Otherwise, the consequences of an innovation can be neither properly assessed nor predicted.

A panel study* in which the same respondents are interviewed both before and after the innovation is introduced could yield desired information about consequences. Firm data about consequences could also come from carefully conducted field experiments in which an innovation is introduced on a pilot basis in a system and its results evaluated under realistic conditions prior to its widespread diffusion and adoption. These studies over time, like the panel study and the pilot field experiment, can provide quantifiable, "firm" data about expected consequences. Such data can lead to generalizations about consequences, rather than simply description. And they are predictive to a future point in time, rather than being simply post mortems of consequences that have already occurred.

3. Consequences are difficult to measure. Clients of change are often not fully aware of all the consequences of their adoption of a given innovation. Therefore, any attempt to study consequences which rests only on respondents' reports may lead to incomplete and misleading conclusions.**

Judgments concerning consequences are almost unavoidably subjective and value laden, whoever makes them. Cultural norms, personal preferences, and bias are an integral part of the frame of reference of every observer of the social scene, in spite of efforts to be rid of such prejudicial attitudes. To some degree every judge of the functionality (or dysfunctionality) of innovations is influenced by his personal experiences, his educational background, his philosophical viewpoint, and the like. A researcher from a more developed country may find it hard to make completely objective judgments of the desirability of an innovation in a less developed country such as Thailand, Nigeria, or Pakistan.

The concept of *cultural relativism* is the viewpoint that each culture should be judged in light of its own specific circumstances and needs. No culture is actually "best" in an absolute sense; each culture works out its own set of norms, values, beliefs, and attitudes that are best for itself. Conditions in a less developed country may therefore seem strange and unsuitable to a Western

*Which is really a double survey in that the respondents are interviewed more than once.
**This is one advantage of the observation method of data-gathering, which does not depend so much (as the survey) upon the receivers' perceptions of consequences.

observer, when many of these conditions result from centuries of experiment, trial and error, and evolution. Most are quite reasonable, given the conditions in which they exist.

The concept of cultural relativism has implications for the measurement of consequences, for whether data about the results of an innovation are gathered from clients, change agents, or from scientific observers, their view of an externally introduced innovation is likely to be subjectively flavored by their own cultural beliefs. Consequences should be judged as to their function-ality in terms of the *clients'* culture, without imposing outsiders' normative beliefs about the needs of the client system. And this is extremely difficult to accomplish.

A further problem in measuring the consequences of an innovation is that they are often confounded with other effects. In assessing the results of chemical fertilizer on crop yields, one cannot ignore the consequences caused by natural events like droughts, volcanic eruptions, and others. And this confounding is difficult or impossible to avoid completely, even with care-fully conducted field experiments with before-and-after measurements and a control system.

Two Examples of the Consequences of Innovation

We now consider in detail two different examples of the consequences of innovation—one dealing with the results of farm mechanization in a less developed country and the other with the consequences of innovation in education. In perusing these accounts notice the rather extensive and compli-cated chain of events precipitated by the innovations. Many of these conse-quences, both positive and negative, were not expected by change agencies.

From Hoes to Diesels in Turkey*

The social effects of farm mechanization were studied in twenty Turkish villages ten years after a program of planned change was initiated. Prior to mechanization the people of the little valley in which the study villages were located lived a slow-moving, isolated, peaceful kind of life. Communication,

*Adopted from Karpat (1960) and used by permission.

both from the outside and between villages, was extremely limited. Although the Turkish government had planned improved travel connections and had opened experiment stations in the valley, peasant life remained static. The roads were crude and dangerous, and few motorized vehicles ventured over them. The agricultural innovations devised at the research stations failed to spread to the villagers because of a lack of proper communication between government workers and the peasants.

With the help of a change agency now known as the U.S. Agency for International Development, the Turkish government began a massive change program which included the introduction of 40,000 tractors, a variety of other farm equipment, a huge road-building program, and large numbers of motor vehicles. The impact was immediate and dramatic. The communication system was greatly improved, and local agriculture was rapidly transformed.

Under the new program the amount of land owned by each village was the primary factor in determining the kind of crop to be grown and the rate of mechanization. If a village had large areas of cultivated land, villagers started using tractors and concentrated exclusively on raising only one crop, such as cotton. One village of 700 people, for example, purchased eighty tractors to cultivate about 5,200 acres planted entirely to cotton. Over four million pounds of cotton were produced, worth about four million Turkish *liras,* a very high income for the village. On the other hand another village with about 2,000 cultivated acres could buy only eighteen tractors and was required to retain a diversified type of agriculture. Smaller villages were hard hit by economic competition.

The development of an improved communication system with the outside world produced an abrupt break with the isolated tradition of the villages. The construction of better roads brought villagers closer together, both psychologically and geographically. Peasants in small settlements began commuting to nearby towns by regular bus service and on tractors, and some peasants found work in the city, while continuing to live in the villages. Curiosity regarding new products, plus the new-found ability to purchase them, produced a heightened eagerness to travel to urban centers, resulting ultimately in a migration toward the cities. New houses sprang up, house styles and construction materials changed, and families began spending money on radios and new furniture. Not only was cosmopoliteness increased, but the old way of life was considered backward and doomed to disappear.

Food habits also changed. Bakery-produced bread replaced homemade bread, and foods like canned goods, sugar, and especially alcoholic beverages increased greatly. New grocery stores opened in each village, and a cash economy sprang up. Commercial farming replaced subsistence living.

The extensive changes wrought by the mechanization program were also attended by a number of undesirable consequences. The newly acquired farm equipment lasted only a relatively short time, because of inexperience in mechanical maintenance and carelessness in use, resulting from the ease with which the equipment had been acquired. Many tractors were idle—as many as 70 percent in some villages—because of lack of parts and improper maintenance.*

Social status differences were emphasized by the new economic and social changes. Poor villagers with little land could not buy farm equipment and were forced to continue using old tools and traditional farming methods. This caused a major social problem: How to insure equal benefits to the huge number of villagers at the bottom of the social scale who had become keenly aware of the possibilities for economic development but were little affected by it. Their wants increased and outran their ability to acquire, leading to frustration.

Further, a conspicuous desire arose for new avenues of pleasure and enjoyment made accessibile through extra time and income. Many of the newly enriched villagers were observed to squander their money on food and drink in bars and restaurants and on costly transportation to and from urban centers.

A feeling of hopelessness and sorrow arose among those not fortunate enough to share in the higher incomes. In speaking of the death of the old ways, one patriarch said, "There is no future left to the [nomadic] life. We failed to see it in time and insisted on remaining ignorant. Look at me. I am fifty-five years old and not as capable as a small boy. Give me a piece of paper with something written on it, and I am blind because I cannot read it."

The mechanization of agriculture produced a complex network of results. It appears to have been initiated without due regard or understanding of village social structure and without envisioning the social consequences likely to result from improperly planned mass mechanization. The process directly benefited only a limited number of villages and a limited number of individuals in each village, mainly those already in the upper income group.**

*The tractors were also utilized by the peasants in unanticipated ways, such as for travel to the cities. One villager drove his family by tractor from Turkey to Berlin, a distance of over 2,000 miles.

**This point is expressed in Generalization 11-3, which is discussed later in this chapter.

Programmed Learning and Teachers' Need to Perform*

An equally dramatic and unexpected series of consequences occurred in the adoption of programmed learning by a U.S. school system. Programmed instruction is an educational innovation designed to improve the learning process by allowing students to learn at their own rates. However, in the Pittsburgh schools that were studied, adoption of programmed instruction produced results far different from those anticipated.

Most teachers seem to possess a compelling need to capture and hold the attention of their students and to serve continuously as the chief mediator between the students and the material being taught. This is what teachers seem to define as "teaching," with all other schoolroom activities assessed as supportive acts. The introduction of programmed instruction frustrated the teachers' need to perform; the student is oriented toward the programmed instructional materials (such as a programmed textbook) instead of the teacher. Various devices were created by the teachers to bring themselves back into a position of classroom prominence. In one such arrangement students worked on the programmed material part of the time but also met with an instructor using regular teaching procedures. Other programs were stretched over a considerable period of time, so that the teacher could serve as a "director of learning" and restore his position relative to the student.

Educational experts speak of the need for "teaching children, not subjects," and "starting the learning experience where the child is." Yet, when programmed instruction was introduced in the school system, permitting students to work at their own rates, a host of practices emerged to keep students working at approximately the *same rate*. Teachers either consciously or unconsciously restricted the output of students who were proceeding more quickly. In addition special provisions were made for the slower students. A previous decision had been made not to allow students to work on their programs at home. However, this original decision was reversed for slow students, but average and fast learners continued to have access to the programs only during class time. Other students took two semesters to complete a program designed for one semester.

Although such practices were justified in the eyes of the teachers and administrators by pragmatic considerations, the net effect was to minimize the range of student progress. And the advantages of individual instruction, which was possible with programmed instruction, were almost totally lost.

*Adopted from Carlson (1965) and used by permission.

Supervision of teachers by school administrators became quite difficult under the new system of programmed instruction. In most cases the assessment of teaching prowess had been conducted by classroom observation. But programmed instruction cut down on the number of oral presentations by the teacher, making evaluation by the administrator a problem. Consequently in many cases teacher observation was abandoned entirely. Other principals who continued to perform this supervisory function threw out the conventional procedures they had used previously. For example, some began asking students to tell them how the teacher was doing, others merely peeked into the classroom to see if the teacher was helping the pupils. Still others judged the teachers' ability by noting their skill in storing the instructional material and keeping the machines and programs in good working order.

The introduction of programmed instruction had unanticipated consequences, results that went far beyond its intended impact on student achievement. Its ability to permit students to move at their own rate of learning proved inconvenient for the teachers, so that various steps were taken to reduce the variability in student progress. Programmed instruction tended to replace the teacher as director of learning. Having a need to perform, teachers introduced their own "innovations" to enable them to recapture some of their role as director of learning, which was lost to programmed instruction. Finally, the use of the new materials rendered inadequate the principals' usual methods of classroom evaluation of teachers.

Classifications of Consequences

So far in our discussion, we have used a simple breakdown of consequences as either desirable or undesirable. However, consequences may also be analyzed according to several other dimensions. With a more comprehensive classification system, we can better understand the nature of different types of consequences. We begin with a more adequate discussion of desirable and undesirable effects of an innovation.

Functional Versus Dysfunctional Consequences

Functional consequences are desirable effects of an innovation in a social system. On the contrary, *dysfunctional consequences* are undesirable effects of an innovation in a social system. The degree to which consequences are desirable or undesirable ultimately depends, of course, on how the innovation affects the members of the system. The determination of whether consequences are functional or dysfunctional depends on how the innovations affect the adopters.

Any social system has certain qualities which should not be destroyed if the welfare of the system is to be maintained. These might include family bonds, respect for human life and property, maintenance of individual respect and dignity, and appreciation for others, including appreciation for contributions made by ancestors. Other sociocultural elements are more trivial and can be modified, discontinued, or supplanted with little impact, either positive or negative.

An innovation may be functional for a system but not functional for certain individuals in the system. Consider the example of the adoption of "miracle" varieties of rice and wheat in India that led to what is called the "Green Revolution." These innovations provide higher crop yields and more income to the farmers who adopt. Yet it also leads to a smaller farm labor force, a hastened farm-city exodus to urban slums, higher unemployment rates in the nation, and political instability. So although certain individuals profit from the adoption of the new seeds, it causes important but unequal conditions for the system. Are the consequences functional or dysfunctional? The answer depends on whether one takes certain individuals or the entire system as his point of reference.

The functionality of consequences also depends on time. Obviously, an innovation's short-range and long-range effects may be quite different. Later in this chapter we shall discuss an illustration of this point: Television in the U.S. In its first years of adoption TV watching was a group activity in the home and it seemed to be contributing to the stability of family life. But in later years television has been blamed for emphasizing violent themes to children and contributing to the generation gap between parents and their offspring.

An innovation may be more functional for some individuals than for others; certain positive consequences may occur for certain members of a system at the expense of others. For instance, laggards are the last to adopt innovations; by the time they adopt a new idea they are often forced to do so by economic pressures. But by being the first in the field innovators frequently secure a kind of economic gain called windfall profits.

Windfall profits are a special advantage earned by the first adopters of a new idea in a social system.* Their unit costs are usually lowered and their ad-

---

*Windfall profits, in a more general sense, could be measured in social as well as economic terms. An example is the prestige that the innovator of consumer products may obtain by being the first to use a new idea.

ditions to total production have little effect on the price of the product. However, when all members of a social system adopt a new idea, total production or efficiency increases, and the price of the product or service often goes down. This offsets the advantage of lowered unit costs.

The innovator must take risks in order to earn windfall profits. All new ideas do not turn out successfully, and occasionally the innovator gets his fingers burned. It is possible that adoption of a noneconomic innovation could result in "windfall losses" for the first individuals to adopt.

Windfall profits are a relative type of economic gain that one individual in a social system receives and others do not. Windfall profits are a reward for innovativeness and a penalty for laggardness. We know that innovators are wealthier than laggards. In one sense new ideas may tend to make the rich richer and the poor poorer.

In order to illustrate the nature of windfall profits, data from the Iowa hybrid seed corn study by Gross (1942) were reanalyzed by Rogers (1962b, p. 276). The innovators of this new idea, who adopted in the late 1920s, earned almost $2,500 more than the laggards, who adopted hybrid seed in 1940–1941. The innovators earned these windfall profits because: (1) of a higher market price for corn which lasted only until most farmers adopted hybrid seed and corn production was increased; (2) of their larger corn acreages (for example, the innovator who adopted in 1927 averaged 124 acres of corn while the typical laggard who adopted in 1941 raised only seventy acres of corn); and (3) of the greater number of years they received the higher yields from hybrid seed.*

Direct Versus Indirect Consequences

Very often the consequences of an innovation are not terminated with the direct impact upon the individual who adopts. Shock waves may be set in motion which reach far beyond the immediate environment of the actual adopter. Ogburn and Gilfillian (1963, p. 153) list 150 effects of the radio in the U.S. These consequences spread out through primary, secondary, and tertiary levels; the most direct effects caused more ultimate consequences. Because of the intricate, often invisible web of interrelationships among cultural elements,

*Gross (1942) presented these data on the basis of average year of trial. It was necessary to adjust the data with a factor measuring the length of the trial-adoption period in order to plot them on the basis of year of adoption.

a change in one part of the system often initiates a chain reaction of consequences.

Direct consequences are those changes in a social system that occur in immediate response to an innovation. *Indirect consequences* are changes in a social system that occur as a result of direct consequences of an innovation. An illustration of the direct and indirect consequences of a new idea is detailed in the anthropological study of the adoption of wet rice farming by a tribe in Madagascar (Linton and Kardiner, 1952, pp. 222–231). The tribe had been a nomadic group that cultivated rice by dry-land methods. After each harvest they would move to a different location. Many social changes resulted in the tribe's culture after the adoption of wet-land rice farming. A pattern of land ownership developed, social status differences appeared, the nuclear family replaced the extended clan, and tribal government changed. The consequences of the technological change were both direct and far-reaching, in that a second generation of consequences from wet rice growing spread from the more direct results. This point is illustrated in Figure 11-2.

Manifest Versus Latent Consequences

Manifest consequences are changes that are recognized and intended by the members of a social system. An example of a manifest consequence is the development of new occupational skills, such as that of tractor repairman, necessitated by the mechanization of Turkish agriculture.

Although they are less discernible to casual observers, latent or "subsurface" consequences may be just as important as manifest consequences. *Latent consequences* are changes that are neither intended nor recognized by the members of a social system. The development of new attitudes about ideal family size because of the introduction of birth control techniques may have significant impact in less developed countries. Another example is the disintegration of respect for their elders among the Yir Yoront, in the case study that follows. This change in familial relations was of tremendous importance to that tribe, even though such a consequence was not readily apparent when steel axes were first introduced.

Almost no innovation comes without any strings attached. The more important, the more advanced, the more "modern" the innovation (and therefore the more desire by the change agent for its rapid adoption), the more likely its introduction is to produce many consequences—some of them

Figure 11-2
Paradigm of the direct and indirect consequences of the adoption of wet rice growing in Madagascar.

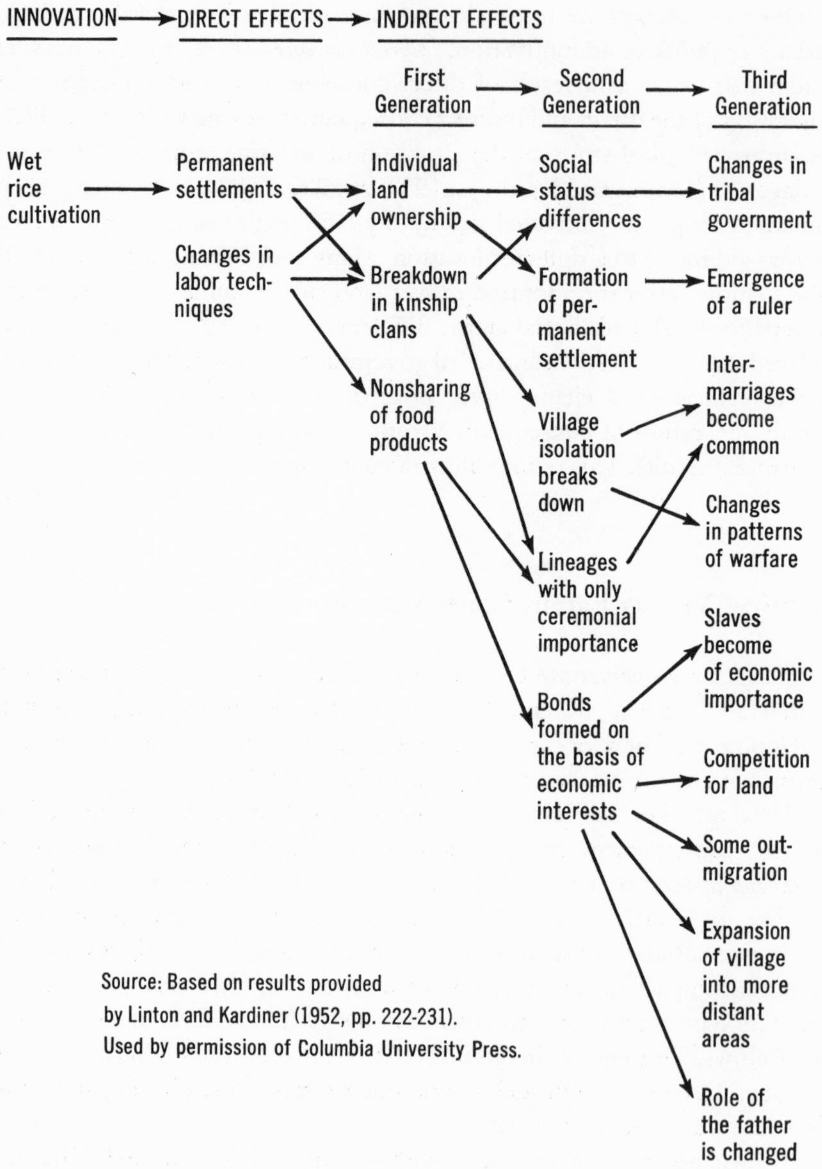

INNOVATION——▶DIRECT EFFECTS ——▶ INDIRECT EFFECTS

| | First Generation | Second Generation | Third Generation |

Wet rice cultivation ——▶ Permanent settlements ——▶ Individual land ownership ——▶ Social status differences ——▶ Changes in tribal government

Changes in labor techniques ——▶ Breakdown in kinship clans ——▶ Formation of permanent settlement ——▶ Emergence of a ruler

Nonsharing of food products

Village isolation breaks down

Inter-marriages become common

Lineages with only ceremonial importance

Changes in patterns of warfare

Slaves become of economic importance

Bonds formed on the basis of economic interests

Competition for land

Some out-migration

Expansion of village into more distant areas

Role of the father is changed

Source: Based on results provided
by Linton and Kardiner (1952, pp. 222-231).
Used by permission of Columbia University Press.

intended and manifest, others unintended and latent.* A system is like a bowl of marbles: Move any one of its elements and the positions of all the others are also changed.

Steel Axes for Stone Age Aborigines**

The consequences of the adoption of steel axes by a tribe of Australian aborigines vividly illustrates the need for consideration of the consequences of an innovation. The tribe was the Yir Yoront, who traveled in small nomadic groups over a vast territory in search of game and other food. The central tool in their culture was the stone ax, which the Yir Yoront found indispensable in producing food, constructing shelters, and obtaining warmth. It is hard to imagine a more complete revolution than that precipitated by the adoption of the steel ax as a replacement for the stone ax.

The method of study used by Sharp (1952) to investigate the Yir Yoront is that of participant observation, in which a scientist studies a culture by taking part in its everyday activities. In the 1930s an American anthropologist was able to live with the Yir Yoront for thirteen months without seeing another white man. Because of their isolation, the natives were relatively unaffected by modern civilization until the establishment of a nearby missionary station in recent years. The missionaries distributed a great many steel axes among the Yir Yoront as gifts and as pay for work performed.

Before the days of the steel ax, the stone ax was a symbol of masculinity and of respect for elders. The men owned the stone axes, but the women and children were the principal users of these tools. The axes were borrowed from fathers, husbands, or uncles according to a system of social relationship prescribed by custom. The Yir Yoront obtained their stone ax heads in exchange for spears through bartering with other tribes, a process which took place as part of elaborate rituals at seasonal fiestas.

*Obviously, the foregoing list of dimensions of consequences does not exhaust all possibilities. Other classifications might include anticipated versus unanticipated, material versus non-material, individual versus system, and perhaps others. However, the dimensions suggested here probably established the point that consequences are not unidimensional, that they take many forms and are expressed in various ways, and that change agencies must recognize the dynamic and complex impact that ensues from the introduction of an innovation in a social system.

**Adapted from Sharp (1952, pp. 69–92) and used by permission.

When the missionaries distributed the steel axes to the Yir Yoront, they hoped that a rapid improvement in living conditions would result. There was no important resistance to the shift from stone to steel axes, because the aborigines were accustomed to securing their tools through trade. Steel axes were more efficient for most tasks, and the stone axes rapidly disappeared among the Yir Yoront.

However, the steel ax contributed little to progress; to the disappointment of the missionaries, the Yir Yoront used their new-found leisure time for sleep, "an act they had thoroughly mastered." The missionaries distributed the steel axes to men, women, and children alike. In fact, the young men were more likely to adopt the new tools than were the elders, who maintained a greater distrust for the missionaries. The result was a disruption of status relations among the Yir Yoront and a revolutionary confusion of age and sex roles. Elders, once highly respected, now became dependent upon women and younger men and were often forced to borrow their steel axes.

The trading rituals of the tribe were also disorganized. Friendship ties among traders broke down, and interest in the fiestas, where the barter of stone axes for spears had formerly taken place, declined. The religious system and social structure of the Yir Yoront became disorganized as a result of inability to adjust to the innovation. Later, the men began the practice of prostituting their daughters and wives in exchange for use of someone else's steel ax.

We see, then, that many of the consequences of the innovation among the Yir Yoront were latent, indirect, and dysfunctional (especially in the eyes of the well meaning missionaries). The case of the steel ax also illustrates a common error made by the change agents in regard to consequences. Agents are able to anticipate the form and function of an innovation but not its meaning for potential adopters. What do we mean by the form, function, and meaning of an innovation?

The Intrinsic Elements of an Innovation

The anthropologist Ralph Linton (1936, pp. 402–404) recognized that every innovation has three intrinsic elements which should ideally be recognized by change agents:

1. *Form,* which is the directly observable physical appearance and substance of an innovation. Both the missionaries and the Yir Yoront recognized the form of the new tool, perhaps in part because of its similarity to the appearance of the stone ax.

2. *Function,* which is the contribution made by the innovation to the way of life of members of the social system. The aborigines immediately perceived the steel ax as a cutting tool, to be used in much the same way as the stone ax had been. The innovation's function was manifest to the receivers.

3. *Meaning,* which is the subjective and frequently unconscious perception of the innovation by members of the social system. Linton (1936) points out that "because of its subjective nature, *meaning* is much less susceptible to diffusion than either *form* or *use* [*function*] . . . A receiving culture attaches new meanings to the borrowed elements or complexes, and these may have little relation to the meanings which the same elements carried in their original setting."*

In view of these intrinsic innovation elements, what mistakes did the missionaries make in the introduction of the steel ax? The change agents seem to have understood the *form* of the innovation. They knew in advance that the Yir Yoront would perceive the steel ax as essentially similar to the stone axes they already possessed.

The missionaries also seem to have partly comprehended the *function* of the steel ax. They believed the Yir Yoront would utilize the new tool in much the same way as the stone ax, such as for brush cutting. But the missionaries made an egregious error in not predicting the *meaning* of the new idea to their clients. They did not anticipate that the steel ax would lead to more hours of sleep, prostitution, and a breakdown of social relationships. Change agents frequently do not sense or understand the social meaning of the innovations they introduce, especially the negative consequences that accrue when an apparently functional innovation is used under different conditions. Change agents are especially likely to make this mistake when they do not empathize fully with the members of the recipient culture, as in cross-cultural contacts or in other heterophilous situations.

So we conclude with Generalization 11-1: *Change agents can more easily anticipate the form and function of an innovation for their clients than its meaning.*

---

*This notion is basically similar to the concept of *reinterpretation,* defined in Chapter 1 as the process which occurs when the receivers use an innovation for different purposes than when it was invented or diffused to them.

The Change Agent's Responsibilities for Consequences

Surely such incidents as the Yir Yoront and the steel ax emphasize that change agencies (as well as social science researchers) must be concerned with the consequences of innovation. If change agents are as familiar with a client system as they should be, they will be able to predict with considerable accuracy the consequences resulting from introduction of an innovation.* How can this ability to predict consequences accurately be developed?

Certainly, trial and error is the most expensive method from the standpoint of wasted human resources and the disastrous disruption of cultures. An improved technique for predicting consequences consists of extensive investigation into the conditions of the receiving system, followed by a test-market pilot program in which the innovation is introduced on a small scale. Such an experimental operation reveals major errors in anticipating the consequences of an innovation. This approach can prove far less costly than the blind introduction of an innovation on a massive scale, based on the vague hope by the change agent that he has correctly guessed the nature of the innovation's consequences.

Change agents should remember that short-range results and long-range effects of innovations may be vastly different and are sometimes contradictory. Is it morally reprehensible to ask clients to tolerate undesirable short-range consequences if the long-range consequences are highly desirable? Is it ethical to introduce an innovation with desirable short-range benefits when long-range effects are highly undesirable? For instance, consider the introduction of television into the U.S. in the late 1940s and early 1950s. This communication innovation had an immediate effect upon family life in that group viewing—especially in the evenings—quickly became a national pastime. But few people anticipated the long-range impact of this fascinating medium of communication on the U.S. population. Almost two decades after its introduction, social commentators assert that violence on the television screen adversely affects the mental and moral welfare of youth, attendance at some spectator sports is damaged whereas others are assisted by the advent of TV, and the technique of national political campaigns has been vastly changed.

*The accuracy with which consequences can be predicted also depends upon such attributes of the innovation as its complexity, relative advantage, and the like.

Change agents have some degree of control over whether certain consequences accrue in the short- or long-range. Certainly, the intensity of promotional effort with which an innovation is introduced into a social system is related to the rapidity with which consequences are felt. A major question which change agents must consider is the ideal rate of change. What rate of change will secure an immediate reaping of benefits and yet not produce a traumatic shock to the client system, followed by more negative consequences?

What Is an Ideal Rate of Change?

Theorists in both the physical and social sciences speak of the concept of equilibrium. *Equilibrium* is the tendency of a system to achieve a balance among the various forces operating within and upon it.* We distinguish three kinds.

Stable equilibrium occurs when there is almost no change in the structure or functioning of the social system. An example is a ball in the bottom of a round-bottomed container. The pressure of gravity forces the ball toward a stationary position. Another example of a stable equilibrium is a traffic circle without any traffic; no vehicles are moving. Perhaps, a completely isolated and traditional peasant village provides a social analogy; the rate of change is almost zero.

Dynamic equilibrium occurs when the rate of change in a social system is commensurate with the system's ability to cope with it. An example is a ball on a flat table-top. A small push moves it to a new position, and it again comes to rest. Or, when an innovation is introduced into a farm community, the new idea is integrated into the pattern of agricultural methods, and a new level of agricultural production is achieved. So there is change occurring in a system in dynamic equilibrium, but it occurs at a rate that allows the system to adapt to it.

Disequilibrium occurs when the rate of change is too rapid to permit the social system to adjust. An analogy is a traffic circle with one too many cars on

---

*Systems analysis, which is an approach that views each part of a system as vitally interrelated with every other part, is especially appropriate to the study of consequences of innovation. Gouldner (1957) discusses ways in which systems analysis may be utilized in applied social science. First, systems models point out the possibility that a change in one part of a system may yield unforeseen and undesirable consequences in another part of the system, due to the interdependence of its elements. Second, systems models suggest that changes may be secured in one part of a system, not only by a direct frontal attack, but also by a circumspect, indirect manipulation of more distantly removed variables.

it; all movement stops. The social disorganization that accompanies disequilibrium marks it as a painful and inefficient way for change to occur in a system.

The long-range goal of planned change is to produce a condition of dynamic equilibrium in the receiving system. Innovations are introduced into the system at a deliberate rate, carefully balancing the system's ability to adjust to the changes. Just as a water container with both a point of entry and a point of exit may receive large amounts of water without spilling over, a social system may be changed from traditional to modern without traumatic disturbances if the proper rate of change is maintained.* But this delicate gauging of the rate of change is extremely difficult.

The Power Elite in a Social System and the Consequences of Innovation

In any social system there is a hierarchy of social statuses. Those at the top, often called the power elite, are mainly responsible for making decisions affecting the entire system. Because of their position of power, the elite are able to act as gatekeepers in determining which innovations enter the system from external sources (Figure 11-3).

The Elite and Restructuring Innovations

The elite are inclined to screen out innovations whose consequences threaten to disturb the *status quo*, for such disruption may lead to a loss of position for the elite. The "dangerous" innovations are often those of a restructuring nature, rather than new ideas which will affect only the functioning of the system. An illustration of this point was presented in Chapter 1, where we suggested that oligarchic leaders in Latin American nations promote technological innovations in agriculture and industry but oppose such re-

---

*The framework used here in describing various kinds of equilibria is borrowed largely from Chin (1966, p. 204). Earlier, Lewin (1958) spoke of balance and equilibrium within social groups.

structuring innovations as land reform and overhauling of the tax system. The leading families in Latin America fear that the consequences of these new ideas, currently promoted by the Alliance for Progress, would change the social structure. Restructuring ideas are usually resisted by the elite.

We conclude with Generalization 11-2: *The power elite in a social system screen out potentially restructuring innovations while allowing the introduction of innovations which mainly affect the functioning of the system.* There is so little empirical support for this generalization that it must be regarded mainly as an hypothesis for future study.

Sometimes the masses, or at least the counter-elite (who are out of power and opposed to the elite), desire the restructuring innovations so much that they overthrow the elite. In a sense this disorganizing event illustrates that consequences can occur as the result of the original rejection of an innovation. We see, then, that the anticipated consequences of innovations can themselves cause consequences.

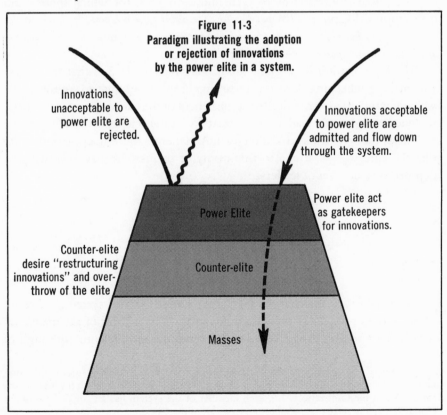

**Figure 11-3
Paradigm illustrating the adoption
or rejection of innovations
by the power elite in a system.**

Innovations unacceptable to power elite are rejected.

Innovations acceptable to power elite are admitted and flow down through the system.

Power Elite

Power elite act as gatekeepers for innovations.

Counter-elite desire "restructuring innovations" and overthrow of the elite

Counter-elite

Masses

The Elite and the Distribution of Good

There is yet another interface between consequences and power: One type of consequences of innovations is a higher *level* of "Good" (such as levels of living) in a system, whereas another consequence is the *distribution* of such Good.* These two types of consequences are often somewhat in conflict. For instance, the tractor mechanization program in Turkey, mentioned earlier in this chapter, raised the *average* level of income in the villages involved, but it also acted to cause a major redistribution of incomes, so that the larger farmers became much richer. Thus, the desired consequences of tractor introduction in Turkey accrued disproportionately to a few individuals. Innovations whose anticipated results raise not only levels of Good but also cause a less equal distribution of the Good are more likely to be favored by the power elite, who act to gain most (Figure 11-4).

This discussion leads us to Generalization 11-3, for which there is yet little empirical support: *The power elite in a social system especially encourage the introduction of innovations whose consequences not only raise average levels of Good but also lead to a less equal distribution of Good.*

The logic behind both Generalizations 11-2 and 11-3 is that the power elite make gatekeeping decisions regarding innovations in terms of their own welfare, rather than altruistically for the benefit of the entire system. Of course, there are enlightened leaders with great social conscience who may put the welfare of the system ahead of their own. And there is the tempering possibility that the elite who continue to think only of themselves may eventually be deposed by a new set of leaders.

Summary

Consequences are the changes that occur within a social system as a result of the adoption or rejection of an innovation. This is third of the main sub-processes of social change, following invention and diffusion. Although of

*One index of the equality of the distribution of a Good in a social system is the Gini-ratio, which expresses the relative degree of concentration of a resource in a few or in many hands. When the Gini-ratio is 1.0, each member of the system has an equal share of the Good. When it is zero, one member possesses all of the Good.

Figure 11-4
Paradigms illustrating two possible consequences of an innovation on the distribution of Good in a system.

1. Before the Innovation

The total amount of income or other Good in the system is relatively small; some portion (say, 10 percent) is held by a wealthy minority.

2. After the Innovation

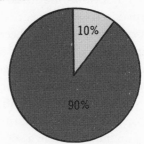

The total amount of Good is now larger, but the proportion held by the wealthy minority remains the same.

I. The Level of Good in a System Increases, but its Distribution Remains the Same.

1. Before the Innovation

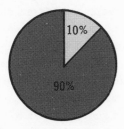

The prior conditions are the same as above.

2. After the Innovation

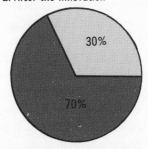

Both the total amount of Good in the system and the proportion of Good held by the wealthy increase as a consequence of the innovation.

II. The Level of Good in a System Increases, and its Distribution Also Changes.

obvious importance, the consequences of innovations have received little attention by change agents or by diffusion researchers, who have concentrated primarily on investigating the correlates of innovativeness. We propose a new model to guide future inquiries of change in which the main dependent variable to be explained is consequences.

Consequences have not been studied because: (1) change agencies have overemphasized adoption *per se,* assuming that the consequences will be positive, (2) survey research methods may be inappropriate for investigating consequences, and (3) consequences are difficult to measure.

Consequences may be classified as (1) functional or dysfunctional, (2) direct or indirect, and (3) manifest or latent. *Functional* consequences are desirable effects of an innovation in a social system, whereas *dysfunctional* consequences are undesirable effects. *Direct* consequences are those changes in a social system that occur in immediate response to an innovation; *indirect* consequences result from the direct consequences. *Manifest* consequences are changes that are recognized and intended by the members of a social system; *latent* consequences are neither intended nor recognized.

The introduction of the steel ax among Australian aborigines brought many dysfunctional and indirect consequences, including breakdown of the family structure, the emergence of prostitution, and "misuse" of the innovation itself. The story of the steel ax illustrates three intrinsic elements of an innovation: (1) *form,* which is the directly observable physical appearance and substance of an innovation, (2) *function,* which is the contribution made by the innovation to the way of life of members of the social system, and (3) *meaning,* which is the subjective and frequently subconscious perception of the innovation by members of the social system. Change agents can more easily anticipate the form and function of an innovation for their clients than its meaning.

In determining an ideal rate of change, the concept of equilibrium must be considered. *Stable equilibrium* occurs when there is almost no change in the structure or functioning of the social system. *Dynamic equilibrium* occurs when the rate of change in a social system is commensurate with the system's ability to cope with it. *Disequilibrium* occurs when the rate of change is too rapid to permit the social system to adjust. Change agents generally wish to achieve a rate of change which leads to dynamic equilibrium, somewhere short of disequilibrium.

There is a concentration of power in the hands of the elite in a system; restructuring innovations are resisted because they are perceived by the elite to have undesirable consequences for their position of advantage. If the positive consequences of an innovation accrue to only a few individuals in a system, the innovation will concentrate the distribution of Good within the system. The elite especially encourage the introduction of innovations whose consequences not only raise average levels of Good but also lead to a less equal distribution of Good.

Appendix A

GENERALIZATIONS

The rate at which empirical results have been adequately digested and integrated into theoretical formulations has not kept pace. If we continue to generate studies at even the present rate, without a major 'leap forward' in terms of integrative theory, we shall drown in our own data.

JOSEPH E. McGRATH AND IRWIN ALTMAN (1966)

ABOUT THE DIFFUSION

OF INNOVATIONS*_.._ ._ ._.___. .___. . _. _..

T HE purpose of this Appendix is to list each of the generalizations contained in this book and to report the empirical diffusion studies that support and that do not support each. A total listing of these generalizations provides a skeleton summary of the major conclusions of what is presently known about the diffusion of innovations.

For each generalization we present the number and percentage of empirical diffusion studies that support and that do not support the generalization. The author(s) and year for each diffusion publication are provided so that the full citation can be located in the bibliography (Appendix B). The page on which the data relevant to the generalization may be found is also provided for each study cited.

1-1: *System effects may be as important in explaining individual innovativeness as such individual characteristics as education, cosmopoliteness, and so on* (7 studies, or 100 percent, support).

Supporting: Davis (1968), Flinn (1961), Flinn (1963), Qadir (1966), Rogers and Burdge (1962), Saxena (1968), van den Ban (1960b).

3-1: *Earlier knowers of an innovation have more education than later knowers* (17 studies, or 71 percent, support; 7 studies do not support).

*This Appendix was prepared by L. Jaganmohan Rao, Research Assistant in the Department of Communication, Michigan State University.

347

> *Supporting:* Abu-Lughod (1963, p. 100), Bauder (1960, p. 74), Bhosale (1960, p. 4), Bogart (1951, p. 771), Feldman (1966, p. 103), Gamson (1961b, p. 6), Glaser (1958, p. 143), Hess (1954, p. 28), Kivlin (1968, p. 37), Martinez (1963, p. 119), Martinez (1964, p. 134), Martinez and others (1964, p. 4), Raina and others (1967, p. 11), Roberts and others (1965, p. 91), Sallan (1966, p. 69), Sinha and Yadav (1964d, p. 19), Zaidi (1968b, p. 28).
>
> *Not supporting:* Celade (1965, p. 110), Danbury and Berger (1965, p. 16), Deutsch-mann and Danielson (1960, p. 16), Deutschmann and others (1965, p. 25), Singh and Akhouri (1966, p. 28), Wilson and Trotter (1933, p. 20), White (1967, p. 80).

3-2: *Earlier knowers of an innovation have higher social status than later knowers** (18 studies, or 64 percent, support; 10 studies do not support).

> *Supporting:* Abu-Lughod (1963, p. 100), Bonjean and others (1965, p. 182), DeFleur and Larsen (1958, p. 164), Glaser (1958, p. 143), Greenberg (1964b, p. 494), Kivlin (1968, p. 37), Larsen and Hill (1954, p. 429), Martinez (1963, pp. 119–120), Martinez (1964, pp. 134–135), Martinez and others (1964, p. 5), Raina and others (1967, p. 36), Sahgal (1966, p. 79), Sallan (1966, p. 69), Sheth (1966, p. 198), Sizer and Porter (1960, p. 5), Spitzer and Denzin (1965, p. 236), Wager (1962, p. 98), Wilson and Trotter (1933, p. 20).
>
> *Not supporting:* Bonjean and others (1965, p. 182), Danbury and Berger (1965, p. 16), Deutschmann and others (1965, p. 25), Dodd (1951, p. 5), Jones (1965, p. 61), Kivlin (1968, p. 38), Larsen and Hill (1954, p. 429), Roberts and others (1965, p. 91), White (1967, p. 81), Winick (1961, p. 388).

3-3: *Earlier knowers of an innovation have greater exposure to mass media channels of communication than later adopters* (18 studies, or 62 percent, support; 11 studies do not support).

> *Supporting:* Aurbach and Kaufman (1956, p. 2), Beal and others (1960, p. 369), Celade (1965, p. 111), Danielson (1956, p. 437), Deutschmann and Danielson (1960, p. 15), Deutschmann (1963, p. 120), Feldman (1966, p. 165), Greenberg and others (1965, p. 41), Katz and Menzel (1954, p. 4), Kivlin (1968, p. 37), Kivlin (1968, p. 38),** Koya (1964, p. 142), Larsen and Hill (1954, p. 428), Lassey (1968, p. 10), Leuthold (1960, p. 2), McNelly (1964, p. 12), Raina and others (1967, p. 41), Spector and others (1963, p. 35).
>
> *Not Supporting:* Deutschmann and Danielson (1960, p. 16), Deutschmann and others (1965, p. 25), De Fleur and Rainboth (1952, p. 736), Fonseca (1966, p.

---

*Different indicators of social status include self-perceived identification with a social class, occupational prestige, income, level of living, and possession of economic resources.
**Notice that it is possible for the same study to be cited twice (often with different page numbers) when two different measures of the same concept were utilized.

70), Jones (1965, p. 63), Keith (1968b, p. 108), Keith (1968a, p. 143), Myren (1962, p. 3), Sinha and Yadav (1964d, p. 19), White (1967, p. 84), White (1967, p. 85).

3-4: *Earlier knowers of an innovation have greater exposure to interpersonal channels of communication than later adopters* (16 studies, or 89 percent, support; 2 studies do not support).

Supporting: Aurbach and Kaufman (1956, p. 2), Bonjean and others (1965, p. 181), Deutschmann and others (1965, p. 34), Inayatullah (1964, p. 14), Katz and Menzel (1954, p. 4), Kivlin (1968, p. 37), Koya (1964, p. 143), Lassey (1968, p. 10), Martinez (1963, p. 120), Martinez (1964, p. 135), Martinez and others (1964, p. 5), McMillion (1960, p. 15), Myren (1962, p. 3), Periera de Melo (1964, p. 208), Rahudkar (1962, p. 101), Requena (1965, p. 82).
Not Supporting: Spitzer and Denzin (1965, p. 236), Troldahl (1963b, p. 14).

3-5: *Earlier knowers of an innovation have greater change agent contact than later knowers* (23 studies, or 81 percent, support; 3 studies do not support).

Supporting: Aurbach and Kaufman (1956, p. 2), Beal and others (1958a, p. 9), Hsieh (1966, p. 17), Inayatullah (1964, p. 15), Katz and Menzel (1954, p. 4), Kivlin (1968, p. 37), Kivlin (1968, p. 38), Koya (1964, p. 143), Leuthold (1960, p. 2), Moulik and others (1966, p. 473), Neurath (1962, p. 279), Polson and Pal (1958, p. 73), Sinha and Yadav (1964d, p. 19).
Not Supporting: Jones (1965, p. 63), Sallan (1966, p. 70), Wilkening and Santopolo (1952, p. 13).

3-6: *Earlier knowers of an innovation have more social participation than late knowers** (11 studies, or 85 percent, support; 2 studies do not support).

Supporting: Delenne and others (1966, p. 214), Erbe (1962, p. 509), Hsieh (1966, p. 3), Kivlin (1968, p. 38), Leuthold (1965, p. 103), Lu and others (1967, p. 174), Martinez (1963, p. 119), Martinez (1964, p. 134), Martinez and others (1964, p. 5), Sheth (1966, p. 2), Sinha and Yadav (1964d, p. 19).
Not Supporting: Sheth (1966, p. 2), White (1967, p. 82).

3-7: *Earlier knowers of an innovation are more cosmopolite than later knowers* (5 studies, or 100 percent, support).

Supporting: Hobbs (1960, p. 64), Kivlin (1968, p. 38), Martinez (1963, p. 120), Martinez (1964, p. 135), Martinez and others (1964, p. 5).

---

*Social participation includes organization membership as well as informal group participation.

3-8: *Later adopters are more likely to discontinue innovations than are earlier adopters* (6 studies, or 100 percent, support).

Supporting: Bishop and Coughenour (1964), Deutschmann and Havens (1965), Johnson and van den Ban (1959), Leuthold (1965), Leuthold (1967), Silverman and Bailey (1961).

3-9: *Innovations with a high rate of adoption have a low rate of discontinuance* (4 studies, or 100 percent, support).

Supporting: Coughenour (1961), Johnson and van den Ban (1959), Leuthold (1965), Silverman and Bailey (1961).

3-10: *Traditional individuals are more likely to skip functions in the innovation-decision process than are modern individuals* (3 studies, or 100 percent, support).

Supporting: Deutschmann and Fals Borda (1962b), Oliver-Padilla (1964), Rahim (1961).

3-11: *There are functions in the innovation-decision process* (13 studies, or 100 percent, support).

Supporting: Beal and others (1957), Beal and Rogers (1960), Coleman and others (1966), Copp and others (1958), Kohl (1966, p. 68), LaMar (1966), Mason (1962), Mason (1963), Mason (1964), Mason (1966a), Mason (1966b), Singh and Pareek (1968), Wilkening (1956).

3-12: *The rate of awareness-knowledge for an innovation is more rapid than its rate of adoption* (2 studies, or 100 percent, support).

Supporting: Beal and Rogers (1960, p. 8), Ryan and Gross (1950, p. 679).

3-13: *Earlier adopters have a shorter innovation-decision period than later adopters* (5 studies, or 84 percent, support; 1 study does not support).

Supporting: Beal and Rogers (1960, p. 14), Petrini (1967), Rogers and Yost (1960, p. 28), Ryan (1948), Wilkening and Santopolo (1952, p. 31).
Not Supporting: Haber (1961).

4-1: *The relative advantage of a new idea, as perceived by members of a social system, is positively related to its rate of adoption* (29 studies, or 67 percent, support; 14 studies do not support).

Supporting: Alers-Montalvo (1957, p. 6), Barnett (1952, p. 359), Erasmus (1952a, p. 20), Erasmus (1952b, p. 418), Erasmus (1962, p. 7), Feliciano (1964,

p. 12), Fliegel and Kivlin (1962a, p. 11), Fliegel and others (1967a, p. 13), Griliches (1957), Griliches (1960a, p. 355), Griliches (1960b), Griliches (1962), Hochstrasser (1963, p. 144), Kivlin (1960, p. 49), Kivlin and Fliegel (1968), Mansfield (1960, p. 9), Mansfield (1961, p. 763), Mansfield (1961, p. 5), Mansfield (1963, p. 12), Mansfield (1964, p. 358), Mansfield (1968, p. 153), McCorkle (1964, p. 12), Miller (1957, p. 272), Petrini (1966a, p. 60), Singh (1966, p. 3), Suttles (1951, p. 280), Tajima (1959, p. 7), Toussaint and Stone (1960, p. 242), Tully and others (1954, pp. 307–309).

Not Supporting: Brandner (1960, p. 150), Brandner and Straus (1959, p. 382), Brandner and Kearl (1964, p. 298), Carlson (1965, p. 73), Coughenour (1962, p. 5), Danda and Danda (1968, p. 192), Fliegel and Kivlin (1962b, p. 366), Fliegel and Kivlin (1966, p. 202), Kivlin and Fliegel (1967, p. 87), Mansfield (1961, p. 6), Massey (1960, p. 183), Tucker (1961, p. 53), Wilkening and Johnson (1958, p. 14), Wilkening (1961, p. 6).

4-2: *The compatibility of a new idea, as perceived by members of a social system, is positively related to its rate of adoption* (18 studies, or 67 percent, support; 9 studies do not support).

Supporting: Alers-Montalvo (1957, p. 6), Barnett (1952, p. 360), Brandner (1960, p. 150), Brandner and Kearl (1964, p. 301), Brandner and Strauss (1959, p. 382), Danda and Danda (1968, p. 197), Erasmus (1962, p. 8), Feliciano (1964, p. 12), Fliegel and Kivlin (1962a, p. 332), Fliegel and Kivlin (1962b,) Fliegel and others (1967a, p. 17), McCorkle (1961, p. 22), Magdub (1964, p. 12), Mahoney (1960, p. 465), Mead (1955, pp. 258, 262), Suttles (1951, p. 283), Tajima (1959, p. 7), Tully and others (1964, p. 301), Zwerman (1956, p. 127).

Not Supporting: Carlson (1965, p. 73), Fliegel and Kivlin (1962, p. 366), Fliegel and Kivlin (1966, p. 202), Fliegel and Kivlin (1966a, p. 241), Havens (1961, p. 414), Hodgdon and Singh (1963, p. 37), Kivlin (1960, p. 51), Kivlin and Fliegel (1967, p. 89), Pemberton (1936, p. 231).

4-3: *The complexity of an innovation, as perceived by members of a social system, is not related to its rate of adoption* (9 studies, or 56 percent, support; 7 studies do not support).

Supporting: Carlson (1965, p. 73), Danda and Danda (1968, p. 200), Feliciano (1964, p. 12), Fliegel and Kivlin (1966, p. 241), Hsu (1955, p. 149), Prundeanu and Zwerman (1958a, p. 440), Singh (1965, p. 117), Tajima (1959, p. 7), Tucker (1961, p. 54).

Not Supporting: Erasmus (1952, p. 21), Fliegel and Kivlin (1962a, p. 10), Fliegel and Kivlin (1962b, p. 360), Kivlin (1960, p. 39), Kivlin and Fliegel (1967, p. 88), Petrini (1966, p. 60), Singh and Warlow (1966, p. 3).

4-4: *The trialability of an innovation, as perceived by members of a social system, is positively related to its rate of adoption* (9 studies, or 69 percent, support; 4 studies do not support).

> *Supporting:* Fliegel and Kivlin (1962a, p. 10), Fliegel and Kivlin (1962b, p. 360), Fliegel and Kivlin (1966a, p. 202), Fliegel and Kivlin (1966b, p. 241), Fliegel and others (1967a, p. 15), Kivlin (1960, p. 70), Kivlin and Fliegel (1967, p. 87), Polgar (1963), Singh and Warlow (1966, p. 3).
>
> *Not Supporting:* Carlson (1955, p. 73), Danda and Danda (1968, p. 201), Tucker (1961, p. 53), Wilkening and Johnson (1961, p. 6).

4-5: *The observability of an innovation, as perceived by members of a social system, is positively related to its rate of adoption* (7 studies, or 78 percent, support; 2 studies do not support).

> *Supporting:* Fliegel and others (1967a, p. 16), Havens and Rogers (1961b), Kivlin and Fliegel (1967, p. 88), Mansfield (1961a), Polgar (1963, p. 110), Rogers and Havens (1962b), Singh and Warlow (1966, p. 3).
>
> *Not Supporting:* Carlson (1965, p. 73), Tucker (1961, p. 54).

4-6: *The degree of communication integration in a social system is positively related to the rate of adoption of innovations* (6 studies, or 100 percent, support).

> *Supporting:* Coleman and others (1966), Coughenour (1962), Guimarães (1968), Mendez (1964), Mendez (1968), Yadav (1967).

5-1: *Earlier adopters are no different from later adopters in age* (44 studies, or 19 percent, show that earlier adopters are younger; 108 studies, or 48 percent, show no relationship; and 76 studies, or 33 percent, show that earlier adopters are older).

> *Younger:* Arndt (1966a, p. 138), Arndt (1968c, p. 4), Barnabas (1960, p. 26), Beal and Sibley (1962, p. 102), Beal and others (1967a, p. 15), Belcher (1968, p. 166), Bhasin (1966, p. 63), Bhatia (1966, p. 135), Blackmore and others (1955, p. 4), Bose (1960, p. 1), Buley (1947, p. 67), Bylund (1963, p. 9), Clark and Abell (1966, p. 4), Copp (1958, p. 110), Dhaliwal (1963, p. 3), Eibler (1965, p. 3), Fougeyrollas (1967, p. 31), Gupta (1966, p. 110), Havens (1963, p. 58), Jones (1960b, p. 1), Jones (1962b, p. 10), Kahlon and Kaushal (1967, p. 94), King (1964, p. 331), Koontz (1958, p. 74), Larsen and De Fleur (1954, p. 596), Martinez (1963, p. 121), Martinez (1964, p. 6), Martinez and others (1964, p. 135), Nichols (1959, p. 126), Rao (1966 p. 15), Robertson and Rossiter (1968, p. 15), Rogers (1961b, p. 64), Ross (1966a, p. 11), Russell (1964, p. 2), Sabri (1966, p. 103), Schindler (1962, p. 46), Singh (1966, p. 82), Sprunger (1968, p. 85), Takeshita (1966, p. 701), van den Ban (1953, p. 320),

Wilkening and others (1962, p. 164), Wish (1967, p. 164), Yang and others (1965, p. 246).

No relationship: Ahmad (1964, p. 13), Arndt (1968a, p. 6), Bang (1966, p. 8), Barnabas (1957, p. 2), Barnabas (1960, p. 216), Bauder (1961, p. 24), Beal and others (1967b, p. 74), Bell (1962, p. 1), Bertrand and others (1956, p. 9), Bhosale (1960, p. 4), Bonser (1958b, p. 9), Booth (1966, p. 8), Bose and Saxena (1965, p. 141), Bostian and Oliveira (1965, p. 9), Brady and Adams (1962, p. 6), Brandner and Kearl (1964, p. 296), Burleson (1950b, p. 34), Celade (1965, p. 110), Chopde and others (1959, p. 2), Chow (1965, p. 158), Christiansen and Taylor (1966, p. 20), Copp (1956, p. 13), Correa (1965, p. 40), Cummings (1950, p. 63), Danda and Danda (1968, p. 174), Dantre (1963, p. 3), Davis (1965, p. 109), Deutschmann and others (1965, p. 24), Dhara (1965, p. 26), Dickerson (1967, p. 56), Dubey (1967, p. 2), Fallding (1957, p. 36), Fischer and Timmons (1959, p. 437), Fosen (1956, p. 52), Frank and Tietze (1965, p. 123), Goldsen and Ralis (1957, pp. 30, 32), Goldstein and others (1961, p. 28), Forman (1967b, p. 18), Marsh and Coleman (1954, p. 113), Havens (1966, p. 150), Hay and Lowry (1957, p. 6), Hay and Lowry (1958, p. 9), Hicks (1942, p. 79), Igbal (1963, p. 82), Inayatullah (1964, p. 15), Ingman (1963, p. 55), International Research Associates (1963, p. 5), Jain (1965, p. 68), Johnson and others (1967, p. 27), Jones (1960b, p. 1), Juliano (1967, p. 199), Junghare (1962, p. 294), Junghare and Roy (1963, p. 397), Kahlon and Kaushal (1967, p. 48), Kar (1967, p. 36), Katz (1956, p. 206), Katz and Menzel (1954, p. 26), Kim (1967, p. 92), Klonglan and others (1964, p. 175), Letourneau (1963, p. 73), Leuthold (1960a, p. 3). Leuthold (1960b, p. 40), Leuthold (1965, p. 68), Lionberger (1956a, p. 10), Loomis (1967b, p. 885), Lowry and Hay (1957, p. 5), Lowry and others (1958, p. 201), Mahajan (1966, p. 2), Mansfield (1960, p. 20), Mansfield (1968, p. 172), Marsh and Coleman (1955b, p. 290), Mauldin (1967, p. 2), McMillion (1960, p. 19), Metraux (1959, p. 6), Miller (1965, p. 20), Oakley (1965, p. 113), Paul (1965, p. 30), Prasada (1966, p. 61), Presser and Russell (1965, p. 155), Presser and Russell (1965, p. 156), Purohit (1963, p. 86), Rahim (1961a, p. 34), Rahudkar (1958, p. 123), Rahudkar (1961, p. 81), Rahudkar (1962, p. 83), Rao (1961, p. 16), Reddy (1962, p. 28), Reddy and Kivlin (1968, p. 20), Rizwani (1964, p. 94), Rogers and Burdge (1961, p. 18), Shetty (1966, p. 196), Singh (1967, p. 150), Singh (1965, p. 136), Sizer and Porter (1960, p. 5), Stapel (1960, p. 14), Stickley (1964, p. 76), Stycos (1965b, p. 130), Tajima (1959, p. 18), Takes (1963, p. 89), Thorat (1968, p. 8), Ullah (1962, p. 79), Verma (1966, p. 114), Verner and Millerd (1966, p. 12), Voget (1948, p. 645), Wells and Andapia (1966, p. 485), Westoff and others (1961, p. 69), Wilkening (1952a, p. 61), Wilson and Trotter (1933, p. 20), Yeracaris (1961, p. 7).

Older: Afaruala (1961, p. 114), Alves (1962, p. 29), Anastasio (1960, p. 38), Bakshi (1962, p. 2), Barnett (1941, p. 162), Beal and Rogers (1960, p. 13), Bell

(1964, p. 90), Bertrand (1951, p. 48), Bogue and Palmore (1966, p. 14), Bonser (1958a, p. 11), Burleson (1950b, p. 18), Carlson (1965, p. 56), Chaparro (1955, p. 183), Christiansen (1965, p. 165), Coughenour (1960, p. 297), DeFleur and Larson (1958, p. 230), Denis (1961, p. 4), Eastmond (1953, p. 326), Ellenbogen (1964, p. 19), Ellenbogen and others (1961, p. 23), Emery and Oeser (1958, p. 58), Erasmus (1952a, p. 21), Fallding (1958, p. 39), Feaster (1966, p. 10), Fliegel (1962, p. 3), Freedman and Takeshita (1965, p. 247), Geiger and Sokol (1960, p. 51), Gorman (1966, p. 160), Gross (1942, p. 88), Gross (1949, p. 152), Gross and Taves (1952, p. 326), Horn and others (1959, p. 511), Hildenbrand and Partenhiemer (1958, p. 447), Johnson (1966, p. 36), Jones (1960a, p. 6), Kaufman and Bryant (1958, p. 1), Keenan (1964, p. 20), Kivlin (1968, p. 37), Khan (1964, p. 128), Klietsch (1961, p. 74), Klonglan (1963, p. 89), Klonglan and others (1960, p. 20), Kunstadter (1961, p. 676), Lackey (1958, p. 28), Lionberger (1958, p. 4), Louisiana Agricultural Extension Service (1950, p. 113), Madigan (1962, p. 207), Magdub (1964, p. 12), Marion (1962, p. 192), Mort and Cornell (1938, p. 42), Planning Research and Action Institute (1961, p. 29), Ploch (1960, p. 8), Polgar and others (1963, p. 110), Richardson (1964, p. 74), Rogers (1959b, p. 12), Rogers (1962a, p. 66), Rohrer (1955, p. 1), Ross and Bang (1965, p. 20), Roy (1967, p. 205), Sahgal (1966, pp. 124, 158), Sicinski (1964, p. 57), Sill (1958, p. 42), Stock and Johnson (1966, p. 8), Stuby (1962, p. 65), Suchman (1966, p. 13), Tajima (1959, p. 18), van den Ban (1956, p. 42), van den Ban (1957a, p. 207), van den Ban (1958, p. 52), White and Heighton (1968, p. 65), White (1968, p. 34), Whyte (1964, p. 142), Wilkening and others (1963, p. 7), Wilkinson and Bailey (1964, p. 13), Yeracaris (1961a, p. 301), Young (1959, p. 2).

5-2: *Earlier adopters have more years of education than do later adopters* (203 studies, or 74 percent, support; 72 do not support).

Supporting: Advertest Research (1958, p. 6), Alves (1962, p. 29), Anastasia (1960, p. 41), Andrus (1965, p. 97), Arell (1960, p. 8), Arias (1967, p. 51), Baden-horst (1962, p. 289), Bakshi (1962, p. 2), Barg (1966, p. 8), Barnabas (1960, p. 214), Beal and others (1959b, p. 23), Beal and others (1967a, p. 8), Beal and Sibley (1967, p. 102), Belcher and Hay (1959, p. 18), Bell (1962, p. 1), Bell (1964, p. 90), Berelson (1966, p. 662), Bhasin (1966, p. 63), Bhatia (1966, p. 119), Bhosal (1960, p. 4), Blackmore and others (1955, p. 4), Bonser (1958a, p. 13), Bonser (1958b, p. 10), Booth (1966, p. 8), Bose (1960, p. 1), Bostian and Oliveira (1965, p. 5), Brady and Adams (1962, p. 6), Beeley (1947, p. 67), Burleson (1950a, p. 11), Burleson (1950b, p. 18), Bylund (1963, p. 7), Campbell and Lionberger (1963, p. 28), Carlson (1965b, p. 26), Cassel (1963, p. 744), Castillo (1964, pp. 6, 11), Chopde and others (1959, p. 3), Chow (1965,

p. 160), Christiansen and Taylor (1966, p. 20), Christiansen (1965, p. 166), Copp (1956, p. 14), Copp (1958, p. 110), Coughenour (1960, p. 296), Coughenour and Kothari (1962, p. 6), Cummings (1950, p. 127), Danda and Danda (1968, p. 175), Dantre (1963, p. 3), Dasgupta (1963b, p. 7), Daspurohit (1963, p. 85), Denis (1961, p. 4), Deshmukh (1960, p. 8), Dhaliwal (1963, p. 3), Diswath (1964, p. 6), Dodd (1934, p. 78), Dubey (1966, p. 2), Eastmond (1953, p. 19), Eibler (1965, p. 185), Ellenbogen (1964, p. 16), Ellenbogen and and others (1961, p. 23), Erasmus (1952b, p. 419), Fallding (1958, p. 39), Feaster (1966, p. 10), Finley (1965, pp. 35, 44), Finley (1968, p. 13), Fonseca (1966, p. 64), Fosen (1956, p. 50), Freedman and others (1964, p. 5), Geiger and Sokol (1960, p. 51), Georgiopoulas (1967, p. 76), Gillis (1958, p. 6), Glaser (1958, p. 143), Goldstein and others (1961, p. 16), Gorman (1966, p. 161), Gorman (1967b, p. 6), Graham (1956, p. 96), Gross (1949, p. 151), Gross and Taves (1952, p. 326), Graham and Gibson (1967, p. 6), Gupta (1966, p. 110), Havens and Rogers (1961a, p. 14), Hay and Lowry (1958, p. 9), Higgins (1962, p. 75), Hoffer and Strangland (1958b, p. 15), Igbal (1963, p. 83), Ingman (1963, p. 55), International Research Associates (1963, p. 5), Johnson (1966, p. 36), Jones (1960, p. 1), Jones (1960a, p. 6), Jones (1962b, p. 10), Julian (1967, p. 199), Junghare (1962, p. 294), Kahlon and Kaushal (1967, p. 96), Kar (1967, p. 36), Katz (1956, p. 206), Katz (1961, p. 26), Keenan (1964, p. 20), Kivlin (1968, p. 37), Klonglan and others (1960, p. 20), Klonglan and others (1963, p. 90), Klonglan and others (1964, p. 175), Koontz (1958, p. 74), Lackey (1958, p. 28), Letourneau (1963, p. 73), Leuthold (1960a, p. 3), Leuthold (1960b, p. 40), Leuthold (1965, p. 68), Leuthold and Wilkening (1965, p. 8), Lionberger (1950, p. 205), Loomis (1967b, p. 885), Louisiana Agricultural Extension Service (1950, p. 12), Lowry and Hay (1958, p. 12), Lowry and others (1958, p. 201), Loy (1967, p. 7), Loy (1967, p. 153), Loy (1968b, p. 8), Maalout (1965, p. 117), Madigan (1962, p. 207), Magdub (1964, p. 12), Mahajan (1966, p. 2), Makarczyk (1965, p. 51), Marsh and Coleman (1954b), Marsh and Coleman (1955b, p. 290), Martinez (1963, p. 122), Martinez (1964, p. 135), Martinez and others (1964, p. 6), McMillion (1960, p. 16), Miro and Rath (1965, p. 58), Mookherjee and Singh (1966, p. 11), Morgan and others (1966, p. 209), Morrison (1956, p. 283), Mort and Cornell (1938, p. 54), Mort and Cornell (1941, p. 94), Nichols (1959, p. 126), Oakley (1965), Parish (1954, p. 211), Ploch (1960, p. 8), Prasada (1966, p. 62), Presser and Russell (1965, p. 155), Rahudkar (1961, p. 81), Rahudkar (1962, p. 83), Rajagopalan and Singh (1967, p. 89), Reddy (1962, p. 28), Research Division, U.S.AID/Bangkok (1966, p. 1), Richardson (1964, p. 73), Roberts and others (1965, p. 91), Rogers (1959a, p. 133), Rogers (1959b, p. 12), Rogers (1961b, p. 64), Rogers (1962, p. 62), Rogers and Burdge (1961, p. 18), Rogers and Cartano (1963, p. 9), Rogers and Pitzer (1960, p. 17), Rohwer (1949, p. 1), Ross and Bang (1965, p. 20), Ross (1955, p. 176), Roy and others (1968, p. 64),

Sabri (1966, p. 108), Sawhney (1962, p. 38), Sicinski (1964, p. 58), Sill (1958, p. 44), Singh (1967, p. 151), Schindler (1962, p. 46), Singh (1965, p. 3), Singh (1965, p. 135), Singh (1966, p. 81), Sinha and Yadav (1964c, p. 15), Sinha and Yadav (1964e, p. 23), Sinha and Yadav (1964f, p. 26), Sizer and Porter (1960, p. 5), Smith (1964, p. 299), South and others (1965, p. 9), Spector and others (1963, p. 46), Spencer (1958, p. 68), Sprunger (1968, p. 85), Straus and Estep (1959, p. 13), Tajima (1959, p. 18), Takes (1963, p. 89), Takeshita (1964, p. 2), Thorat (1968, p. 9), van den Ban (1953, p. 320), van den Ban (1956, p. 42), van den Ban (1957a, p. 209), van den Ban (1958, p. 63), van den Ban (1958, p. 52), van den Ban (1963b, p. 239), Waisanen (1952), Wang (1962, p. 81), Watson (1946), Westermarck (1963, p. 62), Wheelock (1964), Wilkening (1949b, p. 2), Wilkening (1952a, p. 61), Wilkinson and Bailey (1964, p. 13), Wilson (1928, p. 20), Wilson and Gallup (1955, p. 23), Wish (1967, p. 164), Yeracaris (1961a, p. 301), Yeracaris (1961, p. 8), Young (1959, p. 2), Young and Marsh (1956, p. 5).

Not supporting: Ahmad (1964, p. 13), Armstrong (1959, p. 36), Barnabas (1957, p. 2), Bauder (1961, p. 25), Beal and others (1967b, p. 75), Belcher (1958, p. 167), Bertrand and South (1963, p. 11), Bogue and others (1966, p. 26), Brandner and Kearl (1964, p. 297), Card (1960, p. 116), Celade (1965, p. 110), Chaparro (1955, p. 184), Correa (1965, p. 40), Couch (1965, p. 14), Dasgupta (1966, p. 11), Dasgupta (1966, p. 23), Dernburg (1958, p. 42), Dhara (1965, p. 26), Fliegel (1965, p. 288), Frank and others (1964, p. 13), Frank and Tietze (1965, p. 123), Gillespie (1965, p. 32), Graham (1954, p. 168), Graham (1956, p. 95), Havens (1965, p. 156), Havens (1966, p. 150), Hay and Lowry (1957, p. 9), Hochstrasser (1963, p. 146), Horn and others (1959, p. 505), Hildebrand and Partenheimer (1958, p. 447), Jain (1965, p. 48), Junghare and Roy (1963, p. 397), Klietsch (1961, p. 72), Kim (1967, p. 100), Khan and Choldin (1965, p. 4), Kunstadter (1961, p. 680), Leuthold (1965, p. 68), Lionberger (1951, p. 19), Lionberger (1956a, p. 10), Lionberger (1958, p. 4), Maamary (1965, p. 24), Marion (1966, p. 190), Miller (1965, p. 20), Nichols (1959, p. 126), Paul (1965, p. 31), Planning Research and Action Institute (1961, p. 28), Presser and Russell (1965, p. 156), Rao (1966, p. 19), Reddy and Kivlin (1968, p. 20), Rijwani (1964, p. 103), Sepulveda and Carter (1964, p. 2), Shetty (1966, p. 196), Sills and Gill (1959, p. 247), Singh and Pardasani (1967, p. 146), Stock and Johnson (1966, p. 13), Straus (1960, p. 223), Stuby (1965, p. 64), Tajima (1959, p. 18), Takeshita (1964, p. 2), Takeshita (1966, p. 702), van den Ban (1960, p. 308), Verma (1966, p. 114), Verner and Millerd (1966, p. 13), Weldon (1966, p. 68), Wells and Andapia (1966, p. 485), Wells and MacLean (1962, p. 16), Westoff and others (1961, p. 69), White (1965, p. 41), Whittenbarger and Maffei (1966, p. 3), Wilkening and others (1962, p. 163), Wilkening and others (1963, p. 7), Wilson and Trotter (1933, p. 20).

5-3: *Earlier adopters are more likely to be literate than are later adopters* (24 studies, or 63 percent, support; 14 do not support).

> *Supporting:* Barnabas (1960, p. 218), Berelson (1966), Bose and Saxena (1965, p. 141), Bose (1964a, p. 15), Bose (1964c, p. 10), Bose and Dasgupta (1962, p. 10), Bostian and Oliveira (1965, p. 9), Choudhary and Maharaja (1966, p. 163), Dickerson (1967, p. 57), Goldsen and Ralis (1957, p. 33), Goldsen and Ralis (1957, p. 18), Kivlin (1968, p. 37), Madigan (1962, p. 207), Rahim (1961a, p. 34), Ramos (1966, pp. 50, 60), Reddy and Kivlin (1968, p. 20), Rhoads and Piper (1963, p. 662), Roy (1967, p. 205), Singh and Reddy (1965b, p. 267), Spector and others (1963, p. 53), Spector and others (1964, p. 60), Stickley (1964, p. 86), Stickley (1964, p. 87), Sycip (1964, p. 597).

> *Not supporting:* Beal and others (1967a, p. 102), Beal and Rogers (1962a, p. 8), Bose (1961, p. 144), Bostian and Oliveira (1965, p. 10), Dasgupta (1966, p. 23), Deutschmann and others (1965, p. 24), Desai and Sharma (1966, p. 145), Goldsen and Ralis (1957, p. 18), Hrabouszky (1966, p. 12), Inayatullah (1964, p. 15), Rao (1961, p. 16), Rogers and Herzog (1966, p. 20), Rogers and Ramos (1965, p. 7), Ullah (1962, p. 79).

5-4: *Earlier adopters have higher social status than later adopters* (275 studies, or 68 percent, support; 127 studies do not support).

> *Supporting:* Abell and Larsen (1960, p. 8), Adams and others (1963, p. 82), Advertest Research (1958, p. 6), Ahmad (1964, p. 13), Andrus (1965, pp. 94, 97), Arias (1967, p. 51), Arndt (1968c, p. 4), Ayer (1952, p. 24), Badenhorst (1961, p. 292), Bang (1966, p. 8), Barber and Lobel (1952, p. 126), Barnabas (1957, p. 2), Barnabas (1960, pp. 20, 205, 214), Bauder (1961, p. 26), Beal and others (1958a, p. 13), Beal and others (1959b, p. 23), Beal and others (1967a, p. 9), Belcher and Hay (1959, p. 16), Belcher and Hay (1959, p. 18), Bell (1962, p. 1), Bertrand (1951, p. 48), Bertrand and others (1956, p. 28), Biggar and Sauer (1966, p. 16), Bhatia (1966, p. 119), Bhasin (1966, p. 63), Bose (1964a, pp. 12, 13), Bose and Dasgupta (1962, p. 9), Bose (1960, p. 1), Bose (1961, p. 473), Bose and Saxena (1965, p. 141), Bonser (1958a, p. 18), Burleson (1950a, p. 11), Burleson (1950b, pp. 18, 32), Bylund (1963, p. 9), Campbell and Bennett (1961, pp. 111, 119), Campbell and Lionberger (1963, p. 28), Carlson (1964, pp. 334, 336), Carlson (1965, pp. 26, 55, 56, 64), Cassel (1963, p. 744), Cawelt (1967, p. 60), Central Treaty Organization (1962, p. 143), Chaparro (1955, p. 184), Chaudhary and Maharaja (1966, p. 163), Chopde and others (1959, p. 4), Clark (1958, p. 2), Copp (1956, pp. 16, 19, 20), Copp (1958, pp. 106, 110), Correa (1965, p. 40), Coughenour (1960, p. 296), Coughenour (1966, pp. 7, 9), Danda and Danda (1968, p. 179), Dantre (1963, pp. 4, 5), Dasgupta (1963a, p. 30), Dasgupta (1963b, p. 7), Dasgupta (1966, p. 23), Deasy (1956, p. 187), Denis (1961, p. 4), Dernburg (1958, p. 43), Deshmukh

(1960, p. 8), Dhaliwal (1963, pp. 2, 3), Dickersen (1967, p. 77), Diswath (1964, p. 6), Dodd (1934, p. 78), Dubey (1966, p. 2), Dubey (1967, p. 115), Duncan and Kreitlow (1954, p. 355), Ellenbogen and others (1961, p. 23), Erasmus (1952b, pp. 420, 421), Fallding (1958, p. 37), Fals Borda (1960, p. 21), Feaster (1966, p. 10), Feliciano (1964, p. 12), Finley (1965, pp. 35, 44), Finley (1968, pp. 13, 14), Fischer and Timmons (1959, p. 434), Fliegel (1955, p. 21), Fliegel (1956, p. 289), Fliegel (1957, p. 3), Fliegel (1960, p. 349), Fliegel (1962, p. 3), Fliegel (1967a, p. 15), Flinn (1961, p. 33), Flinn (1963, p. 9), Fonseca (1966, p. 63), Freedman and others (1964, p. 6), Geiger and Sokol (1960, p. 51), Georgiopoulos and Potter (1967, p. 72), Georgiopoulos and Potter (1967, p. 74), Gillis (1958, p. 6), Glasser (1958, pp. 143, 144), Goldsen and Ralis (1957, p. 16), Gorman (1966, pp. 160, 161), Gorman (1967b, pp. 6, 7), Graham (1951, p. 292), Graham (1956, p. 96), Graham and Gibson (1967, p. 6), Gross (1942, p. 94), Gross (1947, p. 153), Gruen (1955, p. 197), Gupta (1966, pp. 110, 111), Gugman (1964, p. 9), Hall (1964, p. 434), Harris (1956, p. 82), Havens (1962, p. 84), Havens (1963, p. 58), Havens (1965, p. 158), Hay and Lowry (1957a, pp. 6, 7, 9), Hay and Lowry (1957b, pp. 3, 7, 9), Higgins (1962, p. 75), Hochstrasser (1963, p. 136), Inayatullah (1964, p. 15), International Research Associates (1963, p. 5), Jain (1965, p. 48), Jones (1960a, p. 5), Jones (1960b, p. 1), Jones (1962b, p. 10), Junghare and Roy (1963, p. 398), Katz, (1956, p. 206), Katz (1961, p. 26), Katz (1967, pp. 36, 37), Katz and Menzel (1954, p. 2), Kaufman and Bryant (1958, p. 1), Kim (1967, p. 100), King (1964, p. 331), Kivlin (1968, p. 37), Klietsch, (1961, p. 69), Klietsch (1961, p. 71), Klonglan (1963, p. 89), Klonglan and others (1964, p. 175), Kronus (1967, p. 152), Kienstadter (1961, p. 684), Lackey (1958, p. 28), Larsen (1962, p. 32), Letourneau (1963, p. 73), Leuthold (1960a, p. 3), Leuthold (1960b, p. 40), Leuthold and Wilkening (1965, p. 8), Lionberger (1950, pp. 204, 205), Lionberger (1951, p. 19), Lionberger (1956a, p. 10), Lionberger (1958, p. 4), Lionberger and Coughenour (1957, p. 89), Lionberger and Chang (1965, p. 55), Lionberger and Hassinger (1954b, p. 10), Lipman (1964, pp. 51, 56), Louisiana Agricultural Extension Service (1950, p. 12), Lowry and Hay (1957, pp. 5, 25), Lowry and Hay (1958a, p. 12), Lowry and Hay (1958b, p. 201), Loy (1967a, p. 153), Loy (1967b, p. 8), Loy (1968a, p. 7), Maddala and Knight (1965, p. 42), Madigan (1962, p. 207), Maffei (1966, p. 25), Magdub (1964, p. 12), Marsh and Coleman (1955b, p. 290), Martinez (1963, pp. 123, 124), Martinez (1964b, p. 136), Martinez and others (1964, p. 6), Matthew (1942, p. 161), Miller (1957, p. 40), Mookherjee and Singh (1966, p. 21), Morgan and others (1966, p. 221), Mort and Cornell (1938, pp. 43, 48, 53), Mort and Cornell (1941, pp. 93, 99), Mort and Pierce (1947, p. 2), Oakley (1965, p. 115), Opinion Research Corporation (1959b, p. 75), Parish (1954, p. 206), Pessemier and others (1967, p. 15), Pierce and Rowntree (1961, pp. 141, 145), Plant (1959, p. 222), Ploch (1960, p. 17), Polgar and others

(1963, p. 110), Polson and Pal (1964, p. 73), Prasada (1966, pp. 59, 62), Rahim (1961a, p. 34), Rahudkar (1958, p. 125), Rahudkar (1961, p. 81), Rajagopalan and Singh (1967, pp. 89, 91, 92), Rao (1961, p. 16), Reddy (1962, p. 28), Reddy and Kivlin (1968, p. 20), Richardson (1964, p. 73), Riley and Riley (1940, p. 903), Rigwani (1964, p. 103), Robertson (1966, p. 19), Robertson (1966a, p. 128), Robertson (1967a, p. 14), Robertson (1967b, p. 13), Rogers (1957a, p. 132), Rogers (1958a, p. 144), Rogers (1959a, p. 133), Rogers (1959b, p. 12), Rogers (1961b, p. 64), Rogers (1961b, p. 65), Rogers (1962, pp. 68, 72), Rogers and Burdge (1962, p. 17), Rogers and Havens (1962, p. 38), Rogers and Pitzer (1960, p. 17), Rohrer (1949, p. 1), Ross (1955, pp. 175, 176), Rowntree and Pierce (1961, p. 17), Roy (1967, p. 205), Roy and others (1968, pp. 66, 67, 69), Sabri (1966, p. 107), Sahgal (1966, pp. 78, 161, 273), Savale (1966, p. 206), Sawhney (1962, p. 38), Schindler (1962, p. 46), Schwieder (1966, p. 130), Sheth (1966, p. 188), Sill (1958, p. 51), Singh (1965, pp. 2, 12), Singh (1965, p. 135), Singh (1966, p. 134), Singh (1967, p. 151), Singh and Reddy (1965b, p. 267), Sinha and Yadav (1964b, p. 11), Sinha and Yadav (1964c, p. 16), Sinha and Yadav (1964e, p. 23), Sinha and Yadav (1964f, p. 23), Sizer and Porter (1960, p. 5), Smith (1958, p. 56), Smith (1964, p. 299), South and others (1965, pp. 10, 17), Spencer (1958, p. 70), Spencer (1958, p. 72), Spencer (1958, p. 73), Stickley (1964, p. 113), Stickley (1964, p. 508), Stickley and others (1967, p. 20), Stock and Johnson (1966, p. 9), Straus (1960, p. 226), Straus and Estep (1959, pp. 16, 32), Stuby (1965, p. 51), Suchman (1966, p. 40), Sycip (1964, p. 598), Tajima (1959, p. 18), Taylor (1955, p. 27), Thorat (1968, p. 9), Ullah (1962, p. 79), van den Ban (1953, p. 323), van den Ban (1957a, p. 207), van den Ban (1960a, p. 63), van den Ban (1960b, p. 308), van den Ban (1960c, p. 239), Verma (1966, p. 115), Verner and Gubbels (1967, p. 25), Verner and Millerd (1966, p. 23), Vincent (1945, p. 36), von Blanckenburg (1962a, p. 12), von Blanckenburg (1962b, p. 14), Waisanen (1952, pp. 26, 29, 30, 36), Wang (1963, p. 20), Watson (1946, p. 8), Weldon (1966, p. 72), Wellin (1955, pp. 89, 94), Wells and MacLean (1962, p. 16), Westoff and others (1961, p. 69), White (1968, p. 34), White and Heighton (1968, p. 65), Whittenbarger and Maffei (1966, p. 5), Whyte (1954, p. 142), Wilkening (1949b, p. 2), Wilkening (1952a, p. 61), Wilkening (1952b, p. 61), Wilkinson and Bailey (1964, p. 13), Wilson (1928, p. 20), Wilson and Trotter (1933, p. 20), Yeracaris (1961, pp. 10, 11), Yeracaris (1961a, p. 302), Young and Marsh (1956, p. 6), Zwerman and Prundeanu (1956, p. 128).

Not supporting: Adams (1951, p. 189), Alves (1962, p. 28), Arndt (1966a, p. 138), Badenhorst and Unterhalter (1961, p. 299), Bauder (1961, p. 25), Beal and others (1967a, p. 9), Belcher (1958, p. 167), Belcher (1958, p. 169), Ben-David (1960, p. 90), Berelson (1966, p. 660), Bertrand and others (1956, p. 28), Bertrand and South (1963, p. 7), Bhatia (1966, p. 119), Blair (1967, p. 22), Bogue and others (1966, p. 26), Bonjean and others (1965, p. 182), Bonser

(1958a, p. 30), Bose (1961, p. 142), Bose (1964c, p. 10), Cancian (1967, p. 25), Card (1960, p. 116), Carlson (1965, p. 62), Carter and Williams (1959, p. 104), Carter and Williams (1957, p. 190), Celade (1965, p. 110), Chaparro (1956, p. 582), Christiansen and Taylor (1966, p. 18), Choudhary and Maharaja (1966, p. 163), Clark and Abell (1966, p. 4), Copp (1956, p. 13), Coughenour (1966, p. 8), Dasgupta (1966, p. 7), Dernburg (1958, p. 42), Desai and Sharma (1966, p. 147), Deutschmann and others (1965, p. 25), Dickerson (1967, p. 53), Dickerson (1967, p. 62), Dubey (1966, pp. 2, 119), Fliegel (1967a, p. 15), Fosen (1956, p. 52), Frank and Massey (1964, p. 105), Frank and others (1964, p. 13), Frank and Tietze (1965, p. 123), Georgiopoulos and Potter (1967, p. 75), Glasser (1958, p. 144), Goldsen and Ralis (1957, p. 32), Goldstein and others (1961, p. 18), Graham (1954, p. 168), Graham (1954, p. 170), Graham (1956, pp. 94, 95, 97, 291), Gross (1942, p. 94), Gross (1949, p. 154), Gupta (1966, p. 110), Havens (1962, pp. 84, 85), Havens (1965, pp. 156, 158), Havens and Rogers (1961a, p. 14), Higgin (1962, p. 77), Hrabouszky and Moulik (1966, p. 10), Hildebrand and Partenheimer (1958, p. 448), International Research Associates (1963, p. 15), Jones (1960b, p. 1), Juliano (1967, p. 202), Junghare (1962, p. 2), Katz (1967, p. 36), Khan and Choldin (1965, p. 6), Kim (1967, p. 96), Klietsch (1961), Klonglan and others (1964, p. 175), Kunstadter (1961, p. 686), Larsen (1962, p. 33), Letourneau (1963, p. 73), Leuthold (1960a, p. 3), Loewy (1961, p. 25), Louisiana Agricultural Extension Service (1950, p. 13), Madigan (1962, p. 207), Maamary (1965, p. 29), Marsh and Coleman (1955b, p. 290), Massy (1960, p. 240), Matthew and others (1942, p. 1), McKain and others (1958, p. 13), McMillion (1960, p. 20), Menzel (1957, p. 402), Miller (1965, p. 17), Mookherjee and Sinha (1966, p. 11), Newbold (1957, p. 350), Opler (1964, p. 214), Paul (1965, p. 32), Pessemier and others (1967, p. 15), Planning Research and Action Institute (1961, pp. 25, 27), Polson and Pal (1964, p. 72), Presser and Russell (1965, pp. 155, 156), Prundeanu and Zwerman (1958a, p. 904), Rahudkar (1962, p. 84), Rao (1966, p. 20), Reddy and Kivlin (1968, p. 18), Rogers (1961b, p. 65), Rosner (1968, p. 623), Savale (1966, p. 204), Sepulveda and Carter (1964, p. 2), Shetty (1966, p. 196), Sills and Gill (1959, p. 247), Singh (1967, p. 150), Singh and Pardasani (1967, p. 139), Smith (1951, p. 16), Stapel (1960, p. 14), Stickley (1964, p. 84), Stock and Johnson (1966, p. 16), Stuby (1965, p. 62), Stycos (1955, p. 196), Stycos (1965a, p. 12), Tajima (1959, p. 18), Takeshita (1966, p. 702), Thomas (1966, p. 88), van den Ban (1957a, pp. 207, 212), van den Ban (1960, p. 308), Verma (1966, p. 114), von Blanckenburg (1962, pp. 20, 22), Wells and Andapia (1966, p. 485), Wheelock (1964, p. 23), Wicher (1958, p. 46), Wilkening (1952a, p. 61), Wilkening and others (1962, p. 163), Wilkening and others (1963, p. 7), Wilson and Gallup (1955, p. 24), Winick (1961, p. 388), Yeracaris (1961a, p. 301), Yeracaris (1961a, p. 8), Zaidi (1968b, p. 36).

5-5: *Earlier adopters have a greater degree of upward social mobility than do later adopters* (5 studies, or 100 percent, support).

Supporting: Chaparro (1956, p. 583), Cohen (1962, p. 46), Robertson (1966a, p. 124), Robertson (1967a, p. 13), Robertson (1967b, p. 13).

5-6: *Earlier adopters have larger sized units (farms, and so on) than do later adopters* (152 studies, or 67 percent, support; 75 studies do not support).

Supporting: Alves (1962, p. 27), Bakshi (1962, p. 2), Barnabas (1960, p. 214), Bauder (1961, p. 26), Beal and Rogers (1960, p. 14), Beal and others (1959b, p. 23), Beal and others (1967a, p. 9), Beal and others (1967b, p. 75), Beal and Sibley (1962, p. 105), Berelson (1966, p. 661), Berelson and Freedman (1964, p. 37), Bertrand (1951, p. 48), Bertrand and others (1956, p. 28), Bertrand and South (1963, p. 7), Bhasin (1966, p. 62), Blackmore and others (1955, p. 4), Bonser (1958b, p. 11), Bose and Saxena (1965, p. 141), Brady and Adams (1962, p. 6), Brandner and Kearl (1964, p. 298), Bylund (1963, p. 11), Caird and Moisley (1961, p. 100), Campbell and Lionberger (1963, p. 28), Card (1960, p. 113), Carlson (1965, p. 55), Cawelti (1967, p. 60), Chopde and others (1959, p. 6), Choudhary and Marahaja (1966, p. 163), Chow (1965, p. 159), Christiansen (1965, p. 166), Copp (1956, pp. 17, 18), Copp (1958, p. 60), Coughenour (1966, p. 7), Coughenour and Kothari (1962, p. 6), Crain (1962a, p. 8), Crain (1965, p. 12), Cummings (1950, p. 127), Dantre (1963, p. 4), Dasgupta (1963a, p. 30), Dasgupta (1966, p. 7), Dasgupta (1966, p. 23), Dhaliwal (1963, p. 3), Dhara (1965, p. 26), Dubey (1966, p. 2), Emery and Oeser (1958, p. 66), Fallding (1958, p. 39), Finley (1965, pp. 35, 44), Finley (1968, p. 13), Fischer and Timmons (1959, p. 433), Fliegel (1962, p. 5), Flinn (1961, p. 34), Flinn (1963, p. 9), Freedman and Takeshita (1965, p. 247), Fund for Advancement of Education (1964, p. 10), Gillis (1958, p. 6), Gorman (1966, p. 160), Gross (1941, p. 94), Gross (1949, p. 152), Gross and Taves (1952, p. 326), Gupta (1966, p. 156), Havens (1962, p. 84), Havens (1966, p. 150), Havens (1965, p. 158), Hochstrasser (1963, p. 135), Inayatullah (1964, p. 15), International Research Associates (1963, p. 5), Jain (1965, p. 48), Johnson (1966, p. 36), Jones (1960a, p. 5), Junghare (1962, p. 294), Kahlon and Kaushal (1967, p. 98), Katz (1967, p. 37), Kaufman and Bryant (1958, p. 1), Keenan (1964, p. 20), Khan and Choldin (1965, p. 4), Kivlin and Fliegel (1967, p. 82), Klietsch (1961, pp. 67, 68), Klonglan (1963, p. 89), Klonglan and others (1960, p. 20), Koontz (1958, p. 74), Leuthold (1965, p. 68), Lindstrom (1960, p. 74), Lionberger (1958, p. 4), Loomis (1967b, p. 885), Madigan (1962, p. 207), Magdub (1964, p. 12), Mahajan (1966, p. 2), Makarczyk (1965, p. 51), Maamary (1965, p. 28), Mansfield (1960, p. 9), Mansfield (1968, p. 171), Marsh and Coleman (1954, p. 5), Marsh and Coleman (1955b, p. 290), Martinez (1963, p. 121), Martinez (1964, p. 135), Martinez and others (1964a, p. 6), McMillion (1960,

p. 18), Morrison (1956, p. 283), Mort and Cornell (1938, p. 41), Mulay and Ray (1965, p. 108), Oakley (1965, p. 114), Presser and Russell (1965, p. 156), Rahudkar (1958, p. 124), Rahudkar (1961, p. 81), Rajagopalan and Singh (1967, p. 90), Ramos (1966, p. 60), Rao (1966, p. 17), Reddy (1962, p. 28), Reddy and Kivlin (1968, p. 16), Rogers (1959a, p. 133), Rogers (1961b, p. 65), Rogers (1962, p. 68), Rogers and Burdge (1962, p. 17), Rogers and Cartano (1963, p. 9), Rogers and Havens (1962, p. 38), Rogers and Pitzer (1960, p. 17), Rohrer (1949, p. 1), Ross and Bang (1965, p. 16), Roy (1967, p. 205,) Roy and others (1968, p. 56), Sagi and others (1961, p. 292), Sawhney (1962, p. 38), Schindler (1962, p. 46), Sepulveda and Carter (1964, p. 1), Sheppard (1960, p. 28), Shetty (1966, p. 198), Singh (1965, p. 3), Singh (1965, p. 135), Singh (1966, p. 81), Singh (1967, p. 151), Singh and Pardasani (1967, p. 143), Singh and Reddy (1965b, p. 267), Sinha and Yadav (1964c, p. 15), Sinha and Yadav (1964e, p. 23), Sinha and Yadav (1964f, p. 26), Sill (1958, p. 53), Smith (1954, p. 86), Smith (1958, p. 54), Stickley (1964, pp. 17, 80, 106), Stock and Johnson (1966, p. 9), Straus and Estep (1959, p. 17), Tajima (1959, p. 17), Takeshita (1964, p. 2), Takeshita (1966, p. 700), Thorat (1968, p. 10), van den Ban (1953, p. 321), van den Ban (1956, p. 42), van den Ban (1958, p. 52), van den Ban (1958, p. 63), van den Ban (1963b, p. 239), Verner and Gubbels (1967, p. 24), von Blanckenburg (1962b, p. 15), Young (1959, p. 2).

Not supporting: Aberle (1966, p. 311), Advertest Research (1958, p. 21), Ahmad (1964, p. 13), Armstrong (1959, p. 33), Ascroft (1966, p. 95), Badenhorst and Unterhalter (1961, p. 288), Bhasin (1966, p. 63), Bhosale (1960, p. 4), Bogue and Palmore (1966, p. 14), Bonser (1958a, p. 10), Bonser (1958a, p. 29), Bostian and Oliveira (1965, p. 9), Chopde and others (1959, p. 150), Correa (1965, p. 40), Cummings (1950, p. 127), Danda and Danda (1968, p. 178), Dernburg (1958, p. 43), Desai and Sharma (1966, p. 143), Feaster (1966, p. 10), Fliegel (1955, p. 21), Fliegel (1956, p. 155), Fosen (1956, p. 52), Frank and Tietze (1965, p. 123), Goldsen and Ralis (1957, p. 30), Gupta (1966, pp. 110, 111), Hage (1963, p. 103), Havens (1963, p. 58), Havens (1965, pp. 156, 158), Hoffer and Stangland (1958b, p. 17), Hrabouszky and Moulik (1966, p. 10), Hildebrand and Partenheimer (1958, p. 448), Ingman (1963, p. 55), Jain (1965, p. 48), Johnson and others (1967, p. 31), Juliano (1967, pp. 200, 202), Junghare and Roy (1963, p. 397), Khan and Choldin (1965, p. 4), Kim (1967, p. 93), Kivlin (1968, p. 38), Klonglan and others (1964, p. 175), Kunstadter (1961, p. 689), Larsen (1962, pp. 31, 33), Letourneau (1963, p. 73), Mansfield (1963a, p. 358), Mansfield (1968, p. 107), Mauldin (1967, p. 2), Menzel (1959, p. 199), Miller (1965, p. 20), Mort and Cornell (1941, p. 81), Parish (1954, p. 79), Presser and Russell (1965, p. 155), Rahudkar (1961, p. 81), Rahudkar (1962, p. 84), Ross (1966a, p. 11), Roy and others (1968, p. 65), Prundeanu and Zwerman (1958a, p. 907), Singh (1965, p. 3), Singh (1965, p. 135), Singh (1967, p. 150), Singh and Pardasani (1967, p. 142), Stickley (1964, pp. 78, 82),

Stock and Johnson (1966, p. 9), Stuby (1965, p. 62), Taylor (1955, p. 27), Thomas (1966, p. 88), Ullah (1962, p. 79), van den Ban (1958, p. 209), Verner and Gubbels (1967, p. 25), Verner and Millerd (1966, p. 22), Wells and Andapia (1966, p. 485), Wilkening and others (1963, pp. 4,7), Wilson and Trotter (1933, p. 20), Zwerman and Prundeanu (1956, pp. 95, 128).

5-7: *Earlier adopters are more likely to have a commercial (rather than a subsistence) orientation than are later adopters* (20 studies, or 71 percent, support; 8 studies do not support).

Supporting: Armstrong (1959, p. 18), Beal and others (1967b, p. 72), Bertrand and others (1956, p. 29), Carter and Williams (1957, p. 179), Carter and Williams (1959, p. 90), Coughenour (1966, p. 9), Coughenour and Armstrong (1963, p. 6), Coward (1967, p. 13), Dasgupta (1966, p. 23), Emery and Oeser (1958, p. 89), Fals Borda (1960, p. 21), Fund for Advancement of Education (1964, p. 10), Jones (1960b, p. 1), Letourneau (1963, p. 73), Moulik and others (1966, p. 473), Presser and Russel (1965, pp. 155, 156), Sill (1958, p. 46), White (1968, p. 34), White and Heighton (1968, p. 65), Wilkinson and Bailey (1964, p. 13).

Not supporting: Bertrand and South (1963, p. 9), Coward (1967, p. 13), Dasgupta (1966, p. 9), Fosen (1956a, pp. 47, 48, 52), Fosen (1956b, p. 41), Voget (1948, p. 645), Ramsey and others (1959, p. 43), Russell (1964, p. 2).

5-8: *Earlier adopters have a more favorable attitude toward credit (borrowing money) than later adopters* (19 studies, or 76 percent, support; 6 studies do not support).

Supporting: Anastasio (1960, p. 37), Beal and others (1967a, p. 8), Beal and others (1967b, p. 2), Beal and Sibley (1962, p. 100), Bose (1962, p. 6), Carter and Williams (1957, p. 179), Carter and Williams (1959, p. 90), Copp (1956, p. 24), Dasgupta (1966, p. 23), Goldsen and Ralis (1957, p. 29), Havens (1962, p. 84), Havens (1965, p. 158), Hess and Miller (1954, p. 28), Hoffer and Stangland (1958a, p. 118), Hoffer and Stangland (1958b, p. 21), Ramsey and others (1959, p. 42), Whittenbarger and Maffei (1966, p. 5), Whyte (1954, p. 206), Yacoub (1963, p. 4).

Not supporting: Goldsen and Ralis (1957, pp. 29, 36), Havens (1962, pp. 84, 85), Havens (1965, p. 158), Ingman (1963, p. 55), Letoureneau (1963, p. 73), Reddy and Kivlin (1968, p. 8).

5-9: *Earlier adopters have more specialized operations than later adopters* (9 studies, or 60 percent, support; 6 studies do not support).

Supporting: Armstrong (1959, p. 18), Campbell and Lionberger (1963, p. 28), Hage (1963, p. 104), Proctor (1956), Rogers (1959b, p. 12), Rogers (1961b,

p. 65), Rogers (1962, p. 72), South and others (1965, p. 10), Sutherland (1959, p. 122).

Not supporting: Copp (1956, p. 21), Fliegel and Kivlin (1962b, p. 366), Fosen (1956, p. 52), Hrabouszky and Moulik (1966, p. 10), Reddy and Kivlin (1968, p. 16), Stock and Johnson (1966, p. 13).

5-10: *Earlier adopters have greater empathy than later adopters* (9 studies, or 64 percent, support; 5 studies do not support).

Supporting: Kivlin (1968, p. 39), Loomis (1967b, p. 885), Maamary (1965, p. 29), Portocarrero (1966, p. 51), Roy (1967, p. 205), Roy and others (1968, p. 112), Stickley (1964, p. 98), Thorat (1968, p. 14), Whiting (1967, p. 91).

Not supporting: Herzog (1967, p. 43), Narang (1966, p. 55), Reddy and Kivlin (1968, p. 27), Rogers and Ramos (1965, p. 7), Yaukey and others (1967, p. 725).

5-11: *Earlier adopters are less dogmatic than later adopters* (17 studies, or 47 percent, support; 19 studies do not support).

Supporting: Baur (1957, p. 4), Brim (1954, p. 485), Chattopadhyaya (1963, p. 3), Chattopadhyaya (1967, p. 57), Copp (1956, p. 26), Copp (1958, p. 116), Fathi (1965, p. 212), Goldstein and others (1961, p. 39), Hoffer and Stangland (1958a, p. 118), Hoffer and Stangland (1958b, p. 21), Jamias (1964, pp. 63, 78), Leuthold (1960a, p. 3), Loy (1967, p. 153), Marion (1966, p. 190), Opler (1964, p. 214), Pitzer (1959, p. 71), Singh and Beal (1967, p. 18).

Not supporting: Bhasin (1966, p. 62), Chattopadhyaya (1967, p. 57), Chattopadhyaya and Pareek (1967, p. 328), Childs (1965, p. 71), Fals Borda (1960, p. 21), Farnsworth (1940, p. 100), Havens (1962, pp. 84, 85), Havens (1965, p. 157), Hudspeth (1966, p. 81), Jamias and Troldahl (1964, p. 14), Kivlin (1968, p. 39), Klonglan and others (1964, p. 81), Lin (1966, p. 63), Mulay and Roy (1965, p. 111), Presser and Russell (1965, p. 4), Robertson (1967b, p. 14), Rogers (1957b, p. 6), Sprunger (1968, p. 85), Stuby (1965, p. 56).

5-12: *Earlier adopters have a greater ability to deal with abstractions than do later adopters* (5 studies, or 63 percent, support; 3 studies do not support).

Supporting: Benvenuti (1961, p. 353), Loy (1967, p. 7), Rogers (1961a, p. 81), South and others (1965, p. 11), Tajima (1959, p. 17).

Not supporting: Narang (1966, p. 55), Rogers (1957b, p. 7), Rogers (1961b, p. 65).

5-13: *Earlier adopters have greater rationality than later adopters* (11 studies, or 79 percent, support; 3 studies do not support).

Supporting: Bose (1962, p. 6), Bose (1963, p. 5), Bose (1964a, p. 21), Dean and others (1958b, p. 133), Gross (1942, p. 104), Hobbs (1960, p. 73), Klietsch (1961, p. 70), McMillion (1960, p. 21), Rieck and Pulver (1962, p. 15), Spencer (1958, p. 83), Westermarck (1963, p. 62).

Not supporting: Bemiller (1960, p. 1), Ramsey and others (1959, p. 40), Wilkening and others (1962, p. 191).

5-14: *Earlier adopters have greater intelligence than later adopters* (5 studies, or 100 percent, support).

Supporting: Benvenuti (1961, p. 353), Booth (1966, p. 9), Brien and others (1965, p. 16), Daspurohit (1963, p. 83), Rogers (1962, p. 76).

5-15: *Earlier adopters have a more favorable attitude toward change than later adopters* (43 studies, or 75 percent, support; 14 studies do not support).

Supporting: Ascroft (1966, p. 95), Beal and Rogers (1960, p. 14), Bauder (1961, p. 31), Beal and others (1958b, p. 7), Beal and others (1959b, p. 23), Beal and others (1967a, p. 18), Beal and others (1967b, p. 73), Bonser (1958a, p. 28), Bonser (1958b, p. 21), Bose (1962, p. 6), Bylund (1964, p. 203), Carter and Williams (1959, p. 104), Carter and Williams (1957, p. 190), Castillo and others (1963, p. 70), Celade (1965, p. 111), Chaparro (1955, p. 185), Chattopadhyaya (1963, p. 3), Chattopadhyaya (1967, p. 57), Chattopadhyaya and Pareek (1967, p. 328), Chow (1965, p. 157), Coughenour (1960, p. 297), Coughenour (1966, p. 7), Coughenour and Kothari (1962, p. 6), Dasgupta (1966, p. 16), Denis (1961, p. 4), Deutschmann and Mendez (1962, p. 39), Eibler (1965, p. 128), Eichholz (1963, p. 268), Evans (1968, p. 138), Fliegel (1955, p. 21), Fliegel (1956, p. 289), Fliegel and Oliveira (1963, p. 2), Frank and others (1964, p. 14), Fund for the Advancement of Education (1964, p. 10), Gupta (1966, p. 110), Hains (1964, p. 1), Hoffer (1942, p. 32), Hudspeth (1966, p. 81), Johnson and others (1967, p. 17), Keith (1968a, p. 143), Keith (1968b, p. 108), Kimball (1960, p. 155), Klonglan and others (1960, p. 20), Klonglan and others (1964, p. 177), Koontz (1958, p. 74), Leuthold (1960b, p. 42), Mathen (1962, p. 37), Morgan and others (1966, p. 314), Moulik and others (1966, p. 473), Newell (1943, p. 53), Nichols (1959, p. 129), Pessemier and others (1967, pp. 8, 9), Plaut (1959, p. 222), Ploch (1960, p. 37), Rao (1966, p. 28), Reddy and Kivlin (1968, p. 27), Robertson (1965, p. 93), Robertson (1966, p. 26), Rogers (1957a, p. 130), Rogers (1958a, p. 144), Rogers (1961a, p. 80), Rosecrance (1964, p. 294), Rowntree and Pierce (1961, p. 14), Schindler (1962, p. 45), Schwieder (1966, p. 131), Sheppard (1961, p. 13), Singh and Pardasani (1967, p. 143), South and others (1965, pp. 16, 17), Spencer (1958, p. 96), Suchman (1966, p. 41), Sycip (1964, p. 595), Weintraub and Bernstein (1966, p. 519), Wilkening (1949b, p. 1), Wilkening (1950b, p. 361), Wish

(1966, p. 893), Wish (1967, p. 164), Yacoub (1963, p. 4), Yeracaris (1961a, p. 303).

Not supporting: Bonser (1958a, p. 30), Bose (1963, p. 5), Havens (1965, p. 157), Jain (1965, p. 48), Jamias (1964, p. 79), Kivlin (1967, pp. 5, 6), Marion (1966, p. 190), Mathen (1962, p. 42), Miller (1965, p. 18), Oakley (1965, p. 117), Roberts and others (1965, p. 94), Rogers (1957b, p. 7), Rohrer (1955, p. 303), White (1967).

5-16: *Earlier adopters have a more favorable attitude toward risk than later adopters* (27 studies, or 73 percent, support; 10 studies do not support.)

Supporting: Arias (1967, p. 51), Arndt (1966b, pp. 8, 9), Beal and others (1959b, p. 23), Beal and others (1967a, p. 8), Beal and others (1967b, p. 74), Beal and Sibley (1962, p. 98), Fonseca (1966, p. 80), Gorman (1967b, p. 7), Loy (1968a, p. 5), Pessemier and others (1967, p. 10), Potthoff and Rheinwald (1958, p. 63), Ramsey and others (1959, p. 44), Rao (1966, p. 24), Reddy and Kivlin (1968, p. 6), Robertson (1966, p. 16), Robertson (1966a, p. 108), Robertson (1967a, p. 10), Robertson (1967b, p. 12), Robertson and Rossiter (1968, p. 15), Rogers (1961b, p. 10), Sinha and Yadav (1964e, p. 23), Sinha and Yadav (1964f, p. 28), Strassman (1959, p. 184), Thorat (1968, p. 14), Westermarck (1963, p. 56), Wish (1967, p. 164), Yeracaris (1961a, p. 303).

Not supporting: Arndt (1967a, p. 294), Cancian (1967, p. 26), Desai and Sharma (1966, p. 150), Klonglan and others (1964, p. 177), Loy (1967, p. 153), Loy (1968b, p. 8), Narang (1966, p. 55), Prasada (1966, p. 58), Rogers (1957b, p. 7), Singh and Beal (1967, p. 18).

5-17: *Earlier adopters have a more favorable attitude toward education than later adopters* (25 studies, or 81 percent, support; 6 studies do not support).

Supporting: Advertest Research (1958, p. 6), Bertrand (1951, p. 48), Bohlen and others (1959, p. 30), Borman (1965, p. 11), Caplow (1952, p. 111), Fallding (1958, p. 32), Goldsen and Ralis (1957, p. 12), Hage (1963, p. 104), Havens (1962, p. 84), Havens (1965, p. 160), Havens and Rogers (1961a, p. 14), Holmberg (1952, p. 121), Klietsch (1961, p. 84), Klonglan (1963, p. 91), Klonglan and others (1964, p. 178), Leuthold (1960a, p. 3), Lindstrom (1958, p. 169), Marsh and Coleman (1954a, p. 181), Mort and Cornell (1941, p. 254), Rogers and Beal (1958b, p. 77), Rogers and Havens (1961a, p. 4), Rogers and Havens (1961a, p. 19), Ross and Bang (1965, p. 16), Sinha and Yadav (1964a, p. 8), Yacoub (1963, p. 4).

Not supporting: Goldsen and Ralis (1957, p. 12), Greenwald (1964, p. 601), Havens (1962, pp. 84, 85), Havens (1965, p. 160), Stuby (1965, p. 51), Voget (1950, p. 645).

5-18: *Earlier adopters have a more favorable attitude toward science than later adopters* (20 studies, or 74 percent, support; 7 studies do not support).

Supporting: Arias (1967, p. 51), Beal and others (1967a, p. 8), Beal and others (1967b, p. 73), Beal and Sibley (1962, p. 99), Bose (1963, p. 5), Carter and Williams (1957, p. 183), Carter and Williams (1959, p. 97), Clark and Abell (1966, p. 4), Junghare and Roy (1963, p. 399), Katz (1961, p. 26), Katz and Menzel (1954, p. 2), Kivlin (1967, p. 5), Loy (1967, p. 153), Loy (1968b, p. 8), Rogers (1959b, p. 12), Rogers and Beal (1958b, p. 77), Singh and Beal (1967, p. 18), Singh and Reddy (1965b), Spencer (1958, p. 83), Yacoub (1963, p. 4).

Not supporting: Anand (1966, p. 77), Beal and others (1967b, p. 73), Bose (1963, p. 5), Levitt (1965, p. 166), Ramsey and others (1959, p. 40), Reddy and Kivlin (1968, p. 6), Voget (1948, p. 645).

5-19: *Earlier adopters are less fatalistic than later adopters* (14 studies, or 82 percent, support; 3 studies do not support).

Supporting: Arias (1967, p. 51), Beal and Sibley (1962, p. 98), Castillo (1964, p. 16), Chattopadhyaya (1963, p. 3), Chattopadhyaya (1967, p. 57), Chattopadhyaya and Pareek (1967, p. 328), Dubey (1967, p. 122), Ingman (1963, p. 55), Klonglan and others (1964, p. 176), Mather (1962, p. 38), Niehoff and Anderson (1964a, p. 6), Rogers and Cartano (1963, p. 9), Singh and Beal (1967, p. 18), Suchman (1966, p. 41).

Not supporting: Cohen (1961, p. 32), Goldsen and Ralis (1957, p. 30), Smith (1966, p. 254).

5-20: *Earlier adopters have higher levels of achievement motivation than later adopters* (14 studies, or 61 percent, support; 9 studies do not support).

Supporting: Carter and Williams (1959, p. 95), Carter and Williams (1957, p. 180), Chattopadhyaya (1963, p. 3), Christiansen and Taylor (1966, p. 20), Harris (1956, p. 83), Hess and Miller (1954, p. 28), Hoffer and Stangland (1958a, p. 118), Hoffer and Stangland (1958b, p. 21), Morgan and others (1966, p. 314), Morrison (1964, p. 377), Ramsey and others (1958, p. 39), Roy (1967, p. 205), Stickley (1964, p. 102), Yacoub (1963, p. 3).

Not Supporting: Beal and others (1967b, p. 74), Beal and Sibley (1962, p. 99), Morrison (1964, p. 377), Neill and Rogers (1963, p. 12), Pitzer (1959, p. 74), Rogers (1964, p. 16), Roy and others (1968, p. 112), Smith (1966, p. 254), Spencer (1958, p. 83).

5-21: *Earlier adopters have higher aspirations (for education, occupations, and so on) than later adopters* (29 studies, or 74 percent, support; 10 studies do not support).

Supporting: Anastasio (1960, p. 38), Castillo (1964, pp. 13, 15), Chattopadhyaya (1967, p. 57), Chattopadhyaya and Pareek (1967, p. 328), Dubey (1967, p. 124),

Erasmus (1961, p. 309), Gupta (1966, p. 110), Hrabouszky and Moulik (1966, p. 12), Johnson and van den Ban (1959, p. 5), Loomis (1967b, p. 885), Palmore and Freedman (1968, p. 15), Pitzer (1959, p. 72), Ramsey and others (1959, p. 41), Roy (1967, p. 205), Roy and others (1968, p. 112), Sheppard (1960, p. 69), Sill (1958, p. 45), Singh (1966, p. 80), Spector and others (1963, p. 75), Spector and others (1964, p. 681), Spencer (1958, p. 88), Stickley (1964, p. 74), Sycip (1964, p. 602), Wichers (1958, p. 47), Wilkening (1949b, p. 2), Wilkening (1950b, p. 359), Wilkening (1953, p. 5), Wilkening and others (1963, p. 7), Wilkening and others (1966, p. 163).

Not supporting: Feaster (1966, p. 10), Fliegel and others (1967a, pp. 11, 12), Kivlin (1968, p. 38), Maamary (1965, p. 29), Mort and Cornell (1941, p. 96), Narang(1966, p. 55), Paul (1965, p. 34), Reddy and Kivlin (1968, p. 27), Singh (1966, p. 134), Tully and others (1964, p. 318).

5-22: *Earlier adopters have more social participation than later adopters* (109 studies, or 73 percent, support; 40 studies do not support).

Supporting: Abell and Larsen (1960, p. 8), Advertest Research (1958, p. 134), Bauder (1961, p. 33), Belcher (1958, p. 196), Bhasin (1966, p. 63), Bonser (1958a, p. 16), Bonser (1958b, p. 15), Bose (1964a, p. 16), Bose and Dasgupta (1962, p. 10), Bose and Saxena (1965, p. 141), Brady and Adams (1962, p. 6), Bylund (1964, p. 203), Campbell and Holik (1960, p. 10), Carlson (1965, p. 64), Cassel (1963, p. 744), Chopde and others (1959, p. 6), Clark and Abell (1966, p. 4), Coleman and others (1966, p. 134), Copp (1956, p. 15), Copp (1958, p. 110), Cummings (1950, p. 128), Dasgupta (1963a, p. 22), Dasgupta, (1963b, p. 9), Dasgupta (1966, p. 14), Denis (1961, p. 4), Dhaliwal (1963, p. 3), Dubey (1967, p. 150), Fallding (1958, p. 39), Feaster (1966, p. 10), Finley (1965, p. 35), Finley (1965, p. 44), Finley (1968, p. 14), Fosen (1956, p. 51), Gallagher (1949, p. 8), Gallagher (1949, p. 22), Gillis (1958, p. 6), Gittell and Hollander (1968, p. 197), Gorman (1966, p. 162), Graham (1951, p. 96), Gross (1942, p. 134), Gross and others (1956, p. 326), Havens (1966, p. 150), Hay and Lowry (1957, p. 9), Hay and Lowry (1958, p. 9), Hoffer and Stangland (1958b, p. 18), Jain (1965, p. 48), Junghare (1962, p. 294), Junghare and Roy (1963, p. 398), Kahlon and Kaushal (1967, p. 99), Kaufman and Bryant (1958, p. 2), King (1965, p. 431), King (1964, p. 336), Kivlin (1968, p. 38), Koontz (1958, p. 44), Koontz (1958, p. 74), Lackey (1958, p. 101), Letourneau (1963, p. 73), Leuthold (1960a, p. 3), Leuthold (1960b, p. 41), Leuthold (1965, p. 68), Leuthold and Wilkening (1965, p. 8), Lin (1966, p. 67), Lionberger (1956a, p. 10), Lionberger and Coughenour, (1957, p. 89), Lowry and Hay (1958, p. 12), Lowry and others (1958, p. 201), Loomis (1967b, p. 885), Maamary (1965, p. 29), Marsh and Coleman (1954b, p. 7), Marsh and Coleman (1955b, p. 290), Martinez (1963, p. 122), Martinez (1964, p. 135), Martinez and others

(1964, p. 6), McMillion, (1960, p. 17), Mort and Cornell (1938, p. 55), Mulford (1959, p. 26), Oakley (1965, p. 113), Purohit (1963, p. 96), Rahim (1961a, p. 34), Rahudkar (1961, p. 82), Reddy and Kivlin (1968, p. 20), Rizwani (1964, p. 104), Rogers (1959b, p. 6), Rogers (1961b, p. 64), Rohrer (1949, p. 1), Roy (1967, p. 6), Russell (1964, p. 2), Ryan and Gross (1950, p. 701), Sawhney (1962, p. 38), Sill (1958, p. 62), Singh (1957, p. 151), Singh (1965, p. 151), Singh (1967, p. 14), Singh and Reddy (1965b, p. 267), Sinha and Yadav (1964c, p. 15), Sizer and Porter (1960, p. 5), Spector and others (1963, p. 64), Spector and others (1964, p. 679), Sprunger (1968, p. 85), Straus (1960, p. 225), Straus and Estep (1959, p. 19), Suchman (1966, p. 42), van den Ban (1956, p. 42), van den Ban (1958, p. 52), Waisanen (1956, p. 2), Wilkening (1952a, p. 61), Wilkening (1953, p. 5), Wilkening and others (1962, p. 163), Yacoub (1963, p. 4).

Not supporting: Bertrand and South (1963, p. 9), Bhatia (1966, p. 119), Bigger and Sauer (1966, p. 18), Bonser (1958a, p. 29), Booth (1966, p. 10), Bose (1961, p. 145), Bose (1964c, p. 10), Buley (1947, p. 69), Bylund (1963, p. 14), Christiansen, (1965, p. 164), Coleman and others (1957, p. 268), Dasgupta (1966, p. 23), Dasgupta (1966a, p. 9), Gorman (1967b, p. 8), Graham (1954, p. 170), Gupta (1966, p. 111), Hage (1963, p. 103), Horn and others (1959, p. 507), Jones (1960a, p. 6), Jones (1960b, p. 1), Klietsch (1961, p. 74), Kunstedter (1961, p. 682), Littunen (1959, p. 22), Menzel (1959, p. 216), Opler (1964, p. 215), Paul (1965, p. 33), Paul (1965, p. 34), Presser and Russell (1965, pp. 155, 156), Reddy, (1962, p. 28), Rizwani (1964, p. 104), Robertson and Rossiter (1968, p. 21), Roy and others (1968, pp. 70, 71), Sheth (1966, p. 2), Singh (1965, p. 136), Singh (1966, p. 81), Stuby (1965, p. 67), Verma (1966, p. 115), Verner (1966, p. 28), Whittenbarger and Maffei (1966, p. 3), Wilkening (1953, p. 5),

5-23: *Earlier adopters are more highly integrated with the social system than later adopters* (6 studies, or 100 percent, support).

Supporting: Coleman and others (1966), Coughenour (1962), Guimarães (1968), Mendez (1964), Mendez (1968), Yadav (1967).

5-24: *Earlier adopters are more cosmopolite than later adopters** (132 studies, or 76 percent, support; 42 studies do not support).

Supporting: Aberle (1966, p. 95), Adler (1955, p. 1), Advertest Research (1958, p. 6), Anastasio (1960, p. 40), Arias (1967, p. 51), Armstrong (1959, p. 18), Atamian (1966, p. 337), Bauer and Wortzel (1966, p. 40), Beal and Rogers

---

*Notice that the measures of cosmopoliteness include trips to cities, exposure to cosmopolite communication channels, and the like.

(1958, p. 5), Beal and others (1967a, p. 18), Beal and Sibley (1962, p. 103), Ben-David (1960, p. 567), Berelson and Freedman (1964, p. 37), Bertrand and South (1963, p. 10), Bhatia (1966, p. 119), Bose (1961, p. 473), Bowden (1964, p. 5), Bowden (1965, p. 85), Bowers (1937, p. 834), Brien and others (1965, p. 17), Carlson (1965, p. 64), Carter and Williams (1957, p. 179), Carter and Williams (1959, p. 94), Celade (1965, p. 110), Christiansen (1965, p. 167), Christiansen and Taylor (1966, p. 19), Cocking (1951, p. 54), Coleman and others (1956, p. 1), Coleman and others (1966, p. 134), Copp (1956, p. 22), Coughenour and Armstrong (1963, p. 6), Crain (1962a, p. 7), Crain (1962b, p. 167), Crain (1966, p. 472), Danda and Danda (1968, p. 181), Dasgupta (1963a, p. 31), Dasgupta (1963b, p. 8), Dasgupta (1965, p. 333), Dasgupta (1966, p. 23), Dasgupta (1966a, p. 10), Denis (1961, p. 4), Dhaliwal (1963, p. 3), Dhara (1965, pp. 26, 27), Dickerson (1967, p. 77), Dubey (1966, p. 2), Eastmond (1953, pp. 19, 28), Emery and Oeser (1958, p. 58), Evans (1968, p. 93), Fallding (1958, p. 39), Fals Borda (1960, p. 11), Flinn (1961, p. 33), Francis (1960, p. 5), Freedman and others (1964, p. 6), Freedman and Takeshita (1965, p. 244), Gillis (1958, p. 6), Gittell and Hollander (1968, p. 199), Glaser (1966, pp. 1, 41), Goldsen and Ralis (1957, pp. 25, 27), Gorman (1966, p. 567), Graham (1951, p. 293), Gross (1949, p. 154), Gross and Taves (1956, p. 326), Gunther (1950, p. 178), Hage (1963, p. 103), Hägerstrand (1952, p. 18), Hägerstrand (1953, p. 262), Hägerstrand (1965, p. 262), Hay and Lowry (1957, p. 5), Hicks (1946, p. 80), Hills (1955, p. 163), Hobbs (1966, p. 75), Jones (1962b, p. 10), Jones (1957, p. 113), Katz (1956, p. 204), Katz (1961, p. 23), Katz and Menzel (1954, p. 2), Keith (1968b, pp. 108, 143), King (1965, p. 431), Kivlin (1967, p. 7), Kivlin (1968, p. 38), Kienstadter (1961, p. 678), Larsen and DeFleur (1954, p. 596), Leuthold and Wilkening (1965, p. 8), Loomis (1967b, p. 885), Lowry and Hay (1957, p. 5), Loy (1967, p. 153), Loy (1968b, p. 8), Magdub (1964, p. 12), Makarczyk (1965, p. 49), Marion (1966, p. 190), Martinez (1963, p. 124), Martinez (1964b, p. 136), Martinez and others (1964, p. 6), McMillion (1960, p. 17), McVoy (1940, p. 222), Menzel (1959, p. 213), Mort and Cornell (1938, p. 44), Mort and Cornell (1941, p. 118), Newbold (1957, p. 348), Opinion Research Corporation (1959b, p. 75), Opinion Research Corporation (1960, p. 65), Patnaik (1963, p. 741), Pemberton (1938, p. 251), Ploch (1960, p. 12), Polson and Pal (1955, p. 158), Rahudkar (1966, p. 84), Rizwari (1964, p. 104), Robertson (1966, p. 18), Rogers (1959b, p. 5), Rogers (1962, p. 80), Rogers and Burdge (1961, p. 18), Ross (1955, p. 176), Roy and others (1968, pp. 72, 112), Ryan and Gross (1950, p. 706), Sabri (1966, p. 109), Schindler (1962, p. 46), Simon and Golembo (1967, p. 387), Singh (1965, p. 3), Singh (1965, p. 135), Singh (1967, p. 171), Singh and Beal (1967, p. 18), Sinha and Yadav (1964g, p. 30), Smith (1964, p. 299), Stickley (1964, p. 109), Thorat (1968, p. 13), van den Ban (1963a, p. 231), Verma (1966, p. 144), Weintraub and Bernstein (1966, p. 519), Wellin (1955,

p. 79), Wish (1967, p. 164), Wright and others (1967, p. 17), Yacoub (1963, p. 4).

Not supporting: Beal and others (1967a, p. 8), Beal and others (1967b, p. 75), Booth (1966, pp. 9, 10), Brandner and Kearl (1964, p. 299), Camaren (1966, p. 84), Card (1960, p. 116), Carlson (1965, p. 57), Christiansen and Taylor (1966, p. 19), Cummings (1950, p. 105), Dasgupta (1965, p. 333), Dickerson (1967, p. 71), Dodd (1955, p. 393), Donohew and Singh (1967, p. 16), Eibler (1965, p. 69), Fliegel (1965, p. 285), Francis and Rogers (1960, p. 5), Goldsen and Ralis (1957, p. 27), Goldsen and Ralis (1957, p. 36), Gross (1949, p. 155), Gorman (1967b, p. 8), Hartman and Brown (1965, p. 12), Havens (1965, p. 157), Herzog, (1967, p. 43), Kivlin (1967, p. 4), Madigan (1962, p. 207), Maamary (1965, p. 29), Menzel (1959, p. 213), Menzel (1960, p. 711), Miller (1965, p. 14), Mulford (1959, p. 27), Prasada (1966, p. 60), Robertson (1966a, p. 121), Robertson (1967a, p. 12), Robertson (1967b, p. 13), Rogers (1957a, p. 132), Rogers and Ramos (1965, p. 7), Sapolsky (1967, p. 508), Spaulding (1955, p. 4), Thorat (1968, p. 12), Takes (1963, p. 89), van den Ban (1960, p. 308), Whittenbarger and Maffei (1966, p. 3).

5-25: *Earlier adopters have more change agent contact than later adopters* (135 studies, or 87 percent, support; 21 studies do not support).

Supporting: Alves (1962, p. 30), Australia Bureau of Agricultural Economics (1956, p. 33), Beal and others (1960, p. 14), Beal and others (1967a, p. 9), Bhasin (1966, p. 63), Bhatia (1966, p. 119), Bittner (1959, p. 43), Bonser (1958a, p. 21), Bonser (1958b, p. 15), Berelson (1950a, p. 11), Berelson (1950b, pp. 18, 32), Carter and Williams (1957, p. 179), Carter and Williams (1959, p. 94), Copp (1956, pp. 11, 12), Correa (1965, p. 39), Couch (1965, p. 9), Coughenour (1960, p. 287), Coughenour and Kothari (1962, p. 8), Cummings (1950, p. 88), Dasgupta (1965, p. 333), Dasgupta (1966, p. 23), Dean and others (1958b, p. 133), Dhaliwal (1963, p. 4), Dhaliwal and Sohal (1965, p. 59), Dhara (1965, p. 26), Dobyns (1951, p. 32), Dubey (1966, p. 1), Dubey (1967, p. 150), Elliot (1965, p. 43), Elliott and Couch (1965, pp. 2, 3), Emery and Oeser (1958, p. 66), Fallding (1958, p. 43), Feaster (1966, p. 10), Feliciano (1964, p. 12) Fliegel (1955, p. 21), Fliegel (1961, p. 350), Fliegel (1965, p. 287), Francis and Rogers (1960, p. 5), Glaser (1966, p. 2), Hardee (1963a, p. 86), Hardee (1963b, p. 27), Hobbs (1964, p. 155), Hoffer (1942, p. 33), Johnson and Wilkening (1961, p. 27), Jones (1962b, p. 10), Junghare (1962, p. 294), Junghare and Roy (1963, p. 398), Kahlon and Kaushal (1967, p. 98), Katz and Menzel (1954, p. 2), Kaufman and Bryant (1958, p. 2), Kivlin (1967, p. 11), Kivlin (1968, pp. 37, 38), Klietsch (1961, pp. 88, 89), Lamar (1966, p. 57), Leuthold (1960b, p. 49), Leuthold and Wilkening (1963, p. 9), Leuthold and Wilkening (1965, p. 8), Lindstrom (1958, p. 179), Lionberger and Chang (1965, p. 55),

Loomis (1967b, p. 885), Louisiana Agricultural Extension Service (1950, p. 12), Magdub (1964, p. 12), Maamary (1965, p. 28), Marsh and Coleman (1954b, p. 6), Marsh and Coleman (1955b, p. 290), McCarthy and Tagby (1966, p. 2), McMillion (1960, p. 15), Menzel (1959, p. 213), Mookherjee and Singh (1966, p. 13), Moulik and others (1966, p. 473), Naquin (1957, p. 4), Neurath (1962, p. 280), Nielson and Crosswhite (1959, p. 26), Oakley (1957, p. 116), Parish (1954, p. 207), Photiadis (1961, p. 21), Polgar and others (1963, p. 110), Polson and Pal (1964, p. 73), Presser and Russell (1965, pp. 155, 156), Rahim (1961a, p. 34), Rahudkar (1961, p. 71), Rahudkar (1962, p. 84), Rahudkar (1963, p. 101), Reddy and Kivlin (1968, p. 24), Rizwani (1964, p. 104), Robertson (1964, p. 10), Robertson (1966a, p. 147), Rogers (1959a, p. 133), Rogers (1959b, p. 6), Rogers (1961a, p. 80), Rogers (1961b, pp. 65,66), Rogers (1962, pp. 77, 83), Rogers and Beal (1958b, p. 77), Rogers and Burdge (1961, p. 20), Rogers and Cartano (1963, p. 9), Rogers and Havens (1962, p. 38), Rogers and Pitzer (1960, p, 17), Roy (1967, p. 205), Roy and others (1968, p. 88), Sawhney (1962, p. 38), Savale (1966, p. 205), Scantland and others (1956, p. 40), Sill (1958, p. 62), Silverman and Bailey (1961, p. 8), Singh (1965, p. 3), Singh (1965, p. 135), Singh (1967, p. 15), Sinha and Yadav (1964a, p. 7), Sinha and Yadav (1964e, p. 23), Sinha and Yadav (1964g, p. 30), Stock and Johnson (1966, p. 17), Stone (1952, p. 16), Straus (1960, p. 223), Straus and Estep (1959, p. 9), Tajima (1959, p. 17), Thorat (1968, p. 17), Tiedeman and Van Doren (1964, p. 6), Ullah (1962, p. 80), van den Ban (1958, p. 64), van den Ban (1963a, p. 231), van den Ban (1963b, p. 240), van den Ban (1965, p. 4), Verma (1966, p. 115), Verner and Gebbels (1967, p. 25), Wellin (1955, p. 86), Wichers (1958, p. 47), Wilkening (1952a, p. 61), Wilkening (1953, p. 4), Wilkening and others (1962, pp. 170, 171), Wilkening and others (1963, p. 7), Wilkening and Santopolo (1952, pp. 12, 13, 16, 25), Wilson and Gallup (1955, p. 24), Young and Marsh (1956, p. 7), Zwerman and Prundeanu (1956, p. 129).

Not supporting: Armstrong (1959, p. 26), Bertrand and South (1963, p. 9), Bhasin (1966, p. 61), Bose (1964c, p. 10), Christiansen and Taylor (1966, p. 18), Dasgupta (1966, p. 9), Diswath (1964, p. 59), Elliott and Couch (1965, p. 3), Fonseca (1966, p. 73), Francis and Rogers (1960, p. 5), Jones (1962b, p. 70), Letourneau (1963, p. 73), Marsh and Coleman (1955b, p. 290), McMillion (1960, p. 15), Ramos (1966, p. 64), Saxena (1963, p. 94), Stickley (1964, p. 94), Stuby (1965, p. 65), Verner and Gubbels (1967, p. 31), White (1965, p. 41), Wilkening and Santopolo (1952, p. 16).

5-26: *Earlier adopters have greater exposure to mass media communication channels than later adopters* (80 studies, or 69 percent, support; 36 studies do not support).

Supporting: Anderson (1955, p. 7), Arndt (1968c, pp. 4, 8), Beal and others (1967a,

p. 9), Bowers (1938, p. 28), Bylund (1963, p. 26), Cassel (1963, p. 744), Christiansen and Taylor (1966, p. 19), Cohen (1962, p. 46), Coleman and others (1959, p. 5), Copp (1956, p. 11), Coughenour (1960, p. 287), Cummings (1950, p. 89), Deutschmann (1961, p. 28), Deutschmann (1963, pp. 32, 33), Deutschmann and Fals Borda (1962a, pp. 8, 11), Deutschmann and others (1963, p. 29), Dubey (1966, p. 1), Dubey (1967, p. 113), Ellenbogen and others (1961, p. 23), Emery and Oeser (1958, p. 66), Fallding (1958, p. 39), Fliegel (1966, p. 12), Fonseca (1966, p. 66), Freedman and others (1964, p. 6), Goldsen and Ralis (1957, pp. 20, 33), Gorman (1966, p. 162), Gorman (1967b, p. 7), Graham (1954, p. 168), Gross (1942, p. 137), Gross (1949, p. 152), Gross and Taves (1952, p. 326), Havens (1963, p. 58), Jones (1960a, p. 7), Katz and Menzel (1954, p. 2), Kivlin (1967, pp. 8, 9), Kivlin (1968, pp. 32, 37), Klonglan (1963) p. 58), Klonglan and others (1960, p. 13), Klonglan and others (1964, p. 178), Leuthold and Wilkening (1965, p. 8), Loomis (1967b, p. 885), Magdub (1964, p. 12), Mahajan (1966, p. 2), Martinez (1963, p. 124), Martinez (1964b, p. 136), Martinez and Myren (1964a, p. 79), Martinez and others (1964, p. 6), McMillion (1960, p. 14), McNelly (1964, p. 12), Menzel (1959, p. 88), Oakley (1965, p. 116), Paul (1965, p. 32), Polgar and others (1963, p. 107), Presser and Russell (1965, pp. 155, 156), Purohit (1963, pp. 93, 94), Ramos (1966, p. 56), Reddy and Kivlin (1968, p. 21), Rhoads and Piper (1963, p. 60), Rizwani (1964, p. 105), Robertson and Rossiter (1968, p. 19), Robinson (1963, p. 31), Rogers (1961b, p. 66), Rogers (1962, p. 77), Roy (1967, p. 205), Roy and others (1968, p. 88), Spector and others (1963, pp. 28, 32), Spector and others (1964, p. 678), Stickley (1964, p. 91), Suchman (1966, p. 42), Singh (1967, p. 171), Sycip (1964, p. 601), Tajima (1959, p. 17), Takeshita (1964, p. 2), Thomas (1966, p. 86), van den Ban (1963a, p. 231), van den Ban (1964, p. 248), van den Ban (1965, p. 5), Watson (1946, p. 12), Wells and MacLean (1962, p. 16), Wichers (1958, p. 47), Wish (1967, p. 164).

Not supporting: Armstrong (1959, p. 26), Arndt (1968c, p. 6), Beal and Rogers (1960, p. 14), Belcher (1958, p. 168), Bhasin (1966, p. 61), Bhatia and others (1966, p. 15), Bowers (1938, p. 28), Celade (1965, p. 111), Christiansen and Taylor (1966, p. 21), Couch (1965, p. 14), Deutschmann and others (1965, p. 25), Donohew and Singh (1967, p. 15), Fliegel (1966, p. 12), Fonseca (1966, p. 73), Goldsen and Ralis (1957, p. 20), Gorman (1967b, pp. 8, 9), Herzog (1967, p. 43), Jain (1963, p. 64), Keith (1968a, p. 143), Keith (1968b, p. 108), King (1965, p. 429), Kivlin (1967, p. 8), Klietsch (1961, p. 89), Mahajan (1966, p. 2), Maamary (1965, p. 29), Marsh and Coleman (1955b, p. 290), Oakley (1965, p. 116), Rao (1966, p. 42), Reddy and Kivlin (1968, p. 21), Robertson (1967b, p. 15), Rogers and Beal (1958b, p. 77), Sokol (1959, p. 4), Spector and others (1964, p. 678), Takeshita (1964, p. 3), Torres and Spector (1964, p. 6), Wilkening and others (1962, p. 167).

5-27: *Earlier adopters have greater exposure to interpersonal communication channels than later adopters* (46 studies, or 77 percent, support; 14 studies do not support).

> *Supporting:* Aberle (1966, p. 103), Arndt (1967a, p. 292), Arndt (1968d, p. 17), Arroyo (1965, p. 39), Bonser (1958a, p. 21), Campbell and Holik (1960, p. 15), Cohen (1962, p. 46), Correa (1965, p. 40), Crain (1966, p. 470), De Fleur and Larson (1958, p. 229), Dhaliwal (1963, p. 4), Dubey (1966, p. 1), Evans (1968, p. 153), Feliciano (1964, p. 12), Fonseca (1966, p. 78), Glaser (1966, p. 19), Graham (1951, p. 292), Graham (1956, p. 96), Haines (1964, p. 61), Havens and Rogers (1961a, p. 17), Hay and Lowry (1957, p. 19), Jain (1963, p. 63), Jones (1957, p. 113), Kim (1967, p. 86), King and Summers (1968, p. 21), Kivlin (1968, p. 37), Lewin (1947, p. 464), Lewin (1963, p. 62), McMillion (1960, p. 76), Palmore and Freedman (1968, p. 11), Rahudkar (1958, p. 133), Reddy and Kivlin (1968, p. 27), Rehder (1961, p. 4), Rogers (1962, p. 77), Rogers and Burdge (1962, p. 3), Sabri (1966, p. 109), Satyanarayana (1967, p. 23), Scantland and others (1952, p. 39), Spector and others (1964, p. 679), Suchman (1966, p. 42), Takeshita (1964, p. 2), van den Ban (1963a, p. 231), van den Ban (1964, p. 248), Welch and Verner (1962, p. 236), Whyte (1954, p. 204), Wilson and Gallup (1955, p. 19).
>
> *Not supporting:* Arndt (1968c, p. 6), Bonser (1958b, p. 20), Booth (1966, p. 10), Deutschmann (1961, p. 28), Gupta (1966, p. 110), Jones (1960b, p. 1), Letourneau (1963, p. 73), Marsh and Coleman (1955b, p. 290), Rizwani (1964, p. 105), Spencer (1958, p. 88), Verner and Gubbels (1967, p. 31), Wilkening and others (1962, pp. 172, 176), Wilkening and others (1963, p. 7), Winick (1961, p. 388).

5-28: *Earlier adopters seek information about innovations more than later adopters* (12 studies, or 86 percent, support; 2 studies do not support).

> *Supporting:* Arias (1967, p. 51), Beal and Rogers (1959, p. 562), Carlson (1965, p. 64), Carter and Williams (1957, p. 180), Carter and Williams (1959, p. 94), Dubey (1967, p. 158), Eastmond (1953, p. 19), Fliegel (1965, p. 285), Klietsch (1961, p. 89), Whittenbarger and Maffei (1966, p. 6), Wilson (1927, p. 1), Wilson (1928, p. 20).
>
> *Not supporting:* Carlson (1952, p. 57), Oakley (1965, p. 117).

5-29: *Earlier adopters have greater knowledge of innovations than later adopters* (61 studies, or 76 percent, support; 19 studies do not support).

> *Supporting:* Arndt (1968b, p. 5), Bauder (1961, p. 30), Beal and others (1958b, p. 7), Beal and others (1959b, p. 23), Beal and others (1967a, p. 8), Beal and Rogers (1960, p. 14), Beal and Sibley (1962, p. 102), Bhasin (1966, p. 63), Bohlen and others (1959, p. 39), Bose (1964c, p. 10), Campbell and Lionberger (1963, p. 28), Cassel (1963, p. 744), Castillo (1964, p. 7), Central Treaty Organization

(1962, p. 146), Eichholz (1963, p. 267), Farnsworth (1940, p. 100), Fischer and Timmons (1959, p. 447), Glaser (1958, p. 143), Goldsen and Ralis (1957, p. 35), Graham (1956, p. 99), Haines (1964, p. 1), Havens (1966, p. 150), Hess and Miller (1954, p. 28), Hobbs (1960, p. 74), Hsieh (1966, p. 18), Iqbal (1966, p. 83), Kar (1967, p. 37), Keith (1968b, p. 108), Kivlin (1968, p. 39), Klonglan (1963, p. 90), Klonglan and others (1960, p. 20), Klonglan and others (1964, p. 176), Koya (1962, p. 6), Kronus (1967, p. 156), Leuthold (1960a, p. 3), Leuthold (1960b, p. 42), Leuthold (1965, pp. 68, 103), Lin (1966, p. 67), Loomis (1967b, p. 885), Maamary (1965, p. 24), Moulik and others (1966, p. 473), Mulay and Roy (1965, p. 111), Reddy and Kivlin (1968, p. 27), Rogers (1957b, p. 5), Rogers and Havens (1961a, p. 4), Rose-crance (1964, p. 294), Roy and others (1968, p. 112), Sabri (1966, p. 132), Sawale (1966, p. 206), Schwieder (1966, pp. 132, 134), Sheth (1966, p. 186), Singh (1965, p. 2), Sinha and Yadav (1964a, p. 9), Sizer and Porter (1960, p. 9), Stickley (1964, p. 111), Suchman (1966, p. 41), Thomas (1966, p. 86), Whittenbarger and Maffei (1966, p. 6), Wilkening and Santopolo (1952, p. 15), Wish (1966, p. 893), Wish (1967, p. 163).

Not supporting: Agarwala (1961, p. 113), Barnett (1964b, p. 367), Beal and Sibley (1966, pp. 67, 100), Beal and others (1967a, p. 8), Beaton and Maddox (1962, p. 185), Chaparro (1955, p. 184), Diswath (1964, p. 54), Eichholz (1961, p. 167), Francis (1960, p. 54), Goldsen and Ralis (1957, p. 30), Graham (1951, p. 291), Havens and Rogers (1961a, p. 14), Herzog (1967, p. 43), Klietsch (1961, p. 86), Klonglan and others (1964, p. 175), Rahim (1964, p. 7), Robertson (1967b, p. 16), Rogers (1957b, p. 5), Wilkening and Johnson (1958, p. 14).

5-30: *Earlier adopters have a higher degree of opinion leadership than later adopters* (42 studies, or 76 percent, support; 13 studies do not support).

Supporting: Advertest Research (1958, p. 50), Arndt (1967a, p. 293), Arndt (1968a, p. 6), Arndt (1968b, p. 5), Arndt (1968d, p. 12), Barnabas (1957, p. 2), Bertrand and South (1963, p. 12), Carlson (1965, p. 22), Christiansen and Taylor (1966, pp. 17, 21), Chaparro (1955, p. 185), Coleman and others (1957, p. 268), Coleman and others (1959, p. 19), Couch (1965, p. 19), Crain (1966, p. 472), Dasgupta (1966, p. 12), Flinn (1963, p. 9), Inayatullah (1964, p. 15), Katz (1961, p. 15), Leuthold (1960a, p. 3), Leuthold (1960b, p. 40), Leuthold (1965, p. 68), Leuthold and Wilkening (1965, p. 8), Lin (1968, p. 22), Loomis (1967a, p. 11), Loomis (1967b, p. 883), Marion (1966, p. 190), Marsh and Coleman (1954a, p. 181), Marsh and Coleman (1956, p. 593), Rahudkar (1960, p. 8), Robertson (1966a, p. 158), Robertson (1967b, p. 17), Robertson and Rossiter (1968, p. 15), Rogers (1961b, p. 64), Rogers and Burdge (1962, p. 17), Rogers and Havens (1966, p. 38), Roy (1967, p. 205), Sheppard (1960, p. 44), Sheppard (1963, pp. 5, 12), Singh (1967, p. 181), Stickley (1964, pp. 104, 105), van den Ban (1963b, p. 240), Wilkening and others (1962, p. 132).

Not supporting: Booth (1966, p. 11), Couch (1965, p. 10), King (1964, p. 125), Menzel and Katz (1963, p. 127), Narang (1966, p. 55), Putney and Putney (1962, p. 551), Robertson (1966, p. 27), Rohrer (1949, p. 1), Ryan and Gross (1950, p. 699), Taylor (1966, p. 121), Wilkening (1952b, p. 272), Wilkening and others (1966, p. 134), Yadav (1967, p. 171).

5-31: *Earlier adopters are more likely to belong to systems with modern rather than traditional norms than are later adopters* (32 studies, or 70 percent, support; 14 studies do not support).

Supporting: Atamian (1966, p. 44), Bose (1961, p. 473), Bose (1962, p. 6), Bose (1964a, p. 21), Caird and Moisley (1961, p. 95), Dasgupta (1966, p. 19), Davis (1965, p. 109), Flinn (1961, p. 35), Flinn (1963, p. 19), Hage (1963, p. 104), Hill and others (1959, p. 58), Lin (1966, p. 67), Makarczyk (1965, p. 54), Marion (1966, p. 190), Marsh and Coleman (1954b, p. 9), Marsh and Coleman (1954d, p. 387), Marsh and Coleman (1956, p. 590), Marriott (1955, p. 245), Menzel and Katz (1955, p. 346), Morrison (1956, p. 284), Mort and Cornell (1941, p. 87), Oliver (1964, p. 3), Pedersen (1951, p. 49), Rogers and Burdge (1962, p. 17), Rogers and Havens (1966, p. 38), Sharma and Potti (1966, p. 56), van den Ban (1958, p. 52), van den Ban (1956, p. 42), van den Ban (1963b, p. 239), Verner and Millerd (1966, p. 28), Wish (1967, p. 164), Young and Marsh (1956, p. 8).

Not supporting: Beal and Rogers (1958, p. 7), Booth (1966, p. 23), Crain (1962, p. 9), Gordon and Marquis (1966, p. 201), Jamias and Troldahl (1964, p. 14), Keith (1968a, p. 135), Keith (1968b, p. 154), Letourneau (1963, p. 72), Niehoff (1964, p. 112), Prasada (1966, p. 58), Rahim (1961b, p. 23), Rogers and Ramos (1965, p. 7), Takeshita (1966, p. 702), Welikala (1959, p. 2).

5-32: *Earlier adopters are more likely to belong to well integrated systems than are later adopters* (8 studies, or 53 percent, support; 7 studies do not support).

Supporting: Carman (1966), Madigan (1962, p. 207), Oberg and Rios (1955, p. 370), Press (1966, p. 4), Robertson (1966a, p. 113), Robertson (1967a, p. 10), Robertson (1967b, p. 12), Weintraub and Bernstein (1966, p. 518).

Not supporting: Eibler (1965, p. 157), Fliegel (1967a, p. 16), Kreitlow and Duncan (1956, p. 3), Mead (1955, p. 13), Menzel (1959, p. 216), Raphael (1964, p. 358), van den Ban (1960, p. 308).

6-1: *Interpersonal diffusion is mostly homophilous** (22 studies, or 63 percent, support; 13 studies do not support).

Supporting: Allingham (1964, p. 56), Beal and others (1960, pp. 140, 143), Booth

---

*On such variables as social status, education, mass media exposure, cosmopoliteness, change agent contact, and innovativeness.

(1966, p. 35), Bose and Basu (1963, p. 44), Carlson (1965, pp. 17, 35, 47), Castillo (1964, p. 7), Chaparro (1956, pp. 580, 591), Chou (1966, pp. 48, 49), Feldman and Spencer (1965, p. 12), Kahlon and Kaushal (1967, p. 115), Larsen and Hill (1958, p. 504), Lionberger (1959), Lionberger and Campbell (1963a, p. 11), Marsh and Coleman (1954c, p. 1), Menzel and Katz (1955, p. 340), Patel (1964, p. 10), Patel (1966, p. 206), Rogers and Leuthold (1962, p. 4), Rogers with Svenning (1969, p. 237), Rogers and van Es (1964), Troldahl and Van Dam (1965, p. 628), Warland (1963, p. 136).

Not supporting: Carlson (1965, p. 37), Chou (1966, p. 50), Coleman and others (1959, p. 19), Hartman (1964, p. 10), Inayatullah (1964, p. 13), Larsen and Hill (1958, p. 504), Lionberger (1955, pp. 30, 31), Lionberger (1956a, pp. 3, 9), Lionberger and Campbell (1963b, p. 15), Menzel and Katz (1955, p. 343), Patel (1966, pp. 13, 207, 208), Wager (1962, p. 95), Yadav (1967).

6-2: *When interpersonal diffusion is heterophilous, followers seek opinion leaders of higher social status* (11 studies, or 100 percent, support).

Supporting: Allingham (1964), Carlson (1965), Hartman (1964), Inayatullah (1964), Kahlon and Kaushal (1967), Larsen and Hill (1958), Lionberger (1959), Lionberger and Campbell (1963b), Patel (1966), Rogers and van Es (1964), Rogers with Svenning (1969, p. 237).

6-3: *When interpersonal diffusion is heterophilous, followers seek opinion leaders with more education* (6 studies, or 75 percent, support; 2 studies do not support).

Supporting: Carlson (1965), Hartman (1964), Patel (1964), Rogers and Leuthold (1962), Rogers and van Es (1964), Rogers with Svenning (1969).
Not supporting: Troldahl and Van Dam (1965), Warland (1963).

6-4: *When interpersonal diffusion is heterophilous, followers seek opinion leaders with greater mass media exposure* (5 studies, or 100 percent, support).

Supporting: Chou (1966), Patel (1966), Rogers and Leuthold (1962), Rogers with Svenning (1969), Troldahl and Van Dam (1965).

6-5: *When interpersonal diffusion is heterophilous, followers seek opinion leaders who are more cosmopolite* (1 study, or 100 percent, support).

Supporting: Rogers and Leuthold (1962).

6-6: *When interpersonal diffusion is heterophilous, followers seek opinion leaders with greater change agent contact* (2 studies, or 100 percent, support).

Supporting: Lionberger (1955), Patel (1964).

6-7: *When interpersonal diffusion is heterophilous, followers seek opinion leaders who are more innovative* (10 studies, or 91 percent support; 1 study does not support).

Supporting: Carlson (1965), Hartman (1964), Inayatullah (1964), Kahlon and Kaushal (1967), Lionberger (1955), Lionberger and Campbell (1963b), Patel (1966), Rogers and Leuthold (1962), Rogers and van Es (1964), Rogers with Svenning (1969).
Not supporting: Carlson (1965).

6-8: *Interpersonal diffusion is characterized by a greater degree of homophily in traditional than in modern systems* (3 studies, or 60 percent, support; 2 studies do not support).

Supporting: Rogers and van Es (1964), Rogers with Svenning (1969, pp. 237–238), Yadav (1967).
Not supporting: Ho (1969), van den Ban (1963b).

6-9: *In traditional systems followers interact with opinion leaders less (or no more) technically competent than themselves, whereas in modern systems opinion leaders are sought who are more technically competent than their followers* (1 study, or 100 percent, support).

Supporting: Rogers with Svenning (1969, p. 238).

6-10: *Opinion leaders have greater exposure to mass media than their followers* (9 studies, or 90 percent, support; 1 study does not support).

Supporting: Abu-Lughod (1963, p. 102), Deutschmann (1962a, p. 11), Lionberger (1953, p. 333), Menzel and Katz (1955), Rahim (1961, p. 58), Rahim (1965, p. 36), Rogers (1958, p. 8), Rogers with Svenning (1969, p. 228), Summers (1968, p. 229).
Not supporting: Rahim (1965, p. 34).

6-11: *Opinion leaders are more cosmopolite than their followers* (10 studies, or 77 percent, support; 3 studies do not support).

Supporting: Katz (1957), Lionberger (1953, p. 332), Rahim (1965, p. 36), Rahudkar (1960), Rogers and Burdge (1961, p. 21), Rogers and Burdge (1962), Rogers and Leuthold (1961), Rogers with Svenning (1969, p. 228), Sepulveda and Carter (1964, p. 1), van den Ban (1963b, p. 240).
Not supporting: Attah (1968, p. 74), Summers (1968, p. 190), van Es (1964, p. 98).

6-12: *Opinion leaders have greater change agent contact than their followers* (10 studies, or 77 percent, support; 3 studies do not support).

Supporting: Emery and Oeser (1958, p. 51), Kahlon and Kaushal (1967), Loomis (1967a, p. 15), Loomis (1967b, p. 884), Rogers (1958c, p. 8), Rogers and Burdge

(1961, p. 21), Savale (1966, p. 202), Troldahl and others (1965, p. 6), van den Ban (1961, p. 129), van den Ban (1965, p. 4).

Not supporting: Lionberger and Campbell (1963a, p. 3), Menzel and Katz (1955, p. 343), Troldahl and others (1965, p. 408).

6-13: *Opinion leaders have greater social participation than their followers* (11 studies, or 73 percent, support; 4 studies do not support).

Supporting: Attah (1968, p. 74), Inayatullah (1964, p. 9), Lionberger (1953, pp. 329–334), Narang (1966, p. 46), Rahim (1961, p. 58), Rahudkar (1960, p. 7), Rogers (1958c, pp. 6–8), Sheppard (1960, p. 74), Sicinski (1964, p. 64), Summers (1968, pp. 199, 208), van den Ban (1963b).

Not supporting: Lionberger (1956a, p. 9), Robertson and Rossiter (1968, pp. 21–22), Troldahl and Van Dam (1965, p. 631), Yadav (1967, p. 171).

6-14: *Opinion leaders have higher social status than their followers* (20 studies, or 74 percent, support; 7 studies do not support).

Supporting: Barnabas (1957, p. 6), Barnabas (1958, p. 2), Campbell (1965, p. 117), Carlson (1965, p. 42), Chaparro (1956), Emery and Oeser (1958, pp. 49–53), Fliegel (1957, p. 3), Jones (1964, p. 12), Kahlon and Kaushal (1967, p. 117), Lionberger (1953, pp. 333–334), Lionberger and Coughenour (1957), Narang (1966, p. 44), Rahim (1961, p. 58), Rahudkar (1960, pp. 6, 34), Robertson and Rossiter (1968, p. 15), Rogers (1958c, p. 9), Rogers and Burdge (1962, p. 11), Rogers with Svenning (1969, p. 228), Stickley and others (1967, p. 622), Summers (1968, p. 192).

Not supporting: Abu-Lughod (1963, p. 102), Attah (1968, p. 74), Larsen and Hill (1958, p. 501), Lionberger (1956a, p. 9), Rahim (1965, p. 34), Rogers and van Es (1964, p. 28), Summers (1968, p. 191).

6-15: *Opinion leaders are more innovative than their followers* (24 studies, or 86 percent, support; 4 studies do not support).

Supporting: Attah (1968, p. 74), Brandner (1960), Chaparro (1956), Christiansen (1965, p. 335), Jones (1964, p. 12), Katz (1957), Lionberger (1953, pp. 335–337), Lionberger (1956a, p. 9), Lionberger (1959, p. 116), Marsh and Coleman (1954a), Rahim (1961, p. 58), Rahudkar (1960), Robertson and Rossiter (1968, p. 14), Rogers (1958c, p. 8), Rogers (1961, p. 21), Rogers (1962, p. 86), Rogers and Burdge (1962, p. 11), Rogers with Svenning (1969, p. 228), Summers (1968, p. 230), van den Ban (1963b), Wilkening (1952b), Wilkening (1958d), Wilkening (1961), Young and Coleman (1959).

Not supporting: Chaparro (1956, p. 133), Gross (1942, p. 134), Rogers and van Es (1964, p. 24), van Es (1964, pp. 96, 104).

6-16: *When the system's norms favor change, opinion leaders are more innovative; but when the norms are traditional, opinion leaders are not especially innovative* (7 studies, or 78 percent, support; 2 studies do not support).

Supporting: Herzog and others (1968, p. 72), Marek (1966), Marsh and Coleman (1954a), Rogers and Burdge (1962), Rogers with Svenning (1969, pp. 230–231), Sen (1969), van den Ban (1963b).
Not supporting: Lionberger (1960, p. 61), Rahudkar (1960).

6-17: *When the norms of a system are more modern, opinion leadership is more monomorphic* (4 studies, or 80 percent, support; 1 study does not support).

Supporting: Attah (1968), Sen (1969), Sengupta (1968), Yadav (1967).
Not supporting: Rogers with Svenning (1969, p. 227).

7-1: *Change agent success is positively related to the extent of change agent effort* (16 studies, or 84 percent, support; 3 studies do not support).

Supporting: Armstrong (1959), Deutschmann and Fals Borda (1962b), Fliegel and others (1967b), Hoffer (1944), Hursh and others (1969), Niehoff (1964), Niehoff (1966a), Nye (1952, p. 36), Petrini (1966b), Petrini (1967), Planning Research and Action Institute (1961, p. 17), Rogers and others (forthcoming), Ross (1952), Singh (1952, p. 67), Stone (1952, p. 16), Whiting and others (1968).
Not supporting: Barnabas (1955, p. 2), Preiss (1954, p. 3), Saigaonkar (1967, p. 2).

7-2: *Change agent success is positively related to his client orientation rather than change agency orientation* (6 studies, or 100 percent, support).

Supporting: Dobyns (1951, p. 31), Erasmus (1961, p. 25), Gans (1962), Preiss (1954), Saunders and Samora (1955, p. 396), Straus (1953, p. 251).

7-3: *Change agent success is positively related to the degree to which his program is compatible with clients' needs* (10 studies, or 100 percent, support).

Supporting: Adams (1955, p. 454), Allahabad Agricultural Institute (1957, p. 36), Dobyns (1951, p. 31), Erasmus (1961, p. 25), Hoffer (1946, p. 31), Niehoff (1964, p. 3), Rogers (1966a), Sasaki (1956, p. 309), Saunders and Samora (1955, pp. 391, 396), Singh (1952, p. 61).

7-4: *Change agent success is positively related to his empathy with clients.*
There are no empirical studies supporting or not supporting this generalization; hence, it should be regarded as an hypothesis for future research.

7-5 : *Change agent contact is positively related to higher social status among clients* (37 studies, or 86 percent, support; 6 studies do not support).

Supporting: Anderson (1955, p. 8), Burleson (1950b, pp. 27, 43), Campbell and Bennett (1961, p. 10), Coleman (1951, p. 214), Coleman and Marsh (1955, p. 99), Coughenour (1958, p. 2), Coughenour (1959, p. 16), Dasgupta (1965, p. 335), Deutschmann (1961, pp. 10, 12), Dhaliwal and Sohal (1965, p. 61), Dhara (1965, p. 22), Feldman (1966, p. 10), Hardee (1963a, p. 66), Hardee (1963b, p. 27), Hoffer (1944, p. 23), Harstendahl (1966, p. 11), Jones (1963, p. 12), Lionberger (1949, pp. 5, 10), Lionberger (1955, pp. 18, 20), Lionberger (1956b, p. 8), Louisiana Agricultural Extension Service (1950, p. 18), Photiadis (1961, pp. 3, 24), Photiadis (1962, p. 321), Quesada (1965, pp. 33, 41), Rahudkar (1961, p. 80), Rahudkar (1963, p. 10), Rogers and Capener (1960, pp. 3, 4), Rogers and Havens (1961b, p. 18), Rogers and Leuthold (1962, p. 4), Scantland and others (1952, p. 29), Slocum and others (1958, p. 27), Stickley and others (1967, p. 19), van den Ban (1963b, p. 239), Verma (1966, p. 115), Wilkening (1949b, p. 3), Wilkening (1950a, p. 28), Wilkening (1952a, p. 60).

Not supporting: Dasgupta (1965, p. 335), Kapoor (1966, p. 76), Lionberger (1950, p. 193), Photiadis (1962, p. 321), Stickley (1964, p. 95), Stuby (1965, p. 60).

7-6 : *Change agent contact is positively related to greater social participation among clients* (18 studies, or 90 percent, support; 2 studies do not support).

Supporting: Campbell and Bennett (1961, p. 10), Coleman (1951, p. 215), Coughenour (1958, p. 2), Coughenour (1959, p. 16), Hardee (1963a, p. 66), Hardee (1963b, p. 27), Hoffer (1944, p. 14), Kapoor (1966, p. 76), Lionberger (1949, pp. 9, 10), Lionberger (1955, p. 20), McCarthy and Tugby (1962, p. 9), Photiadis (1961, p. 19), Photiadis (1962, p. 321), Rahudkar (1961, p. 80), Robertson (1964, p. 16), Rogers and Leuthold (1962, p. 4), Slocum and others (1958, p. 27), Verma (1966, p. 115).

Not supporting: Quesada (1965, pp. 28, 30), Singh (1966, p. 81).

7-7 : *Change agent contact is positively related to higher education and literacy among clients* (32 studies, or 74 percent, support; 11 studies do not support).

Supporting: Anderson (1955, p. 10), Burleson (1950b, pp. 28, 43), Campbell and Barnett (1961, p. 10), Coleman (1951, p. 213), Coleman and Marsh (1955, p. 99), Coughenour (1958, p. 2), Coughenour (1959, p. 16), Coughenour and Patel (1962, p. 3), Deutschmann (1961, p. 10), Dhaliwal (1963, p. 4), Dhaliwal and Sohal (1965, p. 60), Kapoor (1966, p. 76), Lionberger (1949, p. 9), Lionberger (1965, p. 9), Louisiana Agricultural Extension Service (1950, p. 18), McCarthy and Tugby (1962, p. 5), Maddox (1965, p. 5), Photiadis (1961, p. 23), Rahudkar (1961, p. 80), Rahudkar (1963, p. 101), Rogers and Capener (1960,

p. 3), Rogers and Havens (1961b, p. 18), Rogers and Leuthold (1962, p. 4), Satyanarayana (1966, p. 18), Satyanarayana (1967, p. 26), Saxena (1963, p. 95), Scantland and others (1952, p. 32), Singh (1966, p. 81), Slocum and others (1958, p. 27), van den Ban (1963b, p. 239), Verma (1966, p. 15), Wilkening and Santopolo (1952, p. 24).

Not supporting: Fliegel (1967b, p. 96), Hardee (1963a, p. 64), Hoffer (1944, p. 24), Parish (1956, p. 229), Photiadis (1962, p. 321), Quesada (1965, p. 35), Ramos (1966, p. 63), Robertson (1964, p. 16), Stuby (1965, p. 60), Takeshita and others (1964, p. 227), White (1965, p. 38).

7-8: *Change agent contact is positively related to cosmopoliteness* (5 studies, or 100 percent, support).

Supporting: Campbell and Barnett (1961, p. 10), Emery and Oeser (1958, p. 14), Jones (1963, p. 12), Photiadis (1961, p. 3), Maddox (1965, p. 5).

7-9: *Change agent success is positively related to his homophily with clients* (2 studies, or 100 percent, support).

Supporting: Allahabad Agricultural Institute (1957), Rahudkar (1959).

7-10: *Change agent success is positively related to the extent that he works through opinion leaders* (3 studies, or 100 percent, support).

Supporting: Alers-Montalvo (1957, p. 16), Bliss (1952, p. 30), Niehoff (1964, p. 5).

7-11: *Change agent success is positively related to his credibility in the eyes of his clients* (1 study, or 100 percent, support).

Supporting: Alers-Montalvo (1957, p. 6).

7-12: *Change agent success is positively related to his efforts in increasing his clients' ability to evaluate innovations* (4 studies, or 100 percent, support).

Supporting: Alers-Montalvo (1957), Castillo (1967, p. 71), Rogers (1966a), Sasaki (1956, p. 310).

8-1: *Mass media channels are relatively more important at the knowledge function and interpersonal channels are relatively more important at the persuasion function in the innovation-decision process* (18 studies, or 90 percent, support; 2 studies do not support).

Supporting: Beal and Rogers (1957), Beal and Rogers (1960, p. 8), Copp and others (1958), Deutschmann and Fals Borda (1962a), Jain (1965), Mason (1962a), Mason (1962b), Mason (1963a), Mason (1963b), Mason (1964), Myren (1962), Rahim (1961, p. 48), Rogers and Beal (1958a), Rogers and

Pitzer (1960), Singh and Jha (1965), Wilkening (1952b, p. 16), Wilkening (1956), van den Ban (1963b).

Not supporting: Rogers and Meynen (1965), Rogers with Svenning (1969, p. 129).

8-2: *Cosmopolite channels are relatively more important at the knowledge function and localite channels are relatively more important at the persuasion function in the innovation-decision process* (6 studies, or 86 percent, support; 1 study does not support).

Supporting: Beal and Rogers (1957a), Katz (1961), Jain (1965), Rogers with Svenning (1969, p. 132), Ryan and Gross (1943), Wilkening and others (1960).

Not supporting: Sawhney (1966).

8-3: *Mass media channels are relatively more important than interpersonal channels for earlier adopters than for later adopters* (8 studies, or 80 percent, support; 2 studies do not support).

Supporting: Beal and Rogers (1960, p. 16), Bowers (1938), Jain (1965), Rahim (1961, p. 43), Rogers and Beal (1958a), Ryan and Gross (1943), Wilkening (1952b, p. 19), van den Ban (1963b).

Not supporting: Rogers and Meynen (1965), Rogers with Svenning (1969, p. 129).

8-4: *Cosmopolite channels are relatively more important than localite channels for earlier adopters than for later adopters* (9 studies, or 100 percent, support).

Supporting: Campbell (1959), Carter and Williams (1959), Coleman and others (1966), Jain (1965), Rogers and Burdge (1961), Rogers and Burdge (1962), Rogers and Leuthold (1962), Rogers and Meynen (1965), Rogers with Svenning (1969, pp. 132–133).

8-5: *The effects of mass media channels, especially among peasants in less developed countries, are greater when these media are coupled with interpersonal channels in media forums* (3 studies, or 100 percent, support).

Supporting: Menefee and Menefee (1967), Neurath (1960), Neurath (1962).

9-1: *Stimulators of collective innovation-decisions are more cosmopolite than other members of the social system* (1 study, or 100 percent, support).

Supporting: Hage (1963).

9-2: *Initiators of collective innovation-decisions in a social system are unlikely to be the same individuals as the legitimizers* (3 studies, or 100 percent, support).

Supporting: Agger and others (1964, p. 47), Chandler (1962), Dahl (1961, p. 128).

9-3: *Rate of adoption of a collective innovation is positively related to the degree to which the social system's legitimizers are involved in the decision-making process* (1 study, or 100 percent, support).

Supporting: Rosenthal and Crain (1965).

9-4: *Legitimizers of collective innovation-decisions possess higher social status than other members of the social system* (4 studies, or 100 percent, support).

Supporting: Agger and others (1964, p. 279), Dahl (1961, p. 169), Edwards (1963, p. 112), Miller (1953, p. 22).

9-5: *The rate of adoption of collective innovations is positively related to the degree of power concentration in a system* (6 studies, or 86 percent, support; 1 study does not support).

Supporting: Fliegel and others (1967), Gamson (1968), Hawley (1962), Hursh and others (1969), Rosenthal and Crain (1968), Whiting and others (1968). *Not supporting:* Clark (1968).

9-6: *Satisfaction with a collective innovation-decision is positively related to the degree of participation of members of the social system in the decision* (5 studies, or 100 percent, support).

Supporting: French and others (1958), French and others (1960), Hamblin and others (1961), Morse and Reimer (1956), Seashore and Bowers (1963).

9-7: *Member acceptance of collective innovation-decisions is positively related to the degree of participation in the decision by members of the social system* (10 studies, or 91 percent, support; 1 study does not support).

Supporting: Bennett (1952), Coch and French (1948), Davis (1965), Giffin and Erlich (1963), Levine and Butler (1952), Lewin (1943), Queeley and Street (1965), Radke and Klisurich (1947), Wallach and Kogan (1965), Wallach and others (1967). *Not supporting:* Crain and others (1969).

9-8: *Member acceptance of collective innovation-decisions is positively related to member cohesion with the social system* (2 studies, or 100 percent, support).

Supporting: Eibler (1965), Kelley and Volkhardt (1952).

10-1: *A supportive relationship between the adoption unit (a subordinate) and the decision unit (a superior) leads to more upward communication about the innovation* (3 studies, or 100 percent, support).

Supporting: Likert (1961), Likert (1967), Read (1962).

10-2: *An individual's acceptance of an authority innovation-decision is positively related to his participation in innovation decision-making* (4 studies, or 100 percent, support).

Supporting: Coch and French (1948), Davis (1965), Giffin and Erlich (1963), Queeley and Street (1965).

10-3: *An individual's satisfaction with an authority innovation-decision is positively related to his participation in innovation decision-making* (4 studies, or 100 percent, support).

Supporting: French and others (1958), French and others (1960), Morse and Reimer (1956), Seashore and Bowers (1963).

10-4: *When an individual's attitudes are dissonant with the overt behavior demanded by the organization, the individual will attempt to reduce the dissonance by changing either his attitudes or his behavior.*
There are no empirical studies supporting or not supporting this generalization; hence, it should be regarded as an hypothesis for future research.

10-5: *The rate of adoption of authority innovation-decisions is faster by the authoratitative approach than by the participative approach.*
There are no empirical studies supporting or not supporting this generalization; hence, it should be regarded as an hypothesis for future research.

10-6: *Changes brought about by the authoritative approach are more likely to be discontinued than those brought about by the participative approach.*
There are no empirical studies supporting or not supporting this generalization; hence, it should be regarded as an hypothesis for future research.

11-1: *Change agents can more easily anticipate the form and function of an innovation for their clients than its meaning* (1 study, or 100 percent, support).
Supporting: Sharp (1952).

11-2: *The power elite in a social system screen out potentially restructuring innovations while allowing the introduction of innovations which mainly affect the functioning of the system.*
There are no empirical studies supporting or not supporting this generalization; hence, it should be regarded as an hypothesis for future research.

11-3: *The power elite in a social system especially encourage the introduction of innovations whose consequences not only raise average levels of Good but also lead to a less equal distribution of Good* (1 study, or 100 percent, support).

Supporting: Karpat (1960).

Appendix B

BIBLIOGRAPHY

T HE references included in this bibliography are organized under three headings: (1) empirical diffusion studies, (2) diffusion publications that do not report empirical results, and (3) general references cited for their relevance to the diffusion of innovations but which do not deal with diffusion *per se*. The first two categories include only those publications which deal with an *innovation* which is *communicated* through certain *channels* over *time* among the members of a *social system*. Unless a publication includes consideration of these elements of diffusion, it is not considered as a diffusion report. The distinction between the first two categories is that the second (nonempirical) includes bibliographies, summaries of diffusion findings reported in other publications, and theoretical writings.* There are approximately 1,200 empirical reports and about 300 non-empirical reports in this bibliography.

Both of these first two categories are coded to indicate the diffusion research tradition of the author, as indicated by his institutional affiliation:

Author's Diffusion Research Tradition	Code for Tradition
Anthropology	A
Agricultural Economics	AE
Communication	C
Education	E

*The reader should note that the text refers to an author's works as (1960a), (1960b), (1960c), and so on. Yet (1960a) may be in the empirical part of the Bibliography; and (1960b) and (1960c) in the non-empirical part.

Early Sociology	ES
Extension Education*	EX
Geography	G
General Economics	GE
General Sociology	GS
Industrial Engineering	I
Journalism**	J
Marketing	MR
Medical Sociology	MS
Psychology	P
Public Administration	PA
Rural Sociology	RS
Statistics	S
Speech	SP
Others	O
Unclassifiable	U

*This category of extension education is frequently combined with the rural sociology tradition in the analyses presented in this book, as explained in Chapter 2.
**This category of journalism is frequently combined with the communication tradition in this book.

Empirical Diffusion Research Publications

Abell, Helen C. (1951), *The Differential Adoption of Homemaking Practices in Four Rural Areas of New York*. Ph.D. thesis. Ithaca, N.Y.: Cornell Univ.—RS

——— (1952), "The Use of Scale Analysis in a Study of the Differential Adoption of Homemaking Practices," *Rural Soc.* 17: 161–167.—RS

——— (1965), *Farm Radio Forum Project: Ghana, 1964-65*, Ontario, Canada: Agricultural College, Univ. of Guelph, Mimeo Rept.—RS

Abell, Helen C. and Larson, Olaf F. (1960), *Homemaking Practices and the Communication of Homemaking Information: Four Rural New York Areas*. Ithaca, N.Y.: Cornell Univ. Agri. Exp. Sta., Dept. of Rural Soc., Mimeo Bul. 56.—RS

Abell, Helen C. and others (1957), *Communication of Agricultural Information in a South-Central New York County*. Ithaca, N.Y.: Cornell Univ. Agri. Exp. Sta., Mimeo Bul. 49.—RS

Aberle, David F. (1966), *The Peyote Religion among the Navaho*. Chicago: Aldine.—A

Aberle, David F. and Stewart, Omer C. (1957), *Navaho and Ute Peyotism: A Chronological and Distributional Study*. Boulder: Univ. of Colorado Studies, Series in Anthro. 6.—A

Abu-Lughod, Ibraham (1963), "The Mass Media and Egyptian Village Life," *Social Forces*, 42: 97–104.—RS

Adams, F. G. and others (1963), "The Diffusion of New Durable Goods and Their Impact on Consumer Expenditures," *Proceedings of the Business and Economic Statistics Section of the Amer. Stat. Assn.*—S

Adams, Richard N. (1951), "Personnel in Culture Change: A Test of a Hypothesis," *Social Forces*, 30: 185-189.—A

——— (1955), "A Nutritional Research Program in Guatemala," in Benjamin D. Paul (Ed.), *Health, Culture and Community*. New York: Russell Sage Fdn.—A

Adler, David (1955), *An Analysis of Quality in the Associated Public School Systems through a Study of the Patterns of Diffusion of Selected Educational Practices*. D. Ed. thesis. New York: Teachers College, Columbia Univ.—E

Advertest Research (1958), *Colortown: A Profile of TV Set Owners*, New Brunswick, N.J.:Batten, Barton, Durstine and Osborn and the National Broadcasting Company. —MR

Agarwala, S. N. (1961), "A Family Planning Survey in Four Delhi Villages," *Population Stud.,* 15: 110–120.—MS

Agnew, Paul C. and Hsu, Francis L. K. (1960), "Introducing Change in a Mental Hospital," *Human Organization,* 19: 195–198.—MS

Ahmad, Bashir (1964), *A Study of Some Factors in Relation to the Acceptance of Modern Agricultural Practices*, M.A. thesis. Lahore: Univ. of Punjab.—RS

Ahmed, Mohiuddin and Ahmed, Fatema (1965), "Male Attitudes Toward Family Limitation in East Pakistan," *Eugenics Qtrly.*, 12: 209–226.—GE

Albaum, Gerald S. (1962), *A New Approach to the Information Function in Marketing*, Ph.D. thesis. Madison:Univ. of Wisconsin.—MR

Alers-Montalvo, Manuel (1953), *Cultural Change in a Costa Rican Village,* Ph.D. thesis. East Lansing: Michigan State Univ.—A

——— (1957), "Cultural Change in a Costa Rican Village," *Human Organization,* 15: 2–7.—A

Alexander, Frank D. and others (1963), "A Field Experiment in Diffusion of Knowledge of Dairy Cattle Feeding Through a TV School," *Rural Soc.,* 28: 400–404.—RS

Allahabad Agricultural Institute (1957), *Extension Evaluation*. Allahabad, India: Allahabad Agri. Inst.—EX

Allen, Harley E. (1956), *The Diffusion of Educational Practices in the School Systems of the Metropolitan Study Council*. D. Ed. thesis. New York: Columbia Univ., Teachers College.—E

Allen, Irving L. and Colfax, J. David (1968), "The Diffusion of News of LBJ's March 31 Decision," *Journalism Qtrly.*, 45: 321–324.—C

Allingham, James R. (1964), *A Descriptive Study of Communication Networks and Rumor Diffusion Aboard a Navy Ship*. M.S. thesis. Boston Univ.—C

Alves, Eliseu R. A. (1962), *Adocão de Pratica: Area Atingida Pelo Escritorio Local de Viçosa,* Belo Horizonte, Brazil: Extensão Rural e Credito Supervisionado.—AE

Anand, Asha (1966), *A Study of Selected Value-Orientations of Rural Women as Related to the Adoption of Recommended Health Practices*. M.S. thesis. India: Univ. of Delhi.—EX

Anastasio, Angelo (1960), *Port Haven: A Changing Northwestern Community*. Pullman, Wash.: Agri. Exp. Sta., Bul. 616.—A

Anderson, Marvin A. (1955), *Information Sources Important in the Acceptance and Use of Fertilizer in Iowa*. Knoxville: Tennessee Valley Authority, Rept. P 55–1.—RS

Anderson, Marvin A. and others (1955), *An Appraisal of Factors Affecting the Acceptance and Use of Fertilizer in Iowa, 1953*. Ames: Iowa Agri. Exp. Sta., Spec. Rept. 16.—RS

Andrews, John H. M. and Greenfield, T. Barr (1967), "Organizational Themes Relevant to Change in Schools," *Ont. J. of Edu. Res.*, 9: 81–99.—E

Andrus, Roman R. (1965), *Measures of Consumer Innovative Behavior*. Ph.D. thesis. Ann Arbor: Univ. of Michigan.—MR

Anello, Michael (1966), "The Cure for Illiteracy: Adult Education in Italy," *Adult Leadership*, 15: 159–179.—GS

Apodaca, Anacleto (1952), "Corn and Custom: Introduction of Hybrid Corn to Spanish American Farmers in New Mexico," in Edward H. Spicer (Ed.), *Human Problems in Technological Change*. New York: Russell Sage Fdn.—A

Arce, Antonio M. (1959), *Rational Introduction of Technology on a Costa Rican Coffee Hacienda: Sociological Implications*. Ph.D. thesis. East Lansing: Michigan State Univ.—A

Arias, Emil (1967), *Relationship between Perceptions of Sources, Adoption Behavior, and Value Orientations of Mexican Dairymen.* M.S. thesis. Ames: Iowa State Univ.—C

Armstrong, Joseph B. (1959), *County Agent Activities and the Adoption of Soil-Building Practices.* M.S. thesis. Lexington: Univ. of Kentucky.—RS

Arndt, Johan (1966a), *Word of Mouth Advertising: The Role of Product-Related Conversations in the Diffusion of a New Food Product.* D. B. A. thesis. Cambridge, Mass.: Harvard Univ.—MR

——— (1966b), "Word of Mouth Advertising: The Role of Product-Related Conversations in the Diffusion of a New Food Product." Paper presented at the American Marketing Association. Bloomington, Ind.—MR

——— (1966c), "Perceived Risk, Sociometric Integration, and Word of Mouth Advertising in the Adoption of a New Food Product," in Raymond M. Haas (Ed.), *Science, Technology and Marketing.* Chicago: Amer. Mktg. Assn.—MR

——— (1967a), "A Study of Word-of-Mouth Advertising," *Markedskommunikasjon,* 4: 94–117.—MR

——— (1967b), "Role of Product-Related Conversation in the Diffusion of a New Product," *J. of Mktg. Res.* 4: 291–295.—MR

——— (1968a), "Selective Processes in Word-of-Mouth Advertising," *J. of Adv. Res.,* 8: 19–22.—MR

——— (1968b), "Profiling Consumer Innovators," in Johan Arndt (Ed.), *Insights into Consumer Behavior.* Boston: Allyn and Bacon.—MR

——— (1968c), "Word-of-Mouth Advertising and Perceived Risk," in Harold H. Kassarjian and Thomas S. Robertson (Eds.), *Perspectives in Consumer Behavior.* Glenwood, Ill.: Scott, Foresman.—MR

——— (1968d), "Profiling Consumer Innovators." New York: Columbia Univ., Grad. School of Bus., unpublished paper. —MR

——— (1968e), "New Product Diffusion: The Interplay of Innovativeness, Opinion Leadership, Learning, Perceived Risk and Product Characteristics." New York: Columbia Univ., Grad. School of Bus., unpublished paper.—MR

——— (1968f), "A Cold-Blooded Analysis of Movie-Going as a Diffusion Process." New York: Columbia Univ., Grad. School of Bus., unpublished paper.—MR

——— (1968g), "A Test of the Two Step Flow in Diffusion of a New Product," *Journalism Qtrly.,* 45: 457–465.—MR

Arroyo, Ricardo (1965), *Estudio Sobre el Proceso de Difusión y Adopción de los Maíses Híbridos en la Vanguardia.* Pergamino, Buenos Aires: Argentina Agri. Exp. Sta., Bul. 33.—RS

Ascroft, Joseph (1966), *A Factor Analytic Investigation of Modernization among Kenya Villagers.* M.A. thesis. East Lansing: Michigan State Univ.—C

Atamian, Sarkis (1966), "The Anaktuvuk Mask and Cultural Innovation," *Science,* 151: 1337–1345.—A

Attah, Efiong Ben (1968), *An Analysis of Polymorphic Opinion Leadership in Eastern Nigerian Communities.* M.A. thesis. East Lansing: Michigan State Univ.—C

Aurbach, Herbert A. and Kaufman, Harold F. (1956), *Knowledge and Use of Recommended Farm Practices.* State College: Miss. Agri. Exp. Sta. Information Sheet 540.—RS

Australia Bureau of Agricultural Economics (1952), *Report on the Agricultural Extension Services in the Murrumbidgee Irrigation Area.* Canberra, A.C.T.: Bureau of Agri. Eco., Dept. of Commerce and Agriculture.—AE

Ayer, Frederic L. (1952), *An Analysis of Certain Community Characteristics Related to the Quality of Education.* Ph.D. thesis. New York: Columbia Univ.—E

Back, Kurt W. (1958), "The Change-Prone Person in Puerto Rico," *Pub. Opin. Qtrly.,* 22: 330–340.—MS

Badenhorst, L. T. (1962), "Family Limitation and Methods of Contraception in an Urban Population," *Population Stud.,* 16: 286–301.—MS

Badenhorst, L. T. and Unterhalter, B. (1961), "A Study of Fertility in an Urban

African Community," *Population Stud.*, 15: 70–87.—MS

Baker, Jasper N. (1954), *A Study of the Relative Effectiveness of Sources from Which Farmers Get Information Regarding Agricultural Station Results.* Ph.D. thesis. Minneapolis: Univ. of Minnesota.—EX

Bakshi, B. (1962), *Differential Acceptance of Alkathene Storage Bin as Related to Levels of Exposure and Socio-Economic Characteristics of the Farmers in Kanjhawla Block.* M.S. thesis. New Delhi: Indian Agri. Res. Inst.—EX

Bang, Sook (1966), "The Koyang Study: Results of Two Action Programs," *Stud. in Farm. Plng.*, 11: 5–12.—O

Barber, Bernard and Lobel, Lyle S. (1952), "Fashion in Women's Clothes and the American Social System," *Social Forces*, 31: 124–131.—GS

Bareiss, G. and others (1962), *Problem des Beispiels Betriebes.* Stuttgart, Germany: Eugen Ulmer.—RS

Barnabas, A. P. (1955), *Relationship between Area of Assignment and Accomplishment of Individual Extension Workers.* Allahabad, India: Allahabad Agri. Inst. Rept.—RS

——— (1957), "Who Are the Village Leaders?" *Kurukshetra.*—RS

——— (1958), "Characteristics of 'Lay Leaders' in Extension Work," *J. of the M.S. Univ. of Baroda*, 7: 1–21.—RS

——— (1960), *Social Change in a North Indian Village.* Ph.D. thesis. Ithaca, N.Y.: Cornell Univ.—RS

Barnett, Homer G. (1941), "Personal Conflicts and Culture Change," *Social Forces*, 20: 160–171.—A

———(1953), *Innovation: The Basis of Cultural Change.* New York: McGraw-Hill.—A

——— (1964a), "Diffusion Rates," in Robert A. Manners (Ed.), *Process and Pattern in Culture.* Chicago: Aldine.—A

——— (1964b), "The Acceptance and Rejection of Changes," in George K. Follschan and Walter Hirsch (Eds.), *Explorations in Social Change.* Boston: Houghton Mifflin.—A

Barrington, Thomas M. (1953), *The Introduction of Selected Educational Practices into Teachers Colleges and Their Laboratory Schools.* New York: Teachers College, Columbia Univ., Bureau of Publications. —E

Bass, Frank M. (1969), "A New Product Growth Model for Consumer Durables," *Mgmt. Sci. Theory*, 15: 215–227.—MR

Bass, Frank M. and King, Charles W. (1968), "The Theory of First Purchase of New Products," in Keith Cox and Ben M. Enis (Eds.), *A New Measure of Responsibility for Marketing: Proceedings of the 1968 June Conference of the American Marketing Association.* Chicago: American Mktg. Assn.—MR

Basu, R. N. (1964), "Experiences with a Poorly Effective Oral Contraceptive in an Indian Village," *Demography*, 1: 106–110.—MS

Basu, Sunil K. (1964), "On Diffusion and Adoption of Farm Traits," *Bul. of the Cultural Res. Inst.* 3: 47–51.—A

Bauder, Ward W. (1960), *Iowa Farm Operators' and Farm Landlords' Knowledge of, Participation in and Acceptance of the Old Age and Survivors Insurance Program.* Ames, Iowa: Agri. and Home Eco. Exp. Sta., Res. Bul. 479.—RS

——— (1961), *Influences on Acceptance of Fertilizer Practices in Piatt County.* Champaign, Ill.: Agri. Exp. Sta. Res. Bul. 679. —RS

Bauer, Raymond A. (1965), "Risk Handling in Drug Adoption: The Role of Company Preference," *Pub. Opin. Qtrly.*, 25: 546–559.—MS

Bauer, Raymond A. and Wortzel, Lawrence H. (1966), "Doctor's Choice: The Physician and His Sources of Information about Drugs," *J. of Mktg. Res.*, 3: 40–47.—MS

Baur, E. Jackson (1957), "Cultural Factors Affecting the Use of Research by Welfare Agencies." Unpublished paper. Lawrence: Univ. of Kansas.—GS

Bauwens, A. L. G. M. and Heunks, F. J. (1963), *De Landbouw in Sprang-Capelle een Economisch-Sociologisch Onderzoek.* The Hague, Netherlands: Agri. Eco. Inst. Rept. 66.—RS

Bauwens, A. L. G. M. and others (1963), *De Landbouw in Rucphen: Economisch-*

Sociologische Gezichtspunten. The Hague, Netherlands: Agri. Eco. Inst., Rept. 67.—RS

Beal, George M. and John, M. J. (1967), "Role Performance of Change Agents." Paper presented at the Rural Sociological Society. San Francisco.—RS

Beal, George M. and others (1957), "Validity of the Concept of Stages in the Adoption Process," *Rural Soc., 22:* 166–168.—RS

———— (1958a), *The Fertilizer Dealers: Attitudes and Activity.* Ames: Iowa State Univ., Rural Soc. Mimeo Rept.—RS

———— (1958b), *The Role of the Retail Dealer in the Adoption of Commercial Fertilizer by the Ultimate Consumers.* Ames: Iowa State Univ., Rural Soc. Mimeo Rept.—RS

———— (1959a), *The Effectiveness of a Free Sample Coupon Promotional Technique.* Ames: Iowa State Univ., Rural Soc. Mimeo Rept. 6.—RS

———— (1959b), *Agricultural Chemical Use Patterns.* Ames: Iowa State Univ., Rural Soc. Mimeo Rept. 17.—RS

———— (1960), *Communication Patterns among Farmers Related to a Free Sample Coupon Technique.* Ames: Iowa State Univ., Rural Soc. Mimeo Rept. 17.—RS

———— (1966), "Adoption of an Abstract Idea: Psychological Adoption of Public Fallout Shelters." Paper presented at the Rural Soc. Society.—RS

———— (1967a), *Adoption of Agricultural Technology by the Indians of Guatemala.* Ames: Iowa State Univ., Rural Soc. Rept. 71.—RS

———— (1967b), *Emerging Patterns of Commercial Farming in a Subsistence Farm Economy: An Analysis of Indian Farmers in Guatemala.* Ames: Iowa State Univ., Dept. of Soc. and Anthrop., Rural Soc. Report 68.—RS

Beal, George M. and Rogers, Everett M. (1957), "Informational Sources in the Adoption Process of New Fabrics," *J. of Home Eco., 49:* 630–634.—RS

———— (1958), "The Communication Process in the Purchase of New Products: An Application of Reference Group Theory." Paper presented at the American Assn. of Pub. Opin. Res. Chicago, Ill.—RS

———— (1959), "The Scientist as a Referent in the Communication of New Technology," *Pub. Opin. Qtrly., 22:* 555–563.—RS

———— (1960), *The Adoption of Two Farm Practices in a Central Iowa Community.* Ames: Iowa Agri. and Home Eco. Exp. Sta., Spec. Rept. 26.—RS

Beal, George M. and Sibley, Donald N. (1966), "Adoption of Agricultural Technology among the Indians of Guatemala." Paper presented at the Rural Soc. Society, Miami Beach.—RS

Beaton, Leonard and Maddox, John (1962), *The Spread of Nuclear Weapons.* New York: Praeger.—U

Becker, Howard S. (1953), "Becoming a Marijuana User," *Amer. J. of Soc., 59:* 235–242.—MS

Becker, Marshall H. (1968a), "Factors Affecting Diffusion of Innovations among Health Professionals." Paper presented at the Amer. Public Health Assn. Detroit.—MS

———— (1968b), *Patterns of Interpersonal Influence and Sources of Information in the Diffusion of Two Public Health Innovations.* Ann Arbor: Univ. of Michigan, Pub. Health Practice Res. Program, Rept.—MS

Becker, Selwyn W. and Stafford, Frank (1967), "Some Determinants of Organizational Success," *J. of Bus., 40:* 511–578.—P

Belcher, John C. (1958), "Acceptance of the Salk Polio Vaccine," *Rural Soc., 23:* 158–170.—RS

Belcher, John C. and Hay, Donald G. (1959), *Use of Health Care Services and Enrollment in Voluntary Health Insurance in Hancock County, Georgia, 1956.* Athens: Ga. Agri. Exp. Sta. Bul., N.S. 72.—RS

———— (1960), *Use of Health Care Services and Enrollment in Voluntary Health Insurance in Habersham County, Georgia, 1957.* Athens: Ga. Agri. Exp. Sta. Bul., N.S. 73.—RS

Bell, William E. (1962), *Consumer Innovation: An Investigation of Selected Characteristics of*

Innovators. D. B.A. thesis. East Lansing: Michigan State Univ.—MR

——— (1964), "Consumer Innovators: A Unique Market for Newness," in Stephen A. Greyser (Ed.), *Toward Scientific Marketing: Proceedings of the 1963 Winter Conference of the American Marketing Association.* Chicago: Amer. Mktg. Assn.—MR

Bemiller, James L. (1960), *Development of a Scale to Measure the Rationality Element of Farm Management Ability.* M.S. thesis. Columbus: Ohio State Univ.—RS

Ben-David, Joseph (1960), "Roles and Innovations in Medicine," *Amer. J. of Soc.*, 65: 557–578.—MS

Benvenuti, B. (1961), *Farming in Cultural Change.* Assen, Netherlands: Van Gorcum, and New York: The Humanities Press.—RS

Berelson, Bernard (1964c), "Turkey: National Survey on Population," *Stud. in Fam. Plng.*, 5: 1–5.—MS

——— (1966), "KAP Studies on Fertility," in Bernard Berelson and others (Eds.), *Family Planning and Population Programs.* Chicago: Univ. of Chicago Press.—MS

Berelson, Bernard and Freedman, Ronald (1964), "A Study in Fertility Control," *Scientific American,* 210: 29–37.—MS

Bertrand, Alvin L. (1951), *Agricultural Mechanization and Social Change in Rural Louisiana.* Baton Rouge: La. Agri. Exp. Sta., Bul. 458.—RS

Bertrand, Alvin L. and others (1956), *Factors Associated with Agricultural Mechanization in the Southwest Region.* Fayetteville: Ark. Agri. Exp. Sta., Bul. 567.—RS

Bertrand, Alvin, L. and South, D. R. (1963). "The Acceptance of New and Improved Forestry Practices by Nonindustrial Forests Landowners," in T. Hansborough (Ed.), *Southern Forests and Southern People.* Baton Rouge: Louisiana State Univ. Press.—RS

Bhasin, H. S. (1966), "An Investigation into Some Factors Influencing the Low Adoption of Selected Farm Practices in Samana Block, Patiala, Punjab," in Aditya N. Shukla and Iqbal Singh Grewal (Eds.), *Summaries of Extension by Post-*

Graduate Students. Ludhiana, India: Punjab Agricultural Univ.—EX

Bhatia, Brajesh and others (1966), *A Study in Family Planning Communication: Direct Mailing.* New Delhi, India: Central Fam. Plng. Inst., CFPI Monograph Series 1.—U

Bhatia, Rajendra P. (1966), "A Study of Some Factors Affecting Adoption of Poultry Farming in Hissar District, Punjab," in Aditya N. Shukla and Iqbal Singh Grewal (Eds.), *Summaries of Extension by Post-Graduate Students.* Ludhiana, India: Punjab Agricultural Univ.—EX

Bhosale, R. J. (1960), *Relative Effectiveness of Extension Techniques for Popularising Improved Vegetable Growing in the I.C. Scheme Area, Nangloi.* M.S. thesis. New Delhi: Indian Agri. Res. Inst.—EX

Bienenstok, Theodore (1965), "Resistance to an Educational Innovation," *Elem. School J.,* 65: 420–428.—E

Bigelow, Merrill A. (1947), *Discovery and Diffusion in Pioneer Schools.* D.Ed. thesis. New York: Teachers College, Columbia Univ.—E

Biggar, Jeanne C. and Sauer, Howard M. (1966), *Evaluation of the Farm and Home Development Program in Duel County, 1958 to 1964.* Brookings: S. Dak. Agri. Exp. Sta., Bul. 535.—RS

Binion, Stuart D. (1954), *An Analysis of the Relationship of Pupil-Teacher Ratio to School Quality.* D.Ed. thesis. New York: Teachers College, Columbia Univ.—E

Bishop, Rowland and Coughenour, C. Milton (1964), *Discontinuance of Farm Innovations.* Columbus: Ohio State Univ., Dept. of Agri. Eco. and Rural Soc., Mimeo Bul. AE 361.—RS

Bittner, Ruford F. (1959), *Farm Practice Adoption as Related to Extension Participation and Importance of Enterprise.* M.S. thesis. East Lansing: Michigan State Univ.—AE

Blackmore, John and others (1955), *Test-Demonstration Farms and the Spread of Improved Farm Practices in Southwest Virginia.* Knoxville: Tennessee Valley Authority, Report P 55–3.—RS

Blair, Annie O. (1967), "A Comparison of Negro and White Fertility Attitudes," in

Donald J. Bogue (Ed.), *Sociological Contributions to Family Planning Research.* Chicago: Univ. of Chicago Press.—MS

Blake, Judith (1961), *Family Structure in Jamaica.* New York: The Free Press.—GS

Bliss, Wesley (1952), "In the Wake of the Wheel: Introduction of the Wagon to the Papago Indians of Southern Arizona," in Edward H. Spicer (Ed.), *Human Problems in Technological Change.* New York: Russell Sage Fdn.—A

Bodiguel, M. (1968), "Trois Innovations Techniques: Trois Societes Rurales" (Three Technical Innovations in Three Rural Communities). Paper presented at the Second World Congress of Rural Soc. Drienerlo, Netherlands.—MS

Bogart, Leo (1951), "The Spread of News on a Local Event: A Case History," *Pub. Opin. Qtrly.,* 15: 769–772.—MR

Bogue, Donald J. and others (1966), *The Rural South Fertility Experiments.* Univ. of Chicago, Community and Family Study Center.—MS

Bogue, Donald J. and Palmore, James A. (1966), "The Eastern Kentucky Private Physician-Plus-Education Program," in Donald J. Bogue (Ed.), *The Rural South Fertility Experiments.* Univ. of Chicago, Community and Family Study Center.—MS

Bohlen, Joe M. and others (1959), *The Iowa Farmer and Agricultural Chemicals: Attitudes, Level of Knowledge and Patterns of Use.* Ames: Iowa State Univ., Rural Soc. Rept. 8.—RS

Bonilla, Elssy (1964), *La Predicíon de la Adopción de Hortalizas en Tres Comunidades Colombianas.* Bogotá, Colombia: Monografia para Licenciatura, Universidad Nacional de Colombia, Facultad de Sociología.—RS

Bonjean, Charles M. and others (1965), "Reactions to the Assassination in Dallas," in Bradley S. Greenberg and Edwin B. Parker (Eds.), *The Kennedy Assassination and the American Public: Social Communication in Crisis.* Stanford, Cal.: Stanford Univ. Press.—C

Bonser, Howard J. (1958a), *Better Farming Practices through Rural Community Organizations.* Knoxville: Tenn. Agri. Exp. Sta., Bul. 286.—RS

——— (1958b), *Better Homemaking Practices through Rural Community Organizations.* Knoxville: Tenn. Agri. Exp. Sta., Bul. 287.—RS

Booth, Alan (1966), *Factors Which Influence Participation in Adult Education Conferences and Programs by Members of Professional Associations.* Lincoln: Univ. of Nebraska, Adult Edu. Res. Rept.—E

Borman, Leonard D. (1965), "The Marginal Route of a Mental Hospital Innovation." Paper presented at the Society for Applied Anthro. Lexington, Ky.—A

Bose, A. B. and Saxena, P. C. (1965), "The Diffusion of Innovations in a Village in Western Rajasthan," *Eastern Anthro.,* 18: 138–151.—RS

Bose, Nirmal K. (1961), "Impact of Changing Technology on Society," *Eco. Wkly.,* 473–474.—A

Bose, S. P. (1960), *Relative Influence of Socio-Economic Factors on the Acceptance of the Cow-Dung Gas Plant by the Farmers When Exposed to Extension Teaching.* M.S. thesis. New Delhi: Indian Agri. Res. Inst.—EX

——— (1961), "Characteristics of Farmers Who Adopt Agricultural Practices in Indian Villages," *Rural Soc.,* 26: 138–145.—RS

——— (1962), "Peasant Values and Innovation in India," *Amer. J. of Soc.,* 67: 552–556.—RS

——— (1963), "Aims and Methods of Agricultural Extension and Their Adaptation to the Human Factor in Developing Countries." Paper presented at Rehovoth Conference on Comprehensive Planning of Agriculture in Developing Countries. Rehovoth, Israel.—RS

——— (1964a), *The Adopters.* Calcutta, India: West Bengal Dept. of Agri., Ext. Bul. 2.—RS

——— (1964b), "The Diffusion of a Farm Practice in Indian Villages," *Rural Soc.,* 29: 53–66.—RS

——— (1964c), *Social and Cultural Factors in Farm Management Efficiency.* Calcutta, India: West Bengal Dept. of Agri., Socio-

Agro-Economic Res. Org., unpublished paper.—RS

——— (1967), "Social Interaction in an Indian Village," *Sociología Ruralis*, 7: 156–175.—RS

Bose, S. P. and Basu, Sunil Kumar (1963), "Influence of Reference Groups on Adoption Behavior of Farmers," *Cultural Res. Inst.*, 2: 62–65.—RS

Bose, S. P. and Dasgupta, Satadal (1962), *The Adoption Process*. Calcutta, India: West Bengal Dept. of Agri., *Soico-Agro-Economic Res. Org., Ext. Bul.* 1.—RS

Bostian, Lloyd R. and Oliveira, Fernando (1965), "Relationships of Literacy and Education to Communication and to Social and Economic Conditions on Small Farms in Two Municipios of Southern Brazil." Paper presented at the Rural Soc. Society. Chicago.—C

Bowden, Leonard W. (1964), "Simulation and Diffusion of Irrigation Wells in the Colorado Northern High Plains." Paper presented at the Working Conference on Spatial Simulation Systems. Pittsburgh. —G.

——— (1965a), *Diffusion of the Decision to Irrigate: Simulation of the Spread of a New Resource Management Practice in the Colorado Northern High Plains*. Chicago: Univ. of Chicago, Dept. of Geography, Res. Paper 97.—G

——— (1965b), *Pump Irrigation in the Colorado Northern High Plains*. Ph.D. thesis. Worcester, Mass.: Clark Univ.—G

Bowers, Raymond V.(1937),"The Direction of Intra-Societal Diffusion," *Amer. Soc. Rev.* 2: 826–836.—ES

——— (1938), "Differential Intensity of Intra-Societal Diffusion," *Amer. Soc. Rev.* 3: 21–31.—ES

Boyle, R. P. (1963), *The Diffusion and Adoption of Innovations: A Study of an Indian Student Community*. M.A. thesis. Seattle: Univ. of Washington.—C

Brady, Dorothy S. and Adams, F. Gerald (1962), *The Diffusion of New Products and Their Impact on Consumer Expenditures*. Philadelphia: Univ. of Pennsylvania, Eco. Res. Serv. Unit, Mimeo Rept.—GE

Brandner, Lowell (1960), *Evaluation For Congruence as a Factor in Accelerated Adoption of an Agricultural Innovation*. Ph.D. thesis. Madison: Univ. of Wisconsin.—C

Brandner, Lowell and Kearl, Bryant (1964), "Evaluation for Congruence as a Factor in Adoption Rate of Innovations," *Rural Soc.*, 29: 288–303.—RS

Brandner, Lowell and Straus, Murray A. (1959), "Congruence Versus Profitability in the Diffusion of Hybrid Sorghum," *Rural Soc.*, 24: 381–383.—RS

Brien, J. P. and others (1965), "A Study of Some Personal and Social Factors in Relation to Farmer Performance," *Mktg. and Agri. Eco.* 1-23.—EX

Brim, Orville (1965), "The Acceptance of New Behavior in Child-Rearing," *Human Relations*, 7: 473–492.—GS

Brooks, Robert C., Jr. (1957), "Word-of-Mouth Advertising in Selling New Products," *J. of Mktg.*, 22: 154–161.—MR

——— (1963), "Relating the Selling Effort to Patterns of Purchase Behavior," *Bus. Topics*, 11: 73–79.—MR

Budd, Richard W. and others (1966), "Regularities in the Diffusion of Two Major News Events," *Journalism Qtrly.*, 43: 221–230.—C

Buley, Hilton C. (1947), *Personnel Characteristics and Staff Patterns Associated with the Quality of Education*. D.Ed. thesis. New York: Teachers College, Columbia Univ. —E

Burdge, Rabel J. (1961), *Development of a Scale to Measure Leisure-Orientation*. M.S. thesis. Columbus: Ohio State Univ.—RS

Burleson, G. L. (1950a), *Studying Extension Work with Farmers and Farm Homemakers in Washington Parish, Louisiana: Negro Families*. Baton Rouge: La. Agri. Ext. Svce., Mimeo Bul.—EX

——— (1950b), *Studying Extension Work with Farmers and Farm Homemakers in Washington Parish, Louisiana: White Families*. Baton Rouge: La. Agri. Ext. Svce., Mimeo Bul.—EX

Burns, Tom and Stalker, O. M. (1961), *The Management of Innovation*. London,

England: Tavistock, and Garden City, New York: Alcuin Press.—I

Buzzell, Robert D. and Nourse, Robert E. M. (1967), *Product Innovation in Food Processing, 1954–1964.* Cambridge, Mass.: Div. of Res., Grad. School of Bus. Adm., Harvard Univ.—MR

Bylund, H. Bruce (1963), *Social and Psychological Factors Associated with Acceptance of New Food Products.* University Park: Pa. Agri. Exp. Sta., Bul. 708.—RS

—— (1964), "Predicting Adoption of New Food Products by the Optical Coincidence Method," *Rural Soc., 29:* 199–203.—RS

Caird, J. B. and Moisley, H. A. (1961), "Leadership and Innovation in the Crofting Communities of the Outer Hebrides," *Soc. Rev., 9:* 85–102.—A

Camaren, Reuben James (1966), *Innovation as a Factor Influencing the Diffusion and Adoption Process.* D.Ed. thesis. Berkeley: Univ. of California.—E

Campbell, Herbert L. (1959), *Factors Related to Differential Use of Information Sources.* M.S. thesis. Ames: Iowa State Univ.—RS

Campbell, Rex R. (1965), *Prestige of Farm Operators in Two Rural Missouri Communities.* Ph.D. thesis. Columbia: Univ. of Missouri.—RS

Campbell, Rex R. and Bennett, John (1961), *Your Audience: What It's Like.* Columbia, Mo.: Agri. Exp. Sta., Bul. 771.—RS

Campbell, Rex R. and Holik, John S. (1960), "The Relationship between Group Structure and the Perception of Community's Willingness to Change." Paper presented to Rural Soc. Society. University Park, Pa.—RS

Campbell, Rex R. and Lionberger, Herbert F. (1963), "Adopters and Non-Adopters of an Idea in an Uninstitutionalized Communication System." Paper presented at the Rural Soc. Society, Northridge, Cal. —RS

Cancian, Frank (1967), "Stratification and Risk-Taking: A Theory Tested on Agricultural Innovation," *Amer. Soc. Rev., 32:* 912–927.—A

Caplow, Theodore (1952), "Market Attitudes: A Research Report from the Medical Field," *Harvard Bus. Rev., 30:* 105–112.—MS

Caplow, Theodore and Raymond, John J. (1954), "Factors Influencing the Selection of Pharmaceutical Products," *J. of Mktg., 19:* 18–23.—MS

Card, B. Y. (1960), "The Diffusion of Educational Sociology to Western Canada Through Textbooks and Periodicals," *Alberta J. of Edu. Res., 6:* 110–118.—GS

Carlson, Richard O. (1964), "School Superintendents and Adoption of Modern Math: A Social Structure Profile," in Matthew B. Miles (Ed.), *Innovation in Education.* New York: Teachers College, Columbia Univ.—E

—— (1965a), *Adoption of Educational Innovations.* Eugene: Univ. of Oregon, Center for the Adv. Study of Edu. Adm. —E

Carman, James M. (1966), "The Fate of Fashion Cycles in Our Modern Society," in Raymond M. Haas (Ed.), *Science, Technology, and Marketing.* Chicago: Amer. Mktg. Assn.—MR

Carroll, Jean (1967), "A Note on Departmental Autonomy and Innovation in Medical Schools," *J. of Bus., 40:* 531–534.—MS

Carroll, Tom W. and Hanneman, Gerhard J. (1968), "Two Models of Innovation Diffusion." Paper presented at the Second Conference on Application of Simulation. New York.—C

Carstairs, G. Morris (1955), "Medicine and Faith in Rural Pakistan," in Benjamin D. Paul (Ed.), *Health, Culture and Community.* New York: Russell Sage Fdn.—A

Carter, C. F. and Williams, B. R. (1957), *Industry and Technical Progress: Factors Governing the Speed of Application of Science.* London: Oxford Univ. Press.—I

—— (1959), "The Characteristics of Technically Progressive Firms," *J. of Ind. Eco., 7:* 87–104.—I

Cassel, John (1955), "Comprehensive Health Program among South African Zulus," in Benjamin D. Paul (Ed.), *Health, Culture and Community.* New York: Russell Sage Fdn.—A

—— (1963), "Social and Cultural Considerations in Health Innovations," *Annals of the N.Y. Acad. of Sci.,* 107: 739–747.—MS

Castillo, Gelia T. (1964), "Some Insights on the Human Factor in Overcoming Barriers to Adequate Food Supply," *Philippine J. of Nutr.,* 17: 134–147.—RS

Castillo, Hernan and others (1963), *Occopata: The Relectant Recipient of Technological Change.* Ithaca, N.Y.: Cornell Univ. Socio-Economic Dev. of Andean Communities, Report 2.—A

Cawelti, Gordon (1967), "Innovative Practices in High Schools: Who Does What—and Why—and How," *Nation's Schools,* 79: 56–88.—E

Central Treaty Organization (1962), *Travelling Seminar for Increased Agricultural Production, Regional Tour, April 7–May 30, 1962.* Ankara, Turkey: Office of U.S. Economic Coordinator for CENTO Affairs —U

Centro Latinoamericano de Demografia (1965), "La Fecundidad Rural en Latinoamerica: Una Encuesta Experimental para Medir Actitudes, Conocimiento y Comportamiento," *Demography,* 2: 97–114.—GS

Champion, Phyllis (1967), "A Pilot Study of the Success or Failure of Low Income Negro Families in the Use of Birth Control," in Donald J. Bogue (Ed.), *Sociological Contributions to Family Planning Research.* Chicago: Univ. of Chicago Press.—MS

Chaparro, Alvaro (1955), *Role Expectation and Adoption of New Farm Practices.* Ph.D. thesis. University Park: Pennsylvania State Univ.—RS

—— (1956), "Soziale Aspekte des Kulturellen Wandels: Die Diffusion Neuer Techniken in der Landwirtschaft," *Kolner Zeitschrift F. Soziologie und Sozialpsychologie,* 8: 567–594.—RS

Chapin, F. Stuart (1928), *Cultural Change.* New York: Century.—ES

Chattopadhyaya, S. N. (1963), *A Study of Some Psychological Correlates of Adoption of Innovation in Farming.* Ph.D. thesis. New Delhi: Indian Agri. Res. Inst.—EX

—— (1967), "Psychological Correlates and Adoption of Innovations," in T. P. S. Chawdhari (Ed.), *Selected Readings on Community Development.* Hyderabad, India: Natl. Inst. of Community Dev.—EX

Chattopadhyaya, S. N. and Pareek, Udai (1967), "Prediction of Multi-Practice Adoption Behavior from Some Psychological Variables," *Rural Soc.* 32: 324–333.—RS

Childs, John W. (1965), *A Study of the Belief Systems of Administrators and Teachers in Innovative and Non-Innovative School Districts.* Ph.D. thesis. East Lansing: Michigan State Univ.—E

Chopde, S. R. and others (1959), "Role of Demonstration Trials on Cultivators' Fields in the Spread of Improved Farm Practices," *Nagpur Agri. College Mag.,* 33: 38–45.—RS

Chou, Teresa Kang Mei (1966), *Homophily in Interaction Patterns in the Diffusion of Innovations in Colombian Villages.* M.S. thesis. East Lansing: Michigan State Univ.—C

Choudhary, K. M. (1965), *Factors Affecting Acceptance of Improved Agricultural Practices.* Rajasthan, India: Agro-Economic Research Centre.—AE

Choudhary, K. M. and Maharaja, Madhukar (1966), "Acceptance of Improved Practices and Their Diffusion among Wheat-Growers in the Pali District of Rajasthan," *Indian J. of Agri. Eco.,* 21: 161–165.—AE

Chow, L. P. (1965), "A Programme to Control Fertility in Taiwan: Settings, Accomplishment and Evaluation," *Population Stud.,* 19: 155–166.—U

Christiansen, James Edward (1965), *The Adoption of Educational Innovations among Teachers of Vocational Agriculture.* Ph.D. thesis. Columbus: Ohio State Univ.—E

Christiansen, James E. and Taylor, Robert E. (1966), *The Adoption of Educational Innovations among Teachers of Vocational Agriculture.* Columbus: Ohio State Univ., Dept. of Agri. Ed., Bul.—E

Christie, Samuel G. and Scribner, Jay D. (1969), "A Social System Analysis of Innovations in Sixteen School Districts."

Paper presented at the Conference of the Amer. Edu. Res. Assn. Los Angeles.—E

Clark, Lincoln H. (1958), *Consumer Behavior*. New York: Harper.—MR

Clark, Orion T. A. and Abell, Helen C. (1966), *The Relevancy of Certain Social and Psychological Variables as Related to the Adoption of Recommended Farming Practices among Dutch Dairy Farmers*. Guelph. Ontario: Univ. of Guelph, Agri. Ext. Edu., Rept. 12.—EX

Clark, Robert C. and Akinbode, I. A. (1968), *Factors Associated with Adoption of Three Farm Practices in the Western State, Nigeria*. Ile-Ife, Nigeria: Univ. of Ife Press.—RS

Clark, Terry N. (1968), "Community Structure, Decision-Making, Budget Expenditures, and Urban Renewal in 51 American Communities," *Amer. Soc. Rev.*, 33: 576–593.—GS

Clendenen, H. Franklin (1961), "Novelty Selection and Some Television Phenomena." Paper presented to the Northwest Anthro. Conf. Vancouver, B.C., Canada. —A

Cocking, Walter (1951), *The Regional Introduction of Educational Practices in Urban School Systems in the United States*. New York: Teachers College, Columbia Univ., Inst. of Adm. Res. Study 6.—E

Coe, Rodney M. and Barnhill, Elizabeth A. (1967), "Social Dimensions of Failure in Innovation," *Human Organization*, 26: 149–156.—MS

Cohen, Reuben (1962), "A Theoretical Model for Consumer Market Prediction," *Soc. Inquiry*, 32: 43–50.—MR

Cohen, Ronald (1961), "The Success that Failed: An Experiment in Culture Change in Africa," *Anthropologica*, 3: 21–36.—A

Coleman, A. Lee (1951), "Differential Contact with Extension Work in a New York Rural Community," *Rural Soc.*, 16: 207–216.—RS

Coleman, A. Lee and Marsh, C. Paul (1955), "Differential Communication among Farmers in a Kentucky County," *Rural Soc.* 20: 93–101.—RS

Coleman, James and others (1956), "Social Processes in the Diffusion of a Medical Innovation among Physicians." Paper presented at the Amer. Soc. Society. Detroit, Mich.—MS

——— (1957), "The Diffusion of an Innovation among Physicians," *Sociometry*, 20: 253–270.—MS

——— (1959), "Social Processes in Physicians' Adoption of a New Drug," *J. of Chronic Dis.*, 59: 1–19.—MS

——— (1966), *Medical Innovation: A Diffusion Study*. New York: Bobbs-Merrill.—MS

Copp, James H. (1956), *Personal and Social Factors Associated with the Adoption of Recommended Farm Practices among Cattlemen*. Manhattan: Kan. Agri. Exp. Sta., Tech. Bul. 83.—RS

——— (1958), "Toward Generalization in Farm Practice Research," *Rural Soc.*, 23: 103–111.—RS

Copp, James H. and others (1958), "The Function of Information Sources in the Farm Practice Adoption Process," *Rural Soc.*, 23: 146–157.—RS

Correa, Heli (1965), *Eficacia Relativa dos Meios de Comunicacão em uma Campanha Agricola (Relative Efficiency of Two Communication Media in an Agricultural Campaign)*. Turrialba, Costa Rica: Instituto Interamericano de Ciencias Agricolas de la OEA.—C

Couch, Carl J. (1965), *Agent Contact and Community Position of Farmers Related to Practice Adoption and NFO Membership*. East Lansing: Michigan State Univ., Inst. of Ext. Personnel Dev., Pub. 16.— EX

Coughenour, C. Milton (1958), "Who Uses the County Extension Agent?" *Ky. Farm and Home Sci.*, 4: 4–8.—RS

———(1959), *Agricultural Agencies as Information Sources for Farmers in a Kentucky County, 1950–1955*. Lexington: Ky. Agri. Exp. Sta., Prog. Rept. 82.—RS

——— (1960), "The Functioning of Farmers' Characteristics in Relation to Contact with Media and Practice Adoption," *Rural Soc.*, 25: 183–297.—RS

———(1964c), "The Rate of Technological Diffusion Among Locality Groups," *Amer. J. of Soc.*, 69: 325–339.—ES

—— (1966), "Group Factors and the Adoption of Agricultural Innovations in Seven Commercial Farming Localities." Paper presented at the Rural Soc. Society. Miami Beach, Fla.—RS

Coughenour, C. Milton and Armstrong, Joseph B. (1963), *County Agents' Activity and Farmers' Use of Soil Building Practices.* Lexington: Ky. Agri. Exp. Sta., Prog. Rept. 130.—RS

Coughenour, C. Milton and Kothari, K. B. (1962), *What Bluegrass Farmers Think Are the Conditions for Using Soil-Conservation and Soil-Building Practices.* Lexington: Ky. Agr. Exp. Sta., Prog. Rept. 120.—RS

Coughenour, C. Milton and Patell, N. B. (1962), *Trends in Use of Recommended Farm Practices and Farm Information Sources in 12 Kentucky Neighborhoods.* Lexington: Ky. Agri. Exp. Sta., Prog. Rept. 111.—RS

Coward, E. Walter, Jr. (1967), "Dimensions of Adoption in the Commercialization of Agriculture in a Developing Nation." Paper presented at the Rural Soc. Society. San Francisco.—RS

Crain, Robert L. (1962a), *Inter-City Influence in the Diffusion of Fluoridation.* Ph.D. thesis. Univ. of Chicago.—GS

—— (1962b), "Inter-City Influence in the Diffusion of Fluoridation." Paper presented at the Amer. Soc. Society. Washington, D.C.—GS

—— (1965), "The Diffusion of an Innovation among Cities." Chicago: Univ. of Chicago, Nat. Opin. Res. Center, unpublished paper.—GS

—— (1966), "Fluoridation: The Diffusion of an Innovation among Cities," *Social Forces,* 44: 467–475.—GS

Crain, Robert L. and others (1969), *The Politics of Community Conflict: The Fluoridation Decision.* Indianapolis: Bobbs-Merrill.—MS

Crain, Robert L. and Rosenthal, Donald B. (1966), "Structure and Values in Local Political Systems: The Case of Fluoridation Decisions," *J. of Politics,* 28: 169–196.—MS

Crane, Diana (1969), "The Diffusion of Innovations in Science: A Case Study."

Baltimore: Johns Hopkins Univ., unpublished paper.—GS

Cummings, Gordon J. (1950), *The Differential Adoption of Recommended Farm Practices among Dairymen in a New York Community.* M.S. thesis. Ithaca, New York: Cornell Univ.—RS

Cunningham, Scott M. (1964), "Perceived Risk as a Factor in Product-Oriented Word-of-Mouth Behavior: A First Step," in L. G. Smith (Ed.), *Reflections on Progress in Marketing.* Chicago: Amer. Mktg. Assn.—MR

—— (1966), "Perceived Risk as a Factor in the Diffusion of New Product Information," in Raymond M. Haas (Ed.), *Science, Technology, and Marketing.* Chicago: Amer. Mktg. Assn.—MR

Dahling, Randall L. (1962), "Shannon's Information Theory: The Spread of an Idea," in Wilbur Schramm (Ed.), *Studies of Innovation and of Communication to the Public.* Stanford, Cal.: Stanford Univ. Inst. for Com. Res.—C

Danbury, Thomas and Berger, Charles (1965), "The Diffusion of Political Information and Its Effect on Candidate Preference." Paper presented at the Assn. for Edu. in Journalism. Syracuse Univ.—C

Danda, Ajit K. and Danda, Dipali Ghosh (1968), *Development and Change in a Bengal Village.* Hyderabad, India: Natl. Inst. of Community Dev., Res. Rept. 20.—A

Danielson, Wayne (1956), "Eisenhower's February Decision: A Study of News Impact, "*Journalism Qtrly.,* 33: 433–441. —J

Dantre, M. P. (1963), *Adoption of Improved Farm Practices and Related Factors.* M.S. thesis. Gwalior, India: Col. of Agriculture.—EX

Dasgupta, Satadal (1962), "Sociology of Innovation," *Bul. of the Cultural Res. Inst.,* 1: 13–15.—RS

—— (1963a), "Innovation and Innovators in Indian Villages," *Man in India,* 43: 27–34.—RS

—— (1963b), *The Innovators.* Calcutta, India: Dept. of Agri., Gov. of West Bengal.—RS

Dasgupta, Satadal (1965a), "Communication and Innovation in Indian Villages," *Social Forces*, 43: 330–337.—RS

——— (1966), "Village (or Community) Factors Related to the Level of Agricultural Practice." Paper presented at Southern Soc. Society. New Orleans.—RS

Davis, Alice (1940), "Technicways in American Civilization," *Social Forces*, 18: 317–330.—ES

Davis, Alva L. and McDavid, Raven I., Jr. (1949), "'Shivaree': An Example of Cultural Diffusion," *Amer. Speech*, 24: 249–255.—SP

Davis, Burl E. (1968), *System Variables and Agricultural Innovativeness in Eastern Nigeria*. Ph.D. thesis. East Lansing: Michigan State Univ., Dept. of Communication.—C

Davis, Morris (1959), "Community Attitudes toward Fluoridation," *Pub. Opin. Qtrly.*, 23: 474–482.—MS

Davis, Richard H. (1965), *Personal and Organizational Variables Related to the Adoption of Educational Innovations in a Liberal Arts College*. Ph.D. thesis. Univ. of Chicago.—E

Davis, Roncisco W. (1952), *An Explanatory Study of Factors Governing the Practical Use of Research Findings in Local School Systems*. D.Ed. thesis. New York: Teachers College, Columbia Univ.—E

Dean, Alfred and others (1958), "Some Factors Related to Rationality in Decision Making among Farm Operators," *Rural Soc.*, 23: 121–135.—RS

Deasy, Leila C. (1956), "Socio-Economic Status and Participation in the Poliomyelitis Vaccine Trial," *Amer. Soc. Rev.*, 21: 185–191.—MS

DeFleur, Melvin L. (1962b), "Mass Communication Theory and the Study of Rumor," *Soc. Inquiry*, 32: 51–70.—GS

——— (1966a), "Mass Communication and Social Change," *Social Forces*, 44: 314–326.—GS

DeFleur, Melvin L. and Larson, Otto W. (1958), *The Flow of Information: An Experiment in Mass Communication*. New York: Harper.—GS

DeFleur, Melvin L. and Rainboth, Edith Dyer (1952), "Testing Message Diffusion in Four Communities: Some Factors in the Use of Airborne Leaflets as a Communication Medium," *Amer. Soc. Rev.*, 17: 737–743.—GS

Delenne, Michel and others (1966), "Les Agriculteurs du Sudest Face au Progres Technique" (Southeast Farmers and Technical Progress), *La Revue de Geographie de Lyon*, 41: 177–286.—GS

Dempsey, Richard A. (1963), *An Analysis of Teachers' Expressed Judgments of Barriers to Curriculum Change in Relation to the Factor of Individual Readiness to Change*. Ph.D. thesis. East Lansing: Michigan State Univ.—E

Denis, Elmer C. (1961), *Adoption Status of the Countrywide Farm Panel*. St. Louis: Doane Agri. Svce.—MR

Dernberg, Thomas F. (1958), "Consumer Response to Innovation: Television," in Thomas F. Dernberg and others (Eds.), *Studies in Household Economics Behavior*. New Haven: Yale Univ. Press.—MR

Desai, D. K. and Sharma, B. M. (1966), "Technological Change and Rate of Diffusion," *Indian J. of Agri. Eco.* 21: 141–154.—AE

Deshmukh, B. R. (1960), *Relative Influencing Capacity of Different Agencies in Persuading Rural Youths to Adopt Improved Farm Practices*. M.S. thesis. New Delhi: Indian Agri. Res. Inst.—EX

Deutschmann, Paul J. (1961), "Debate Viewing, Conversation and Changes in Voting Intentions in Lansing, Michigan." East Lansing: Michigan State Univ., unpublished paper.—C

——— (1963), "The Mass Media in an Underdeveloped Village," *Journalism Qtrly.*, 40: 27–35.—C

Deutschmann, Paul J. and Danielson, Wayne A. (1960), "Diffusion of Knowledge of the Major News Story," *Journalism Qtrly.*, 37: 345–355.—C

Deutschmann, Paul J. and Fals Borda, Orlando (1962a), *La Comunicación de las Ideas entre Los Campesinos Colombianos*. Bogotá: Univ. Nacional de Colombia, Monografias Sociologicas 14.—C

—— (1962b), *Communication and Adoption Patterns in an Andean Village.* San José, Costa Rica: Programa Interamericano de Información Popular.—C

Deutschmann, Paul J. and Havens, A. Eugene (1965), "Discontinuances: A Relatively Uninvestigated Aspect of Diffusion." Madison: Univ. of Wisconsin Dept. of Rural Soc., unpublished paper. —C

Deutschmann, Paul J. and Mendez, Alfredo (1962), *Adoption of Foods and Drugs in Cholena: A Preliminary Report.* San José, Costa Rica: Programa Interamericano de Información Popular.—C

Deutschmann, Paul J. and others (1965), *Adoption of New Foods and Drugs in Five Guatemalan Communities.* San José, Costa Rica: Instituto de Nutrición de Centro America and Programa Interamericano de Información Popular, mimeographed report.—C

Dhaliwal, A. J. Singh (1963), "A Study of Some Important Factors Affecting the Adoption of a Few Selected Agricultural Practices by the Cultivators of Ludhiana Community Development Block," in Aditya N. Shukla and Iqbal Singh Grewal (Eds.), *Summaries of Extension Research by Post-Graduate Students.* Ludhiana, India: Punjab Agricultural Univ.—EX

Dhaliwal, A. J. Singh and Sohal, T. S. (1965), "Extension Contacts in Relation to Adoption of Agricultural Practices and Socio-Economic Status of Farmers," *Indian J. of Ext. Edu.,* 1: 58–62.—EX

Diaz, Juan B. (1964a), "Bonito and Timbauba: Exploratory Study of the Leaders of Two Towns of the Brazilian Northeast at Different Levels of Development." Paper presented at the Amer. Society for Applied Anthro. San Juan, Puerto Rico.—C

—— (1964b), "Sociological and Psychological Factors Related to the Search for Instrumental Information among Farmers of the Brazilian Northeast." Paper presented to the First Interamer. Res. Symposium on the Role of Communications in Agri. Dev. Mexico City.— C

—— (1966), *The Search for Instrumental Information among Farmers of the Brazilian Northeast.* Ph.D. thesis. East Lansing: Michigan State Univ.—C

Dickerson, Thomas B. (1967), *Some Characteristics of High and Low Adopters in Karen Village Society, North Thailand.* M.S. thesis. Ithaca, N.Y.: Cornell Univ.—RS

Dimit, Robert M. (1954), *Diffusion and Adoption of Approved Farm Practices in 11 Counties in Southwest Virginia.* Ph.D. thesis. Ames: Iowa State Univ.—RS

Diswath, Bencha (1964), *The Adoption of a Sanitary Practice in Five Thai Villages.* M.S. thesis. Ithaca, N.Y.: Cornell Univ. —RS

Div. of Vocational Education (1967), *A Study of the Diffusion Process of Vocational Education Innovations.* East Lansing: Michigan State Univ., Dept. of Edu.—E

Dobyns, Henry F. (1951), "Blunders with Bolsas: A Case Study of Diffusion of Closed-Basin Agriculture," *Human Organization,* 10: 25–32.—A

Dodd, Stuart C. (1934), *A Controlled Experiment on Rural Hygiene in Syria.* New York: Oxford Univ. Press.—GS

—— (1951), "A Measured Wave of Interracial Tension," *Social Forces,* 29: 281–289.—GS

—— (1952), "Testing Message Diffusion from Person to Person," *Pub. Opin. Qtrly.,* 16: 247–262.—GS

—— (1955a), "Diffusion Is Predictable: Testing Probability Models for Laws of Interaction," *Amer. Soc. Rev.,* 20: 392–401.—GS

Donohew, Lewis and Singh, B. K. (1967), "Poverty 'Types' and Their Sources of Information about New Practices." Paper presented at the Assn. for Edu. in Journalism. Boulder, Colo.—C

Dube, S. C. (1967), "Communication, Innovation, and Planned Change in India," in Daniel Lerner and Wilbur Schramm (Eds.), *Communication and Change in the Developing Countries.* Honolulu: East-West Center Press.—C

Dubey, D. C. (1967), *Adoption of a New Contraceptive in Urban India: Analysis of Communication and Family Decision-Making*

Processes. Ph.D. thesis. East Lansing: Michigan State Univ.—GS

Dubey, D. C. and Choldin, Harvey M. (1967), "Communication and Diffusion of the IUD: A Case Study in Urban India." East Lansing: Michigan State Univ., Dept. of Soc., unpublished paper. —MS

Dubey, D. C. and others (1962), *Village Level Workers: Their Work and Result Demonstrations.* Hyderabad, India: Natl. Inst. of Community Dev.—RS

Dubey, D. N. (1966), *Study of Factors Influencing Community Adoption Process in Relation to Improved Farm Practices in Anand Taluka of Gujarat State.* M.S. thesis. Anand, India: Inst. of Agriculture.— EX

Duncan, James A. and Kreitlow, Burton W. (1954), "Selected Cultural Characteristics and the Acceptance of Educational Programs and Practices," *Rural Soc.,* 19: 349–357.—E

Eastmond, Jefferson N. (1953), *An Analysis of Elementary School Staff Characteristics Related to the Quality of Education.* D.Ed. thesis. New York: Teachers College, Columbia Univ.—E

Edling, Jack V. and others (1964), "Four Case Studies of Programmed Instruction." New York: Fund for the Adv. of Edu.—E

Eibler, Herbert J. (1965), *A Comparison of the Relationships Between Certain Aspects or Characteristics of the Structure of the High School Faculty and the Amount of Curriculum Innovation.* Ph.D. thesis. Ann Arbor: Univ. of Michigan.—E

Eichholz, Gerhard C. (1961), *Development of a Rejection Classification for Newer Educational Media.* Ph.D. thesis. Columbus: Ohio State Univ.—E

——— (1963), "Why Do Teachers Reject Change?" *Theory Into Practice,* 2: 264–268.—E

Eichholz, Gerhard C. and Rogers, Everett M. (1964), "Resistance to the Adoption of Audio-Visual Aids by Elementary School Teachers: Contrast and Similarities to Agricultural Innovation," in Matthew B. Miles (Ed.), *The Nature of Educational Innovation.* New York: Teachers College, Columbia Univ.—E

Ellenbogen, Bert L. (1964), *Age, Status and Diffusion of Preventive Health Practices.* Ithaca, N.Y.: Cornell Univ. Dept. of Rural Soc., Bul. 64.—RS

Ellenbogen, Bert L. and Lowe, George D. (1968), "Health Care 'Styles' in Rural and Urban Areas," *Rural Soc.,* 33: 300–312.—RS

Ellenbogen, Bert L. and others (1961), *Factors in Farmers' Subscription to Health Insurance.* Ithaca, N.Y.: Cornell Univ. College of Agriculture, Bul. 60.—RS

Ellinghaus, Dieter and others (1968), "Die Verbreitung von Neuerungen Bei Klein Und Mittelunternehmern (The Diffusion of Innovations among Small and Medium Size Entrepreneurs)," *Institut fur Mittelstandsforchung, Soziologische Abteilung,* working paper.—RS

Elliott, John G. (1965), *The Diffusion of Farm Practices as a Function of the Role of Elevator Operators (Feed and Grain Dealers) as Reference Others to Their Farmer Clientele.* M.S. thesis. East Lansing: Michigan State Univ.—EX

——— (1968), *Farmers' Perceptions of Innovations as Related to Self-Concept and Adoption.* Ph.D. thesis. East Lansing: Michigan State Univ.—O

Elliott, John G. and Couch, Carl J. (1965), "Operators of Grain Elevators as Diffusers of Farm Practices." East Lansing: Michigan State Univ., Inst. for Ext. Personnel Dev., Pub. 20.—EX

Emery, F. E. and Oeser, O. A. (1958), *Information, Decision and Action: A Study of the Psychological Determinants of Changes in Farming Techniques,* New York: Cambridge Univ. Press.—P

Engel, James F. and others (1966), "Sources of Influence in the Acceptance of New Products for Self-Medication: Preliminary Findings," in Raymond M. Haas (Ed.), *Science, Technology, and Marketing.* Chicago: Amer. Mktg. Assn.—MR

Enos, John L. (1958), "A Measure of the Rate of Technological Progress in the Petroleum Refining Industry," *J. of Ind. Eco.,* 6: 180–197.—I

—— (1960), "Invention and Innovation in the Petroleum Refining Industry." Paper presented at the Conference on the Economic and Social Factors Determining the Rate and Direction of Inventive Activity. New York.—I

Erasmus, Charles J. (1952a), "Agricultural Changes in Haiti: Patterns of Resistance and Acceptance," *Human Organization*, 11: 20–26.—A

—— (1952b), "Changing Folk Beliefs and the Relativity of Empirical Knowledge," *Southwestern J. of Anthro.*, 8: 411–428.—A

—— (1961), *Man Takes Control: Cultural Development and American Aid*. Minneapolis: Univ. of Minnesota Press.—A

—— (1962), "Introducing New Agricultural Practices in Latin America," *Migration News*, 11: 7–12.—A

Erbe, William (1962), "Gregariousness, Group Membership, and the Flow of Information," *Amer. J. of Soc.*, 67: 502–516.—GS

Eteng, William I. A. (1968), *Factors Related to Farm Practice Adoption among Cocoa Farmers of Western Nigeria*. M.S. thesis. Madison: Univ. of Wisconsin.—RS

Evan, William M. and Black, Guy (1967), "Innovation in Business Organizations: Some Factors Associated with Success or Failure of Staff Proposals," *J. of Bus.*, 40: 519–530.—GS

Evans, Richard I. (1968), *Resistance to Innovation in Higher Education*. San Francisco: Jossey-Bass.—P

Fallding, Harold (1957), *Social Factors in Serrated Tussock Control: A Study of Agricultural Extension*. Sydney, Australia: Univ. of Sydney, Dept. of Agri. Eco., Res. Bul. 1.—RS

—— (1958), *Precept and Practice on North Coast Dairy Farms*. Sydney, Australia: Univ. of Sydney, Dept. of Agri. Eco., Res. Bul. 2.—RS

Fals Borda, Orlando (1960), *Facts and Theory of Socio-Cultural Change in a Rural Social System*. Bogotá: Univ. Nacional de Colombia, Monografias Sociologicas 2 Bis.—RS

Farnsworth, Philo T. (1940), *Adaptation Processes in Public School Systems as Illustrated by a Study of Five Selected Innovations in Educational Service in New York, Connecticut, and Massachusetts*. New York: Teachers College, Columbia Univ.—E

Fathi, Asghar (1965), "Leadership and Resistance to Change: A Case from an Underdeveloped Area," *Rural Soc.*, 30: 204–212.—RS

Feaster, J. Gerald (1966), "The Measurement of Innovativeness among Primitive Agriculturists." Paper presented at the Rural Soc. Society. Miami Beach, Fla.—AE

—— (1968), "Measurement and Determinants of Innovativeness among Primitive Agriculturists," *Rural Soc.*, 33: 339–348.—AE

Feldman, Jacob J. (1966), *The Dissemination of Health Information*. Chicago: Aldine.—MR

Feldman, Sidney P. (1966), "Some Dyadic Relationships Associated with Consumer Choice," in Raymond M. Haas (Ed.), *Science, Technology, and Marketing*. Chicago: Amer. Mktg. Assn.—MR

Feldman, Sidney P. and Spencer, Merlin C. (1965), "The Effect of Personal Influence in the Selection of Consumer Services," in Peter Bennett (Ed.), *Marketing and Economic Development*. Chicago: Amer. Mktg. Assn.—MR

Feliciano, Gloria D. (1964), "The Human Variable in Farm Practice Adoption: Philippine Setting." Paper presented at Philippine Soc. Society. Manila.—C

Finley, James R. (1965), *Prediction of the Adoption of Improved Farm Practices*. M.S. thesis. Columbus: Ohio State Univ. —RS

—— (1968), "Farm Practice Adoption: A Predictive Model," *Rural Soc.*, 33: 5–18.—RS

Fischer, Loyd K. and Timmons, John F. (1959), *Progress and Problems in the Iowa Soil Conservation Districts Program*. Ames: Iowa Agri. and Home Eco. Exp. Sta., Res. Bul. 466.—AE

Fisk, George (1959), "Media Influence Reconsidered," *Pub. Opin. Qtrly.*, 23: 83–91.—C

Fliegel, Frederick C. (1955), *A Multiple Correlation Analysis of Factors Associated with Adoption of Farm Practices*. Ph.D. thesis. Madison: Univ. of Wisconsin.—RS

—— (1956), "A Multiple Correlation Analysis of Factors Associated with Adoption of Farm Practices," *Rural Soc.,* 21: 284–292.—RS

—— (1957), "Farm Income and the Adoption of Farm Practices," *Rural Soc.,* 22: 159–162.—RS

—— (1959), "Aspirations of Low-Income Farmers and Their Performance and Potential for Change," *Rural Soc.,* 24: 205–214.—RS

—— (1960), "Obstacles to Change for the Low-Income Farmer," *Rural Soc.,* 25: 347–351.—RS

—— (1962), "Traditionalism in the Farm Family and Technological Change," *Rural Soc.,* 27: 70–76.—RS

—— (1965), "Differences in Prestige Standards and Orientation to Change in a Traditional Agricultural Setting," *Rural Soc.,* 30: 279–290.—RS

—— (1966), "Literacy and Exposure to Instrumental Information among Farmers in Southern Brazil," *Rural Soc.,* 31: 15–28.—RS

—— (1967a), "Community Organization and Acceptance of Change in Rural India." Paper presented at the Rural Soc. Society. San Francisco.—RS

—— (1967b), "Community Structure and the Success or Failure of Agricultural Change Programs." Paper presented at the Rural Soc. Society. San Francisco.—RS

—— (1967c), "Literacy and Exposure to Agricultural Information: A Comparison of Some Indian and Brazilian Data," *Behav. Sci. and Com. Dev.,* 1: 89–99.—C

Fliegel, Frederick C. and Kivlin, Joseph E. (1962a), *Differences among Improved Farm Practices as Related to Rate of Adoption.* University Park: Pa. Agri. Exp. Sta. Res. Bul. 691.—RS

—— (1962b), "Farm Practice Attributes and Adoption Rates," *Social Forces,* 40: 364–370.—RS

—— (1966a), "Farmers' Perception of Farm Practice Attributes," *Rural Soc.,* 31: 197–206.—RS

—— (1966b), "Attributes of Innovations as Factors in Diffusion," *Amer. J. of Soc.,* 72: 235–248.—RS

Fliegel, Frederick C. and Oliveira, Fernando C. (1963), "Receptivity to New Ideas and Rural Exodus in an Area of Small Farms, Rio Grande do Sul." Paper presented at the Rural Soc. Society. Northridge, Cal. —RS

Fliegel, Frederick C. and others (1967a), "A Cross-National Comparison of Farmers' Perceptions of Innovations as Related to Adoption Behavior." East Lansing: Michigan State Univ., unpublished paper.—RS

—— (1967b), *Innovation in India: The Success or Failure of Agricultural Development Programs in 108 Indian Villages.* Hyderabad, India: Natl. Inst. of Community Dev., Res. Rept. 9.—C

—— (1968), "A Cross-National Comparison of Farmers' Perceptions of Innovations as Related to Adoption Behavior," *Rural Soc.,* 33: 437–449.—RS

Fliegel, Frederick C. and Sekhon, Gurmeet S. (1969), "Balance Theory and the Diffusion of Innovations: An Empirical Test." Paper presented at the Rural Soc. Society. San Francisco.—RS

Flinn, William L. (1961), *Combined Influence of Group Norms and Personal Characteristics on Innovativeness.* M.S. thesis. Columbus: Ohio State Univ.—RS

—— (1963), "Community Norms in Predicting Innovativeness." Paper presented at the Rural Soc. Society. Northridge, Cal.—RS

Fonseca, Luiz (1966), *Information Patterns and Practice Adoption among Brazilian Farmers.* Ph.D. thesis. Madison: Univ. of Wisconsin.—C

Fosen, Robert H. (1956a), *Structural and Social Psychological Factors Affecting Differential Acceptance of Recommended Agricultural Practices: A Study in the Application of Reference Group Theory.* M.S. thesis. Ithaca, N.Y.: Cornell Univ.—RS

——— (1956b), *Social Solidarity and Differential Adoption of a Recommended Agricultural Practice*. Ph.D. thesis. Ithaca, N.Y.: Cornell Univ.—RS

Fougeyrollas, Pierre (1967), *Television and the Social Education of Women*. Paris: UNESCO Bul. 50.—P

Fourt, Louis A. and Woodlock, Joseph W. (1960), "Early Prediction of Market Success for New Grocery Products," *J. of Mktg.*, 25: 31–38.—MR

Francis, David G. (1960), *Communication Credibility of a Non-Recommended Innovation*. M.S. thesis. Columbus: Ohio State Univ.—RS

Francis, David G. and Rogers, Everett M. (1960), "Adoption of a Nonrecommended Innovation: The Grass Incubator." Paper presented at the Rural Soc. Society. University Park, Pa.—RS

Frank, Richard and Tietze, Christopher (1965), "Acceptance of an Oral Contraceptive Program in a Large Metropolitan Area," *Amer. J. of Obs. and Gyn.*, 93: 122–127.—MS

Frank, Ronald E. and Massy, William F. (1964), "Innovation and Brand Choice: The Folger's Invasion," in Stephen A. Greyser (Ed.), *Toward Scientific Marketing: Proceedings of the 1963 Winter Conference of the American Marketing Association*. Chicago: Amer. Mktg. Assn.—MR

Frank, Ronald E. and others (1964), "The Determinants of Innovative Behavior with Respect to a Branded, Frequently Purchased Food Product." Paper presented at the Amer. Mktg. Assn. Winter Conf. Chicago.—MR

Fraser, Thomas M., Jr. (1963), "Sociocultural Parameters in Directed Change," *Human Organization*, 22: 95–104.—A

Freedman, Ronald (1964), "Sample Surveys for Family Planning Research in Taiwan," *Pub. Opin.Qtrly.*, 28: 374–382.—MS

Freedman, Ronald and others (1964), "Fertility and Family Planning in Taiwan: A Case Study in Demographic Transition," *Amer. J. of Soc.*, 70: 16–27.—MS

Freedman, Ronald and Takeshita, John Y. (1965), "Studies of Fertility and Family

Limitation in Taiwan," *Eugenics Quarterly*, 12: 233–250.—GS

Freedman, Ronald and Takeshita, John Y. (1969), *Family Planning in Taiwan: An Experiment in Social Change*. Princeton, N.J.: Princeton University Press.—MS

Funkhouser, G. Ray (1968), "A General Mathematical Model of Information Diffusion." Stanford, Cal.: Stanford Univ., Inst. for Communication Res.—C

Gaffin, Ben and Associates (1956), *Effectiveness of Promotion in a Medical Marketing Area*. Chicago: Amer. Medical Assn.—MR

Gaikwad, V. R. (1968), "Location of Contributions of Variables in Adoption Process." Hyderabad, India: Natl. Inst. of Community Dev., unpublished paper. —RS

Galjart, B. F. (1968), *Itaguai: Old Habits and New Practices in a Brazilian Land Settlement*. Wageningen, Netherlands: Centre for Agricultural Publishing and Documentation.—RS

Gallagher, Ralph P. (1949), *Some Relationships of Symbiotic Groups to Adaptability in Public Schools and to Other Related Factors*. D.Ed. thesis. New York: Teachers College, Columbia Univ.—E

Gallup International (1966), *School Board Members' Reactions to Educational Innovations*, Princeton, N.J.—O

Gamson, William A. (1961a), "The Fluoridation Dialogue: Is It an Ideological Conflict?" *Pub. Opin. Qtrly.*, 25: 526–537.—MS

——— (1961b), "Public Information in a Fluoridation Referendum," *Health Edu. J.*, 19: 47–54.—MS

——— (1968), "Rancorous Conflict in Community Politics," in Terry N. Clark (Ed.), *Community Structure and Decision-making: Comparative Analyses*. San Francisco: Chandler.—MS

Ganorkar, P. L. (1961), *An Appraisal of the Factors Affecting the Acceptance and Use of Agricultural Practices in College Extension Block*. M.S. thesis. Nagpur, India: Col. of Agriculture.—EX

Garabedian, Peter G. and Dodd, Stuart C. (1962), "Clique Size as a Factor in

Message Diffusion," *Soc. Inquiry*, 32: 71–81.—GS

Garcia, Julio Cesar Borelli (1960), *Determinación de Algunas Necesidades de Caficultores en Tres Localidades de Costa Rica (Identification of Some Needs of Coffee Farmers in Three Costa Rican Localities)*. Turrialba, Costa Rica: Instituto Interamericano de Ciencias Agricolas.—EX

Gaviria, Hernan (1960), *Determinación Tecnica de Las Necesidades de Los Caficultores de San Ignacio, Costa Rica (Technical Identification of Coffee Farmers' Needs, San Ignacio, Costa Rica)*. M.S. thesis. Turrialba, Costa Rica: Instituto Interamericano de Ciencias Agricolas.—EX

Geiger, Kent and Sokol, Robert (1960), "Educational Television in Boston," in Wilbur Schramm (Ed.), *The Impact of Educational Television*. Urbana: Univ. of Illinois.—GS

Georgiopoulos, Ioannis P. (1967), *Adoption of County Zoning by the Counties of the State of Indiana as Related to Social Change*. M.S. thesis. Lafayette, Ind.: Purdue Univ.—RS

Georgiopoulos, Ioannis P. and Potter, Harry R. (1967), "Ideology and Social Change: The Diffusion of County Zoning." Lafayette, Ind.: Purdue Univ., working paper II.—RS

Ghildyal, U. C. (1967b), "Innovations in Educational Methodology," *Behav. Sci. and Com. Dev.*, 1: 114–134.—E

Gill, Dhara S. (1965), *Socio-Economic and Educational Aspects of Rural Life in Ako, Mallam Madure, and Ebeto Districts*. Kaduna: No. Nigeria Ministry of Agri.—EX

Gillespie, Robert W. (1965), *Family Planning on Taiwan, 1964–1965*. Taichung, Taiwan: Population Council, Mimeo Report.—U

Gillis, Willie Mae (1958), *The Adoption of Recommended Farm Practices in Alcorn County and Its Relationship to Other Variables*. State College: Miss. Agri. Exp. Sta., Mimeo Bul.—RS

Gilmore, J. S. and others (1967), *The Channels of Technology Acquisition in Commercial Firms, and the NASA Dissemination Program*. Denver. Colo.: Denver Res. Inst., Univ. of Denver.—U

Gittell, Marilyn and Hollander, T. Edward (1968), *Six Urban School Districts: A Comparative Study of Institutional Response*. N.Y.: Praeger.—E

Glaser, Edward M. (1966), *Utilization of Applicable Research and Demonstration Results*. Los Angeles: Human Interaction Res. Inst., Mimeo Report.—P

Glaser, Melvin A. (1958), "A Study of the Public's Acceptance of the Salk Vaccine Programs," *Amer. J. of Pub. Health*, 48: 141–146.—MS

Goldsen, Rose K. and Ralis, Max (1957), *Factors Related to Acceptance of Innovations in Bang Chan, Thailand*. Ithaca, N.Y.: Cornell Univ., Southeast Asia Program, Data Paper 25.—A

Goldstein, Bernice and Eichhorn, Robert L. (1961), "The Changing Protestant Ethic: Rural Patterns in Health, Work, and Leisure," *Amer. Soc. Rev.*, 26: 557–565.—GS

Goldstein, Marshall N. and others (1961), *Educational Television Project, Preliminary Report Number One*. Eugene, Ore.: Inst. for Community Stud., Studies in Resistances to Cultural Innovation.—GS

Gordon, Gerald and Marquis, Sue (1966), "Freedom, Visibility of Consequences, and Scientific Innovation," *Amer. J. of Soc.*, 72: 195–202.—GS

Gorman, Walter, P., III (1966), *Market Acceptance of a Consumer Durable Good Innovation: A Socio-Economic Analysis of First and Second Buying Households of Color Television Receivers in Tuscaloosa, Alabama*. Ph.D. thesis. University: Univ. of Alabama.—MR

——— (1967b), "The Diffusion of Color Television Sets into a Metropolitan Fringe Area Market." Paper presented at the Southern Mktg. Assn. New Orleans. —MR

Gottlieb, David and Brookover, Wilbur B. (1967), *Acceptance of New Educational Practices by Elementary School Teachers*. East Lansing: Michigan State Univ., Educ. Res. Series 33.—E

Graham, L. Saxon (1951), *Selection and Social Stratification: Factors in the Acceptance and Rejection of Five Innovations by Social Strata in New Haven, Connecticut.* Ph.D. thesis. New Haven: Yale Univ.—GS

——— (1954), "Cultural Compatibility in the Adoption of Television," *Social Forces,* 33: 166–170.—GS

——— (1956), "Class and Conservatism in the Adoption of Innovations," *Human Relations,* 9: 91–100.—GS

Graham, L. Saxon and Gibson, Robert (1967), "Acceptance and Rejection of a Decremental Innovation: Cessation of Smoking." Paper presented at Amer. Soc. Assn., San Francisco.—MS

Greenberg, Bradley S. (1964a), "Diffusion of News of the Kennedy Assassination," *Pub. Opin. Qtrly.,* 28: 225–232.—C

——— (1964b), "Person-to-Person Communication in the Diffusion of News Events," *Journalism Qtrly.,* 41: 490–494.—C

Greenberg, Bradley S. and others (1965), "Diffusion of News about an Anticipated Major News Event," *J. of Broadcasting,* 9: 129–142.—C

Greenberg, Joseph H. (1951), "Social Variables in Acceptance or Rejection of Artificial Insemination," *Amer. Soc. Rev.,* 16: 86–91.—GS

Greenwald, Anthony G. (1964), "Effects of Prior Commitment on Behavior Change after a Persuasive Communication," *Pub. Opin. Qtrly.,* 29: 595–601. —C

Grewal, I. S. (1965), "Differential Characteristics of Farmers of Predominantly Refugee, a Native and a Mixed Village, Affecting the Adoption of Improved Agricultural Practices in a Block in Ludhiana District of the Punjab State," in Aditya N. Shukla and Iqbal Singh Grewal (Eds.), *Summaries of Extension by Post-Graduate Students.* Ludhiana, India: Punjab Agri. Univ.—EX

Griliches, Zvi (1957), "Hybrid Corn: An Exploration in the Economics of Technological Change," *Econometrica,* 25: 501–522.—AE

——— (1960a), "Congruence Versus Profitability: A False Dichotomy," *Rural Soc.,* 25: 354–356.—AE

——— (1960b), "Hybrid Corn and the Economics of Innovation," *Science,* 132: 275–280.—AE

Gross, Neal C. (1942), *The Diffusion of a Culture Trait in Two Iowa Townships.* M.S. thesis. Ames: Iowa State Univ.—RS

——— (1949), "The Differential Characteristics of Acceptors and Non-Acceptors of an Approved Agricultural Technological Practice," *Rural Soc.,* 14: 148–156.—RS

Gross, Neal C. and Taves, Marvin J. (1952), "Characteristics Associated with Acceptance of Recommended Farm Practices, *Rural Soc.,* 17: 321–328.—RS

Gruen, Fred H. G. (1955), "Incomes of Dairy Farmers in the Richmond-Tweed Region," *Rev. of Mktg. and Agri. Eco.,* 23: 177–206.—AE

Guimarães, Lytton L. (1968), *Matrix Multiplication in the Study of Interpersonal Communication.* M.A. thesis. East Lansing: Michigan State Univ.—C

Gunther, Erna (1950), "The Westward Movement of Some Plains Traits," *Amer. Anthro.,* 52: 174–180.—A

Gupta, Ashok K. (1966), "A Study of Some Factors Affecting Adoption of Poultry Farming in Moga Tehsil, Punjab," in Aditya N. Shukla and I. S. Grewal (Eds.), *Summaries of Extension Research by Post-Graduate Students.* Ludhiana, India: Punjab Agricultural Univ.—EX

Guzman, Leopoldo P. de (1964), "The Rice Farmers' Response to Technological Change." Paper presented at Intl. Rice Res. Inst. Laguna, Philippines.—RS

Haber, Ralph N. (1963), "The Spread of an Innovation: High School Langlue Laboratories," *J. of Exp. Edu.* 31: 359–369.—E

Hage, Jerald T. (1963), *Organizational Response to Innovation: A Case Study of Community Hospital.* Ph.D. thesis. N.Y.: Columbia Univ.—GS

Hägerstrand, Torsten (1952), *The Propagation of Innovation Waves.* Lund, Sweden: Lund Studies in Geog. 4.—G

——— (1953), *Innovations for Loppet ur Korologisk Synpunkt (Innovation Diffusion as a Spatial Process).* Lund, Sweden: Univ. of Lund, Dept. of Geog., Bul. 15, and Univ. of Chicago Press (1968).—G

Hägerstrand, Torsten (1965b), "Quantitative Techniques for Analysis of the Spread of Information and Technology," in C. A. Anderson and M. J. Bowman (Eds.), *Education and Economic Development*. Chicago: Aldine.—G

———— (1968), *Innovation Diffusion as a Spatial Process*. Chicago and London: Univ. of Chicago Press.—G

Haines, George H., Jr. (1964a), "Change in Small Groups: An Experimental Study of a Management Game as a Research Tool," *Ind. Mgmt. Rev.*, 5: 61–65.—MR

———— (1966), "A Study of Why People Purchase New Products," in Raymond M. Haas (Ed.); *Science, Technology, and Marketing*. Chicago: Amer. Mktg. Assn. —MR

Hall, M. Francoise (1964), "Birth Control in Lima, Peru: Attitudes and Practices," *Milbank Memorial Fund Qtrly.*, 43: 409–438.—U

Hamuy, Eduardo and others (1958), *El Primer Satelite Artificial: Sus Efectos en la Opinion Publica (The First Artificial Satellite: Its Effects on Public Opinion)*. Santiago de Chile: Editorial Universitaria.—GS

Hanks, L. M. and others (1955), "Diphtheria Immunization in a Thai Community," in Benjamin D. Paul (Ed.), *Health, Culture and Community*. N.Y.: Russell Sage Fdn.—A

Hanneman, Gerhard J. (1969), *A Computer Simulation of Information Diffusion in a Peasant Community*. M.A. thesis. East Lansing: Michigan State Univ.—C

Hanson, John O. (1966), *A Descriptive Study of Basic Data and the Educational Innovations Found in Twenty-two Selected North Dakota Small Schools*. Ph.D. thesis. Grand Forks: Univ. of North Dakota.—E

Hardee, J. Gilbert (1963a), *Evaluation of an Educational Program with Part-Time Farm Families: Transylvania County, North Carolina, 1955–1960*. Raleigh: North Carolina State Col., N.C. Ext. Evaluation Stud. 5.—RS

———— (1963b), *Program Development with Part Time Farm Families: A Five-Year*

Evaluation. Raleigh: N.C. Agri. Exp. Sta., Bul. 420.—RS

Hardenbrook, Robert F. (1967), *Identification of Processes of Innovation in Selected Schools in Santa Barbara County*. Ph.D. thesis. Los Angeles: Univ. of So. California.—E

Haring, Ardyce E. (1965), *Attitude Change as a Function of Communicator and Audience Types in a Traditional and a Modern Farming Community*. Ph.D. thesis. Madison: Univ. of Wisconsin.—C

Harris, Ruth A. (1956), *Certain Socio-Economic Factors and Value Orientation as Related to the Adoption of Home Practices*. M.A. thesis. Madison: Univ. of Wisconsin.—RS

Hartman, Joel A. (1964), "Validity of Using Sociometric Questions in Determining Characteristics of Personal Information Sources." Paper presented at the Rural Soc. Society. Montreal, Canada.—RS

Hartman, Joel A. and Brown, Emory J. (1965), "Ecological Patterns of Farm Practice Adoption." Paper presented at the Rural Soc. Society. Chicago.—RS

———— (1967), "Influence of Demonstration Farms on Diffusion of Practices." Paper presented at the Rural Soc. Society. San Francisco.—RS

Havens, A. Eugene (1962b), *Social Psychological Factors Associated with Differential Adoption of New Technologies by Milk Producers*. Ph.D. thesis. Columbus: Ohio State Univ.—RS

———— (1963a), "El Cambio en la Tecnologia Agricola de Subachoque," in *Factores Sociales que Inciden en le Desarrollo Economico de la Hoya del Rio Subachoque*." Bogotá: Facultad de Sociología, Univ. Nacional de Colombia.—RS

———— (1965a), "Increasing the Effectiveness of Predicting Innovativeness," *Rural Soc.*. 30: 151–165.—RS

———— (1966), *Támesis: Estructura y Cambio, Estudio de una Comunidad Antioqueña (Támesis: Structure and Change, Study of an Antioqueña Community)*. Bogotá: Ediciones Tercer Mundo y Facultad de Sociología, Univ. Nacional de Colombia.—RS

Havens, A. Eugene and Rogers, Everett M. (1961a), "A Campaign That Failed: A

Reason Why." Paper presented at Ohio Valley Soc. Society, Cleveland.—RS

——(1961b), "Adoption of Hybrid Corn: Profitability and the Interaction Effect," *Rural Soc.*, 26: 409–414.—RS

Hawley, Amos H. (1962), "Community Power and Urban Renewal Success," *Amer. J. of Soc.*, 68: 422–431.—GS

Hawley, Florence (1946), "The Role of Pueblo Social Organization in the Dissemination of Catholicism," *Amer. Anthro.*, 48: 407–415.—A

Hay, Donald G. and Lowry, Sheldon G. (1957), *Acceptance of Voluntary Health Insurance in Scotland Neck Community, North Carolina, 1955.* Raleigh: N.C. Agri. Exp. Sta., Prog. Rept. 27.—RS

—— (1958), *Use of Health Care Services and Enrollment in Voluntary Health Insurance in Montgomery County, North Carolina, 1956.* Raleigh: N.C. Agri. Exp. Sta., Prog. Rept. RS-31.—RS

Herzog, William A. (1967a), *Literacy Training and Modernization: A Field Experiment.* East Lansing: Michigan State Univ., Dept. of Communication, Tech. Rept. 3.—C

Herzog, William A. and others (1968), *Patterns of Diffusion in Rural Brazil.* East Lansing: Michigan State Univ., Dept. of Communication, Res. Rept. 10.—C

Hess, C. V. and Miller, L. F. (1954), *Some Personal, Economic, and Sociological Factors Influencing Dairymen's Actions and Success.* University Park: Pa. Agri. Exp. Sta., Bul. 577.—AE

Hicks, Alvin W. (1942), *A Plan to Accelerate the Process of Adaptation in a New York City School Community-Teachers College.* D.Ed. thesis. New York: Teachers College, Columbia Univ.—E

Higgins, Edward (1962), "Some Fertility Attitudes among White Women in Johannesburg," *Population Stud.*, 16: 73–78.—U

Hildebrand, Peter E. and Partenheimer, Earl J. (1958), "Socio-Economic Characteristics of Innovators," *J. of Farm Eco.*, 40: 446–449.—AE

Hilfiker, Leo R. (1969), *The Relationship of School System Innovativeness to Selected Dimensions of Interpersonal Behavior in Eight School Systems.* Madison: Univ. of Wisconsin, Wisc. Res. and Dev., Tech. Rept. 70.—E

Hill, Reuben and others (1959), *The Family and Population Control: A Puerto Rican Experiment in Social Change.* Chapel Hill: Univ. of North Carolina Press.—GS

Hill, Richard J. (1955), *Temporal Aspects of Person-to-Person Message Diffusion.* Ph.D. thesis. Seattle: Univ. of Washington.—GS

Hill, Richard J. and Bonjean, Charles M. (1964), "News Diffusion: A Test of the Regularity Hypothesis," *Journalism Qtrly.*, 41: 336–342.—GS

Ho, Yung Chang (1969), *Homophily in the Diffusion of Innovations in Brazilian Communities.* M.A. thesis. East Lansing: Michigan State Univ.—C

Hobbs, Daryl (1960), *Factors Related to the Use of Agricultural Chemicals on Iowa Farms.* M.S. thesis. Ames: Iowa State Univ.—RS

—— (1964), *The Relation of Farm Operator Values and Attitudes to Their Economic Performance.* Ames: Iowa State Univ., Rural Soc. Rept. 33.—RS

Hochstrasser, Donald L. (1963), *Possum Ridge Farmers: A Study in Cultural Change.* Ph.D. thesis. Eugene: Univ. of Oregon. —A

Hodgdon, Linwood L. and Singh, Harpal (1963), *The Adoption of Agricultural Practices in Two Villages of Madhya Pradesh.* Delhi, India: Ford Foundation Printed Report.—RS

Hoffer, Charles R. (1942), *Acceptance of Approved Farming Practices among Farmers of Dutch Descent.* East Lansing: Mich. Agri. Exp. Sta., Spec. Bul. 316.—RS

——(1944), *Selected Social Factors Affecting Participation of Farmers in Agricultural Extension Work.* East Lansing: Mich. Agri. Exp. Sta., Spec. Bul. 331.—RS

—— (1946), *Social Organization in Relation to the Extension Service in Eaton County, Michigan.* East Lansing: Mich. Agri. Exp. Sta., Spec. Bul. 338.—RS

Hoffer, Charles R. and Gibson, D. L. (1941), *The Community Situation as It Affects*

Agricultural Extension Work. East Lansing: Mich. Agri. Exp. Sta., Spec. Bul. 312.—RS

Hoffer, Charles R. and Stangland, Dale (1958a), "Farmers' Attitudes and Values in Relation to Adoption of Approved Practices in Corn Growing," *Rural Soc.,* 23: 112–120.—RS

—— (1958b), *Farmers' Reactions to New Practices.* East Lansing: Mich. Agri. Exp. Sta., Tech. Bul. 264.—RS

Holdaway, E. A. and Seger, John E. (1968), "The Development of Indices of Innovativeness," *Canadian Edu. and Res. Digest,* 8: 366–379.—E

Holmberg, Allan R. (1952), "The Wells That Failed: An Attempt to Establish a Stable Water Supply in the Viree Valley, Peru," in Edward H. Spicer (Ed.), *Human Problems in Technological Change.* New York: Russell Sage Fdn.—A

—— (1959), "Land Tenure and Planned Social Change: A Case from Vicos, Peru," *Human Organization,* 18: 7–10.—A

Holmes, John H. (1967), *Dogmatism as a Predictor of Communication Behavior in the Diffusion of Consumer Innovations.* Ph.D. thesis. East Lansing: Michigan State Univ.—C

Horn, Daniel and others (1959), "Cigarette Smoking among High School Students," *Amer. J. of Pub. Health,* 49: 1497–1511.—MS

Hong, Sung-Bong and Yoon, Joong-Hi (1962), "Male Attitudes Toward Family Planning on the Island of Kangwha-Gun, Korea," *Milbank Mem. Fund Qtrly.,* 40: 443–452.—MS

Hrabovszky, J. K. and Moulik, T. K. (1966), *A Study of Economic Returns to the Olpad Thresher in Two Delhi Villages: Some Psychological and Social Factors Influencing Their Adoption and the Process of Diffusion among Cultivators.* New Delhi, India: Agri. Res. Inst.—AE

Hruschka, E. and Rheinwald, H. (1965), "The Effectiveness of German Pilot Farms," *Sociologia Ruralis,* 5: 101–111.—EX

Hsieh, S. C. (1966), "Management Decisions on Small Farms in Taiwan." New York: Agri. Dev. Council.—AE

Hsu, Francis, L. K. (1955), "A Cholera Epidemic in a Chinese Town," in Benjamin D. Paul (Ed.), *Health, Culture and Community.* New York: Russell Sage Fdn.—A

Hudspeth, Delayne R. (1966), *A Study of Belief System and Acceptance of New Educational Media with Users and Non-Users of Audiovisual Graphics.* Ph.D. thesis. East Lansing: Michigan State Univ.—E

Hughes, Larry W. (1965), *A Study of Administrative Arrangements in Different Types of School Districts.* Ph.D. thesis. Columbus: Ohio State Univ.—E

Hursh, Gerald D. and others (1969), *Innovation in Eastern Nigeria: Success and Failure of Agricultural Programs in 71 Villages.* East Lansing: Michigan State Univ., Dept. of Communication, Diffusion of Innovations Res. Rep. 8.—C

Hvistendahl, J. K. (1966), "Mass Media Usage in South Dakota Hutterite Colonies." Paper presented at the Assn. for Edu. in Journalism. Iowa City, Iowa.—C

Inayatullah (1962), *Diffusion and Adoption of Improved Practices.* Peshawar: West Pakistan Acad. for Rural Dev.—RS

—— (1964), "Communication and Innovation in a Pakistani Village." Hawaii: East-West Center, Intl. Dev. Inst., unpublished paper.—PA

Ingman, Stanley R. (1963), *Private Outdoor Recreational Development: Factors Associated with Adoption.* M.S. thesis. Columbus: Ohio State Univ.—RS

International Research Associates (1963), *A Study of the Market for Fertilizers in West Pakistan.* New York: Esso Standard Eastern.—MR

Iqbal, Mohammad (1963), *The Study of Farmers' Attitudes towards Adoption of Modern Agricultural Practices.* M.A. thesis. Lahore: Univ. of Punjab.—RS

Jackson, Rita (1965), *Psychological Correlates of Innovative Behavior.* M.A. thesis. Logan: Utah State Univ.—RS

Jain, Navin C. (1965), *The Relation of Information Source Use to the Farm Practice*

Adoption and Farmers' Characteristics in Waterloo County. M.S. thesis. Guelph, Canada: Univ. of Guelph.—EX

——— (1969a), *Some Social Psychological Factors Related to the Effectiveness of Radio Forums*. East Lansing: Michigan State Univ., Dept. of Communication, Diffusion of Innovations, Tech. Rept. 11.—C

——— (1969b), *An Experimental Investigation of the Effectiveness of Group Listening, Discussion, Decision, Commitment, and Consensus in Indian Radio Forums*. Ph.D. thesis. East Lansing: Michigan State Univ.—C

Jain, Nemi C. (1963), *A Study of Relative Effectiveness of Extension Methods in Adoption of Improved Agricultural Practices in Development Block Amraudha, District Kanpur*. M.S. thesis. Kanpur, India: Agra Univ.—EX

Jamias, Juan F. (1964), *The Effects of Belief System Styles on the Communication and Adoption of Farm Practices*. Ph.D. thesis. East Lansing: Michigan State Univ.—C

Jamias, Juan F. and Troldahl, Verling C. (1964), "Dogmatism, Tradition, and General Innovativeness." East Lansing, Michigan State Univ., unpublished paper. —C

Jarnagin, Robert A. (1964), *Farmer Attitudes toward the Cooperative Extension Service in Illinois: A Basis for Communication Strategy*. Ph.D. thesis. East Lansing: Michigan State Univ.—C

Jha, P. N. and Singh, B. N. (1966), "Utilization of Sources of Farm Information as Related to Characteristics of Farmers," *Indian J. of Ext. Edu.*, 1: 294–302.—EX

John, Melathathil J. (1966), *Social Psychological Variables Related to the Role Performance of the Gram Sevaks*. Ph.D. thesis. Ames: Iowa State Univ.—RS

Johnson, David W. (1966), "Racial Attitudes of Negro Freedom School Participants and Negro and White Civil Rights Participants," *Social Forces*, 45: 266–273. —GS

Johnson, Donald E. and van den Ban, Anne W. (1959), "The Dynamics of Farm Practice Change." Paper presented to Midwest Soc. Society. Lincoln, Nebraska. —RS

Johnson, Donald E. and Wilkening, E. A. (1961), *Five Years of Farm and Home Development in Wisconsin: A Comparison of Participating and Control Families*. Madison: Wisc. Agri. Exp. Sta., Res. Bul. 228.— RS

Johnson, Homer M. and Marcum, R. Laverne (1969), "Organizational Climate and the Adoption of Educational Innovations." Paper presented at the Conf. of the Amer. Edu. Res. Assn. Los Angeles. —E

Johnson, Homer M. and others (1967), *Personality Characteristics of School Superintendents in Relation to Their Willingness to Accept Innovation in Education*. Logan: Utah State Univ., Res. Rept.—E

Johnson, Ronald L. (1966), *Goal Aspiration and Practice Adoption for Wisconsin Farm Operators*. Ph.D. thesis. Madison: Univ. of Wisconsin.—RS

Jones, Garth N. (1965a), "Planned Change: An Illustrative Case," *Natl. Inst. of Pub. Adm. Reporter*, 4: 57–72.—PA

Jones, Gwyn E. (1960a), *The Differential Characteristics of Early and Late Adopters of Farming Innovations*. Loughborough, England: Univ. of Nottingham, School of Agri., Rept.—RS

——— (1960b), *Factors Affecting the Adoption of New Farm Practices, with Particular Reference to Central Wales and the East Midlands of England*. B. Litt. thesis. Oxford Univ.—RS

——— (1962a), *Bulk Milk Handling: An Investigation into the Adoption of a New Dairy Technique in Lindsey*. Loughborough, England: Univ. of Nottingham, School of Agri.—RS

——— (1962b), "The Diffusion of Agricultural Innovations." Paper presented at the Agri. Eco. Society. London, England. —ES

——— (1963), "Sources of Information and Advice Available to United Kingdom Farmers: Description and Appraisal," *Sociologia Ruralis*, 3: 1–17.—RS

——— (1964), "Leadership among Farmers." Paper presented at a Conf. on the Communication of Scientific and Technological Ideas in Farming under the auspices of the Agri. Edu. Assn., Seale Hayne Agri. College. England.—RS

Jones, Gwyn E. and Howell, Jeremy (1966), *A Farm 'Open Day' on Grain Drying and Storage*. England: Univ. of Reading, Agri. Ext. Centre.—EX

Jones, William O. (1957), "Manioc: An Example of Innovation in African Economics," *Eco. Dev. and Culture Change*, 5: 99–117.—AE

Juliano, Clemente P., Jr. (1967), *The Relationship between Some Characteristics of Rice Farm Operators and Adoption of Some Recommended Practices in Rice Production in Seventeen Barrios of Laguna*. M.S. thesis. Laguna: Univ. of the Philippines.—O

Junghare, Y. N. (1962), "Factors Influencing the Adoption of Farm Practices," *Indian J. of Social Work*, 23: 291–296.—RS

Junghare, Y. N. and Rahudkar, W. B. (1962), "The Influence of Family Members on Decision-Making in Farm Operations," *Nagpur Agri. Col. Mag.*, 36: 21–24.—RS

Junghare, Y. N. and Roy, Prodipto (1963), "The Relation of Health-Practice Innovations to Social Background Characteristics and Attitudes," *Rural Soc.*, 28: 394–400.—RS

Kahlon, A. S. and Kaushal, Mohinder P. (1967), *Adoption of Improved Agricultural Practices in the Plains of the Punjab*. Ludhiana, India: Punjab Agricultural Univ. —RS

Kantner, John F. (1964), "Pakistan: The Medical Social Research Project at Lulliani," *Stud. in Family Plng.*, 4: 5–10.—MS

Kapoor, J. M. and Roy, Prodipto (1969), "Role of Mass Media and Interpersonal Communication in the Diffusion of a News Event in India." East Lansing: Michigan State Univ., Dept. of Communication, unpublished paper.—C

Kapoor, Rajinder P. (1966), "Relative Effectiveness of Information Sources in the Adoption Process of Some Improved Farm Practices," in A. N. Shukla and I. S. Grewal (Eds.), *Summaries of Extension by Post-Graduate Students*. Ludhiana, India: Punjab Agricultural Univ.—EX

Kar, Mati Lal (1967), "The Adoption Process and the Adopter Categories of Some Improved Practices in Paddy Cultivation in a West Bengal Village," in *Research Studies in Extension Education*. Rajendranagar, India: Andhra Pradesh Agricultural Univ.—EX

Karpat, Kemal H. (1960), "Social Effects of Farm Mechanization in Turkish Villages," *Social Research*, 27: 83–103.—AE

Katz, Elihu (1956a), *Interpersonal Relations and Mass Communications: Studies in the Flow of Influence*. Ph.D. thesis. New York: Columbia Univ.—GS

—— (1961), "The Social Itinerary of Technical Change: Two Studies on the Diffusion of Innovation," in Wilbur Schramm (Ed.), *Studies of Innovation and of Communication to the Public*. Stanford, Cal.: Stanford Univ., Inst. for Communication Res.; and (1962), *Human Organization*, 20: 70–82.—GS

Katz, Elihu and Menzel, Herbert (1954), *On the Flow of Scientific Information in the Medical Profession*. New York: Columbia Univ., Bur. of App. Soc. Res.—MS

Katz, Elihu and others (1964), *The Fluoridation Decision: Community Structure and Innovation*. Philadelphia: Natl. Analysts, Mimeo Rept.—GS

Kaufman, Harold F. and Bryant, Ellen M. (1958), *Characteristics of Farmers Following Recommended Practices*. State College: Miss. Agri. Exp. Sta., Inform. Sheet 608.—RS

Keenan, P. J. (1965), "Evaluating Agricultural Advisory Work in Ireland," *Sociologia Ruralis*, 5: 238–266.—EX

Keith, Robert F. (1968a), *An Investigation of Information and Modernization among Eastern Nigerian Farmers*. East Lansing: Michigan State Univ. Dept. of Communication, Tech. Rept. 4.—C

—— (1968b), *Information and Modernization: A Study of Eastern Nigerian Farmers*. Ph.D. thesis. East Lansing: Michigan State Univ.—C

Kelly, Robert F. (1966), "The Diffusion Model as a Predictor of Ultimate Patronage Levels in New Retail Outlets," in Raymond M. Haas (Ed.), *Science, Tech-*

nology, and Marketing. Chicago: Amer. Mktg. Assn.—MR

——— (1967), "The Role of Information in the Patronage Decision Process: A Diffusion Phenomenon," in *Marketing for Tomorrow Today.* Chicago: Amer. Mktg. Assn.—MR

Khan, Anwar K. (1968), *Adoption and Internalization of Educational Innovations among Teachers in the Pilot Secondary Schools of West Pakistan.* Ph.D. thesis. East Lansing: Michigan State Univ.—E

Khan, A. Majeed (1963), *Pilot Project in Family Planning.* Comilla, East Pakistan: Pakistan Acad. for Rural Dev.—MS

——— (1964), "Population Control: A Two Year Rural Action Experience," *Demography,* 1: 126–129.—MS

Khan, A. Majeed and Choldin, Harvey M. (1965), "New 'Family Planners' in Rural East Pakistan," *Demography,* 2: 1–7.—MS

Kim, Han Young (1967), *Structural Balance and Adoption and Diffusion of an Innovation: A Study of Adoption and Diffusion of the Intrauterine Contraceptive Device.* Ph.D. thesis. Seattle: Univ. of Washington.—MS

Kimball, William J. (1960), *The Relationship between Personal Values and the Adoption of Recommended Farm and Home Practices.* Ph.D. thesis. Univ. of Chicago.—RS

King, Charles W. (1964a), "Fashion Adoption: A Rebuttal to the 'Trickle Down' Theory," in Stephen A. Greyser (Ed.), *Toward Scientific Marketing: Proceedings of the 1963 Winter Conference of the American Marketing Association.* Chicago: Amer. Mktg. Assn.—MR

——— (1964b), "The Innovator in the Fashion Adoption Process," in *Reflections on Progress in Marketing.* Chicago: Amer. Mktg. Assn.—MR

——— (1965), "Communicating with the Innovator in the Fashion Adoption Process," in Peter D. Bennett (Ed.), *Marketing and Economic Development.* Chicago: Amer. Mktg. Assn.—MR

King, Charles W. and Summers, John O. (1967a), "Interaction Patterns in Interpersonal Communication." Lafayette,

Ind.: Herman C. Krannert Grad. School of Ind. Adm., Purdue Univ., Paper 168.—MR

——— (1968), *Technology, Innovation and Consumer Decision Making.* Lafayette, Ind.: Herman C. Krannert Grad. School of Ind. Adm., Purdue Univ., Mimeo Paper.—MR

Kivlin, Joseph E. (1960), *Characteristics of Farm Practices Associated with Rate of Adoption.* Ph.D. thesis. University Park: Pennsylvania State Univ.—RS

——— (1967), "Communication and Development Success-Failure in Indian Villages." Paper presented at the Rural Soc. Society. San Francisco.—C

——— (1968), *Correlates of Family Planning in Eight Indian Villages.* East Lansing: Michigan State Univ. Diffusion of Innovations Res. Rept. 18.—C

Kivlin, Joseph F. and Fliegel, Frederick C. (1967a), "Differential Perceptions of Innovations and Rate of Adoption," *Rural Soc.,* 32: 78–91.—RS

——— (1967b), "Orientations to Agriculture: A Factor Analysis of Farmers' Perceptions of New Practices," *Rural Soc.,* 33: 127–140.—C

Klietsch, Ronald (1961), *Decision-Making in Dairy Farming: A Sociological Analysis.* Ph.D. thesis. Minneapolis: Univ. of Minnesota.—RS

Klonglan, Gerald E. (1962), *Message Sharpening in a Multi-Step Communication Situation.* M.S. thesis. Ames: Iowa State Univ.—RS

——— (1963), *Role of a Free Sample Offer in the Adoption of a Technological Innovation.* Ph.D. thesis. Ames: Iowa State Univ.—RS

Klonglan, Gerald E. and others (1960), "The Role of a Free Sample in the Adoption Process." Paper presented at the Midwest Soc. Society. St. Louis.—RS

——— (1963), "Message Sharpening in a Multi-Step Communication Situation." Paper presented at the Rural Soc. Society. Northridge, Cal.—RS

——— (1964a), *Adoption of Public Fallout Shelters.* Ames: Iowa State Univ., Rural Soc. Rept. 49.—RS

Klonglan, Gerald E. and others (1964b), *Family Adoption of Public Fallout Shelters: A Study of Des Moines, Iowa*. Ames: Iowa State Univ., Rural Soc. Rept. 20.—RS

—— (1967), *Factors Related to Adoption Progress*. Ames: Iowa State Univ., Rural Soc. Rept. 64.—RS

—— (1968), "Conceptualizing and Measuring Extent of Diffusion: The Concept of Adoption Progress." Paper presented at the Rural Soc. Society. Boston.—RS

Knedlik, Stanley M. (1967), *The Effect of Administrative Succession Pattern upon Educational Innovation in Selected Secondary Schools*. Ph.D. thesis. New York: New York Univ.—E

Kohl, John W. (1966), *Adoption Stages and Perceptions of Characteristics of Educational Innovations*. D.Ed. thesis. Eugene: Univ. of Oregon.—E

Koontz, Donald H. (1958), *Decision-Making in a Rural Church*. M.S. thesis. Ames: Iowa State Univ.—RS

Koya, Yoshio (1962), "Lessons from Contraceptive Failure," *Population Stud.*, 16: 4–11.—MS

—— (1964), "Does the Effect of a Family Planning Program Continue?" *Eugenics Qtrly.*, 11: 141–147.—MS

Kraus, Sidney and others (1966), "Fear-Threat Appeals in Mass Communication: An Apparent Contradiction," *Speech Monographs*, 33: 23–29.—C

Kreitlow, Burton W. and Duncan, James A. (1956), *The Acceptance of Educational Programs in Rural Wisconsin*. Madison: Wisc. Agri. Exp. Sta., Bul. 525.—RS

Kronus, Sidney (1967), "Fertility Control in the Rural South: A Pretest," in Donald J. Bogue (Ed.), *Sociological Contributions to Family Planning Research*. Chicago: Univ. of Chicago Press.—MS

Krug, Larry L. (1961), *Content of a Farm Magazine Related to Farmer Adoption of Selected Practices*. M.S. thesis. Madison: Univ. of Wisconsin.—J

Kunstadter, Peter (1961), *Culture Change, Social Structure, and Health Behavior: A Quantitative Study of Clinic Use among the Apaches of the Mescalero Reservation*. Ph.D. thesis. Ann Arbor: Univ. of Michigan.—A

Kuswaha, B. S. (1963), *Effectiveness of the Result Demonstration in the Adoption of the Improved Farm Practices*. M.S. thesis. Gwalior, India: Col. of Agriculture.—EX

Lackey, Alvin S. (1958), *The Consistency of Sociological Variables in Predicting the Adoption of Farm Practices*. Ph.D. thesis. Ithaca, N.Y.: Cornell Univ.—RS

Lamar, Ronald V. (1966), *In-Service Education Needs Related to the Diffusion of an Innovation*. Ph.D. thesis. Berkeley: Univ. of California.—E

Lantis, Margaret (1952), "Eskimo Herdsmen: Introduction of Reindeer Herding to the Natives of Alaska," in Edward H. Spicer (Ed.), *Human Problems in Technological Change*. New York: Russell Sage Fdn.—A

Larsen, Otto N. (1962), "Innovators and Early Adopters of Television," *Soc. Inquiry*, 29: 16–33.—GS

Larsen, Otto N. and De Fleur, Melvin L. (1954), "The Comparative Role of Children and Adults in Propaganda Diffusion," *Amer. Soc. Rev.*, 19: 593–602.—GS

Larsen, Otto N. and Hill, Richard J. (1954), "Mass Media and Interpersonal Communication in the Diffusion of a News Event," *Amer. Soc. Rev.*, 19: 426–433.—GS

—— (1958), "Social Structure and Interpersonal Communication," *Amer. J. of Soc.*, 63: 497–505.—GS

Lassey, William R. (1968), "Communication Behavior and Change Orientation in Rural Development: A Study in Guatemala." Paper presented at the Conf. of the Natl. Society for the Study of Communication. New York.—RS

Lazer, William and Bell, William E. (1966), "The Communications Process and Innovation." *J. of Adv. Res.*, 6: 2–7.—MR

Leadley, Samuel M. (1968), "Organizational Innovation for Development in Multi-County Regions." Paper presented at the Rural Soc. Society. Boston.—RS

Letourneau, Marcel (1963), *Relationship of Culture Patterns to the Adoption of Recommended Farm Practices in the St. John Valley (Maine)*. M.S. thesis. Orono: Univ. of Maine.—RS

Leuthold, Frank O. (1960a), "Demonstrators and the Diffusion of Fertilizer Practices." Paper presented at the Rural Soc. Society. University Park, Pa.—RS

——— (1960b), *Demonstrators and the Diffusion of Fertilizer Practices*. M.S. thesis. Columbus: Ohio State Univ.—RS

——— (1965), *Communication and Diffusion of Improved Farm Practices in Two Northern Saskatchewan Farm Communities*. Saskatoon, Sask.: Centre for Community Studies, Mimeo Report.—RS

——— (1967), *Discontinuance of Improved Farm Innovations by Wisconsin Farm Operators*. Ph.D. thesis. Madison: Univ. of Wisconsin.—RS

Leuthold, Frank O. and Wilkening, Eugene A. (1963), "Acceptance of New Farm Technology: A Test of a Theory of Social Interaction." Paper presented at the Rural Soc. Society. Northridge, Cal.—RS

——— (1965), "Measuring the Dimensions of the Adoption Process and Assessing the Factors Associated with the Acceptance and Continued Use of New Farm Technology." Paper presented at the Midwest Soc. Society. Minneapolis.—RS

Levine, Robert and Sangree, Walter H. (1962), "The Diffusion of Age-Group Organization in East Africa: A Controlled Comparison," *J. of the Intl. African Inst.*, 32: 97–110.—A

Levitt, Theodore (1965), *Industrial Purchasing Behavior: A Study of Communications Effects*. Cambridge: Harvard Univ., Grad. School of Bus. Adm. Press.—MR

Lewin, Kurt, "Group Decision and Social Change," in Theodore M. Newcomb and Eugene L. Hartley (Eds.), *Readings in Social Psychology*. New York: Henry Holt.—P

——— (1963), "Forces Behind Food Habits and Methods of Changing," in Report of the Committee on Food Habits, *The Problem of Changing Food Habits*. Washington, D.C.: Natl. Res. Council, Natl. Acad. of Sciences.—P

Lewis, Oscar (1955), "Medicine and Politics in a Mexican Village," in Benjamin D. Paul (Ed.), *Health, Culture and Community*. New York: Russell Sage Fdn.—A

Lin, Nan (1966), *Innovation Internalization in a Formal Organization*. Ph.D. thesis. East Lansing: Michigan State Univ.—C

——— (1968a), "Innovative Methods for Studying Innovation in Education and an Illustrative Analysis of Structural Effects on Innovation Diffusion within Schools." Paper Presented at the Natl. Conf. on Diffusion of Edu. Ideas. East Lansing, Mich.—GS

——— (1968b), "Innovative Methods for Studying Innovation in Education." Paper presented at the Natl. Conf. on Diffusion of Edu. Ideas. East Lansing, Mich.—C

Lin, Nan and others (1966), *The Diffusion of an Innovation in Three Michigan High Schools: Institution Building Through Change*. East Lansing: Michigan State Univ., Proj. on the Diffusion of Edu. Practices in Thailand. Res. Rept. 1.—E

Lindstrom, David E. (1958), "Diffusion of Agricultural and Home Economics Practices in a Japanese Rural Community," *Rural Soc.*, 23: 171–183.—RS

——— (1960), *Community Development in Sekimura*. Urbana: Ill. Agri. Exp. Sta.—RS

——— (1963), "The Communication of New Techniques and Ideas," in Vu Quoc Thuc (Ed.), *Social Research and Problems of Rural Development in South-East Asia*. Place de Fontenoy, Paris, France: UNESCO.—RS

Linton, Ralph and Kardiner, Abram (1952), "The Change from Dry to Wet Rice Cultivation in Tanala-Betsileo," in Guy E. Swanson and others (Eds.), *Readings in Social Psychology*. New York: Henry Holt.—A

Lionberger, Herbert F. (1948), *Low-Income Farmers in Missouri: Situation and Characteristics of 459 Farm Operators in Four Social Area B Countries*. Columbia: Mo. Agri. Exp. Sta. Bul. 413—RS.

Lionberger, Herbert F. (1949), *Low-Income Farmers in Missouri: Their Contacts with Potential Sources of Farm and Home Information.* Columbia: Mo. Agri. Exp. Sta., Res. Bul. 441.—RS

——— (1950), *Reception and Use of Farm and Home Information by Low-Income Farmers in Selected Areas of Missouri.* Columbia: Univ. of Missouri.—RS

——— (1951), *Sources and Use of Farm and Home Information by Low-Income Farmers in Missouri.* Columbia: Mo. Agri. Exp. Sta., Res. Bul. 472.—RS

——— (1953), "Some Characteristics of Farm Operators Sought as Sources of Farm Information in a Missouri Community," *Rural Soc.,* 18: 327–338.—RS

——— (1954), "The Relation of Informal Social Groups to the Diffusion of Farm Information in a Northeast Missouri Farm Community," *Rural Soc.,* 19: 233–243.—RS

——— (1955), *Information Seeking Habits and Characteristics of Farm Operators.* Columbia: Mo. Agri. Exp. Sta., Bul. 581.—RS

——— (1956a), "The Communication of Farm Information in a Missouri Farm County: A Study of Structural Factors and Personal Influence." Paper read to the Society for Social Res. Chicago.—RS

——— (1956b), *Low-Income Farmers in the Good Farming Areas of Missouri: Their Characteristics, Resources, Sources of Information.* Columbia: Mo. Agri. Exp. Sta., Bul. 688.—RS

——— (1958), *Television Viewing in Rural Boone County with Special Reference to Agricultural Shows.* Columbia: Mo. Agri. Exp. Sta., Bul. 702.—RS

——— (1959), "Community Prestige and the Choice of Sources of Farm Information," *Pub. Opin. Qtrly.,* 23: 111–118.—RS

——— (1962c), "Overlap Dispersion of Selected Functions in Adoption Decisions of Farm Operators in Two Missouri Communities." Paper read at the Rural Soc. Society, Washington, D.C.—RS

——— (1963d), *Legitimation of Decisions to Adopt Farm Practices and Purchase Farm Supplies in Two Missouri Farm Communities:* *Ozark and Prairie.* Columbia: Mo. Agri. Exp. Sta., Res. Bul. 826.—RS

Lionberger, Herbert F. and Campbell, Rex R. (1963a), *The Potential of Interpersonal Communicative Networks for Message Transfer from Outside Information Sources: A Study of Two Missouri Communities.* Columbia: Mo. Agri. Exp. Sta., Res. Bul. 842.—RS

——— (1963b), "Segregating and Differentiating Influences of Personal Attitudes on the Choice of Persons as Information Sources and Associates in Two Missouri Communities." Paper presented at the Rural Soc. Society. Northridge, Cal.—RS

Lionberger, Herbert F. and Chang, H. C. (1965), *Comparative Characteristics of Special Functionaries in the Acceptance of Agricultural Innovations in Two Missouri Communities, Ozark and Prairie.* Columbia: Mo. Agri. Exp. Sta., Res. Bul. 885.—RS

——— (1968), *Communication and Use of Scientific Farm Information by Farmers in Two Taiwan Agricultural Villages.* Columbia: Mo. Agri. Exp. Sta., Res. Bul. 940—RS

Lionberger, Herbert F. and Coughenour, C. Milton (1957), *Social Structure and Diffusion of Farm Information.* Columbia: Mo. Agri. Exp. Sta., Res. Bul. 631.—RS

Lionberger, Herbert F. and Hassinger, Edward (1954a), "Neighborhoods as a Factor in the Diffusion of Farm Innovations in a Northeast Missouri Farming Community," *Rural Soc.,* 19: 377–384.—RS

——— (1954b), *Roads to Knowledge.* Columbia: Mo. Agri. Exp. Sta., Bul. 633.—RS

Lionberger, Herbert F. and others (1960), "A Method of Measuring the Communication Potential of Interpersonal Communicative Networks." Paper presented at the Rural Soc. Society. University Park, Pa.—RS

Lipman, Aaron (1966), *El Impresarios de Bogotá.* Bogotá: Univ. Nacional de Colombia, Facultad de Sociología. Monografias Sociologicas 22.—GS

Lippitt, Ronald and Havelock, Ronald (1968), "Needed Research on Research Utilization." Paper presented at the

Natl. Conf. on Diffusion of Edu. Ideas. East Lansing, Mich.—P

Littunen, Yrjo (1959), "Deviance and Passivity in Radio Listener Groups," *Acta Sociologia,* 4: 17–26.—GS

Loewy, Edith B. (1961), *Early Automobile Buying and Status Aspirations.* New York: Columbia Univ., Bur. of Applied Social Res., Mimeo Rept.—MR

Loomis, Charles P. (1967a), "Change in Rural India as Related to Social Power and Sex," *Behav. Sci. and Com. Dev.,* 1 1–27.—RS

—— (1967b), "In Praise of Conflict and Its Resolution," *Amer. Soc. Rev.* 32: 875–890.—RS

—— (1968), "Social Organization and Social Change." Paper presented at the 2nd World Congress of Rur. Soc. Drienerlo, Netherlands.—RS

Loudermilk, Kenneth M. and others (1968), *Open System Theory and Change in Vocational Programs of Idaho Secondary Schools.* Moscow: Univ. of Idaho, State Occupational Res. Unit, Res. Rept.—E

Louisiana Agri. Ext. Svce. (1950), *Extension at Work in Lafourche.* Baton Rouge, La.: Agri. Ext. Svce, Pub. 1054.—EX

Lovos, George J. (1955), *A Description of Educational Practice in Metropolitan School Study Council Systems in 1954: With Special Reference to Elementary Schools.* D.Ed. thesis. New York: Teachers College, Columbia Univ.—E

Lowenstein, Duane E. and others (1967), *A Discussion of the Gap between Knowledge and Use of New Practices.* Washington, D.C.: U.S. Dept. of Health, Edu. and Welfare, Office of Educ.—AE

Lowry, Sheldon G. and Hay, Donald G. (1957), *Acceptance of Voluntary Health Insurance in Sampson County, North Carolina, 1955.* Raleigh: N.C. Agri. Exp. Sta., Rept. RS-28.—RS

—— (1958), *Use of Health Care Services and Enrollment in Voluntary Health Insurance in Stokes County, North Carolina, 1956.* Raleigh: N.C. Agri. Exp. Sta. Prog. Rept. RS-32.—RS

Lowry, Sheldon and others (1958), "Factors Associated with the Acceptance of Health Care Practices among Rural Families," *Rural Soc.,* 23: 198–202.—RS

Loy, John W., Jr. (1967), *Socio-Psychological Attributes of English Swimming Coaches Differentially Adopting a New Technology.* Ph.D. thesis. Madison: Univ. of Wisconsin.—E

—— (1968a), "Competitive Success in Sport," Paper presented at the Natl. Convention of the Amer. Assn. of Health, Phys. Edu., and Recreation, St. Louis.—E

—— (1968b), "Socio-Psychological Attributes Associated with the Early Adoption of a Sport Innovation." Paper presented at the Natl. Convention of the Amer. Assn. of Health, Phys. Edu., and Recreation. St. Louis.—E

—— (1968c), "Socio-Psychological Attributes Associated with the Early Adoption of a Sport Innovation," *J. of Psychology,* 70: 141–147.—E

—— (1969), "Social Psychological Characteristics of Innovators," *Amer. Soc. Rev.,* 34: 73–82.—E

Lu, Laura Pan and others (1967), "An Experimental Study of the Effect of Group Meetings on the Acceptance of Family Planning in Taiwan," *J. of Social Issues,* 23: 171–177.—MS

Lutz, Arien E. (1966), *Change Agents as Predictors of the Rate of Farm Practice Adoption.* D.Ed. thesis. Lincoln: Univ. of Nebraska.—E

Maalouf, Wajih D. (1965), *Factors Associated with Effectiveness of the Result Demonstration Method in Promoting Adoption of Fertilizer Practices by Wheat Farmers in Baaldeck and Akkar Counties, Lebanon.* Ph.D. thesis. Ithaca N.Y.: Cornell Univ.—E

Maamary, Samir M. (1965), *Cross Cultural Comparison of Characteristics of Adopters and Non-Adopters of Farmer Cooperatives Among Villagers.* M.A. thesis. East Lansing: Michigan State Univ.—C

Maccoby, N. and others (1959), "Critical Periods in Seeking and Accepting Information." Paper presented at Amer. Psychological Assn. Cincinnati, Ohio.—C

Maddala, G. S. and Knight, Peter T. (1965), "International Diffusion of Technical Change: A Case Study of the Oxygen

Steel Making Process." Paper presented at the Inter-University Conf. on Microeconomics of Technical Change. Philadelphia.—GS

Maddox, Harry (1965), "Progressive and Traditional Farmers: The Use of N.A.A.S. by Upland Farmers in Radnorshire," *Qtrly Bul. of the Natl. Agri. Advisory Svce. (NAAS),* 67: 1–7.—P

Madigan, Francis C. (1962a), *The Farmer Said No: A Study of Background Factors Associated with Dispositions to Cooperate with or Be Resistant to Community Development Projects.* Diliman: Community Development Research Council, Univ. of the Philippines, Quezon City, Study Series 14.—GS

———— (1962b), "Predicting Receptivity to Community Development Innovations," *Current Anthro.,* 3: 207–208.—A

Maffei, Eugenio (1966), *Innovativeness as Related to Other Factors in a Colombian Community: Contadero, Nariño.* M.S. thesis. Madison: Univ. of Wisconsin.—RS

Magdub, Abdo M. (1964), *La Difusión Adopción del Cultivo de la Soya en el Valle del Yaqui.* Professional thesis. Chapingo, Mexico: Grad. School of Agriculture.—C

Mahajan, Bhagwan S. (1966), "Relative Effectiveness of Selected Extension Methods in Acceptance of Agrosan GN Seed Treatment to Cotton," in *Research Studies in Extension Education.* Rajendranagar, India: Andhra Pradesh Agricultural Univ., Ext. Edu. Inst.—EX

Mahoney, Frank J. (1960), "The Innovation of a Savings System in Truk," *Amer. Anthro.,* 62: 465–482.—A

Makarczyk, Waclaw (1965), "Innovation in Agriculture and the Use of Information Sources," *Polish Soc. Bul.,* 2: 48–60.—RS

———— (1967), "Research on Diffusion of Farm Innovations in Two Rural Communities in Poland." Warsaw, Poland: Inst. of Philos. and Soc., Polish Acad. of Sciences, unpublished paper.—RS

———— (1968b), "The Press as a Source of Information about Farming," *Annals of Rural Soc.,* 50–56.—RS

Malhotra, Mridula (1966), *An Investigation into the Acceptance of the Applied Nutrition Programme in a Selected Village of Punjab.* M.S. thesis. India: Univ. of Delhi.—EX

Mansfield, Edwin (1960), *Acceptance of Technical Change: The Speed of Response of Individual Firms.* Pittsburgh: Carnegie Inst. of Technology, Grad. School of Ind. Adm., Mimeo Rept.—GE

———— (1961), "Technical Change and the Rate of Imitation," *Econometrica,* 29: 741–766.—GE

———— (1963a), "Intrafirm Rates of Diffusion of an Innovation," *Rev. of Econ. and Stat.,* 45: 348–359.—GE

———— (1963b), "Size of Firm, Market Structure, and Innovation," *J. of Pol. Eco.,* 71: 556–576.—GE

———— (1963c), "The Speed of Response of Firms to New Techniques," *Qtrly J. of Eco.,* 77: 290–311.—GE

———— (1968), *Industrial Research and Technological Innovation.* New York: Norton.—GE

Marcum, R. Laverne (1968), *Organizational Climate and the Adoption of Educational Innovations.* Logan: Utah State Univ., Dept. of Edu. Adm.—E

Marcus, Alan S. and Bauer, Raymond A. (1964), "Yes: There Are Generalized Opinion Leaders," *Pub. Opin. Qtrly.,* 28: 628–632.—MR

Marine, Clyde L. (1964), *Testing Institutional Acceptance of New Food Products: A Case Study on Processed Onion Products.* Ph.D. thesis. East Lansing: Michigan State Univ.—AE

Marion, Guy B. (1966), *A Study of Selected Factors Related to the Innovativeness of Elementary School Principals.* Ph.D. thesis. Edmonton: Univ. of Alberta.—E

Marquis, Donald G. and Allen, Thomas J. (1967), "Communication Patterns in Applied Technology," *Amer. Psychologist,* 22: 1052–1060.—MR

Marriott, McKim (1955), "Western Medicine in a Village of Northern India," in Benjamin D. Paul (Ed.), *Health, Culture and Community.* New York: Russell Sage Fdn.—A

Marsh, C. Paul and Coleman, A. Lee (1954a), "Farmers' Practice-Adoption

Rates in Relation to Adoption Rates of 'Leaders'," *Rural Soc.* 19: 180–181.—RS

——— (1954b), *Communication and the Adoption of Recommended Farm Practices.* Lexington: Ky. Agri. Exp. Sta., Prog. Rept. 22.—RS

——— (1954c), "The Relation of Kinship, Exchanging Work, and Visiting to the Adoption of Recommended Farm Practices," *Rural Soc.,* 19: 1–2.—RS

——— (1954d), "The Relation of Neighborhood of Residence to Adoption of Recommended Farm Practices," *Rural Soc.,* 19: 385–389.—RS

——— (1955a), "Differential Communication among Farmers in a Kentucky County," *Rural Soc.,* 20: 93–101.—RS

——— (1955b), "The Relation of Farmer Characteristics to the Adoption of Recommended Farm Practices," *Rural Soc.,* 20: 289–296.—RS

——— (1956), "Group Influences and Agricultural Innovations: Some Tentative Findings and Hypotheses," *Amer. J. of Soc.,* 71: 588–594.—RS

Martinez, Gregorio V. and Myren, Delbert T. (1964), *Alcance e Impacto de la Pagina Agricola de "El Dictamen" de Vera Cruz (Scope and Impact of the Agricultural Page of "El Dictamen": A Newspaper of Veracruz).* Mexico: Secretaria de Agricultura y Ganaderia, Instituto Nacional de Investigaciones Agricolas, Tech. Bul. 47.—C

Martinez, Jesus R. (1963), *La Difusión y Adopción del Maiz Hibrido en Cuatro Municipios de Estado de Guanajuato.* Thesis for Ingeniero Agronomo. Chapingo, Mexico: Grad. School of Agriculture.—C

——— (1964), "Factores Sociales y Economicos que Influyen en la Difusión y Adopción del Maiz Hibrido en el Bajio." Paper presented at the First Inter-American Res. Symposium on the Role of Communications in Agri. Dev. Mexico City.—RS

Martinez, Jesus R. and others (1964a), "Estudio de la Diffusión y Adopción del Maiz Hibrido en Cuartro Municipios del Estado de Guanajuato, *Agricultura Tecnica en Mexico,* 2: 120–125.—C

Mason, Robert G. (1962a), *Information Source Use in the Adoption Process.* Ph.D. thesis. Stanford, Cal.: Stanford Univ.—C

——— (1962b), "An Ordinal Scale for Measuring the Adoption Process," in Wilbur Schramm (Ed.), *Studies of Innovation and of Communication to the Public.* Stanford, Cal.: Stanford Univ., Inst. for Communication Res.—C

——— (1963), "The Use of Information Sources by Influentials in the Adoption Process," *Pub. Opin. Qtrly.,* 27: 455–457.—C

——— (1964), "The Use of Information Sources in the Process of Adoption," *Rural Soc.,* 29: 40–52.—C

——— (1966b), "A Revision of the Two-Step Flow of Communication Hypothesis," *Gazette,* 12: 109–111.—EX

Mason, Robert G. and Halter, Albert N. (1968), "The Application of a System of Simultaneous Equations to an Innovation Diffusion Model," *Social Forces,* 47: 182–195.—C

Massy, William Francis (1960), *Innovation and Market Penetration: A Study in the Analysis of New Product Demand.* Ph.D. thesis. Cambridge: Massachusetts Inst. of Tech.—MR

Mathen, K. K. (1962), "Preliminary Lessons Learned from the Rural Population Control Study of Singur," in Clyde V. Kiser (Ed.), *Research in Family Planning.* Princeton, N.J.: Princeton Univ. Press.—MS

Matthew, Taylor and others (1942), *Attitudes of Edgefield County Farmers Toward Farm Practices and Rural Programs.* Clemson: S.C. Agri. Exp. Sta., Bul. 339.—U

Mauldin, W. Parker (1967), "Retention of IUDS: An International Comparison," *Res. in Fam. Plng.,* 18: 1–12.—O

McCarthy, W. O. and Tugby, Donald J. (1962), "Methods of Management, Sources of Advice and Objectives among a Sample of Queensland Dairy Farmers," *Rev. of Mktg. and Agri. Eco.,* 30: 1–11.—AE

McClellan, George B. (1952), *The Relation of Factors in the Organizational Patterns of School Systems and Adaptability.* Ph.D. thesis. New York: Columbia Univ.—E

McConnell, Douglas and others (1962), *Relative Profitability and Order of Adoption of Soil Conservation Practices.* Madison: Univ. of Wisconsin, Res. Bul. 237.—RS

McCorkle, Thomas (1961), "Chiropractic: A Deviant Theory of Disease and Treatment in Contemporary Western Culture," *Human Organization,* 20: 20–22.—A

McKain, Walter C., Jr., and others (1958), *Campaigns to Increase the Milk Consumption of Older Persons.* Storrs: Conn. Agri. Exp. Sta., Bul. 344.—RS

McKenna, Bernard H. (1955), *Measures of Class Size and Numerical Staff Adequacy Related to a Measure of School Quality.* D.Ed. thesis. New York: Teachers College, Columbia Univ.—E

McLeod, Jack M. and others (1968), "The Mass Media and Political Information in Quito, Ecuador," *Pub. Opin. Qtrly.,* 32: 575–587.—C

McLeod, Jack M. and Swinehart, James W. (1959), "Satellites, Science and the Public." Ann Arbor: Univ. of Michigan, Survey Res. Cen.—P

McMillion, Martin B. (1960), *The Source of Information and Factors which Influence Farmers in Adopting Recommended Practices in Two New Zealand Counties.* Christ Church, N.Z.: Canterbury Agricultural Col., Tech. Pub. 19.—AE

McNelly, John T. (1964), "Mass Communication and the Climate for Modernization in Latin America." Paper presented at the Assn. for Edu. in Journalism. Austin, Texas.—J

McNelly, John T. and Deutschmann, Paul J. (1963), "Media Use and Socioeconomic Status in a Latin American Capital," *Gazette,* 9: 225–231.—C

McNelly, John T. and others (1967), "Cosmopolitan Media Usage in the Diffusion of World Affairs Information." Paper presented at the Assn. for Edu. in Journalism. Boulder, Colo.—J

McVoy, Edgar C. (1940), "Patterns of Diffusion in the United States," *Amer. Soc. Rev.,* 5: 219–227.—ES

Mead, Margaret (1955), *Cultural Patterns and Technical Change.* New York: New American Library.—A

Medalia, Nahum Z. and others (1958), "Diffusion and Belief in a Collective Delusion: The Seattle Windshield Pitting Epidemic," *Amer. Soc. Rev.,* 23: 180–186.—GS

Mendelsohn, Harold (1964), "Broadcast vs. Personal Sources of Information in Emergent Public Crises: The Presidential Assassination," *J. of Broadcasting,* 8: 147–156.—C

Mendez, Alfredo D. (1968), "Social Structure and the Diffusion of Innovation," *Human Organization,* 27: 241–249.—P

Menefee, Selden and Menefee, Audrey (1967), "A Country Weekly Proves Itself in India," *Journalism Qtrly.,* 44: 114–117. —C

Menzel, Herbert (1957), "Public and Private Conformity under Different Conditions of Acceptance in the Group," *J. of Abn. and Soc. Psych.,* 55: 398–402.—MS

——— (1959), *Social Determinants of Physicians' Reactions to Innovations in Medical Practice.* Ph.D. thesis. Madison: Univ. of Wisconsin.—MS

——— (1960), "Innovation, Integration, and Marginality: A Survey of Physicians," *Amer. Soc. Rev.,* 25: 704–713.—MS

Menzel, Herbert and Katz, Elihu (1955), "Social Relations and Innovation in the Medical Profession: The Epidemiology of a New Drug," *Pub. Opin. Qtrly.,* 19: 337–352.—MS

——— (1963), "Comment on Charles Winick, The Diffusion of an Innovation among Physicians in a Large City," *Sociometry,* 26: 125–127.—MS

Menzel, Herbert and others (1959), "Dimensions of Being 'Modern' in Medical Practice," *J. of Chronic Dis.,* 4: 20–40.— MS

Metraux, Alfred (1959), "The Revolution of the Ax," *Diogenes,* 25.—U

Michigan Dept. of Public Instruction (1964), *Five Years of Change in the Public Elementary and Secondary Schools in Michigan.* Lansing: Michigan Dept. of Public Instruction, Rept. 1.—E

Miller, Delbert C. (1945), "A Research Note on Mass Communication," *Amer. Soc. Rev.,* 10: 691–694.—GS

Miller, Genevieve (1957), *The Adoption of Inoculation for Smallpox in England and France*. Philadelphia: Univ. of Pennsylvania Press.—MS

Miller, Richard I. (1968), "Implications for Practice from Research on Educational Change." Paper presented at the Natl. Conf. on Diffusion of Edu. Ideas. East Lansing, Mich.—E

Miller, Robert A. and others (1968), "IUD Rejected Cases: An Example of Clinic and Client Experiences in East Pakistan," *Pakistan J. of Fam. Plng., 2*: 21–27.—MS

Miller, Texton R. (1965), *Teacher Adoption of a New Concept of Supervised Practice in Agriculture*. Raleigh: North Carolina State Univ., Edu. Res. Series 4.—E

Miner, Horace (1960), "Culture Change under Pressure: A Hausa Case," *Human Organization, 19*: 164–167.—A

Miro, Carmen A. and Rath, Ferdinand (1965), "Preliminary Findings of Comparative Fertility Surveys in Three Latin American Cities," *Milbank Mem. Fund Qtrly., 43*: 36–68.—MS

Misra, B. D. (1967), "Correlates of Males' Attitudes Toward Family Planning," in Donald J. Bogue (Ed.), *Sociological Contributions to Family Planning Research*. Chicago: Univ. of Chicago Press.—MS

Montgomery, David B. and Armstrong, J. Scott (1968), "Consumer Response to a Legitimated Brand Appeal," in Johan Arndt (Ed.), *Insights into Consumer Behavior*. Boston: Allyn and Bacon.—MR

Mookherjee, D. K. and Singh, Harjinder (1966), *Utilisation of Tubewell Irrigation in West Bengal: An Analysis*. Kalyani (Nadia), West Bengal: Dept. of Community Dev., Orientation Centre.—RS

Morgan, James N. and others (1966), *Productive Americans: A Study of How Individuals Contribute to Economic Progress*. Ann Arbor: Univ. of Michigan, Survey Res. Center, Monogr. 43.—GE

Morrison, Denton E. (1962), *Achievement Motivation: A Conceptual and Empirical Study in Measurement Validity*. Ph.D. thesis. Madison: Univ. of Wisconsin.—RS

—— (1964), "Achievement Motivation of Farm Operators: A Measurement Study," *Rural Soc., 29*: 367–384.—RS

Morrison, William A. (1956), "Attitudes of Males toward Family Planning in a Western Indian Village," *Milbank Mem. Fund Qtrly., 34*: 262–286.—RS

—— (1957), "Attitudes of Females toward Family Planning in a Maharashtrian Village," *Milbank Mem. Fund Qtrly., 35*: 67–81.—RS

—— (1961), "Family Planning Attitudes of Industrial Workers of Ambarnath, a City of Western India: A Comparative Analysis," *Population Stud., 14*: 235–248.—RS

Mort, Paul R. and Cornell, Francis G. (1938), *Adaptability of Public School Systems*. New York: Teachers College, Columbia Univ.—E

—— (1941), *American Schools in Transition*. New York: Teachers College, Columbia Univ.—E

Mort, Paul R. and Pierce, Truman A. (1947), *A Time Scale for Measuring the Adaptability of School Systems*. New York: Teachers College, Columbia Univ. Metropolitan School Study Council.—E

Mortimore, Fredric J. (1968), *Diffusion of Educational Innovations in the Government Secondary Schools of Thailand*. Ph.D. thesis. East Lansing: Michigan State Univ.—E

Moulik, T. K. and others (1966), "Predictive Values of Some Factors of Adoption of Nitrogenous Fertilizers by North Indian Farmers," *Rural Soc., 31*: 447–467.—RS

Mulay, Sumati and Ray, G. L. (1965), "Caste and Adoption of Improved Farm Practices," *Indian J. of Ext. Edu., 1*: 106–111.—RS

Mulford, Charles Lee (1959), *Relation between Community Variables and Local Industrial Development Corporations*. M.S. thesis. Ames: Iowa State Univ.—GS

Myers, John G. (1966), "Patterns of Interpersonal Influence in the Adoption of New Products," in Raymond M. Haas (Ed.), *Science, Technology, and Marketing*. Chicago: Amer. Mktg. Assn.—MR

Myren, Delbert T. (1962), "The Rural Communications Media as a Determinant of the Diffusion of Information about Improved Farming Practices in Mexico." Paper presented at Rural Soc. Society. Washington, D.C.—C

Nagoke, Jagir S. (1964), "Relative Effectiveness of Extension Methods on the Adoption of Approved Agriculture Practices in Selected N.E.S. Block District Amritsar," in A. N. Shukla and I. S. Grewal (Eds.), *Summaries of Extension by Post-Graduate Students*. Ludhiana, India: Punjab Agricultural Univ.—EX

Naquin, C. J. (1957), *Factors Associated with the Adoption of Selected Dairy Farming Practices, Avoyelles Parish, 1956*. M.S. thesis. Baton Rouge: Louisiana State Univ.—RS

Narang, Swarn (1966), *A Study on the Identification and Characteristics of Rural Women Leaders in a Selected Village, Delhi*. M.S. thesis. India: Univ. of Delhi.—EX

Narayan, J. P. (1963), *Farmers' Attitudes and Beliefs in Relation to Adoption of Improved Agricultural Practices*. M.S. thesis. Bhagalpur, India: Agricultural College Sabour.—EX

National Institute of Community Development (1963), *Perception of National Emergency in Village India*. Mussorie, India.—RS

National Science Foundation (1961), "Diffusion of Technological Change," in *Reviews of Data on Research on Development*. Washington, D.C.: Natl. Science Fdn., Rept. 31.—I

Neill, Ralph F. (1963), *Achievement Motivation among Ohio Farmers*. M.S. thesis. Columbus: Ohio State Univ.—RS

Neill, Ralph E. and Rogers, Everett M. (1963), *Measuring Achievement Motivation among Farmers*. Columbus: Ohio Agri. Exp. Sta., Dept. Series AE 346.—RS

Neurath, Paul (1960), *Radio Farm Forum in India*. Delhi, Govt. of India Press.—GS

——— (1962), "Radio Farm Forum as a Tool of Change in Indian Villages," *Eco. Dev. and Cultural Change*, 108: 275–283.—GS

Newbold, Stokes (1957), "Receptivity to Communist-Fomented Agitation in Rural Guatemala," *Eco. Dev. and Cul. Change*, 5: 338–361.—U

Newell, Clarence A. (1943), *Class Size and Adaptability*. Ph.D. thesis. New York: Teachers College, Columbia Univ.—E

Nichols, Keith M. (1959), *Farming Information Sources Preferred by Iowa Farmers*. M.S. thesis. Ames: Iowa State Univ.—J

Niehoff, Arthur (1964b), "Theravada Buddhism: A Vehicle for Technical Change," *Human Organization*, 23: 108–112.—A

Niehoff, Arthur H. and Anderson, J. Charnel (1964a), "Positive, Negative, and Neutral Factors: The Process of Cross-Cultural Innovations," *Intl. Dev. Rev.*, 6: 5–11.—A

——— (1964b), "Peasant Fatalism and Socio-Economic Innovation." Washington, D.C.: George Washington Univ., HumRRO, unpublished paper.—A

Nielson, James and Bittner, R. F. (1958), *Farm Practice Adoption in Michigan*. East Lansing: Mich. Agri. Exp. Sta., Tech. Bul. 263.—AE

Nielson, James and Crosswhite, William (1959), *Changes in Agricultural Production, Efficiency and Earnings*. East Lansing: Mich. Agri. Exp. Sta., Res. Bul. 274.—RS

Nikhade, D. M. (1963), *Adoption of Agricultural Practices in Relation to Some Aspects of Social Change*. M.S. thesis. Nagpur, India: Col. of Agriculture.—EX

Nye, Ivan (1952), *The Relationship of Certain Factors to County Agent Success*. Columbia: Mo. Agri. Exp. Sta., Res. Bul. 498.—RS

Nylin, Donald W. (1967), *An Investigation of the Relationship between Self-Perceived Traits Associated with Innovators and Assessment of Climate Satisfaction and Limitations on Satisfaction*. Ph.D. thesis. Urbana: Univ. of Illinois.—E

Oakley, John W. (1965), *The Adoption of Recommended Forestry Practices by Small Woodland Owners in Grenada County, Mississippi*. State College: Mississippi State Univ.—RS

Oberg, Kalervo and Rios, Jose A. (1955), "A Community Improvement Project in Brazil," in Benjamin D. Paul (Ed.), *Health, Culture and Community*. New York: Russell Sage Fdn.—A

Oliver-Padilla, Otis (1964), *The Role of Values and Channel Orientations in the Diffusion and Adoption of New Ideas and Practices: A Puerto Rican Dairy Farmers' Study.* Ph.D. thesis. East Lansing: Michigan State Univ.—C

Oloko, Olatunde (1962), *A Study of Socio-Economic Factors Affecting Agricultural Productivity in Annang Province, Eastern Nigeria.* Ibadan: Nigerian Inst. of Social and Eco. Res.—RS

Olson, K. S. (1959a), *4-H and Adoption of Improved Farm Practices.* Fargo: N. Dak. Agri. Ext. Svce. Cir.—EX

—— (1959b), *The Relation of Selected Farmers' 4-H Experience to Their Adoption of Improved Farm Practices.* Ph.D. thesis. Madison: Univ. of Wisconsin.—EX

Opinion Research Corporation (1959a), *America's Taste-Makers: A New Strategy for Predicting Change in Consumer Behavior.* Princeton, N.J.—MR

—— (1959b), *Consumer Values: How They Help Predict Market Change in a Mobile Society.* Princeton, N. J.—MR

—— (1960), *The Initiators.* Princeton, N. J.—MR

Opler, Morris E. (1964), "Cultural Context and Population Control Programs in Village India," in Earl W. Count and Gordon T. Bowles (Eds.), *Fact and Theory in Social Science.* Syracuse, N.Y.: Syracuse Univ. Press.—A

Ortega, Alfonso A. (1964), *Adoption-Proneness and Response to Mail Questionnaires.* M.S. thesis. Ames: Iowa State Univ.—J

Osborne, L. W. (1961), "County Leaflets as an Advisory Tool," *NAAS Qtrly. Rev.* 12: 126–132.—EX

Osborne, L. W. and Dadd, C. V. (1958), "Factors Influencing the Choice of Cereal Varieties," *Outlook on Agri.,* 2: 3–12.—EX

Oyen, Orjar (1953), *The Relationship between Distance and Social Interaction: The Case of Message Diffusion.* M.A. thesis. Seattle: Univ. of Washington.—GS

Oyen, Orjar and De Fleur, Melvin L. (1953), "The Spatial Distribution of an Airborne Leaflet Message," *Amer. J. of Soc.,* 49: 144–149.—GS

Palmore, James A. (1967a), "Awareness Sources and Stages in the Adoption of Specific Contraceptives." Ann Arbor: Univ. of Michigan, Population Studies Center, unpublished paper.—MS

—— (1967b), "The Chicago Snowball: A Study of the Flow and Diffusion of Family Planning Information," in Donald J. Bogue (Ed.), *Sociological Contributions to Family Planning Research.* Chicago: Univ. of Chicago Press.—MS

Palmore, James A. and Freedman, Ronald (1968), "Perceptions of Contraceptive Practice by Others: Effects on Family Planning Acceptance in Taichung, Taiwan." Ann Arbor: Univ. of Michigan, Population Studies Center, unpublished paper.—MS

Pandit, S. (1962), *Study of the Role of Age, Education and Size of Farm in Relation to Adoption of Improved Agricultural Practices.* M.S. thesis. Bhagalpur, India: Agricultural College Sabour.—EX

Papoz, J. C. (1960), "Enquete-Pilote Sur l'Adoption de la Culture du Mais Hybride les Cantons de Nay (Basse-Pyrenees)," *Economic Rurale,* 45: 29–43.—O

Paramdhamayya, N. (1967), "An Analysis of Channels of Communication in the Adoption of Some Improved Poultry Husbandry Practices in Gudivada Panchayat Samithi," in *Research Studies in Extension Education.* Rajendranagar, India: Andhra Pradesh Agricultural Univ.—EX

Parish, Ross (1954), "Innovation and Enterprise in Wheat Farming," *Rev. of Mktg. and Agri. Eco.,* 22: 189–218.—AE

—— (1956), "Extension Services and the Grazier on the South-West Slope," *Rev. of Mktg. and Agri. Eco.,* 24: 222–235.—AE

Patel, Ishwarlal C. (1967), *Communication Behavior of Village Level Workers in Surat and Mehsana Districts, Gujarat State, India.* Ph.D. thesis. Ithaca, N.Y.: Cornell Univ.—EX

Patel, Narsi (1964), "Shift from Primary to Secondary Reference in Interpersonal Communication." Paper presented at the

Rural Soc. Society. Montreal, Canada.—RS

Patel, Narsinhbhai B. (1966), *Status Determinants of Interpersonal Communication: A Dyadic Analysis.* Ph.D. thesis. Lexington: Univ. of Kentucky.—RS

Patnaik, Nityananda (1963), "From Tribe to Caste: The Juangs of Orissa," *The Economic Wkly.*, 741–742.—A

Paul, William J., Jr. (1965), *Psychological Characteristics of the Innovator.* Ph.D. thesis. Cleveland, Ohio: Western Reserve Univ.—P

Pedersen, Harold A. (1951), "Cultural Differences in the Acceptance of Recommended Practices," *Rural Soc.*, 16: 37–49.—RS

Pelto, Perti J. (1960), "Innovation in an Individualistic Society." Paper presented at the Amer. Anthro. Assn.—A

Pelto, Perti J. and others (1969), "The Snowmobile Revolution in Lapland," *J. of the Finno-Ougrienno Society,* 69: 1–42.—A

Pemberton, H. Earl (1936a), "The Curve of Culture Diffusion Rate," *Amer. Soc. Rev.*, 1: 547–556.—ES

——— (1936b), "Culture-Diffusion Gradients," *Amer. J. of Soc.*, 42: 226–233.—ES

——— (1937), "The Effect of a Social Crisis on the Curve of Diffusion," *Amer. Soc. Rev.*, 2: 55–61.—ES

——— (1938), "Spatial Order of Cultural Diffusion," *Sociology and Social Res.*, 22: 246–251.—ES

Pereira de Melo, Gilberto (1964), "Estudio Comparativo de Cuatro Metodos de Extensión en el Salvador," *Turrialba*, 14: 208–211.—EX

Pessemier, Edgar A. and others (1967), "Can New Product Buyers Be Identified?" *J. of Mktg. Res.*, 4: 349–354.—MR

Peterman, Lloyd E. (1966), *The Relationship of In-Service Education to the Innovativeness of the Classroom Teacher in Selected Public Secondary Schools in Michigan.* Ph.D. thesis. Ann Arbor: Univ. of Michigan.—E

Petrini, Frank (1966a), *Upplysningsverksamhetes Effekt I, Innovationernas Spridningstakt (The Rate of Adoption of Selected Agricultural Innovations).* Uppsala: Agricultural College of Sweden, Rept. 53.—RS

——— (1966b), *Upplysningsverksamhetens Effekt II, Vardering av en Upplysningskampanj Mot Dvargskottsjuka (Studies of the Effect of An Extension Campaign Against Disease of Oats).* Uppsala: Agricultural College of Sweden, Rept. 63.—RS

——— (1967a), *Experiment with Different Time Disposition of a Given Extension Programme.* Uppsala: Agricultural College of Sweden, Rept. 85.—RS

——— (1967b), *Upplysningsverksamhetens Effekt V, Forsok Med Stigande Mangd Information (An Experiment with an Increasing Amount of Information).* Uppsala: Agricultural College of Sweden, Rept. 84.—RS

Petrini, Frank and others (1968), *The Information Problem of an Agricultural College.* Uppsala: Nordisk Jordbruksforskning 50.—RS

Photiadis, John D. (1960), "Attitudes of People Toward Experiment Station and Extension Service," *S. Dak. Farm and Home Res.*, 11: 17–20.—RS

——— (1961), *Contacts with Agricultural Agents.* Brookings: S. Dak. Agri. Exp. Sta., Bul. 493.—RS

——— (1962), "Motivation, Contacts, and Technological Change," *Rural Soc.*, 27: 316–326.—RS

Pierce, Rachel M. and Rowntree, Griselda (1961), "Contraceptive Methods Used by Couples Married in the Last Thirty Years," *Population Stud.*, 15: 121–159.—MS

Pierce, Truman M. (1947), *Controllable Community Characteristics Related to the Quality of Education.* Ph.D. thesis. New York: Teachers College, Columbia Univ.—E

Pimpre, D. M. (1963), *Attitude of the Farmers towards Adoption of Improved Agricultural Practices.* M.S. thesis. Nagpur, India: Col. of Agriculture.—EX

Pitzer, Ronald L. (1959), *The Influence of Social Values on the Acceptance of Vertical Integration by Broiler Growers.* M.S. thesis. Columbus: Ohio State Univ.—RS

Planning Research and Action Institute (1961), *Acceptance of Family Planning Methods as a Function of the Nature and Intensity of Contact*. Lucknow, India: PRAI.—U

Planning Research and Action Institute (1966), *Report on Family Planning Communication Action Research Project*. Lucknow, India: PRAI.—U

Plaut, Thomas F. A. (1959), "Analysis of Voting Behavior on a Fluoridation Referendum," *Pub. Opin. Qtrly.*, 23: 213–222.—MS

Ploch, Louis A. (1960), *Social and Family Characteristics of Maine Contract Broiler Growers*. Orono: Me. Agri. Exp. Sta., Bul. 596.—RS

Polgar, Steven and others (1963), "Diffusion and Farming Advice: A Test of Some Current Notions," *Social Forces*, 41: 104–111.—GS

Polson, Robert A. and Pal, Agaton P. (1955), "The Influence of Isolation on the Acceptance of Technological Changes in the Dumaguete City Trade Area, Philippines," *Silliman J.*, 2: 149–159.—RS

——— (1964), *Social Change in the Dumaguete Trade Area: Philippines 1951–1958*. Ithaca, N.Y.: Cornell Univ., Dept. of Rural Soc., Intl. Agri. Dev. Mimeo. 4.—RS

Portocarrero, Cesar A. (1966), *Empathy and Modernization in Colombia*. M.S. thesis. East Lansing: Michigan State Univ.—C

Potter, Robert, G. (1967). "Taiwan: IUD Effectiveness in the Taichung Medical Follow-Up Study," *Stud. in Family Plng.*, 18: 13–20.—MS

Potter, Robert G. and others (1966), "Social and Demographic Correlates of IUCD Effectiveness: The Taichung IUCD Medical Follow-Up Study," *Social Statistics Section Proceedings of the Amer. Stat. Assn.*—MS

Potthoff, Hilda and Rheinwald, Hans (1958), *Beratung in der Schweinmast: Untersuchungen uber ein Gezieltes Beratungs Program* (*Extension among Hog Farmers: An Investigation of the Objectives of an Extension Program*). Stuttgart-Hohenheim, Germany: Institut für Landwirtschaft-

liche Beratung an der Landwirtschaftlicken Hochschule, Bul. 80.—EX

Prasada, Rekha (1966), *A Study of the Process of Acceptance of an Improved Home Practice in One Village of Delhi State*. M.S. thesis. India: Univ. of Delhi.—AE

Preising, Paul P. (1969), "The Relationship of Staff Tenure and Administrative Succession to Structural Innovation." Paper presented at the Conf. of the Amer. Edu. Res. Assn. Los Angeles.—E

Preiss, Jack J. (1954), *Functions of Relevant Power and Authority Groups in the Evaluation of County Agent Performance*. Ph.D. thesis. East Lansing: Michigan State Univ.—GS

Prescott, Raymond B. (1922) "Law of Growth in Forecasting Demand," *J. of Amer. Stat. Assn.*, 18: 471–479.—ST

Press, Irwin (1966), "Role Ambiguity and Innovation: A Maya Case." Paper presented at the Society for App. Anthro. Madison, Wisc.—A

Presser, H. A. and Cornish, J. B. (1968), *Channels of Information and Farmers' Goals in Relation to the Adoption of Recommended Practices*. Sydney, Australia: Univ. of Melbourne, Dept. of Rural Soc., Res. Bul. 1.—RS

Presser, H. A. and Russell, H. M. (1965), "Acceptance of Research Results by Farmers," *Rev. of Mktg. and Agri. Eco.*, 33: 147–165.—RS

Proctor, Charles H. (1956), *Changing Patterns of Social Organization in a Rural Problem Area of Uruguay*. Ph.D. thesis. East Lansing: Michigan State Univ.—RS

——— (1960), "Did Demonstrations Alter the Diffusion of Hybrid Corn Practices in San Ramon?" East Lansing: Michigan State Univ., unpublished paper.—RS

Prundeanu, Julian and Zwerman, Paul J. (1958a), "An Evelution of Some Economic Factors and Farmers' Attitudes That May Influence Acceptance of Soil Conservation Practices," *J. of Farm Eco.*, 40: 903–914.—AE

——— (1958b), "Certain Characteristics of Land in Relation to Tendency of Farmers to Establish Conservation Practices," *Agronom. J.*, 27: 548–551.—AE

Purohit, Bhagwan Das (1963), *The Economic Implications of Human Ability of Finnish Book-Keeping Farms*. Helsinki, Finland: Univ. of Helsinki, Dept. of Agri. Eco.— AE

Putney, Snell and Putney, Gladys J. (1962), "Radical Innovation and Prestige," *Amer. Soc. Rev.,* 27: 548–551.—A

Qadir, Abdul S. (1966), *Adoption of Technological Change in the Rural Philippines: An Analysis of Compositional Effects*. Ph.D. thesis. Ithaca, N.Y.: Cornell Univ.—RS

Queeley, Mary and Street, David (1965), "Innovation in Public Education: The Impact of the 'Continuous Development' Approach." Chicago: Univ. of Chicago, Center for Social Organization Studies, Working Paper 45.—GS

Quesada, Gustavo M. (1965), *Contacts with Professional Services as Related with Social Characteristics in a Rural Area of the State of Rio de Janeiro, Brazil*. M.S. thesis. Madison: Univ. of Wisconsin.—RS

Quesada, Gustavo M. and Lopes, Renato S. (1967), "Contato e Caracteristicas dos Clientes do Servico de Extensão em Minas Gerais, Brazil" (Contact and Client Characteristics of the Extension Service in Minas Gerais, Brazil). Paper presented at the VII Reunion de la ALAF (Asociación Latino-Americano de Fito-Tecnia). Caracas, Venezuela.—C

Quigle, Carroll (1956), "Aboriginal Fish Poisons and the Diffusion Problem," *Amer. Anthro.,* 58: 508–525.—A

Rahim, S. A. (1961a), *The Diffusion and Adoption of Agricultural Practices: A Study in a Village in East Pakistan*. Comilla: Pakistan Acad. for Rural Dev.— C

——— (1961b), *Voluntary Group Adoption of Power Pump Irrigation in Five East Pakistan Villages*. Comilla: Pakistan Acad. for Rural Dev., Res. and Survey Bul. 12.— C.

——— (1964), *Partial and Full Adoption of Improved Practices in Comilla Cooperatives*. Comilla: Pakistan Acad. for Rural Dev., Res. and Survey Bul. 8.—C

——— (1965), *Communication and Personal Influence in an East Pakistan Village*. Comilla: Pakistan Acad. for Rural Dev., Res. and Survey Bul.—C

——— (1968), *Collective Adoption of Innovations by Village Cooperatives in Pakistan: Diffusion of Innovations in a Development System*. East Lansing: Michigan State Univ., Dept. of Communication, Tech. Rept. 5.—C

Rahudkar, Wasudeo B. (1958), "Impact of Fertilizer Extension Programme on the Minds of the Farmers and Their Reactions to Different Extension Methods," *Indian J. of Agron.,* 3: 119–136.—RS

——— (1960), "Local Leaders and the Adoption of Farm Practices," *Nagpur Agri. Col. Mag.,* 34: 1–13.—RS

——— (1961), *Testing a Culturally-Bound Model for Acceptance of Agricultural Practices*. M.A. thesis. Manhattan: Kansas State Univ.—RS

——— (1962), "Farmer Characteristics Associated with the Adoption and Diffusion of Improved Farm Practices," *Indian J. of Agri. Eco.,* 17: 82–85.—RS

——— (1963), "Communication of Farm Innovations in an Indian Community," *Indian J. of Social Work,* 23: 99–103.—RS

——— (1967), "Communication Patterns in the Acceptance of Agricultural Practices," in T. P. S. Chawdhari (Ed.), *Selected Readings on Community Development*. Hyderabad, India: Natl. Inst. of Community Dev.—EX

Rai, B. D. (1961), *Channels of Communication for Improved Farm Practices in Barela Block, M.P.* M.S. thesis. Jabalpur, India: Government Agricultural Col.—EX

Raina, B. L. and others (1967), *A Study in Family Planning Communication—Meerut District*. New Delhi, India: Central Family Planning Inst., Monograph Series 3.—U

Rajagopalan, C. and Singh, Jaspal (1967), *Adoption of Agricultural Innovations: A Sociological Study of the Indo-German Project, Mandi*. Delhi: Indian Inst. of Tech., Dept. of Humanities and Social Sciences.—S

Rajendra, C. (1966), "Adoption of Improved Farm Practices and Caste System," *J. of Ext. Edu.,* 2: 99–101.—S

Ramos, Eduardo L. (1966a), *Client-Change Agent Relationships in Three Colombian Villages*. M.A. thesis. East Lansing: Michigan State Univ.—C

Ramos, Elssy Bonilla de (1966b), *Fatalism and Modernization in Colombia*. M.A. thesis. East Lansing: Michigan State Univ.—C

Ramsey, Charles E. and others (1959), "Values and the Adoption of Practices," *Rural Soc.*, 24: 35–47.—RS

Rangarao, K. (1966), *Influence of Extension Methods on Individual Adoption Process in Relation to Selected Agricultural Practices Around Anand*. M.S. thesis. Anand, India: Inst. of Agriculture.—EX

Rangarao, K. and Patel, A. U. (1966), "Influence of Extension Methods on Individual Adoption Process in Relation to Selected Agricultural Practices around Anand, Gujarat State," *Indian J. of Ext. Edu.*, 1: 282–293.—EX

Rao, C. R. P. (1961), *Relative Influence of Socio-Economic Factors on the Acceptance of Cow-Dung Gas Plant by the Farmers When Exposed to Extension Teaching*. M.S. thesis. New Delhi: India Agri. Res. Inst.—EX

Rao, C. S. S. and Moulik, T. K. (1966), "Influence of Sources of Information on Adoption of Nitrogenous Fertilizers," *J. of Ext. Edu.*, 2: 7–16.—EX

Rao, G. Narayan (1966), "A Study on Extent of Knowledge and Adoption of Hybrid Maize Cultivation in Huzurabad Panchayat Samithi of Karimnagar District," in *Research Studies in Extension Education*. Rajendranager, India: Andhra Pradesh Agricultural Univ., Ext. Edu. Inst.—EX

Rao, Kamala G. (1966), *An Exploratory Study of IUCD Acceptors*. New Delhi, India: Central Family Planning Inst., Monograph Series 2.—P

Rao, L. Jaganmohan (1965), *An Evaluation of Intensive Agricultural District Programme in West Godavary District, A.P.* M.S. thesis. Hyderabad, India: Andhra Pradesh Agricultural Univ.—AE

Raper, Arthur, and Tappan, Pearl W. (1943), "Never Too Old to Learn New Tricks: The Canning Program in Greene County, Georgia," *Applied Anthro.* 2: 3–11.—RS

Raphael, Edna E. (1964), "Community Structure and Acceptance of Psychiatric Aid," *Amer. J. of Soc.*, 69: 340–358.—GS

Rathore, O. S. (1962), *Farmer Typology in Relation to Improved Agricultural Practices*. M.S. thesis. Udaipur, India: Rajashan Col. of Agriculture.—EX

Reddy, S. K. (1962), *A Study of Adoption of Improved Agricultural Practices as a Function of Some Socio-Economic Factors and Sources of Information*. M.S. thesis. New Delhi: Indian Agri. Res. Inst.—EX

Reddy, S. K. and Kivlin, J. E. (1968), *Adoption of High Yielding Varieties in Three Indian Villages*. Hyderabad, India: Natl. Inst. of Com. Dev. Res. Rept. 19.—C

Rehder, Robert R. (1961), *The Role of the Detail Man in the Diffusion and Adoption of an Ethical Pharmaceutical Innovation within a Single Medical Community*. Ph.D. thesis. Palo Alto, Cal.: Stanford Univ.—MR

Requena, Mariano B. (1965), "Studies of Family Planning in the Quinto Normal District of Santiago: The Use of Contraceptives," *Milbank Memorial Fund Qtrly.*, 43: 69–99.—MS

Research Division, U.S. Agency for International Development (1966), *Innovations in Ubol Changwad*. Bangkok, Thailand.—U

Rhoads, William G. and Piper, Anson C. (1963), *Use of Radiophonic Teaching in Fundamental Education*. Williamstown, Mass.: Roper Pub. Opin. Res. Center.—GE

Richardson, Gary A. (1964), *Socio-Economic Factors Associated with Trying New Grocery Products*. M.S. thesis. University Park: Pennsylvania State Univ.—AE

Rieck, Robert E. and Pulver, Glen C. (1962), *An Empirical Measure of Decision Making in Evaluating Farm and Home Development in Wisconsin*. Madison: Wisc. Agri. Exp. Sta. and Coop. Ext. Svce., Res. Bul. 238.—RS

Riley, John W. and Riley, Matilda (1940), "The Use of Various Methods of Contraception," *Amer. Soc. Rev.*, 5: 890–903.—MS

Rizwani, Abdur R. (1964), *Adoption of Recommended Farm Practices in Selected Rural Communities of New York State.* Ph.D. thesis. Ithaca, N.Y.: Cornell Univ. —RS

Roberts, Beryl J. and others (1965), "Family Planning Survey in Dacca, East Pakistan," *Demography*, 2: 74–96.—MS

Robertson, C. A. (1964), *Effectiveness of Crop Demonstrations: A Study of Wheat Demonstrations in Aligarh District.* U.P. Delhi: Univ. of Delhi, Agri. Eco. Centre, Intensive Agri. Distr. Programme Studies 1.—AE

Robertson, Thomas S. (1966a), *An Analysis of Innovative Behavior and Its Determinants.* Ph.D. thesis. Evanston, Ill.: Northwestern Univ.—MR

—— (1966b), *The Touch-Tone Study of Innovative Behavior.* Evanston, Ill.: Northwestern Univ., Mimeo Rept.—MR

—— (1967a), "Characterizing the Consumer Innovator: An Interdisciplinary Approach." Paper presented at the Rural Soc. Society. San Francisco.—MR

—— (1967b), "Determinants of Innovative Behavior." Paper presented at the Amer. Mktg. Assn. Washington, D.C.—MR

—— (1968), "The Effect of the Informal Group Upon Member Innovative Behavior." Paper presented at the Rural Soc. Society. Boston.—MR

Robertson, Thomas S. and Rossiter, John R. (1968), "Fashion Diffusion: The Interplay of Innovator and Opinion Leader Roles in College Social Systems." Los Angeles: Grad. School of Bus. Adm., Univ. of California, unpublished paper.—MR

Robinson, A. V. (1963), *An Evaluation of the Gooroy Extension Group.* Brisbane, Australia: Queensland Dept. of Primary Industries.—U

Rogers, Everett M. (1957a), *A Conceptual Variable Analysis of Technological Change.* Ph.D. thesis. Ames: Iowa State Univ.—RS

—— (1957b), "Personality Correlates of the Adoption of Technological Practices," *Rural Soc.*, 22: 267–268.—RS

—— (1958a), "A Conceptual Variable Analysis of Technological Change," *Rural Soc.*, 23: 136–145.—RS

—— (1958b), "Categorizing the Adopters of Agricultural Practices," *Rural Soc.*, 23: 345–354.—RS

—— (1958c), "Opinion Leaders in the Communication of Agricultural Technology." Paper presented at Amer. Soc. Society. Seattle, Wash.—RS

—— (1959a), "A Note on Innovators," *J. of Farm Eco.*, 41: 132–134.—RS

—— (1959b), "The Role of the Agricultural Innovator in Technological Change." Paper presented at the Amer. Soc. Society. Chicago.—RS

—— (1961a), "The Adoption Period," *Rural Soc.*, 26: 77–82.—RS

—— (1961b), *Characteristics of Agricultural Innovators and Other Adopter Categories.* Wooster: Ohio Agri. Exp. Sta., Res. Bul. 882.—RS

—— (1962), "Characteristics of Agricultural Innovators and Other Adopter Categories," in Wilbur Schramm (Ed.), *Studies of Innovation and Communication to the Public.* Palo Alto, Cal.: Stanford Univ., Inst. for Communication Res.—RS

—— (1964), "Achievement Motivation and the Adoption of Farm Innovations in Colombia." Paper presented at the Society for App. Anthro. San Juan, P.R. —C

—— (1965a), "Mass Media Exposure and Modernization among Colombian Peasants," *Pub. Opin. Qtrly.*, 29: 614–625.—C

—— (1969), "Elements in the Subculture of Traditionalism." Paper presented at the Society for App. Anthro. Mexico City.—C

—— (Forthcoming), "Mass Media and Interpersonal Communication," in Wilbur Schramm and others (Eds.), *Handbook of Communication.* Chicago: Rand-McNally.—C

Rogers, Everett M. and Beal, George M. (1958a), "The Importance of Personal Influence in the Adoption of Technological Changes," *Social Forces*, 36: 329–335.—RS

———— (1958b), *Reference Group Influences in the Adoption of Agricultural Technology.* Ames: Iowa State Univ., Dept. of Eco. and Soc., Mimeo Bul.—RS

Rogers, Everett M. and Burdge, Rabel J. (1961), *Muck Vegetable Growers: Diffusion of Innovations among Specialized Farmers.* Wooster: Ohio Agri. Exp. Sta., Res. Cir. 94.—RS

———— (1962), *Community Norms, Opinion Leadership, and Innovativeness among Truck Growers.* Wooster: Ohio Agri. Exp. Sta., Res. Bul. 912.—RS

Rogers, Everett M. and Capener, Harold R. (1960), *The County Extension Agent and His Constituents.* Wooster: Ohio Agri. Exp. Sta., Res. Bul. 858.—RS

Rogers, Everett M. and Cartano, David, G. (1963a), "Differential Self-Perceptions by Adopter Categories." Paper presented at the Rural Soc. Society. Northridge, Cal.—RS

Rogers, Everett M. and Havens, A. Eugene (1961a), *The Impact of Demonstrations on Farmers' Attitudes toward Fertilizer.* Wooster: Ohio Agri. Exp. Sta., Res. Bul. 891—RS .

———— (1961b), *Extension Contact of Ohio Farm Housewives.* Wooster: Ohio Agri. Exp. Sta., Res. Bul. 890.—RS

———— (1962a), "Predicting Innovativeness." *Soc. Inquiry,* 32: 34–42.—RS

Rogers, Everett M. and Herzog, William (1966), "Functional Literacy among Colombian Peasants," *Eco. Dev. and Cul. Chg.,* 14: 190–203.—C

Rogers, Everett M. and Jain, Nemi C. (1968), "Needed Research on Diffusion within Educational Organizations." Paper presented at the Natl. Conf. on Diffusion of Edu. Ideas. East Lansing, Mich.—C

Rogers, Everett M. and Leuthold, Frank O. (1962), *Demonstrators and the Diffusion of Fertilizer Practices.* Wooster: Ohio Agri. Exp. Sta., Res. Bul. 908.—RS

Rogers, Everett M. and Meynen, Wicky L. (1965), "Communication Sources for 2,4-D Weed Spray among Colombian Peasants," *Rural Soc.,* 30: 213–219.—C

Rogers, Everett M. and Niehoff, Arthur H. (1967), "Diffusion of Agricultural Inno-

vations in Eastern Nigeria: Innovation Characteristics, Motivation, and Communication Bottlenecks." Paper presented at the Conf. on the Nigerian Economy and CSNRD. East Lansing, Mich.—C

Rogers, Everett M. and others (1968), *Diffusion of Educational Innovations in the Government Secondary Schools of Thailand.* East Lansing: Michigan State Univ., Inst. for Intl. Stud. in Edu., Mimeo Rept.—C

Rogers, Everett M. and Pitzer, R. L. (1960), *The Adoption of Irrigation by Ohio Farmers.* Wooster: Ohio Agri. Exp. Sta., Res. Bul. 851.—RS

Rogers, Everett M. and Ramos, E. B. de (1965), "Prediction of the Adoption of Innovations: A Progress Report." Paper presented at the Rural Soc. Society. Chicago.—RS

Rogers, Everett M. and Rogers, L. Edna (1961), "A Methodological Analysis of Adoption Scales," *Rural Soc.,* 26: 325–336.—RS

Rogers, Everett M. with Svenning, Lynne (1969), *Modernization among Peasants: The Impact of Communication.* New York: Holt, Rinehart and Winston.—C

Rogers, Everett M. and van Es, Johannes C. (1964), *Opinion Leadership in Traditional and Modern Colombian Peasant Communities.* East Lansing: Michigan State Univ., Dept. of Communication, Diffusion of Innovations Res. Rept. 2.—RS

Rogers, Everett M. and Yost, M. D. (1960), *Communication Behavior of County Extension Agents.* Wooster: Ohio Agri. Exp. Sta., Res. Bul. 850.—RS

———— (Forthcoming), *Diffusion of Innovations in Brazil, Nigeria, and India.* East Lansing: Michigan State Univ., Dept. of Communication, Diffusion of Innovations Res. Rept. 24.—C

Rohrer, Wayne C. (1955), "On Clienteles of the Agricultural Extension Service," *Rural Soc.,* 20: 299–303.—RS

Rohwer, Robert, A. (1949). "How New Practices Spread," *Iowa Farm Science,* 4: 13–14.—RS

Rollins, Sidney P. and Charters, W. W., Jr. (1965), "The Diffusion of Information within Secondary School Staffs," *J. of Social Psych.*, 65: 167–178.—GS

Rosecrance, R. N. (1964), *The Dispersion of Nuclear Weapons.* New York: Columbia Univ. Press.—PA

Rosenthal, Donald B. and Crain, Robert L. (1965), "Executive Leadership and Community Innovation: The Fluoridation Experience," *Urban Affairs Qtrly.*, 1: 39–57.—MS

——— (1968), "Structure and Values in Local Political Systems: The Case of Fluoridation Decisions," in Terry N. Clark (Ed.), *Community Structure and Decision-Making: Comparative Analyses.* San Francisco: Chandler.—MS.

Rosner, Martin M. (1968), "Economic Determinants of Organizational Innovation," *Adm. Science Qtrly.*, 12: 614–625.—MR

Ross, Donald H. (1952), "Rate of Diffusion for Driver Education," *Safety Edu.*, 32: 16–32.—E

——— (1955), "Measuring Institutional Quality of School Systems," *Teachers College Rec.*, 57: 172–177.—E

Ross, John (1966a), "Predicting the Adoption of Family Planning," *Stud. in Family Plng.*, 9: 8–12.—MS

——— (1966b), "United States: The Chicago Fertility Control Studies," *Stud. in Family Plng.*, 15: 1–8.—MS

Ross, John and Bang, Sook (1965), "The Aid Computer Program Used to Predict Adoption of Family Planning in Koyang." New York: Population Council, unpublished paper.—MS

Rothemich, Vincent J. (1953), *Communication in Educational Change.* D.Ed. thesis. New York: Teachers College, Columbia Univ.—E

Rowntree, Griselda and Pierce, Rachel M. (1961), "Birth Control in Britain," *Population Stud.*, 15: 3–31.—U

Roy, Prodipto (1967), *The Impact of Communications on Rural Development in India.* Hyderabad, India: Natl. Inst. of Community Dev., Mimeo Rept.—RS

Roy, Prodipto and others (1968a) *Patterns of Agricultural Diffusion in Rural India.*

Hyderabad, India: Natl. Inst. of Community Dev., Res. Rept. 12.—C

Roy, Prodipto and others (1968b), *The Impact of Communication on Rural Development: An Investigation in India and Costa Rica.* Hyderabad, India: Natl. Inst. of Community Dev. and Paris: UNESCO.—C

Roy, R. N. (1959), *Study of the Causes of Success and Failure of Improved Farm Practices in an East Bihar Village.* M.S. thesis. Bhagalpur, India: Agricultural College Sabour.—EX

Russell, Hamish M. (1964), *A Survey of the Lands Department Rabbit Eradication Campaign in the Mallee.* M. Agr. Sc. thesis. Melbourne, Australia: Univ. of Melbourne.—RS

Ryan, Bryce (1948), "A Study in Technological Diffusion," *Rural Soc.*, 13: 273–285.—RS

Ryan, Bryce and Gross, Neal C. (1943), "The Diffusion of Hybrid Seed Corn in Two Iowa Communities," *Rural Soc.*, 8: 15–24.—RS.

——— (1950), *Acceptance and Diffusion of Hybrid Corn Seed in Two Iowa Communities.* Ames: Iowa Agri. Exp. Sta., Res. Bul. 372.—RS

Sabri, Medhat M. (1966), *Symbolic Adoption of a New Innovation.* M.S. thesis. Ames: Iowa State Univ.—RS

Sagi, Philip C. and others (1961), "Contraceptive Effectiveness as a Function of Desired Family Size," *Population Stud.*, 15: 291–296.—MS

Saigaonkar, P. B. (1967), *Relationship of Certain Factors with the Success of Gramsevaks in Kaira Districts of Gujarat State.* Anand, India: Sardar Patel Univ.—EX

Sainz, L. R. Collazo and Vargas, Beatriz R. De (1966), *Recomendaciones Adoptadas por los Agricultores de Puerto Rico para Mejorar el Cultivo del Tabaco (Recommended Practices for Puerto Rican Farmers for Improving Tobacco-Growing).* Rio Piedras: Univ. de Puerto Rico, Estacion Exp. Agri.—RS

Salcedo, Rodolfo N. (1968), *A Communication Model of Modernization.* Ph.D. thesis. East Lansing: Michigan State Univ. Dept. of Communication.—C

Salisbury, R. F. (1962), *From Stone to Steel*. New York: Melbourne Univ. Press.—A

Sallan, Vinita (1966), *A Study of the Opinions and Characteristics of 'Discontinuers' of a Selected Family Planning Programme in a Selected Village of Delhi State*. M.S. thesis. Delhi, India: Univ of Delhi.—EX

Sandoval, Rodrigo P. (1966), *El Caso de Candelaria, Valle: La Estructura Social y el Cambio en la Tecnologia Agricola (The Case of Candelaria, Valle: Social Structure and Change in Agricultural Technology)*. Bogotá: Universidad Nacional de Colombia, Facultad de Sociología 21.—RS

Sandria, Y. D. (1962), *Factors Affecting Adoption of Improved Practices in M.P. State*. M.S. thesis. Jabalpur, India: Government Agricultural Col.—EX

Sangle, G. K. (1962), *Knowledge and Acceptance of Recommended Farm Practices among Small Cultivators*. M.S. thesis. Nagpur, India: Col. of Agriculture.—EX

Sapolsky, Harvey M. (1967), "Organizational Structure and Innovation," *J. of Bus.*, 40: 497–510.—PA

Sargent, Harold R. (1965), *A Test of Motivational Appeals Judged Effective by Chief School Administrators to Include Teachers' Acceptance of Educational Innovation*. D.Ed. thesis. University Park: Pennsylvania State Univ.—E

Sarupria, Shanti Lal (1964), "Attitudes towards Family Planning in a Small Urban Community," *Indian J. of Social Work*, 25: 79–87.—GE

Sasaki, Tom T. (1956), "Sociocultural Problems in Introducing New Technology on a Navaho Irrigation Project," *Rural Sociology*, 21: 307–310.—A

Satyanarayana, C. (1967), "A Study of the Sources of Information in the Different Stages of Adoption of Improved Animal Husbandry Practices in Key Village Block Karimnagar, Andhra Pradesh," in *Research Studies in Extension Education*. Rajendranagar, India: Andhra Pradesh Agricultural Univ.—EX

Satyanarayana, Munaganuri (1966), "An Analysis of Channels of Communication in the Adoption of Some Improved Agricultural Practices in an Andhra Village," in *Research Studies in Extension Education*. Rajendranagar, India: Andhra Pradesh Agricultural Univ., Ext. Edu. Inst.—EX

Saunders, Lyle and Samora, Julian (1955), "A Medical Care Program in a Colorado County," in Benjamin D. Paul (Ed.), *Health, Culture and Community*. New York: Russell Sage Fdn.—A

Savale, R. S. (1966), "Technological Change in Agriculture: Study of Sources of Its Diffusion, Efficacy of these Sources and the Economic Factors Affecting the Adoption of Improved Practices," *Indian J. of Agri. Eco.*, 21: 199–208.—AE

Sawhney, M. M. (1962), *Diffusion and Acceptance of Innovations in Agriculture with Special Emphasis of the Role of Rural Leaders and Rural Youth*. Ph.D. thesis. New Delhi: Indian Agricultural Res. Inst.—EX

——— (1966), "Farm Practice Adoption and the Use of Information Sources and Media in a Rural Community in India." Paper presented at the Rural Soc. Society. Miami Beach, Fla.—RS

Saxena, Anant P. (1963), *Communication of Agricultural Information in Mirar Block*. M.S. thesis. Ujjain, India: Vikram Univ.—EX

——— (1968), *System Effects on Innovativeness among Indian Farmers*. Ph.D. thesis. East Lansing: Michigan State Univ., Dept. of Communication.—C

Saxena, A. P. and Warlow, G. L. (1966), *An Analysis of the Effects of Three Television Programs on Selected Ontario Farm Audiences*. Guelph: Ont. Agri. Exp. Sta., Rept. 13.—EX

Scantland, Lois and others (1952), *A Square Look at Spokane County, Washington*. Pullman: Wash. Agri. Ext. Bul. 463.—RS

Schachter, Stanley and Burkick, Harvey (1955), "A Field Experiment on Rumor Transmission and Distortion," *J. of Abn. and Soc. Psycho.*, 50: 363–371.—P

Schindler, Raymond, A. (1962), *Predicting the Adoption or Rejection of Rural Zoning by Township in Ohio*. M.S. thesis. Columbus: Ohio State Univ.—RS

Schwieder, Elmer W. Jr., (1966), *Social Psychological Factors Related to Adoption of Public Fallout Shelters*. Ph.D. thesis. Ames: Iowa State Univ.—RS

Sekhon, G. S. (1968), *Differentials in Perceptions of Attributes of Innovations by Professional Advocates and Their Clientele*. Ph.D. thesis. University Park: Pennsylvania State Univ.—RS

Sen, Lalit K. (1969), *Opinion Leadership in India*. East Lansing: Michigan State Univ. Dept. of Communication, Diffusion of Innovations Res. Rept. 22.—C

Sen, Lalit K. and Roy, Prodipto (1966), *Awareness of Community Development in Village India*. Hyderabad, India: Natl. Inst. of Com. Dev., Prelim. Rept.—RS

Sengupta, T. (1968), "Opinion Leaders in Rural Communities," *Man in India*, 48: 159–166.—RS

Sepulveda, Orlando and Carter, Roy E., Jr. (1964), "La Efectividad de las Comunicaciones de Masas en un Programa de Salud: Una Investigación Evaluativa en Santiago de Chile" (The Effectiveness of Mass Communication in a Health Program: An Investigation in Santiago, Chile). Paper presented to the Seventh Latin-American Cong. of Soc. Bogotá, Colombia.—C

Sharma, S. K. and Potti, V. S. S. (1966), "Differential Adoption of Improved Farm Practices in Relation to Reference Group Influence," *J. of Ext. Edu.*, 2: 51–58.—EX

Sharp, Lauriston (1952), "Steel Axes for Stone Age Australians," in Edward H. Spicer (Ed.), *Human Problems in Technological Change*. New York: Russell Sage Fdn.—A

Shaw, John G., Jr. (1954), *The Relationship of Selected Ecological Variables to Leaflet Message Response*. Ph.D. thesis. Seattle: Univ. of Washington.—GS

Sheatsley, Paul B. and Feldman, Jacob J. (1964), "The Assassination of President Kennedy: A Preliminary Report on Public Reactions and Behavior," *Pub. Opin. Qtrly.*, 28: 189–215.—C

Sheppard, David (1960b), *A Survey among Grassland Farmers*. London: Central Office of Information, Social Survey 274.—P

——— (1961), "Farmers' Reasons for Not Adopting Controversial Techniques in Grassland Farming," *J. of the British Grassland Society*, 16: 6–13.—P

——— (1963), "The Importance of 'Other Farmers,'" *Sociologia Ruralis*, 3: 1–15.—P

Sheth, Naginlal S. (1966), *Formal and Informal Social Participation as Related to Diffusion of Information and Adoption of Farm Practices in a Village in India*. Ph.D. thesis. Columbia: Univ. of Missouri.—RS

Sheth, Jagdish N. (1968), "Perceived Risk and Diffusion of Innovations," in Johan Arndt (Ed.), *Insights into Consumer Behavior*. Boston: Allyn and Bacon.—MR

Shibutani, Tamotsu (1966), *Improvised News: A Sociological Study of Rumor*. New York: Bobbs-Merrill.—P

Sicinski, Andrezej (1964), "'Expert—Innovator—Adviser': Certain Aspects of the Differentiation of Roles in a Factory," *Polish Soc. Bul.*, 9–14: 54–66.—GS

Silk, Alvin J. (1966a), "Monomorphic Versus Polymorphic Opinion Leaders: A Study of Dental Products." Paper presented at the Conf. of the Pacific Chp., Amer. Assn. for Pub. Opin. Res.—MR

——— (1966b), "Overlap among Self-Designated Opinion Leaders: A Study of Selected Dental Products and Services," *J. of Mktg. Res.*, 3: 255–259.—MR

Sill, Maurice L. (1958), *Personal, Situational, and Communicational Factors Associated with the Farm Practice Adoption Process*. Ph.D. thesis. University Park: Pennsylvania State Univ.—RS

Sills, David L. and Gill, Rafael E. (1959), "Young Adults' Use of the Salk Vaccine," *Social Probs.*, 6: 248–253.—MS

Silverberg, James M. (1964), "Something New under the Sun: Applied Ethnographic Research on the Social Use of Solar Energy Devices in Rural Areas." Paper presented to the Seventh Intl. Cong. of Anthro. Ethno. Sciences. Moscow.—A

Silverman, Leslie J. and Bailey, Wilfred C. (1961), *Trends in the Adoption of Recommended Farm Practices*. State College: Miss. Agri. Exp. Sta., Bul. 617.—RS

Simon, Julian L. and Golembo, Leslie (1967), "The Spread of a Cost-Free Business Innovation," *J. of Bus.*, 40: 385–388.—MR

Singh, Avtar, (1964), "Reputational Measure of Leadership: A Study of Two Indian Villages." State College: Mississippi State Univ.—RS

——— (1967a), "Action Measure of Leadership: A Study of Two Indian Villages." Paper presented at the Rural Soc. Society. San Francisco.—RS

Singh, B. N. and Jha, P. N. (1965), "Utilization of Sources of Farm Information in Relation to Adoption of Improved Agricultural Practices," *Indian J. of Ext. Edu.*, 1: 33–42.—EX

Singh, Gurcharn (1965), *The Differential Characteristics of Early and Late Adopters of New Farm Practices, Punjab State, India.* Ph.D. thesis. Ithaca. N.Y.: Cornell Univ.—RS

Singh, K. K. and Pardasani, H. B. (1967), "Effectiveness of Short-Duration Contacts in Popularising Family Planning in Rural India," *Behav. Sci. and Com. Dev.*, 1: 135–149.—U

Singh, K. N. and Akhouri, M. M. P. (1966), "Relationship Between Farmers' Background and Knowledge Gained Through Different Extension Teaching Methods: An Experimental Evidence," *J. of Ext. Edu.*, 2: 22–34.—EX

Singh, Raghubar (1967b), *Adoption of Nitrogenous Fertilizers as Related to Selected Factors and Use of Information Sources in Adoption Process.* Ph.D. thesis. Ithaca, N.Y.: Cornell Univ.—EX

Singh, Ram N. (1966a), *Characteristics of Farm Innovations Associated with the Rate of Adoption.* Guelph: Ont. Agri. Ext. Edu. Rept., 14.—EX

Singh, Ranjit (1966b), "A Correlational Study of the Levels of Aspiration with the Acceptance of Crop Production Plans by the Cultivators," in A. N. Shukla and I. S. Grewal, *Summaries of Extension by Post-Graduate Students.* Ludhiana, India: Punjab Agricultural Univ.—EX

Singh, Rudra D. (1952), "The Village Level: An Introduction of Green Manuring in Rural India," in Edward H. Spicer (Ed.), *Human Problems and Technological Change.* New York: Russell Sage Fdn.—A

Singh, S. K. (1962), *Communication and Diffusion Process Contributing to the Adoption of Improved Farm Practices.* M.S. thesis. Bhagalpur, India: Agricultural College Sabour.—EX

Singh, S. N. and Beal, George M. (1967), "Value Orientations and Adoption Behavior of Indian Cultivators." Paper presented at the Rural Soc. Society. San Francisco.—RS

Singh, S. N. and others (1965), "A Study of Social Values in Relation to Farming," *Indian J. of Ext. Edu.*, 1: 114–118.—RS

Singh, S. N. and Reddy, S. K. (1965a), "Adoption of Improved Farm Practices as Related to Sources of Information," *Indian J. of Agron.*, 10: 100–107.—RS

——— (1965b), "Adoption of Improved Agricultural Practices by Farmers," *Indian J. of Social Work*, 26: 263–269.—RS

Singh, Tej P. (1966c), *The Effect of Aspirational Level on Adoption of Recommended Practices in Rice Cultivation.* M.A. thesis. Manila: Univ. of the Philippines.—EX

Singh, Y. P. and Pareek, Udai (1968), "A Paradigm of Sequential Adoption," *Indian Edu. Rev.*, 3: 89–114.—EX

Singhal, Prem L. (1966), *Diffusion of Technology from Japan to India.* M.A. thesis. Washington, D.C.: Howard Univ.—GE

Sinha, N. K. (1963), *The Adoption Process as Related to Some Socio-Personal Factors.* Ph.D. thesis. New Delhi, India: Agricultural Res. Inst.—EX

Sinha, N. K. and Yadav, D. P. (1964a), *Summary of a Study of the Preparation and Implementation of Intensive and Extensive Farm Plans for Kharif.* Ludhiana, India: State Govt. of Punjab, Operational Res. Rept. 2.—RS

——— (1964b), *Summary of Study on Crop Demonstrations in Ludhiana District.* Ludhiana, India: State Govt. of Punjab, Operational Res. Rept. 3.—RS

Sinha, N. K. and Yadav, D. P. (1964c), *An Evaluation Study Showing Utilization of Medium Term Taccavi Loans for Minor Irrigation to Cultivators in Ludhiana District.* Ludhiana, India: State Govt. of Punjab, Operational Res. Rept. 5.—RS

—— (1964d), *A Study to Determine Cultivators' Awareness and Knowledge of Package Programme and the Nature of Contacts with the Extension Staff, Ludhiana District.* Ludhiana, India: State Govt. of Punjab, Operational Res. Rept. 6.—RS

—— (1964e), *Cultivators' Attitudes toward Adoption of Chemical Fertilizers, Package Programme, Ludhiana.* Ludhiana, India: State Govt. of Punjab, Operational Res. Rept. 7.—RS

—— (1964f), *Attitudes of Farmers towards the Adoption of Insecticides and Pesticides in I.A.D.P., Ludhiana.* Ludhiana, India: State Govt. of Punjab, Operational Res. Rept. 8.—RS

—— (1964g), *Acceptance of Improved Agricultural Implements by Cultivators in I.A.D.P., Ludhiana.* Ludhiana, India: State Govt. of Punjab, Operational Res. Rept. 9.—RS

Sinha, P. R. R. and Parshad, Rajinder (1966), "Sources of Information as Related to Adoption Process of Some Improved Farm Practices," *J. of Ext. Edu.* 2: 86–91.—EX

Sizer, Leonard, M. and Porter, Ward F. (1960), *The Relation of Knowledge to Adoption of Recommended Practices.* Morgantown: W. Va. Agri. Exp. Sta., Bul. 446.—RS

Skogsberg, Alfred H. (1950), *Administrative Operational Patterns.* New York: Teachers College, Columbia Univ., Bureau of Pub.—E

Slocum, Walter L. and others (1958), *Extension Contacts, Selected Characteristics, Practices, and Attitudes of Washington Farm Families.* Pullman: Wash. Agri. Exp. Sta., Bul. 584.—RS

Smith, David H. (1966), "A Psychological Model of Individual Participation in Formal Voluntary Organizations: Application to Some Chilean Data," *Amer. J. of Soc.*, 72: 249–266.—GS

Smith, H. T. E. (1958), "The Communication of Ideas to Farmers," *Agricultural Progress*, 33: 51–57.—EX

Smith, Marian and Sheppard, D. (1959), "A Study of the Dissemination of Information about a New Technique in Dairy Farming," *Farm Economist*, 9: 133–147.—P

Smith, Sampson G. (1951), *Pupil Factors Related to the Quality of Education.* D.Ed. thesis. New York: Teachers College, Columbia Univ.—E

Smith, Stanley V. (1954), *Quality of Education Related to Certain Social and Administrative Characteristics of Well-Financed Rural School Districts.* Ph.D. thesis. New York: Columbia Univ.—E

Smith, William C. (1964), "Hens That Laid Golden Eggs," in Gove Hambridge (Ed.), *Dynamics of Development.* New York: Praeger.—U

Sohoni, A. W. (1963), *Characteristics of Farm Practices Associated with the Rate of Adoption.* M.S. thesis. Nagpur, India: Col. of Agriculture.—EX

Sokol, Robert (1959), "The Television Behavior and Attitudes of Influentials." Paper presented at Amer. Soc. Assn. Chicago.—GS

Sorokin, Pitirim A. (1959), *Social and Cultural Mobility.* New York: Free Press.—GS

South, Donald R. (1968), *A Theoretical and Empirical Analysis of the Adoption-Diffusion of Social Change.* Ph.D. thesis. Baton Rouge: Louisiana State Univ.—RS

South, Donald R. and others (1965), *Factors Related to the Adoption of Woodland Management Practices.* Baton Rouge: La. Agri. Exp. Sta., Bul. 603.—RS

Spaulding, Irving A. (1955), *Farm Operator Time-Space Orientations and the Adoption of Recommended Farming Practices.* Kingston: R.I. Agri. Exp. Sta., Bul. 330.—RS

—— (1960), *Motivation for Communicative Behavior.* Kingston: R.I. Agri. Exp. Sta., Bul. 354.—RS

Spector, Paul and others (1963), *Communication and Motivation in Community Development: An Experiment.* Washington, D.C.: Inst. for Intl. Services.—P

Spector, Paul and others (1964), "The Role of Mass Communications Media in the Adoption of Innovation." Paper presented at the Ninth Cong. of the Interamer. Society of Psych. Miami Beach, Fla.—P

Spencer, George E. (1958), *Value-Orientations and the Adoption of Farm Practices.* Ph.D. thesis. Ithaca, N.Y.: Cornell Univ.—RS

Spicer, Edward H. (1952), "Sheepmen and Technicians: A Program of Soil Conservation on the Navaho Indian Reservation," in E. H. Spicer (Ed.), *Human Problems in Technological Change.* New York: Russell Sage Fdn.—A.

Spitzer, Stephen P. and Denzin, Norman K. (1965), "Levels of Knowledge in an Emergent Crisis," *Social Forces,* 44: 234–237.—GS

Spitzer, Stephen P. and Spitzer, Nancy S. (1965), "Diffusion of News of Kennedy and Oswald Deaths," in Bradley S. Greenberg and Edwin B. Parker (Eds.), *The Kennedy Assassination and the American Public: Social Communication in Crisis.* Stanford, Cal.: Stanford Univ. Press.—C

Splete, A. P. J. (1968), *Substance and Processes of Innovation in an Office of the Vice President for Academic Affairs.* Ph.D. thesis. Syracuse, N.Y.: Syracuse Univ.—E

Sproles, George B. (1967), *A Profile of the Analysis of the Durable Press Clothing Information Communication.* M.S. thesis. Lafayette, Ind.: Purdue Univ.—MR

Sprunger, Benjamin E. (1968), *An Investigation of the Characteristics which Differentiate Innovative from Non-Innovative College Student Personnel Programs.* Ph.D. thesis. East Lansing: Michigan State Univ.—E

Stanfield, J. David (1968), *Interpersonal Trust and Modernization in Rural Brazil.* East Lansing: Michigan State Univ., Dept. of Communication, Diffusion of Innovations Tech. Rept. 9.—C

—— (1969), "Brazilian Farmers' Integration with Their Community and with Life Beyond the Community." Paper presented to the Society for App. Anthro. Mexico City.—C

Stapel, Jan (1960), "The Consumption Pioneers." Paper presented to the European Society for Opin. Surveys and Mkt. Res. (ESOMAR) Conf.—MR

Star, Shirley A. and Hughes, Helen MacGill (1950), "Report on an Educational Campaign: The Cincinnati Plan for the United Nations," *Amer. J. of Soc.,* 55: 389–400.—GS

Starch, Daniel (1958), "Do Ad Readers Buy the Product?" *Harvard Bus. Rev.,* 36: 49–58.—MR

Stewart, John B. (1964), *Repetitive Advertising in Newspapers: A Story of Two New Products.* Boston: Harvard Univ.—MR

Stickley, S. Thomas (1964), "Socio-Economic Correlates of Levels of Living among Farmers in Three Colombian Neighborhoods." M.S. thesis. Columbus: Ohio State Univ.—RS

Stickley, S. Thomas and others (1967), "Levels of Living among Farmers in Three Colombian Neighborhoods." Columbus: Ohio State Univ., Dept. of Agri. Eco. and Rural Soc.—AE

Stock, Garfield R. and Johnson, Donald E. (1966), "Adoption Correlates as Predictors of Organizational Membership over Time." Paper presented at the Rural Soc. Society. Miami Beach, Fla.—RS

Stoeckel, John E. and Choudhury, Moqbul A. (1968), "The Impact of Family Planning on Fertility in Comilla," *Pakistan J. of Family Plng.,* 2: 13–20.—MS

Stone, John T. (1952), *How County Agricultural Agents Teach.* East Lansing, Mich.: Agri. Ext. Svce., Mimeo Bul.—RS

Straits, Bruce C. (1966), "The Discontinuation of Cigarette Smoking: A Multiple Discriminant Analysis." Paper presented at the Amer. Soc. Assn. Miami Beach, Fla.—GS

Strassman, W. Paul (1959), *Risk and Technological Innovation: American Manufacturing Methods during the Nineteenth Century.* Ithaca, N.Y.: Cornell Univ. Press.—GE

Straus, Murray A. (1953), "Cultural Factors in the Functioning of Agricultural Extension in Ceylon," *Rural Soc.,* 18: 249–256.—RS

Straus, Murray A. (1958), *Short Term Effects of Farm and Home Development in Wisconsin*. Madison: Univ. of Wisconsin, Dept. of Rural Soc., Rept. 3.—RS

——— (1960), "Family Role Differentiation and Technological Change in Farming," *Rural Soc.*, 25: 219–228.—RS

Straus, Murray A. and Estep, Allen J. (1959), *Education for Technological Change among Wisconsin Farmers*. Madison: Wisc. Agri. Exp. Sta., Tech. Bul. 214.—RS

Stuby, Richard (1965), *Intermittent Adoption: A Modified Form of Farm Practice Adoption*.

Stycos, J. Mayone (1955), "Birth Control Clinics in Crowded Puerto Rico," in Benjamin D. Paul (Ed.), *Health, Culture, and Community*. New York: Russell Sage Fdn.—A

——— (1965a), "Contraception and Catholicism in Latin America." Ithaca, N.Y.: Cornell Univ., Intl. Population Program, unpublished paper.—MS

——— (1965b), "The Potential Role of Turkish Village Opinion Leaders in a Program of Family Planning," *Pub. Opin. Qtrly.* 29: 120–130.—MS

Suchman, Edward A. (1966), *An Experimental Study of Accident Prevention among Sugar Cane Workers in Puerto Rico*. San Juan, P.R.: Dept. of Health, Office of Res., Mimeo Rept.—RS

Suchman, Edward A. and others (1967), "An Experiment in Innovation among Sugar Cane Cutters in Puerto Rico," *Human Organization*, 26: 214–221.—MS

Summers, John O. (1968), *The Identity of Women's Clothing Fashion Transmitter*. Ph.D. thesis. Lafayette, Ind.: Purdue Univ.—MR

Sutherland, Alistair (1959), "The Diffusion of an Innovation in Cotton Spinning," *J. of Ind. Eco.*, 7: 118–135.—1

Suttles, Wayne (1951), "The Early Diffusion of the Potato among the Coast Salish," *Southwestern J. of Anthro.*, 7: 272–288.—A

Swiney, Kemp L. (1967), *Factors Influencing Members' Selection of 4-H Club Projects, with Special Emphasis on Poultry Projects*. Ph.D. thesis. East Lansing: Michigan State Univ.—O

Sycip, Felicidad (1964), "Factors Related to Acceptance or Rejection of Innovations," in Socorro C. Espiritu and Chester L. Hunt (Eds.), *Social Foundations of Community Development: Readings on the Philippines*. Manila: R. M. Garcia Pub. House. —U

Synergie-Recherche et Organization Commerciales (1961), *Etude Sur le Marche de la Vulgarization*, Volumes I, II. Paris.—MR

Szwengrub, Lili M. (1967), "Diffusion des Nouveautes Techniques dans l'Agriculture" (Diffusion of Technical Innovations in Agriculture). Warsaw: Polish Acad. of Sciences, Inst. of Philo. and Soc.—RS

Tajima, Shigeo (1959), *An Evaluation of Agricultural Extension in Hokkaido*. Obihiro, Japan: Obihiro Zootechnical Univ. —RS

Takes, Charles A. P. (1963b), *Socio-Economic Factors Affecting the Productivity of Agriculture in Okigwi Division (Eastern Division)*. Ibadan: Nigerian Inst. of Social and Eco. Res.—RS

Takeshita, John Y. (1964), "Taiwan: The Taichung Program of Prepregnancy Health," *Stud. in Family Plng.*, 4: 10–12.— MS

——— (1966), "Lessons Learned from Family Planning Studies in Taiwan and Korea," in Bernard Berelson and others (Ed.), *Family Planning and Population Programs*. Chicago: Univ. of Chicago Press.—MS

Takeshita, John Y. and others (1964), "A Study of the Effectiveness of the Prepregnancy Health Program in Taiwan," *Eugenics Qtrly.*, 2: 222–233.—MS

Taylor, Charles H. (1955), *The Relationships of Bonded Indebtedness Measure of the Quality of Education*. D.Ed. thesis. New York: Teachers College, Columbia Univ. —E

Taylor, Robert B. (1966), "Conservative Factors in the Changing Culture of a Zapotec Town," *Human Organization*, 25: 116–121.—A

Thomas, A. Anne Hill (1966), *The Role of Mass Media among Voters and Non-Voters in a Fluoridation Election*. M.A. thesis. Eugene: Univ. of Oregon.—U

Thorat, Sudhakar S. (1966), *Certain Social Factors Associated with the Adoption of Recommended Agricultural Practices by Rural Local Leaders and Ordinary Farmers in India*. Ph.D. thesis. East Lansing: Michigan State Univ.—RS

———(1968), "Some Salient Characteristics of Sarpanchas and the Success or Failure of Agricultural Innovations in India's Villages," *Behav. Sci. and Com. Dev.*, 2: 26–37.—C

Thorat, Sudhakar S. and Fliegel, Frederick C. (1968), "Some Aspects of Adoption of Health and Family Planning Practices in India," *Behav. Sci. and Com. Dev.*, 2: 1–20.—RS

Tiedeman, Clifford E. and Van Doren, Carlton S. (1964), *The Diffusion of Hybrid Seed Corn in Iowa: A Spatial Simulation Model*. East Lansing: Michigan State Univ. Inst. for Com. Dev. and Svces., Tech. Bul. 13–44.—G

Torres, Augusto and Spector, Paul (1964), *Diffusion of Information Through Radio and Supporting Media: Report of Follow-up Interviews*. Washington, D.C.: Inst. for Int. Svces., Phase II Final Rept.—P

Toussaint, W. D. and Stone, P. S. (1960), "Evaluating a Farm Machine Prior to its Introduction," *J. of Farm Eco.*, 42: 241–251.—AE

Troldahl, Verling C. (1963a), *Mediated Communication and Personal Influence: A Field Experiment*. Ph.D. thesis. Minneapolis: Univ. of Minnesota.—C

——— (1963b), *The Communication of Horticultural Information and Influence in a Suburban Community*. Boston: Boston Univ., Com. Res. Center, Rept. 10.—C

——— (1963c), "A Field Experiment Test of a Modified 'Two-Step Flow of Communication' Model." Paper presented at the Assn. for Edu. in Journalism. Lincoln, Nebr.—C

——— (1964), "Communicating to the Suburbs," *J. of Coop. Ext.*, 2: 82–88.—C

——— (1967), "A Field Test of a Modified Two-Step Flow of Communication Model," *Publ. Opin. Qtrly.*, 30: 609–623.—C

Troldahl, Verling C. and others (1965),

"Public Affairs Information-Seeking from Expert Institutionalized Sources," *Journalism Qtrly.*, 42: 403–412.—C

Troldahl, Verling C. and Van Dam, Robert (1965), "Face-to-Face Communication about Major Topics in the News," *Pub. Opin. Qtrly.*, 29: 626–632.—C

Tucker, Carlos F. (1961), *Prediction of Rate of Adoption from Characteristics of Farm Innovations*. M.S. thesis. Columbus: Ohio State Univ.—RS

Tully, Joan (1965), "Towards a Sociological Theory for Extension." Paper presented at the Soc. Assn. of Australia and New Zealand. Melbourne, Australia.—EX

Tully, Joan and others (1964), "Factors in Decision-Making in Farming Problems," *Human Relations*, 17: 295–320.—EX

UNESCO (1953), "Three Experiments in the Spreading of Knowledge about the Universal Declaration of Human Rights: Cambridge, Grenoble, Uppsala," *Intl. Soc. Sci. Bul.*, 5: 583–602.—U

Uno, Yoshiyasu and Aoike, Shinichi (1967a), "A Field Study of Diffusion Process of Innovations in a Japanese Rural Community, Part I: The Case of Diffusion Process of Chrysanthemum Cultivation," *Japanese Annals of Social Psych.*, 8: 205–218.—RS

——— (1967b), "A Field Study of Diffusion Process of Innovations in a Japanese Rural Community, Part II: The Case of Diffusion Process of Gentian Cultivation." *Tetugaku Keio Univ.* 50: 229–258—RS

Vainio-Mattila, Ilkka (1969), *Extension Services as a Source of Farming Information*. Helsinki, Finland: Pellervo Society, Mktg. Res. Inst., Pub. 11.—MR

Van den Ban, Anne W. (1953), "Wie Worden Door de Landbouwvoorlichtingsdienst Bereikt?" (Who Are Influenced by the Agricultural Extension Service?"), *Landbouwkundig Tijdschrift*, 65: 317–327.—RS

——— (1956), *Certain Features and Characteristics of Progressive Farmers*, I. Wageningen, Netherlands: Agricultural Univ., Afdeling Sociologie en Sociografie van de Landbouwhogeschool, Bul. 5.—RS

Van den Ban, Anne W. (1957), "Some Characteristics of Progressive Farmers in the Netherlands," *Rural Soc.,* 22: 205–212. —RS

——— (1958a), *Certain Features and Characteristics of Progressive Farmers, II.* Wageningen, Netherlands: Agricultural Univ., Afdeling Sociologie en Sociografie van de Landbouwhogeschool, Bul. 10.—RS

——— (1958b), *Regionale Verschillen in de Toepassing van Enkele Landbouwmethoden (Regional Differences in the Adoption of Some Farm Practices).* Wageningen, Netherlands: Univ. of Wageningen, Dept. of Rural Soc., Bul. 9.—RS

——— (1960), "Locality Group Differences in the Adoption of New Farm Practices," *Rural Soc.,* 25: 308–320.—RS

——— (1961), "Research in the Field of Advisory Work," *Netherlands J. of Agri. Sci.,* 9: 122–133.—RS

——— (1963a), "Hoe Vinden Nieuwe Landbouwmethodeningand" (How a New Practice Is Introduced), *Landbouwvoorlichting,* 20: 227–239.—EX

——— (1963b), *Boer en Landvoorlichting: De Communicatie over Nieuwe Landbouwmethoden. (Farmers in Change: The Communication of New Ideas).* Assen, Netherlands: Van Gorkum.—RS

——— (1964b), "A Revision of the Two-Step Flow of Communications Hypothesis," *Gazette,* 10: 237–250.—RS

——— (1965), "The Communication of New Farm Practices in the Netherlands," *Sociologia Neerlandica,* 2: 1–18.—RS

Van Es, Johannes C. (1964), *Opinion Leadership in Colombian Veredas with Different Norms on Social Change.* M.S. thesis. Columbus: Ohio State Univ.—RS

Vargas, Fortunato T., (1959), "Testing of Visual Material and Evaluation of an Experience in Health Education Programs." Mexico City, Mexico: Centro Regional de Educación Fundamental Para la America Latina.—U

Velde, Baukje T. (1962), *Waarom Wordt een Melkmachine Aangeschaft (The Diffusion of Milking Machines).* Wageningen, Netherlands: Dairy Research Inst. Rept. 124. —RS

Verma, Ladu Ram (1966), "Study of Some Factors Influencing Adoption of Poultry Farming in Selected Panchayat Samitis (C.D. Blocks) of Jaipur District, Rajasthan," in A. N. Shukla and I. S. Grewal (Eds.), *Summaries of Extension by Post-Graduate Students.* Ludhiana, India: Punjab Agricultural Univ.—EX

Verner, Coolie and Gubbels, Peter M. (1967), *The Adoption or Rejection of Innovations by Dairy Farm Operators in the Lower Fraser Valley.* Vancouver, B.C.: Agri. Eco. Res. Council of Canada.—E

Verner, Coolie and Millerd, Frank W. (1966), *Adult Education and the Adoption of Innovations by Orchardists in the Okanagan Valley of British Columbia.* Vancouver: Univ. of British Columbia, Rural Soc. Monograph 1.—E

Villegas, Leopoldo (1966), *Field Demonstration: Its Effects on Level of Knowledge, Understanding, and Attitudes of Farmers about Choice of a Rice Variety ond Use of Fertilizers.* M.S. thesis. Manila: Univ. of the Philippines.—C

Vincent, William S. (1945), *Emerging Patterns of Public School Practice.* Ph.D. thesis. New York: Teachers College, Columbia Univ.—E

Vitor, Vincente de Paula (1968), *Typologies of Change Agents Based on Their Accuracy in Estimating Community Adoption Level of Innovation.* M.S. thesis. East Lansing: Michigan State Univ.—EX

Voget, Fred (1948), "Individual Motivation in the Diffusion of the Wind River Shoshone Sundance to the Crow," *Amer. Anthro.,* 50: 634–646.—A

——— (1950), "A Shoshone Innovator," *Amer. Anthro.,* 52: 53–63.—A

Von Blanckenburg, Peter (1962a), *Rice Farming in the Abakaliki Area.* Ibadan: Nigerian Inst. of Soc. and Eco. Res.—RS

——— (1962b), *A Study of Some Socio-Economic Factors Influencing Rubber Production.* Ibadan: Nigerian Inst. of Soc. and Eco. Res.—RS

Von Oppenfeld, Horst and others (1962), "Results of a Study of Adoption of Better Farm Practices in the Philippines," *Indian J. of Agri. Eco.,* 17: 23–32.—AE

Voos, Henry (1967), *Organizational Communication: A Bibliography*. New Brunswick, N.J.: Rutgers Univ. Press.—C

Wager, L. Wesley (1962), "Channels of Interpersonal and Mass Communication in an Organizational Setting: Studying the Diffusion of Information about a Unique Organizational Change," *Soc. Inquiry*, 32: 88–107.—S

Waisanen, F. B. (1952), *Television Ownership in an Iowa City: A Study in Some Determinants of Social Innovation*. M.A. thesis. Iowa City: State Univ. of Iowa.—GS

Waisanen, F. B. and Durlak, Jerome T. (1966), *A Survey of Attitudes Related to Costa Rican Population Dynamics*. San José, Costa Rica: Programa Interamericano de Información Popular.—GS

Walker, K. F. (1963), "Psychological Aspects of the Introduction of New Farming Practices: An Australian Study," in Thuc. Vu Quoc (Ed.), *Social Research and Problems of Rural Development in South-East Asia*. Paris, France: UNESCO.—AE

Wang, Inkeun (1962), "Adoption of Recommended Farming Practices and Some Influencing Socio-Economic Factors," *J. of Agri. Eco.*, 5: 67–86.—AE

―――― (1963), "Socioeconomic Status of the Farm Household and Adoption of Agricultural and Home Economics Practices in Several Korean Villages," *J. of Agri. Eco.*, 6: 14–26.—AE

Warland, Rex H. (1963), *Personal Influence: The Degree of Similarity of Those Who Interact*. M.S. thesis. Ames: Iowa State Univ.—RS

Watson, Goodwin (1946), *A Comparison of 'Adaptable' Versus 'Laggard' YMCAs*. New York: YMCA Research Council.—U

Weintraub, D. and Bernstein, F. (1966), "Social Structure and Modernization: A Comparative Study of Two Villages," *Amer. J. of Soc.*, 72: 509–521.—RS

Welch, John M. and Verner, Coolie (1962), "A Study of Two Methods for the Diffusion of Knowledge," *Adult Edu.*, 12: 231–237.—E

Weldon, Peter D. (1966), *Level of Living, Innovativeness, and Social Mobility in the Southern Philippines: A Unidimensional Approach*. M.S. thesis. Ithaca, N.Y.: Cornell Univ.—RS

Welikala, George H. F. (1959), *An Analysis of the Adoption of Some Agricultural, Medical, Public Health and Cooperative Practices in Six Selected Villages of Ceylon*. M.A. thesis. East Lansing: Michigan State Univ.—RS

Wellin, Edward (1955), "Water Boiling in a Peruvian Town," in Benjamin D. Paul (Ed.), *Health, Culture and Community*. New York: Russell Sage Fdn.—A

Wells, Donald E. and Andapia, Alfonso O. (1966), "Adoption Proneness and Response to Mail Questionnaires," *Rural Soc.*, 31: 483–487.—J

Wells, Donald E. and Arias, Emil (1967), "Relationship between Perception of Sources, Adoption Behavior, and Value Orientations of Mexican Dairymen." Paper presented at the Assn. for Edu. in Journalism. Boulder, Colo.—J

Wells, Donald E. and MacLean, Malcolm S. (1962), *A Typological Approach to the Study of Diffusion*. East Lansing: Michigan State Univ., Communication Res. Center, Mimeo Rept.—C

Westermarck, N. (1963), "Human Aspects on the Re-Organization and Reallocation of Resources on Family Farms," *Eripainos* 36: 56–64.—AE

Westoff, Charles F. and others (1961), "Some Estimates of the Reliability of Survey Data on Family Planning," *Population Stud.*, 15: 52–69.—MS

Wheelock, Gerald (1964), *The Farmers' Decision-Making Process in the Purchase of New Tractors*. M.S. thesis. Ames: Iowa State Univ.—RS

White, David M. (1950), "The 'Gate Keeper': A Case Study in the Selection of News," *Journalism Qtrly.*, 27: 383–390.—J

White, Donald J. (1965), *Relationships of Education, Agent Contact and Adoption of Dairy Practices at Recommended Levels for DHIA and Non-DHIA Dairymen*. M.S. thesis. East Lansing: Michigan State Univ.—EX

White, William J. (1967), *Toward the Validation of a Functional Model of the Adoption-Rejection Process*. Ph.D. thesis. East Lansing: Michigan State Univ.—C

—— (1968), "The Adoption of Modern Dairy Practices," *Canadian J. of Agri. Eco.,* 14: 29–30.—AE

White, W. J. and Heighton, V. A. (1968), *The Structure of the Canadian Manufacturing, Milk and Cream Industry*. Ottawa, Ont.: Economics Branch, Canada Dept. of Agri.—AE

Whiting, Gordon C. (1967), *Empathy, Mass Media, and Modernization in Rural Brazil*. East Lansing: Michigan State Univ., Dept. of Comm., Tech. Rept. 1.—C

Whiting, Gordon C. and others (1968), *Innovation in Brazil: Success and Failure of Agricultural Programs in 76 Minas Gerais Communities*. East Lansing: Michigan State Univ., Dept. of Comm., Diffusion of Innovations Res. Rept. 7.—C

Whittenbarger, Robert L. (1966), *Attitudes toward Social Change in a Rural Colombian Community: An Attempt at Measurement*. M.A. thesis. Madison: Univ. of Wisconsin.—RS

Whittenbarger, Robert L. and Maffei, Eugenio (1966), "Innovativeness and Related Factors in a Rural Colombian Community." Paper presented at the Rural Soc. Society. Miami Beach, Fla.—RS

Whyte, William H., Jr. (1954), "The Web of Word of Mouth," *Fortune*, 50: 140–143, 204–212.—U

Wichers, A. J. (1958), *De Evaluatie van een Voorlichtings-Campagne in de Betuwe* (The Evaluation of an Extension Campaign). Wageningen, Netherlands: Univ. of Wageningen, Dept. of Rural Soc., Bul. 11.—RS

Wilkening, Eugene A. (1949a), *The Acceptance of Certain Improved Agricultural Programs and Practices in a Piedmont Community of North Carolina*. Raleigh: N.C. Agri. Exp. Sta., Dept. of Rural Soc., Prog. Rept. 8.—RS

—— (1949b), *The Acceptance of Certain Agricultural Programs and Practices in a Piedmont Community of North Carolina.*

Ph.D. thesis. Chicago: Univ. of Chicago. —RS

—— (1950a), "Sources of Information for Improved Farm Practices," *Rural Soc.,* 15: 19–30.—RS

—— (1950b), "A Socio-Psychological Approach to the Study of the Acceptance of Innovations in Farming," *Rural Soc.,* 15: 352–364.—RS

—— (1951), "Social Isolation and Response of Farmers to Agricultural Programs," *Amer. Soc. Rev.,* 16: 836–837.—RS

—— (1952a), *Acceptance of Improved Farm Practices*. Raleigh: N. C. Agri. Exp. Sta., Tech. Bul. 98.—RS

——(1952b), "Informal Leaders and Innovators in Farm Practices," *Rural Soc.,* 17: 272–275.—RS

—— (1953), *Adoption of Improved Farm Practices as Related to Family Factors*. Madison: Wisc. Agri. Exp. Sta., Res. Bul. 183.—RS

——(1954), "Change in Farm Technology as Related to Familism, Family Decision Making, and Family Integration," *Amer. Soc. Rev.,* 19: 29–37.—RS

—— (1956), "Roles of Communicating Agents in Technological Change in Agriculture," *Social Forces,* 34: 361–367.—RS

—— (1958d), "Joint Decision-Making in Farm Families as a Function of Status and Role," *Amer. Soc. Rev.,* 23: 187–192.— RS

Wilkening, Eugene A. and Johnson, Donald (1958), "A Case Study in Decision-Making among a Farm Owner Sample in Wisconsin." Paper presented at the Rural Soc. Society. Pullman, Wash.—RS

—— (1960), "Why Farmers Quit Doing Things," *Better Farming Methods,* 32: 22–25.—RS

—— (1961), *Goals in Farm Decision-Making as Related to Practice Adoption*. Madison: Wisc. Agri. Exp. Sta., Res. Bul. 225.—RS

Wilkening, Eugene A. and others (1960), "Use and Role of Information Sources among Dairy Farmers of Northern

Victoria." Paper presented at the Rural Soc. Society. University Park, Pa.—RS

——— (1962), "Communication and Acceptance of Recommended Farm Practices among Dairy Farmers of Northern Victoria," *Rural Soc.*, 27: 116–197.—RS

——— (1963), "How Farmers Adopt New Practices," *Rural Res. in CSIRO*, 42: 2–7.—RS

Wilkening, Eugene A. and Santopolo, Frank A. (1952), *The Diffusion of Improved Farm Practices from Unit Test-Demonstration Farms in the Tennessee Valley Counties of North Carolina.* Raleigh: N.C. Agri. Exp. Sta., Mimeo Bul.—RS

Wilkinson, Kenneth P. and Bailey, Wilfrid C. (1964), *Differential Effectiveness of Test-Demonstration Farmers.* State College: Miss. Agri. Exp. Sta., Bul. 714.—RS

Williams, S. K. Taiwo (1968), *Sources of Information on Improved Farming Practices in Some Selected Areas of Western Nigeria.* Ile-Ife: Univ. of Ife, Dept. of Ext., Res. Monograph 5.—EX

Wilson, Meredith C. (1927), *Influence of Bulletins, News Stories, and Circular Letters upon Farm Practice Adoption with Particular Reference to Methods of Bulletin Distribution.* Washington, D.C.: U.S.D.A. Federal Ext. Svce., Circular 495.—EX

——— (1928), *Distribution of Bulletins and Their Use by Farmers.* Washington, D.C.: U.S.D.A. Federal Ext. Svce., Circular 78.—EX

Wilson, Meredith C. and Gallup, Gladys (1955), *Extension Teaching Methods and Other Factors that Influence Adoption of Agricultural and Home Economics Practices.* Washington, D.C.: U.S.D.A. Federal Ext. Svce.—EX

Wilson, Meredith C. and Trotter, Ide P. (1933), *Results of Legume Extension in Three Southeast Missouri Counties Representing Three States of Development of a Statewide Legume Program.* Washington, D.C.: U.S.D.A. Federal Ext. Svce., Circular 188.—EX

Winick, Charles (1961), "The Diffusion of an Innovation among Physicians in a Large City," *Sociometry*, 24: 384–396.—MS

Wish, John (1966), "Some Socio-Psychological Aspects of the Latin American Food Marketing Study," in Raymond Haas (Ed.), *Science, Technology and Marketing.* Chicago: Amer. Mktg. Assn.—MR

——— (1967), *Food Retailing in Economic Development: Puerto Rico, 1950–1965.* Ph.D. thesis. East Lansing: Michigan State Univ.—MR

Wolff, Robert J. (1965), "Modern Medicine and Traditional Culture: Confrontation on the Malay Peninsula," *Human Organization*, 339–345.—MS

Wolpert, Julian (1964), "A Regional Simulation Model of Information Diffusion," *Pub. Opin. Qtrly.*, 30: 597–608. —G

Wood, James R. and Zald, Mayer N. (1966), "Aspects of Racial Integration in the Methodist Church: Sources of Resistance to Organizational Policy," *Social Forces*, 45: 255–265.—GS

Woollatt, Lorne Hedley (1949), *The Cost-Quality Relationship on the Growing Edge.* Ph.D. thesis. New York: Teachers College, Columbia Univ.—E

Wright, Peter C. and others (1967), *The Impact of a Literacy Program in a Guatemalan Ladino Peasant Community.* Tampa: Univ. of South Florida, Mimeo Rept.—E

Wyon, John B. and Gordon, John E. (1962), "A Long Term Prospective Type Field Study of Population Dynamics in the Punjab, India," in Clyde V. Kiser (Ed.), *Research in Family Planning.* Princeton, N.J.: Princeton Univ. Press.—MS

Yacoub, Alah M. (1963), *Sociological Analysis of Factors Related to the Adoption of Recommended Farm Practices among Michigan Apple Growers.* M.A. thesis. East Lansing: Michigan State Univ.—RS

Yadav, Dharam P. (1967), *Communication Structure and Innovation Diffusion in Two Indian Villages.* East Lansing: Michigan State Univ., Dept of Comm., Tech. Rept. 2.—C

Yang, Jae Mo and others (1965), "Fertility and Family Planning in Rural Korea," *Population Stud.*, 18: 237–250.—MS

Yaukey, David and others (1966), "Husbands' vs. Wives' Responses to a

Fertility Study," *Population Stud.*, 19: 29–43.—MS

Yaukey, David and others (1967), "Couple Concurrence and Empathy on Birth Control Motivation in Dacca, East Pakistan,"*Amer. J. of Soc.*, 32:716–726.—MS

Yeracaris, Constantine A. (1961a), "The Acceptance of Polio Vaccine: An Hypothesis," *Amer. Catholic Soc. Rev.*, 22: 299–305.—MS

—— (1961b), "Social Factors Associated with the Acceptance of Medical Innovations: A Pilot Study," *J. of Health and Human Behav.*, 3: 193–198.—MS

Young, James N. (1959), *The Influence of Neighborhood Norms on the Diffusion of Recommended Farm Practices.* Ph.D. thesis. Lexington: Univ. of Kentucky.—RS

Young, James N. and Coleman, A. Lee (1959), "Neighborhood Norms and the Adoption of Farm Practices," *Rural Soc.*, 24: 372–380.—RS

Young, James N. and Marsh, C. Paul (1956), *The Adoption of Recommended Farm Practices and Sources of Farmer Information.* Lexington: Ky. Agri. Exp. Sta., Prog. Rept. 40.—RS

Zaidi, Wiqar Husain (1968a), *Effectiveness of Communication Strategy for Diffusion of Family Planning in West Pakistan.* M.A. thesis. East Lansing: Michigan State Univ.—C

—— (1968b), "PIA—A Comparison of Male and Female Responses Relating to Knowledge, Attitude and Practice Regarding Family Planning," *Pakistan J. Family Plng.*, 2: 28–37.—MS

Zwerman, Paul J. and Prundeanu, Julian (1956), "Obstacles to Soil Conservation," *J. of Soil and Water Conserv.*, 11: 127, 129.—AE

Non-Empirical Diffusion Research Publications

Abelson, Herbert, and Rugg, W. Donald (1958), "Self-Designated Influentiality and Activity," *Pub. Opin. Qtrly.*, 22: 556–567.—U

Abelson, Robert and Bernstein, Alex (1963), "A Computer Simulation Model of Community Referendum," *Pub. Opin. Qtrly.*, 27: 93–123.—P

Albrecht, Hartmut (1963), *Zum Heutigen Stand der Adoption-Forschung in den Vereinieten Staaten* (*The Contemporary Position of Adoption Research in the United States*). Hamburg: Verlag Paul Parey.—RS

—— (1964), "Die Theoretischen Ansatze der Amerikanischen Adoption-Forschung: Eine Kritische Analyze Zor Orienterung der Berantungsforschung" (A Theoretical Analysis of American Adoption Research: A Critical Analysis with an Orientation to Extension Research), in Hans Rheinwald (Ed.), *Probleme de Beratung* (Problems of Extension). Stuttgart: Verlag Eugen Ulmer.—EX

—— (1965), "*Die Bedeutung von Demonstration-Betrieben Als Einer Form der Landwirtschaftlichen Entwicklungshilfe: Herkungsbedingungen und Problembereiche des Demonstrierens*" (*The Importance of Demonstration Farms as a Form of Agricultural Development Aid: Conditions of Efficiency and Problems in Demonstration*). Gottingen, Germany: Univ. of Gottingen, Institut fur Auslandishe Landwirtschaft.—EX

Alers-Montalvo, Manual (1964), "The Study of Communities as a Source of Insight into Communication Problems." Paper presented at the First Interamer. Res. Symposium on the Role of Communications in Agri. Dev. Mexico City.—A

Anderson, C. Arnold (1959), "Trends in Rural Sociology," in Robert K. Merton and others (Eds.), *Sociology Today: Problems and Prospects.* New York: Basic Books.—RS

Anderson, David (1965), *Three Computer Programs for Contiguity Measures.* Evanston, Ill.: Northwestern Univ., Tech. Rept. 5. —G

Anscombe, F. J. (1961), "Estimating a Mixed-Exponential Response Law," *J. of Amer. Stat. Assn.,* 56: 493–502.—ST

Arce, Antonio M. (1961), *Sociología y Desarrollo Rural (Sociology and Rural Development).* Turrialba, Costa Rica: Instituto Interamericano de Ciencias Agricolas de la OEA.—RS

Arensberg, Conrad M. and Niehoff, Arthur H. (1964), *Introducing Social Change.* Chicago: Aldine.—A

Arndt, Johan (1966d), "Word of Mouth Advertising and Informal Communication." Cambridge, Mass.: Harvard Univ., School of Bus. Adm., unpublished paper. —MR

Arndt, Johan (1967c), *Word of Mouth Advertising: A Review of the Literature.* New York: Advertising Res. Fdn.—MR

Arnold, Robert and others (1966), "The Generating and Limiting Function: Dimensions of Innovations and Acceptance of Change in Rural Life." Paper presented at the Rural Soc. Society. Miami Beach, Fla.—RS

Babcock, Jarvis M. (1962), "Adoption of Hybrid Corn: A Comment," *Rural Soc.,* 27: 332–338.—AE

Bailey, Wilfrid C. (1963), "The Dilemma of Demonstrations." Paper presented at the Rural Soc. Society. Northridge, Cal.—RS

Baird, Andrew W. and Bailey, Wilfrid C. (1960), *Test Demonstration and Related Areas: Review of Literature.* State College: Miss. Agri. Exp. Sta., Prelim. Repts. in Soc. and Rural Life 11.—RS

Balfour, M. C. (1961), "Family Planning in Asia," *Population Stud.,* 15: 102–109.—MS

Barbichon, Guy (1968), "La Diffusion des Connaissances Scientifiques et Techniques dans le Public ses Conditions Dans les Pays en Voie de Developpement" (The Diffusion of Scientific Ideas to the Public and the Conditions that Lead to Development), *J. of Social Issues,* 24: 135–159.—U

Barnett, Homer G. (1961), "The Innovative Process," in Alfred L. Kroeber (Ed.), *Alfred L. Kroeber: A Memorial.* Kroeber Anthro. Society Papers.—A

Barton, Glen I. and Loomis, Ralph A. (1957), "Differential Rates of Change in Output per Unit of Input," *J. of Farm Eco.,* 39: 1551–1561.—AE

Bass, Frank M. (1967), "A New Product Growth Model for Consumer Durables." Lafayette, Ind.: Herman C. Krannert Grad. School of Ind. Adm., Purdue Univ., Paper 175.—MR

Bass, Frank M. and King, Charles W. (1968), "The Theory of First Purchase of New Products." Lafayette, Ind.: Herman C. Krannert Grad. School of Ind. Adm., Purdue Univ., Paper 213.—MR

Bauer, Raymond A. (1962), *The Social Cost and Benefits of Promotion: The Case of Ethical Pharmaceuticals.* Cambridge, Mass.: Arthur D. Little.—MR

Beal, George M. and Bohlen, Joe M. (1957), *The Diffusion Process.* Ames: Iowa Agri. Ext. Svce., Spec. Rept. 18.—RS

Becker, Selwyn W. and Whisler, Thomas L. (1967), "The Innovative Organization: A Selective View of Current Theory and Research," *J. of Bus.,* 40: 462–469.—P

Bennis, Warren G. and others (1962), *The Planning of Change: Readings in the Applied Behavioral Sciences.* New York: Holt, Rinehart and Winston.—P

Berelson, Bernard (1964a), "National Family Planning Program: A Guide," *Stud. in Fam. Plng.,* 5: 1–12.—MS

——— (1964b), "On Family Planning Communication," *Demography,* 1: 94–105. —MS

Bertrand, Alvin L. and Von Brock, Robert C. (Eds.), (1968), *Models for Educational Change.* Austin, Tex.: Southwest Edu. Dev. Lab., Monograph 2.—RS

Bessell, J. E. (1964), "Measurement of Leadership Among Farmers," *The Statistician,* 14: 227–273.—AE

Bhola, Harbans S. (1965a), "A Configurational Theory of Innovation Diffusion." Paper presented at the Conf. on Strategies for Edu. Change. Washington, D.C.—E

Bhola, Harbans S. (1965b), *Innovation Research and Theory*. Columbus: Ohio State Univ., School of Edu., Mimeo Rept.—E

—— (1965c), *A Theory of Innovation Diffusion and Its Application to Indian Education and Community Development*. Ph.D. thesis. Columbus: Ohio State Univ.—E

—— (1966), "The Need for Planned Change in Education," *Theory Into Practice*, 5: 7–10.—E

Bienstok, Theodore (1965), "Resistance to an Educational Innovation," *Elem. School J.*, 65: 420–428.—E

Bogue, Donald J. (Ed.), (1967), *Sociological Contributions to Family Planning Research*. Chicago: Univ. of Chicago, Community and Family Study Center.—MS

Bohlen, Joe M. (1964), "The Adoption and Diffusion of Ideas in Agriculture," in James H. Copp (Ed.), *Our Changing Rural Society: Perspective and Trends*. Ames: Iowa State Univ. Press.—RS

—— (1967), "Needed Research on Adoption Models," *Sociologia Ruralis*, 7: 113–129.—RS

Boone, Lalia P. (1949), "Patterns of Innovation in the Language of the Oil Field," *Amer. Speech.* 24: 31–37.—SP

Booth, Alan (1964), "Participation in Occupationally Related Adult Education Programs and the Diffusion of Innovations." Paper presented at the Amer. Edu. Res. Assn. Chicago.—E

Booth, Alan and Knox, Alan B. (1964), "Diffusion of Innovation and Adult Education." Lincoln: Univ. of Nebraska, Adult Edu. Res., Mimeo Paper.—E

Borden, Neil H., Jr. (1968), *Acceptance of New Food Products by Supermarkets*. Cambridge, Mass.: Div. of Research, Grad. School of Bus., Harvard Univ.—MR

Bourne, Francis S (1957), "Group Influence in Marketing and Public Relations," in Samuel P. Hayes and Rensis Likert (Eds.), *Some Applications of Behavioral Research*. Paris, France: UNESCO.—GS

Boyan, Norman J. (1968), "Problems and Issues of Knowledge Production and Utilization," in Terry L. Edill and Joanne

M. Kitchell (Eds.), *Knowledge Production and Utilization in Educational Administration*. Eugene: Univ. of Oregon, Center for the Advanced Study of Edu. Adm.—E

Brickell, Henry M. (1961), *Organizing New York State for Educational Change*. Albany, N.Y.: State Dept. of Edu.—E

—— (1964a), "State Organization for Educational Change: A Case Study and a Proposal," in Matthew B. Miles (Ed.), *Innovation in Education*. New York: Columbia Univ., Teachers College.—E

—— (1964b), "State Organization for Educational Means: New Media as Means and as Ends," in Wesley C. Mierhenry (Ed.), *Media and Educational Innovation*. Lincoln: Univ. of Nebraska, Col. of Ed.—E

Brown, Lawrence A. (1963), *The Diffusion of Innovation: A Markov Chain-Type Approach*. Evanston, Ill.: Northwestern Univ.—G

—— (1965a), "A Bibliography on Spatial Diffusion with Special Empahsis on Methodology and Theory." Evanston, Ill.: Northwestern Univ., Dept. of Geog., Mimeo Paper.—G

—— (1965b), *Models for Spatial Diffusion Research: A Review*. Evanston, Ill.: Northwestern Univ., Dept. of Geog., Tech. Rept. 3.—G

—— (1966), *Diffusion Dynamics: A Review and Revision of the Quantative Theory of the Spatial Diffusion of Innovation*. Ph.D. thesis. Evanston, Ill.: Northwestern Univ.—G

—— (1968), *Diffusion Processes and Location*. Regional Science Res. Inst., Philadelphia, Pa.—G

Brozen, Yale (1954), "Determinants of Entrepreneurial Ability," *Social Res.*, 21: 339–369.—GE

Burger, Henry G. (1965), "Directed Change of the Culture Core," *Amer. Anthro.*, 67: 489–494.—A

—— (1967), *Telesis: Facilitating Directed Cultural Change by Strategically Designing Chain Reactions*. Ph.D. thesis. New York: Columbia Univ.—A

Bushnell, Margaret (1957), "Now We're Lagging Only 20 Years," *School Exec.* 77: 61–63.—E

Byrnes, Francis C. (1966), "Some Missing Variables in Diffusion Research and Innovation." Paper presented at Philippine Soc. Assn. Manila.—C

Campbell, Rex R. (1966), "A Suggested Paradigm of the Individual Adoption Process," *Rural Soc.,* 31: 458–466.—RS

Carlson, Richard O. (1965), "Strategies for Educational Change: Some Needed Research on the Diffusion of Innovations." Paper presented at the Conf. on Strategies for Edu. Change. Washington, D.C.: U.S. Office of Edu.—E

——— (1968), "Summary and Critique of Educational Diffusion Research," Paper presented at the Natl. Conf. on the Diffusion of Edu. Ideas. East Lansing, Mich.—E

Carlson, Richard O. and Kiernan, Owen B. (1966), *A Plan For Curriculum Innovation in Massachusetts.* Boston: Mass.: State Dept. of Edu.—E

Cartano, David G. and Rogers, Everett M. (1963), "The Role of the Change Agents in Diffusing New Ideas," *J. of the Pakistan Acad. for Rural Dev.,* Comilla, 4: 61–65.—RS

Carter, Launor F. (1968), "Knowledge Production and Utilization in Contemporary Organizations," in Terry L. Edill and Joanne M. Kitchell (Eds.), *Knowledge Production and Utilization in Educational Administration.* Eugene: Univ. of Oregon, Center for the Adv. Study of Edu. Adm.—U

Castro Caldas, Eugenio de (1964), "A Difusão de Tecnicas e de Conhecimentos Entre os Agricultores: Aspectos Sociologicos," in Fundacão Calouste Gulbenkian (Ed.), *Analise e Planeamento da exploracão Agricola. (Analysis and Planning of Agricultural Development).* Lisbon, Portugal. —U

Chaudhari, Haider Ali and others (1967), "Social Characteristics of Agricultural Innovators in Two Punjabi Villages in West Pakistan," *Rural Soc.,* 32: 486–473.—RS

Christian, H. L. (1962), "The La Plaine 3–F Campaign," *Com. Dev. Bul.,* 13: 20–23.—E

Churchill, Gilbert A. and Ozanne, Urban (1967), "Adoption and Diffusion Research: A Potential Tool for Improving Technology Transfer," unpublished paper. Philadelphia: University of Pennsylvania.—I

Couch, Carl J. (1964), *Communication and Change.* East Lansing: Michigan State Univ., Inst. for Ext. Pers. Dev., Pub. 9.—EX

Couch, Carl J. and Bebermeyer, James P. (1964), *Some Research Possibilities on Communication and Change Within Extension.* East Lansing: Michigan State Univ., Inst. for Ext. Per. Dev., Bul. 15.—EX

Coughenour, C. Milton (1961), "The Practice-Use Tree and the Adoption, Drop-Out, and Non-Adoption of Recommended Farm Practices: A Progress Report." Paper presented at Rural Soc. Society. Ames, Iowa.—RS

——— (1964a), "Technology, Diffusion, and the Theory of Action." Paper presented at the First World Conf. of Rural Soc. Dijon, France.—RS

——— (1964b), "Toward a Theory of the Diffusion of Technology." Paper presented at the First Interamer. Res. Symposium on the Role of Communications in Agri. Dev. Mexico City.—RS

——— (1965a), "The Problem of Reliability of Adoption Data in Survey Research," *Rural Soc.,* 30: 184–203.—RS

——— (1965b), "Change and Sociological Perspectives," *Bul. of the Bur. of School Svce.,* 38: 28–45.—RS

——— (1965c), "Technology, Diffusion and the Theory of Action," *Indian J. of Ext. Edu.,* 1: 159–184.—RS

——— (1968), "Some General Problems in Diffusion from the Perspective of the Theory of Social Action." Columbia: Mo. Agri. Exp. Sta., North Central Reg. Res. Bul. 186.—RS

Council On Social Work Education (1959), *Interprofessional Training Goals for Technical Assistance Personnel Abroad.* New York: Council on Social Work Edu.—GS

Coutts, Philip (1953), "Five Dams: A Community Development Project in Uganda," *Corona,* 5: 296–299.—A

Cowper, L. T., and others (1958), "Village Sanitation Campaigns on Guam," *S. Pacific Bul.*, 8: 63–65.—U

Crane, Diana (1968a), "Collaboration, Communication, and Influence: A Study of the Effects of Formal Collaboration Among Scientists." Baltimore, Md.: Johns Hopkins Univ., Dept. of Behavioral Sciences, unpublished paper.—GS

Crane, Diana (1968b), "Fashion in Science: Does it Exist?" Baltimore, Md.: Johns Hopkins Univ., Dept. of Behavioral Sciences, unpublished paper.—GS

Crane, Diana (1968c), "Social Structure in a Group of Scientists: A Test of the 'Invisible College' Hypothesis." Baltimore, Md.: Johns Hopkins Univ., Dept. of Behavioral Sciences, unpublished paper.—GS

Crawford, C. Merle (1966), "The Trajectory Theory of Goal Setting for New Products," *J. of Mktg. Res.*, 3: 117–125.—MR

Culbertson, Jack A. (1965), "Organizational Strategies for Planned Change in Education." Paper presented at the Conf. on Strategies for Edu. Change. Washington, D.C.: U.S. Office of Edu.—E

Dalrymple, Dana G. (1969), *Technological Change in Agriculture: Effects and Implications for the Developing Nations*. Washington, D.C.: U.S.D.A., Foreign Agriculture Service, Mimeo Report.—AE

Danhof, Clarence (1949), "Observation on Entrepreneurship in Agriculture," in Harvard Research Center on Entrepreneurship History (Ed.), *Change and the Entrepreneur*. Cambridge, Mass.: Harvard Univ. Press.—GE

Dasgupta, Satadal (1965b), "Studies on Diffusion of Agricultural Innovations in India," State College, Miss.: Mississippi State Univ., Social Science Res. Center, unpublished paper.—RS

Davis, James A. (1963), "Structural Balance, Mechanical Solidarity and Interpersonal Relations," *Amer. J. of Soc.*, 63: 444–462.—GS

DeFleur, Melvin L. (1962a), "The Emergence and Functioning of Opinion Leadership: Some Conditions of Informal Influence Transmission," in Norman F. Washburne (Ed.), *Decisions, Values and Groups*. New York: Pergamon Press.—GS

DeFleur, Melvin L. and Assal, Elaine El (1964), "Innovation to Institution: Patterns of Adoption and Obsolescence in the Mass Media of Communication." Lexington, Univ. of Kentucky, unpublished paper.—GS

Degrazia, Alfred (1961), "Elements of Social Invention," *Amer. Behav. Sci.*, 5: 6–9.—GS

De Jong, Gordon F. and Coughenour, C. Milton (1960), "Reliability and Comparability of Two Instruments for Determining Reference Groups in Farm Practice Decisions," *Rural Soc.*, 25: 298–307.—RS

Desai, D. K. (1966), "Technological Change and Its Diffusion in Agriculture," *Indian J. of Agri. Eco.*, 21: 218–226.—AE

Deutsch, Karl W. (1949), "Innovation Entrepreneurship and the Learning Process," in A. H. Cole (Ed.), *Change and the Enterpreneur: Postulates for Entrepreneurial History*. Cambridge Mass.: Harvard Univ. Press.—PS

—— (1957), "Innovation Curves in Politics and Economics," *Prod.*, 1: 4–7.—PS

Deutschmann, Paul J. (1962a), "A Machine Simulation of Attitude Change in a Polarized Community." San José, Costa Rica: Programma Interamericano de Información Popular, unpublished paper.—C

—— (1962b), "A Machine Simulation of Information Diffusion in a Small Community." San José, Costa Rica: Programma Interamericano de Información Popular, unpublished paper.—C

—— (1962c), "A Model for Machine Simulation of Information and Attitude Flow." San José, Costa Rica: Programma Interamericano de Información Popular, unpublished paper.—C

Dexter, Lewis A. (1966), "Civil Defense Viewed as a Problem in Social Innovation." Paper presented at the Natl.

Meeting of the Society for Appl. Anthro. Milwaukee, Wisc.—A

Dickinson, Elizabeth R. (1955), *A Communication Study: Characteristics of Schuyler County, New York, Farmers Using Eleven Different Media as Sources for Obtaining Information on New Farming Practices.* M.S. thesis. Ithaca, N.Y.: Cornell Univ. —RS

Dillon, John L. and Heady, Earl O. (1958), "Decision Criteria for Innovation," *Australian J. of Agri. Eco.,* 2: 113.—AE

Dodd, Stuart C. (1950), "The Interactance Hypothesis: A Gravity Model Fitting Physical Masses and Human Groups," *Amer. Soc. Rev.,* 15: 245–256.—GS

—— (1953), "Testing Message Diffusion in Controlled Experiments: Charting the Distance and Time Factors in the Interactance Hypothesis," *Pub. Opin. Qtrly.,* 16: 247–262.—GS

—— (1955b), "The Transact Model: A Predictive and Testable Theory of Social Action," *Sociometry,* 18: 432–447.—GS

—— (1956), "A Power of Town Size Predicts an Internal Interacting: A Controlled Experiment Relating the Amount of an Interaction to the Number of Potential Interactors." Seattle: Univ. of Washington, Pub. Opin. Lab., Mimeo Rept.—GS

—— (1957), "The Counteractance Model, *Amer. J. of Soc.,* 53: 483–507.—GS

—— (1958), "Formulas for Spreading Opinions," *Pub. Opin. Qtrly.,* 22: 537–554.—GS

—— (1959), "How Random Interacting Organizes a Population: Exploring a Simple Chance Model to Relate Diffusion Theory to Information Theory." Seattle: Univ. of Washington, Pub. Opin. Lab., unpublished paper.—GS

—— (1961), "The Logistic Law of Interaction When People Pair Off 'At Will'," *J. of Social Psych.,* 53: 143–158.— GS

—— (1962a), "The Momental Models for Diffusing Attributes," *Darshana* (India), 2: 71–85—GS

—— (1962b) "How Momental Laws Can

Be Developed in Sociology by Deducing Testable and Predictive 'Actance' Models from Transacts," *Synthese,* 14: 277–299.—GS

—— (1962c), *Four Countering Laws in Communication.* Seattle: Univ. of Washington, Inst. for Soc. Res.—GS

—— (1965), "A Test of Message Diffusion by Chain Tags," *Amer. J. of Soc.,* 61: 425–432.—GS

Dodd, Stuart C. and McCurtain, Marilyn (1959a), *Logistic Diffusion in Randomly Overlapped Cliques.* Paper read at the Amer. Soc. Assn. Chicago.—GS

—— (1959b), "The Logistic Law in Communication," in *Research Series in Social Psychology.* Washington, D.C.: Natl. Inst. of Social and Behav. Sci., Symposia Studies Series 8.—GS

Dodd, Stuart C. and Pierce, William S. (1962), "Three Momental Models for Predicting Message Diffusion." Seattle: Univ. of Washington, Pub. Opin. Lab., unpublished paper.—GS

Eboch, Sidney (1966), "The Study of Change as a Concept in Education," *Theory into Practice,* 5: 34–38.—E

Eckaus, Richard S. (1966), "Notes on Invention and Innovation in Less Developed Countries," *Amer. Eco. Rev.,* 6: 98–117.—GE

Einsiedel, Luz A. (1960), *Success and Failure in Selected Community Development Projects in Batangas.* Quezon City: Univ. of the Philippines, Comm. Dev. Res. Council, Study Series 3.—U

Engel, James F. and others (1968), "Diffusion of Innovations," in *Consumer Behavior.* New York: Holt, Rinehart and Winston.—MR

Etzioni, Amitai and Etzioni, Eva (1964), *Social Change: Sources, Patterns, and Consequences.* New York: Basic Books.— GS

Fallding, Harold (1960), "The Contribution of Sociology to Effective Agricultural Extension (II)." Sydney, Australia: Univ. of Sydney, Dept. of Agri. Eco.—GS

Fallers, Lloyd A. (1954), "A Note on the 'Trickle Effect'," *Pub. Opin. Qtrly.,* 18: 314–321.—E

Fallon, Berlie J. (1966), *Educational Innovation in the United States.* Bloomington, Indiana: Phi Delta Kappa.—E

Fitzgerald, Kevin (1961), "On Getting Through to the Farmer." Paper presented at the Fertilizer Society. London.—U

Fletcher, D. J. (1967), "Predicted Diffusion of Super 8 MM Sound Films and Hardware." Chicago: Univ. of Chicago, unpublished MBA paper.—MR

Fortune (1954), "Conspicuous Buyers Anonymous: A Fable for Sellers," *Fortune,* 50: 128–130, 181–184.—U

Foster, George M. (1962), *Traditional Cultures and the Impact of Technological Change.* New York: Harper.—A

Foundation For Research On Human Behavior (1956), *Group Influence in Marketing and Public Relations.* Ann Arbor, Mich.—MR

—— (1959), *The Adoption of New Products: Process and Influence.* Ann Arbor, Mich.—MR

—— (1965), *The Obstinate Audience.* Ann Arbor, Mich.—MR

Freedman, Ronald (1967), "The Research Challenge to Social Scientists in the Developing Family Planning Programs: The Case of Taiwan," *J. of Social Issues,* 23: 165–169.—MS

Freedman, Ronald and Slesinger, Doris P. (1961), "Fertility Differentials for the Indigenous Non-Farm Population of the United States," *Population Stud.,* 15: 161–173.—MS

Freymann, Moye W. and Lionberger, Herbert F. (1962), "A Model for Family Planning Action-Research," in Clyde V. Kiser, (Ed.), *Research in Family Planning.* Princeton, N.J.: Princeton Univ. Press.—U

Galeski, Boguslaw (1965), *Sociological Research on the Diffusion of Agricultural Innovations in Poland.* Warsaw: Polish Acad. of Science, Inst. for Philo. and Soc., Center of Rural Soc., Mimeo Rept.—RS

Gallaher, Art, Jr., (1964), "The Role of the Advocate and Directed Change," in Wesley C. Mierhenry (Ed.), *Media and Educational Innovation.* Lincoln: University of Nebraska, Col. of Edu.—A

Gallup, George (1955), "The Absorption Rate of Ideas," *Pub. Opin. Qtrly.,* 19: 234–242.—C

Gamson, William A. (1961c), "Social Science Aspects of Fluoridation: A Summary of Research," Cambridge Mass:. Harvard Univ., School of Pub. Health, Community Aspects of Fluoridation Document 21-S.—MS

Garver, Richard A. (1962), "Communication Problems of Underdevelopment: Cheju-Do, Korea, 1962," *Pub. Opin. Qtrly.,* 26: 613–625.—J

Ghildyal, U.C. (1967a), "Diffusion of Innovations in Educational Methodology." Paper presented at the Seminar on Innovation in Edu. Hyderabad, India: Osmania Univ.—E

Goldhammer, Keith (1968), "Implications for Change in Training Programs," in Terry L. Edill and Joanne M. Kitchell (Eds.), *Knowledge Production and Utilization in Educational Administration.* Eugene: Univ. of Oregon, Center for the Adv. Study of Edu. Adm.—E

Gorman, Walter P., III (1967a), "Analyzing Consumer Markets Through Diffusion Research," *Rural Soc.,* 32: 456–463.—MR

Gould, Peter (1964), "A Note on Research into the Diffusion of Development," *J. of Modern African Stud.,* 2: 123–125.—G

Griliches, Zvi (1958), "Research Costs and Social Returns: Hybrid Corn and Related Innovations," *J. of Pol. Eco.,* 5: 419–431.—AE

—— (1960c), "Congruence Versus Profitability: A False Dichotomy," *Rural Soc.,* 25: 354–356.—AE

—— (1962), "Profitability Versus Interaction: Another False Dichotomy," *Rural Soc.,* 27: 325–330.—AE

Grossman, Gregory (1966), "Knowledge, Information, and Innovation in the Soviet Economy," *Amer. Eco. Rev.,* 66: 118–130.—E

Grunig, James E. (1966), "The Role of Information in Economic Decision Making," *Journalism Monographs,* 3: 1–51.—C

Guba, Egon G. (1965), "Methodological Strategies for Educational Change." Paper presented at the Conf. on Strategies for Edu. Change. Washington, D.C.: U.S. Office of Edu.—E

—— (1967), *The Basis for Educational Improvement*. Bloomington, Ind.: Natl. Inst. for the Study of Edu. Change.—E

—— (1968b), "Diffusion of Innovations," *Edu. Leadership*, 25: 292–295.—E

—— (1968c), "Development, Diffusion and Evaluation," in Terry L. Edill and Joanne M. Kitchell (Eds.), *Knowledge Production and Utilization in Educational Administration*. Eugene: Univ. of Oregon, Center for the Advanced Study of Edu. Adm.—E

Hagen, Everett E. (1961), "The Entrepreneur: A Rebel Against Traditional Society," *Human Organization*, 19: 185–187.—GE

Hägerstrand, Torsten (1965a), "A Monte Carlo Approach to Diffusion," *European J. of Soc.*, 6: 43–67.—G

Haines, George H., Jr. (1964b), "A Theory of Market Behavior After Innovation," *Mgmt. Sci.*, 10: 634–657.—MR

Harp, John (1960), "A Note on Personality Variables in Diffusion Research," *Rural Soc.*, 25: 346–347.—RS

Harris, Jack S. (1940), "The White Knife Shoshoni of Nevada," in Ralph Linton (Ed.), *The Acculturation of Seven American Indian Tribes*. New York: Appleton-Century-Crofts.—A

Harvey, D. W. (1966a), "Geographical Processes and the Analysis of Point Patterns: Testing Models of Diffusion by Quadrat Sampling," *Transactions and Papers*, 40: 81–95.—G

—— (1966b), "Theoretical Concepts and the Analysis of Agricultural Land-Use Patterns in Geography," *Annals of the Assn. of Amer. Geog.*, 56: 361–374.—G

Hassinger, Edward (1959), "Stages in the Adoption Process," *Rural Soc.*, 24: 52–53.—RS

Hauser, Philip M. (1967), "Family Planning and Population Programs: A Book Review Article." Chicago: Univ. of Chicago, unpublished paper.—GS

Havelock, Ronald G. (1968a), "Dissemination and Translation Roles," in Terry L. Edill and Joanne M. Kitchell (Eds.), *Knowledge Production and Utilization in Educational Administration*. Eugene: Univ. of Oregon, Center for the Adv. Study of Edu. Adm.—P

—— (1968b), *Bibliography on Knowledge Utilization and Dissemination*. Ann Arbor: Univ. of Michigan, Inst. for Social Res. —P

Havens, A. Eugene (1962a), *A Review of Factors Related to Innovativeness*. Columbus: Ohio Agri. Exp. Sta., Dept. of Agri. Eco. and Rural Soc., Mimeo Bul. AE 329.—RS

—— (1963b), "La Adopción de Innovaciones: Una Comparación Entre Colombia y Estados Unidos" (The Adoption of Innovations: A Comparison Between Colombia and the United States), in *Memoria del Primer Congress Nacional de Sociología*. Bogotá, Asociación Colombiana de Sociología.—RS

—— (1964), "Some Theoretical and Methodological Considerations for Research on Diffusion in Latin America." Paper presented at the Society for App. Anthro. San Juan, P.R.—RS

—— (1965b), "Some Theoretical and Methodological Considerations for Research on Diffusion in the Hot-Humid Tropics of Latin America." Paper presented at the Conf. on the Potentials of the Hot-Humid Tropics in Latin America Rural Dev. Ithaca, N.Y.: Cornell Univ.—RS

Hawkins, Norman G. (1959a), "Graduate Training for Health Research." Paper presented at Amer. Soc. Assn. Chicago.—MS

—— (1959b), "The Detailman and Preference Behavior," *Southwestern Social Sci. Qtrly.*, 40: 213–224.—MS

Heiskanen, Veronica F. (1967), "A Cross-Cultural Content Analysis of Family Planning Publications," in Donald J. Bogue (Ed.), *Sociological Contributions to Family Planning Research*. Chicago: Univ. of Chicago Press.—MS

Hemphill, H. David (1968), "A General Theory of Innovativeness," *Alberta J. of Edu. Res.*, 14: 101–114.—E

Henderson, Mary S. C. (1965), *Managerial Innovations of John Diebold.* Washington D.C.: Lebaron Fdn.—GE

Herzog, William A., Jr. (1967b), "Mass Media Credibility, Exposure, and Modernization in Rural Brazil." Paper presented at the Assn. for Edu. in Journalism. Boulder, Colo.—C

Hobbs, Daryl J. (1966), "The Study of Change as a Concept." *Theory into Practice,* 5: 20–24.—RS

Hochstrasser, Donald L. (1955), *The Concept of Cultural Diffusion: An Analysis Based on Contemporary Anthropological Usage.* M.A. thesis. Lexington: Univ. of Kentucky.—A

Howard, John A. (1963), "The Process of Marketing Acceptance," in Thomas L. Berg and Abe Shuchmann (Ed.), *Product Strategy and Management.* New York: Holt, Rinehart and Winston.—MR

Hoyos, Hernan M. (1967), "Comunicación e Información" (Communication and Information). Bogotá: Colombian Assn. of Schools of Medicine, Div. of Population Studies.—O

Hoyos, Hernan M. and Torres, Mercedes de (1967), *Planificación Familia* (Family Planning). Bogotá: Colombian Assn. of Schools of Medicine, Div. of Pop. Stud.—O

Hundley, James R. (1962), *The Validity of Farm Practice Adoption Measures.* M.S. thesis. Madison: Univ. of Wisconsin.—RS

Iutaka, Sugiyama (1965), "Social Stratification Research in Latin America," *Latin Amer. Res. Rev.,* 1: 7–16.—GS

Johnson, D. Gale (1966), "The Environment for Technological Change in Soviet Agriculture," *Amer. Eco. Rev.,* 66: 145–153.—GE

Jones, Garth N. (1965b), "Preventive Medicine at Work: A Hypothetical Study of Managed Organizational Change," *Public Adm. Rev.* (Pakistan) 1–31.—PA

Jones, Gwyn E. (1966), "Indexes of Innovativeness." Paper presented at the European Cross-Natl. Res. Proj. on the Diffusion of Tech. Innovations in Agri. Reading, England.—EX

—— (1967a), "The Adoption and Diffusion of Agricultural Practices," *World Agri. Eco. and Rural Soc. Abstracts,* 9: 1–34.—EX

—— (1967b), "Rural Development and Agricultural Extension: A Sociological View," *Comm. Dev.J.* 6: 26–33.—EX

Jordan, P. (1957), "War on a Worm," *Corona,* 9: 369–373.—M

Joshi, Vidya (1962), "Attitude Towards Reception of Technology," *The J. of Social Psych.,* 58: 3–7.—GS

Karlsson, Georg (1958), *Social Mechanisms: Studies in Sociological Theory.* Uppsala, Sweden: Almqvist and Wilkselld, and New York: The Free Press.—GS

Katz, Elihu (1958b), "Communication and Technical Change: On the Convergence of Two Traditions of Social Research." Chicago: Univ. of Chicago, unpublished paper.—GS

—— (1957), "The Two-Step Flow of Communication: An Up-To-Date Report on an Hypothesis," *Public Opin. Qtrly.,* 21: 61–78.—GS

—— (1959), "Review of Information, Decision and Action," *Amer. J. of Soc.,* 65: 321–322.—GS

—— (1960), "Communication Research and the Image of Society: Convergence of Two Traditions," *Amer. J. of Soc.,* 65: 435–440.—GS

—— (1962), "Notes on the Unit of Adoption in Diffusion Research," *Soc. Inquiry,* 32: 3–9.—GS

—— (1963a), "The Characteristics of Innovations and the Concept of Compatibility." Paper presented at Rehovoth Conf. of Comprehensive Planning of Agri. in Dev. Countries. Rehovoth, Israel.—GS

—— (1963b), "The Diffusion of New Ideas and Practices," in Wilbur Schramm (Ed.), *The Science of Human Communication.* New York: Basic Books.—GS

—— (1964), "Communication Research and the Image of Society: Convergence of Two Traditions," in Lewis A. Dexter and David M. White (Eds.), *People, Society and Mass Communications.* New York: The Free Press.—MS

——— (1965), "Diffusion of Innovation," in Donald E. Payne (Ed.), *The Obstinate Audience*. Ann Arbor, Mich.: Fdn. for Res. on Human Behav.—GS

Katz, Elihu and Levin, Martin L. (1959), "Traditions of Research on the Diffusion of Innovation." Paper presented at the Amer. Soc. Assn. Chicago.—GS

Katz, Elihu and others (1963), "Traditions of Research on the Diffusion of Innovations," *Amer. Soc. Rev.*, 28: 237–253.—GS

Kahn, Anwarvzzaman (1963), *Introduction of Tractors in a Subsistence Farm Economy*. Comilla, East Pakistan: Acad. for Rural Dev., Tech. Pub. 14.—AE

King, Charles W. (1966), "Adoption and Diffusion Research in Marketing: An Overview," in Raymond M. Haas (Ed.), *Science, Technology, and Marketing*. Chicago: Amer. Mktg. Assn.—MR

King, Charles W. and Ness, Thomas E. (1969b), "The Adoption and Diffusion of New Architectural Concepts Among Professional Architects: An Overview of the Research Project." Lafayette, Ind.: Purdue Univ., Herman C. Krannert Grad. School of Ind. Adm., unpublished paper.—MR

King, Charles W. and Summers, John O. (1967b), "Dynamics of Interpersonal Communication: The Interaction Dyad," in Donald F. Cox (Ed.), *Risk Taking and Information Handling in Consumer Behavior*. Cambridge, Mass.: Harvard Univ., Grad. School of Bus. Adm.—MR

——— (1967c), "The New Product Adoption Research Project: A Survey of New Product Adoption Behavior Across a Wide Range of New Consumer Products Among Marion County, Indiana Home-makers: A Project Description." Lafayette, Ind.: Purdue Univ., Herman C. Krannert Graduate School of Industrial Administration, Paper 196.—MR

——— (1969a), "Technology, Innovation and Consumer Decision Making," in *Changing Market Systems ... Consumer, Corporate and Government*. Chicago: Amer. Mktg. Assn., 1967 Winter Conf. Proceedings, Series 26.—MR

Kivlin, Joseph E. (1965), "Contributions to the Study of Mail-Back Bias," *Rural Soc.*, 30: 322–326.—RS

Klonglan, Gerald E. (1967), *Radio Listening Groups in Malawi, Africa*. Ames: Iowa State Univ., Rural Soc. Rept. 70.—RS

Klonglan, Gerald E. and others (1966), "Adoption of Public Fallout Shelters." Ames: Iowa State Univ., Rural Soc. Rept. 54.—RS

Knight, Kenneth E. (1967), "A Descriptive Model of the Intra-Firm Innovation Process," *J. of Bus.*, 40: 478–496.—MR

Knop, Edward and Aparicio, Kathryn (1965), *Current Change Literature: An Annotated Classification of Selected Inter-disciplinary Sources*. Grand Forks: Univ. of North Dakota, Dept. of Soc. and Anthro., Monograph L.—GS

Kolbe, F. F. H. (1964), "Vertraagde Verandering in Die Lanbou" (Delayed Change in Agriculture), *Tydskrif Vir Geestes-Wetenskappe*, 4: 333–339.—U

Koval, John P. (1960), *Innovation and Fringe Benefits*. M.S. thesis. Eugene: Univ. of Oregon.—GS

Kroeber, A. L. (1937), "Diffusion," in Edwin R. A. Seligman and Alvin Johnson (Eds.), *The Encyclopedia of the Social Sciences, II*. New York: Macmillan.—A

Kumpf, Carl H. (1952), *The Adaptable School*. New York: Macmillan.—E

Kunstadter, Peter (1957), *Preliminary Report on Use of Clinic Facilities by the Residents of the Mescalero Apache Reservation*. Tucson: Univ. of Arizona, Bur. of Ethnic Res., Mimeo Bul.—A

Kurland, Norman D. (1966), "The Effect of Planned Change on State Departments," *Theory into Practice*, 5: 51–53.—E

Lake, Dale G. (1968), "Concepts of Change and Innovation in 1966," *J. of App. Behav. Sci.*, 4: 3–24.—E

Lancaster, Kelvin (1966). "Allocation and Distribution Theory: Technological Innovation and Progress," *Amer. Eco. Rev.*, 56: 14–23.—GE

Lavidge, Robert J. and Steiner, Gary A. (1961), "A Model for Predicting Measurements of Advertising Effectiveness," *J. of Mktg.*, 25: 59–62.—MR

Leeper, Robert R. (Ed.), *Strategy for Curriculum Change*. Papers from the ASCD Seminar on Strategy for Curriculum Change. Washington, D.C.: Assn. for Supervision and Curriculum Dev.— E

———— (1968), "Most of the Change," *Edu. Leadership,* 25: 283–285.—E

Lerner, Daniel and Schramm, Wilbur (Eds.), (1967), *Communication and Change in the Developing Countries*. Honolulu: East-West Center Press.—C

Leuthold, Frank O. (1966), "Communication and Diffusion of Improved Farm Practices in Two Northern Saskatchewan Farm Communities." Saskatoon: Univ. of Saskatchewan, unpublished paper.— RS

Linton, Ralph (1936), *The Study of Man*. New York: Appleton-Century-Crofts.— A

Lionberger, Herbert F. (1952), "The Diffusion of Farm and Home Information as an Area of Sociological Research," *Rural Soc.,* 17: 132–140.—RS

———— (1960), *Adoption of New Ideas and Practices: A Summary of the Research Dealing with the Acceptance of Technological Change in Agriculture, with Implications for Action in Facilitating Social Change*. Ames: Iowa State Univ. Press.—RS

———— (1961a), "The Role of Communication Media and Agents in the Adoption of New Ideas and Practices (with Special Reference to Family Planning)." Paper presented at the Fourth All India Conf. on Family Plng. Hyderabad, India.—RS

———— (1961b), "Some Observations Regarding the Nature and Scope of Action Research in Family Planning." Columbia: Univ. of Missouri, unpublished paper.— RS

———— (1962a), "Diffusion of Ideas and Practices." Paper presented at the Intl. Seminar on Water and Soil Utilization. Brookings: South Dakota State College. —RS

———— (1962b), "Studies in Hand and in Prospect Relating to the Diffusion of Agricultural Information: American Experience." Paper read at the Organization

for Eco. Coop. and Dev. Seminar on Structure and Orientation of Intellectual Investments in Agriculture in Relation to Economic and Social Development. Paris, France.—RS

———— (1963a), "Individual Adoption Behavior: Applications from Diffusion Research—Part 1, " *J. of Coop. Ext.,* 3: 157–166.—RS

———— (1963b), "Community Adoption: Applications from Diffusion Research— Part 2," *J. of Coop. Ext.,* 4: 201–208.—RS

———— (1963c), "Audiences." Paper presented at the Seminar in Mass Communications. Columbia: Univ. of Missouri.— RS

———— (1964a), "Implications of Diffusion Research in Rural Sociology for Implemented Change in Public Health with Particular Reference to Fluoridation of Water." Paper presented at the Univ. of California Conf. on Fluoridation of Water. Carmel, Cal.—RS

———— (1964b), "Needed Research on the Structure of Interpersonal Communication and Influence in Traditional Rural Societies." Paper presented to the First Interamer. Res. Symposium on the Role of Communications in Agri. Dev. Mexico City.—RS

———— (1964c), "The Diffusion Research Tradition in Rural Sociology and Its Relation to Implemented Change in Public Schools Systems," in Wesley C. Mierhenry (Ed.), *Media and Educational Innovation*. Lincoln: Univ. of Nebraska, Col. of Edu.—RS

———— (1965a), "Some Implications from Diffusion Research for the Adoption of Farm Practices in India," *Indian J. of Ext. Edu.,* 1: 18–25.—RS

———— (1965b),"Diffusion of Innovations in Agricultural Research and in Schools," in Robert R. Leeper (Ed.), *Strategy for Curriculum Change*. Washington, D.C.: Assn. for Supervision and Curriculum Dev.—RS

———— (1966), "Organizing for Implementing Changes in Education: Some Implications from Agriculture and Diffusion Research." Paper presented at the Seminar on Dev. and Coordinating of Res. Columbus, Ohio.—RS

———— (1967), "Differential Views Held of Innovator and Influence Referents as Sources of Farm Information in Adoption Decisions in a South Missouri Farm Community." Columbia: Mo. Agri. Exp. Sta.—RS

Lippitt, Ronald and others (1958), *The Dynamics of Planned Change.* New York: Harcourt, Brace.—E

Little, Arthur D. (1963), *Patterns and Problems of Technical Innovation in American Industry.* Washington, D.C.: U.S. Dept. of Commerce, Natl. Bur. of Stds., Inst. for App. Tech.—U

Loomis, Charles P. and Loomis, Zona K. (1967), "Rural Sociology," in Paul F. Lazarsfeld and others (Eds.), *The Uses of Sociology.* New York: Basic Books.—RS

Lopes, Renato S. (1967), *Evolução Conceptual da Extensão Rural do Brazil* (Conceptual Evolution of Extension in Rural Brazil). Belo Horizonte, Brazil: ACAR.—EX

Luschinsky, Mildred S. (1963), "Problems of Culture Change in the Indian Village," *Human Organization* 22: 66–74.—A

Maccoby, Eleanor and others (1962), "The Communication of Information about Child Care and Development: Summary Report." Stanford Univ., Inst. for Communication Res., Mimeo Paper.—P

MacLaurin, W. Rupert (1950), "The Process of Technological Innovation: The Launching of a New Scientific Industry," *Amer. Eco. Rev.,* 50: 90–112.—GE

Makarczyk, Waclaw (1968a), "An Experiment with Re-Interviewing on Adoption Dates." Warsaw: Polish Acad. of Sciences, Inst. of Philo. and Soc.—RS

Marble, Duane F. and Bowiby, Sophia R. (1968), *Computer Programs for the Operational Analysis of Hägerstrand Type Spatial Diffusion Models.* Evanston Ill.: Northwestern Univ. Dept. of Geog., Tech. Rept. 9.—G

Marsh, Paul E. (1964), "Wellsprings of Strategy: Considerations Affecting Innovations by the PSSC," in Matthew B. Miles (Ed.), *Innovation in Education.* New York: Columbia Univ., Teachers College. —E

Mason, Robert G. (1966a), "Information Sources and the Adoption of Innovations," *Gazette,* 12: 112–116.—EX

Mayo, Selz C. (1960), "An Analysis of the Organizational Role of the Teacher of Vocational Agriculture," *Rural Soc.,* 25: 334–345.—RS

McCormack, William C. (1957), "Mysore Villagers' View of Change," *Eco. Dev. and Cul. Change.* 5: 257–262.—A

McLaughlin, Curtis P. and Penchansky, Roy (1965), "Diffusion of Innovation in Medicine: A Problem of Continuing Medical Education," *J. of Medical Edu.,* 40: 437–447.—MR

McNamara, Robert L. (1957), "The Need for Innovativeness in Developing Societies," *Rural Soc.,* 32: 395–398.—RS

Meadows, Paul (1964), "Novelty and Acceptors: A Sociological Consideration of the Acceptance of Change," in Wesley C. Meierhenry (Ed.), *Media and Educational Innovation.* Lincoln: Univ. of Nebraska, Col. of Edu.—GS

Meierhenry, Wesley C. (1966), "Innovation, Education, and Media," *Audiovisual Communication Rev.,* 14: 451–465.—E

Mendelsohn, Harold (1965), "Comment on Spitzer,…A Comparison of Six Investigations," *J. of Broadcasting,* 9: 51–54.—C

Miles, Matthew B. (1964a), "Innovation in Education: Some Generalizations," in M. B. Miles (Ed.), *Innovation in Education.* New York: Columbia Univ., Teachers College.—E

———— (1964b), "Educational Innovation: Some Generalizations," in Wesley C. Meierhenry (Ed.), *Media Educational Innovation.* Lincoln: Univ. of Nebraska, Col. of Edu.—E

———— (1964c), "Educational Innovation: Resources, Strategies, and Unanswered Questions," *Amer. Behav. Sci.,* 7: 10–13.—E

Millard, I. S. (1950), "The Village Schoolmaster as Community Development Leader," *Mass Edu. Bul.,* 1: 42–45.

Miller, Peggy L. (1968), *Change Agent Strategies: A Study of the Michigan-Ohio*

Regional Educational Laboratory. Ph.D. thesis. East Lansing: Michigan State Univ.—E

Miller, Richard I. (1964a), "Needed Research and Development in the Process of Change," *Bul. of the Bur. of School Svce.,* 38: 72–83.—E

—— (1965b), "Some Current Development in Educational Change," *Bul. of the Bur. of School Svce.,* 32: 7–17.—E

—— (1967), "The Role of Educational Leadership in Implementing Educational Change." Paper presented at the Symposium on System Analysis and Mgmt. Techniques for Edu. Planners. Orange, Cal.—E

Miller, Richard I. and others (1967), *Catalyst for Change: A National Study of ESEA Title III (PACE).* Washington, D.C.: U.S. Govt. Printing Office.—E

MISRA (1966), *Diffusion Studies in Social Sciences: A Bibliography.* India: Univ. of Mysore, Res. Paper 6.—G

Morrill, Richard L. (1968), "Waves of Spatial Diffusion," *J. of Regional Sci.,* 8: 1–18.—G

Morrill, Richard L. and Pitts, Forrest R. (1967), "Marriage, Migration, and the Mean Information Field: A Study in Uniqueness and Generality," *Annals of the Assn. of Amer. Geog.,* 57: 401–422.—G

Mort, Paul R. (1953), "Educational Adaptability," *The School Exec.,* 71: 1–23.—E

—— (1957), *Principles of School Administration.* New York: McGraw-Hill.—E

—— (1964), "Studies in Educational Innovation from the Institute of Administrative Research: An Overview," in Matthew B. Miles (Ed.), *Innovation in Education.* New York: Columbia Univ., Teachers College.—E

Mottur, Ellis (1968), *The Processes of Technological Innovation: A Conceptual Systems Model.* Washington, D.C.: National Bureau of Standards Report 9689.—GE

Mtawali, C. V. (1951), "A Health Campaign in Tanganyika Territory," *Community Dev. Bul.,* 2: 55–58.—PH

Mueller, E. W. and Westrom, Betty (1957), *Planting the Cross on the Contour.* Chicago:

Div. of Amer. Missions, Natl. Lutheran Council.—EX

Myren, Delbert R. (1965), *Bibliography: Communications in Agricultural Development.* Mexico: Rockefeller Fdn.—C

Nelson, Richard R. and others (1967), *Technology, Economic Growth and Public Policy.* Washington, D.C.: The Brookings Institution.—GE

Nelson, Richard R. and Phelps, Edmund S. (1965), "Investment in Humans, Technological Diffusion and Economic Growth," *Amer. Eco. Rev.,* 56: 69–75.—GE

Nicosia, Francesco M. (1966), *Consumer Decision Processes.* Englewood Cliffs, N.J.: Prentice-Hall.—MR

Niederfrank, E. J. (1955), *Main Types of Organization Found in Extension Work and Related Social Factors.* Washington, D.C.: U.S. Dept. of Agriculture, Federal Ext. Svce., Circular 500.—RS

Niehoff, Arthur (1964a), *The Primary Variables in Directed Cross-Cultural Change.* Alexandria, Va.: George Washington Univ., Human Resources Res. Office.—A

—— (1966a), *A Casebook of Social Change.* Chicago: Aldine.—A

—— (1966b), "Food Habits and the Introduction of New Foods." Paper Performed by HumRRO Division 7. Alexandria, Va.—C

—— (1967), "Intra-Group Communication and Induced Change." Paper presented at the Society for App. Anthro. Washington, D.C.—A

Niehoff, Arthur H. and Anderson, J. Charnel (1964c), *A Selected Bibliography of Cross-Cultural Change Projects.* Alexandria, Va.: George Washington Univ. Human Resources Res. Office.—A

—— (1966), "Peasant Fatalism and Socio-Economic Innovation," *Human Organization,* 25: 273–283.—A

Niehoff, Arthur and Niehoff, Juanita (1966), "The Influence of Religion on Socio-Economic Development," *Intl. Dev. Rev.,* 18: 6–62.—A

North Central Rural Sociology Subcommittee for the Study of Diffusion of

Farm Practices (1955), *How Farm People Accept New Ideas*. Ames: Iowa Agri. Ext. Svce., Spec. Rept. 15.—RS

North Central Rural Sociology Subcommittee for the Study of Diffusion of Farm Practices (1959), *Bibliography of Research on Social Factors in the Adoption of Farm Practices*. Ames: Iowa Agri. Ext. Svce., Mimeo Bul.—RS

North Central Rural Sociology Subcommittee for the Study of Diffusion of Farm Practices (1961), *Adopters of New Farm Ideas: Characteristics and Communications Behavior*. East Lansing: Mich. Agri. Ext. Svce., Bul.—RS

North Central Rural Sociology Subcommittee for the Study of Diffusion of Farm Practices (1966), *Diffusion Research Needs*. Ames: Iowa State Univ., North Central Reg. Pub.—RS

Obibuaku, L. O. (1967), "Socio-Economic Problems in the Adoption Process: Introduction of a Hydraulic Palm-Oil Press," *Rural Soc.*, 32: 464–468.—EX

Ogburn, William F. (1922), *Social Change*. New York: Huebsch.—ES

Ogburn, William F. and Gilfillian, Colum S. (1933), "The Influence of Invention and Discovery," in *Recent Social Trends in the United States*. New York: McGraw-Hill.—ES

Palmer, J. E. S. (1962), "Self-Help to Irrigation," *Community Dev. Bul.*, 13: 44–45.—EX

Parameswaran, E. G. and Bhogie, S. (1967), "Psychological Study in Adoption of Farm Innovation and Educational Innovation: A Comparative Study." Paper presented at the Seminar on Innovation in Edu. Hyderabad, India: Osmania Univ.—P

Pareek, Udai (Ed.) (1966), *Behavioral Science Research in India: A Directory*. Secunderabad, India: The Behavioural Science Centre.—P

Pareek, Udai and Chattopadhyaya, S. W. (1966), "Adoption Quotient: A Measure of Multipractice Adoption Behavior," *J. of App. Behav. Sci.*, 2: 95–108.—P

Parker, William and others (1962), *The Diffusion of Technical Knowledge as an Instrument of Economic Development*. Washington, D.C.: Natl. Inst. of Social and Behav. Sci., Symposia Studies Series 13.—U

Patnaik, N. (1967), "Adoption of Agricultural Practices in a Peasant Community in Orissa," in T.P.S. Chawdhari (Ed.), *Selected Readings on Community Development*. Hyderabad, India: Natl. Inst. of Com. Dev.—A

Perry, Astor and others (1967), "The Adoption Process: S-Curve or J-Curve," *Rural Soc.*, 32: 220–222.—EX

Petrini, Frank (1968), "The Effect of Extension Work," *Lantbrukshogskolans Annaler*, 34: 351–376.—RS

Pfaff, Martin and Jambothar, C. G. (1968), *Simulation of Cross-Cultural Transfers*. East Lansing: Michigan State Univ., Computer Inst. for Social Sci. Res., Res. Rept.—MR

Pitts, Forrest R. (1962a), "Problems in Computer Simulation of Diffusion." Paper presented to the Ninth Meeting of the Reg. Sci. Assn. Pittsburgh.—G

—— (1962b), "Computer Simulation of Diffusion in the Japanese Rural Economy." Paper presented at the Northwest Anthro. Conf. Eugene, Oregon.—G

—— (1964a), "A General Computer Program for Spatial Diffusion Processes." Paper presented at the TIMS-ORSA Meeting. Minneapolis.—G

—— (1964b), "Scale and Purpose in Urban Simulation Models." Paper presented at Conf. on Strategy for Reg. Growth. Ames, Iowa.—G

—— (1965), *Hager III and Hager IV: Two Monte Carlo Computer Programs for the Study of Spatial Diffusion Problems*. Evanston, Ill.: Northwestern Univ., Tech. Rept. 4.—G

—— (1967), MIFCAL *and* NONCEL: *Two Computer Programs for the Generalization of the Hägerstrand Models to an Irregular Lattice*. Evanston, Ill.: Northwestern Univ., Tech. Rept. 7.—G

Polgar, Steven (1966), "Sociocultural Research in Family Planning in the United States: Review and Prospects," *Human Organization*, 25: 321–329.—A

Ponsioen, J. A. (1965), *The Analysis of Social Change Reconsidered.* The Hague, Netherlands: Mouton.—GS

Pool, Ithiel de Sola (1960), "The Role of Communication in the Process of Modernization and Technological Change." Paper presented to the North Amer. Conf. on the Social Implications of Industrialization and Technological Change. Chicago.—C

Press, Irwin (1966b), "Innovation in Spite of: A Lamp Factory for Maya Peasants," *Human Organization,* 25: 284–294.—A

Pulschen, Rolf E. (1968), *Effects of Literacy, Informal Leadership and Gatekeeping on Diffusion of Printed Messages: A Field Experiment on Persuasive Mass Communication Strategies in Rural Brazil.* M.S. thesis. Madison: Univ. of Wisconsin.—C

Rahudkar, W. B. (1961), "Measurement Techniques in Agricultural Extension and Rural Sociological Research," *Indian J. of Agron.,* 6: 52–62.—RS

Rainio, Kulleruo (1961), *A Stochastic Model of Social Interaction.* Copenhagen: Munksgaard.—GS

Rapoport, Anatol (1957), "Contributions to the Theory of Random and Biased Nets," *Bul. of Mathematical Biophysics,* 19: 257–277.—ST

Redlich, Fritz (1953), "Ideas, Their Migration in Space and Transmittal over Time," *Kyklos,* 6: 301–322.—U

Rehder, Robert R. (1965), "Communication and Opinion Formation in a Medical Community: The Significance of the Detail Man," *Acad. of Mgmt. J.,* 8: 282–291.—MR

Rein, Martin (1964), "Organization for Social Change," *Social Work,* 9: 32–41.—GS

Ribble, Robert B. (1966), "The Effect of Planned Change on the Classroom," *Theory into Practice,* 5: 41–45.—E

Richardson, Lee (Ed.), (1969), *Dimensions of Communication.* N.Y., Appleton-Century-Crofts.—C

Richman, Barry M. (1962), "A Rating Scale for Product Innovation," *Bus. Horizons,* 5: 37–42.—MR

Robertson, Thomas S. (1967c), "The Process of Innovation and the Diffusion of Innovation," *J. of Mktg.,* 31: 14–19.—MR

Robinson, Ira E. and Bailey, Wilfrid C. (1965), "Consonance and Dissonance in Agricultural Communications," *Rural Soc.,* 30: 332–337.—RS

Rogers, Everett M. (1962a), "How Research Can Improve Practice: A Case Study," *Theory into Practice,* 1: 89–93.—RS

——— (1962b), *Diffusion of Innovations.* New York: The Free Press.—RS

——— (1963a), "The Adoption Process: Part I," *J. of Coop. Ext.,* 1: 16–22.—RS

——— (1963b), "The Adoption Process: Part II," *J. of Coop. Ext.,* 1: 69–75.—RS

——— (1963c), "What Are Innovators Like?" *Theory into Practice,* 2: 252–256.—RS

———(1964a), *Bibliography of Research on the Diffusion of Innovations.* East Lansing: Michigan State Univ., Dept. of Comm., Res. on the Diffusion of Innovations Report 1.—C

——— (1964b), "What Are Innovators Like?," in Richard O. Carlson and others (Eds.), *Change Processes in the Public Schools.* Eugene: Univ. of Oregon, Center for the Adv. Study of Edu. Adm.—C

——— (1965b), "Toward a New Model for Educational Change." Paper presented at the Conf. on Strategies for Edu. Change. Washington, D.C.—C

——— (1966a), *Elementos del Cambio Social en America Latina: Difusión de Innovaciones (Elements of Social Change in Latin America: Diffusion of Innovations).* Bogotá: Ediciones Tercer Mundo and Universidad Nacional de Colombia, Facultad de Sociología, Monografías 23.—C

——— (1966b), "The Communication of Innovations: Strategies for Change in a Complex Institution." Paper presented at the Natl. Conf. on Curricular and Instructional Innovation for Large Colleges and Universities. East Lansing, Mich.—C

——— (1967a), "Developing a Strategy for Planned Change." Paper presented at the Symposium on System Analysis and

Management Techniques for Educational Planners. Orange, Cal.—C

—— (1967b), "Mass Communication and the Diffusion of Innovations: Conceptual Convergence of Two Research Traditions." Paper presented at the Assn. for Edu. in Journalism. Boulder, Colo.—C

—— (1968a), "The Communication of Innovations in a Complex Institution," *Edu. Record*, 49: 67–77.—C

—— (1968b), "Experience with Cross-National Research: The Diffusion Project in Brazil, Nigeria and India." Paper presented at the Second World Cong. of Rural Soc. Drienerlo, Enschede, Netherlands.—C

Rogers, Everett, M. and Beal, George M. (1957a), "An Approach to Measure Reference Group Influences in the Adoption of Farm Practices." Paper presented at the Rural Soc. Society. College Park, Md.—RS

—— (1957b), "Projective Techniques in the Study of Consumer Behavior." Paper presented at the Research Planning Meeting on How Consumer Purchases of Convenience Goods Are Determined. Ann Arbor, Mich., The Fdn. for Res. on Human Behav.—RS

—— (1958c), "Projective Techniques in Interviewing Farmers," *J. of Mktg.*, 23.—RS

—— (1959a), "Projective Techniques: Potential Tools for Agricultural Economists," *J. of Farm Eco.*, 41: 644–648.—RS

—— (1959b), "Projective Techniques and Rural Respondents," *Rural Soc.*, 24: 178–182.—RS

Rogers, Everett M. and Bettinghaus, Erwin P. (1966), "Comparison of Generalizations from Diffusion Research on Agricultural and Family Planning Innovations." Paper presented at the Amer. Soc. Assn. Miami Beach, Fla.—C

Rogers, Everett M. and Cartano, David G. (1962), "Methods of Measuring Opinion Leadership," *Pub. Opin. Qtrly.*, 26: 435–441.—RS

—— (1963b), "Research on the Diffusion of Innovations," *J. of the Pakistan Acad. for Rural Dev.*, 3: 220–224.—RS

Rogers, Everett M. and Havens, A. Eugene (1962b), "Rejoinder to Griliches' 'Another False Dichotomy'," *Rural Soc.*, 27: 330–332.—RS

Rogers, Everett M. and others (1962), *The Construction of Innovativeness Scales.* Columbus, Ohio: Agri. Exp. Sta., Dept. of Agri. Eco. and Rural Soc., Mimeo Bul. AE 330.—RS

Rogers, Everett M. and Safilios, Constantina (1960), "Communication of Agricultural Technology: How People Accept New Ideas," in Everett M. Rogers, *Social Change in Rural Society: A Textbook in Rural Sociology.* New York: Appleton-Century-Crofts.—RS

Rogers, Everett M. and Smith, Leticia (1965), *Bibliography on the Diffusion of Innovations.* East Lansing: Michigan State Univ., Dept. of Comm., Diffusion of Innovations Research Report 3.—C

Rogers, Everett M. and Stanfield, J. David (1968), "Adoption and Diffusion of New Products: Emerging Generalizations and Hypotheses," in Frank M. Bass and others (Eds.), *Application of the Sciences in Marketing Management.* New York: Wiley.—C

Rogers, Everett M. and Svenning, Lynne (1969a), "Change in Small Schools." Paper presented at a Natl. Working Conf. of N-FIRE, Denver.—C

—— (1969b), *Managing Change.* Burlingame, Cal.: Operation PEP, Mimeo Rept.—C

Rogers, Everett M. and van den Ban, Anne W. (1962), "Research on the Diffusion of Agricultural Innovations in the United States and the Netherlands," *Sociologia Ruralis*, 3: 38–51.—RS

Ross, Donald H. (1958), *Administration for Adaptability: A Source Book Drawing Together the Results of More than 150 Individual Studies Related to the Question of Why and How Schools Improve.* New York: Metropolitan School Study Council.—E

Ross, John A. (1966c), "Cost Analysis of the Taichung Experiment," *Stud. in Family Plng.*, 10: 6–15.—MS

Rural Sociological Society Subcommittee on the Diffusion and Adoption of

Farm Practices (1952), *Sociological Research on the Diffusion and Adoption of New Farm Practices: A Review of Previous Research and a Statement of Hypotheses and Needed Research*. Lexington: Ky. Agri. Exp. Sta., Mimeo Rept. RS–2.—RS

Ryan, Bryce (1965), "The Resuscitation of Social Change," *Social Forces*, 44: 1–7.—GS

Saxena, A. P. (1961), "Panchayat Raj in Community Development," *Rural India*, 24: 316–320.—AE

—— (1962), "Programme Planning and Execution in C. D. Blocks," *Rural India*, 25: 217–221.—AE

Schmookler, Jacob (1966), *Invention and Economic Growth*. Cambridge, Mass.: Harvard Univ. Press.—GE

Schmuck, Richard (1968), "Social Psychological Factors in Knowledge Utilization," in Terry L. Edill and Joanne M. Kitchell (Eds.), *Knowledge Production and Utilization in Educational Administration*. Eugene: Univ. of Oregon, Center for the Adv. Study of Edu. Adm.—E

Schon, Donald A. (1967), *Technology and Change: The Impact of Invention and Innovation on American Social and Economic Development*. New York: Delacorte Press.—U

Sen, Lalit K. (1967), "Diffusion of Innovations in a System Model." Paper presented at the Seminar on Innovation in Education. Hyderabad, India: Osmania Univ.—RS

Sen, Lalit K. and others (1967), *People's Image of Community Development and Panchayati Raj*. Hyderabad, India: Natl. Inst. of Comm. Dev.—RS

Sengupta, T. (1967), "A Simple Adoption Scale for Selection of Farmers for High Yielding Varieties Programme on Rice," *Indian J. of Ext. Edu.*, 3: 107–115.—RS

Shapero, Albert (1965), "Diffusion of Innovations Resulting from Research: Implications for Research Program Management." Paper presented at the Second Conf. on Res. Program Effectiveness. Washington, D.C.: Office of Naval Research.—U

Shapiro, Paul S. (1967), *Communications or Transport: Decision-Making in Developing Countries*. Cambridge: Massachusetts Inst. of Tech., Center for Intl. Studies.—GS

Shaw, Steven J. (1965), "Behavioral Science Offers Fresh Insights on New Product Acceptance," *J. of Mktg.*, 29: 9–13.—MR

Shen, T. Y. (1961), "Innovation, Diffusion, and Productivity Changes," *Rev. of Eco. and Stat.*, 43: 175–181.—E

Shepard, Herbert A. (1967), "Innovation-Resisting and Innovation-Producing Organizations," *J. of Bus.*, 40: 470–477. —GE

Sheppard, David (1957), "Studying the Progressiveness of Farmers," *Agri. Progress*, 32: 54–62.—P

—— (1960a), "Neighborhood Norms and the Adoption of Farm Practices," *Rural Soc.*, 25: 356–358.—P

Sherrington, Andrew M. (1965), "An Annotated Bibliography of Studies on the Flow of Medical Information to Practitioners," *Methods of Info. in Med.*, 6: 45–57.—MS

Shetty, N. S. (1966), "Inter-Farm Rates of Technological Diffusion in Indian Agriculture," *Indian J. of Agri. Eco.*, 21: 189–198.—AE

Shirpurkar, G. R. (1964), *Factors Affecting the Adoption of Improved Agricultural Practices*. M.Sc. thesis. Nagpur, India: Col. of Agri.—EX

Sieber, Sam D. (1968), "Organizational Influence on Innovative Roles," in Terry L. Edill and Joanne M. Kitchell (Eds.), *Knowledge Production and Utilization in Educational Administration*. Eugene: Univ. of Oregon, Center for the Adv. Study of Edu. Adm.—E

Singh, Avtar and Kaufman, Harold F. (1965), *A Behavioral Approach to Agricultural Development*. State College: Mississippi State Univ., Social Sci. Res. Center, Preliminary Rept. 7.—RS

Smith, H. T. E. (1959), "Advisory Methods: The Study of How to Get Information to Farmers." Paper presented to the Agri. Edu. Assn.—EX

—— (1964), "Some Notes on the Adoption of Farm Practices: Intensity of

Information," *Farm Econ.*, 10: 345–351.—EX

Smith, M. Brewster (1965), "Motivation, Communications Research and Family Planning." Berkeley: Univ. of California, Dept. of Psych. and Inst. of Human Dev., unpublished paper.—P

Soth, Lauren (1952), *How Farm People Learn New Methods*. Washington, D.C.: Natl. Planning Assn., Pamphlet 79.—AE

South, Donald R. (1966), "Some New Perspectives on the Adoption-Diffusion Process." Paper presented at the Rural Soc. Society. Miami Beach, Fla.—RS

Sparks, Harry M. (1964), *Educational Change in Kentucky Public Schools*. Lexington, Ky.: Dept. of Pub. Instr., Edu. Bul. 32.—E

Spaulding, Irving A. (1956), "Experiences with the Use of Projective Interview Techniques in Farm Practice Acceptance Research." Paper presented at the Rural Soc. Society. East Lansing, Mich.—RS

Spencer, Daniel L. and Woroniak, Alexander (Eds.) (1967), *The Transfer of Technology to Developing Countries*. New York: Praeger.—GE

Spitzer, Stephan P. (1965), "Mass Media vs. Personal Sources of Information about the Presidential Assassination: A Comparison of Six Investigations," *J. of Broadcasting*, 9: 45–50.—C

Stanfield, J. David and others (1965), "Computer Simulation of Innovation Diffusion: An Illustration from a Latin American Village," American Soc. Assn. and the Rural Soc. Society. Chicago.—C

Stoetzel, J. (1965), "General Social Psychology," *Bul. of Psych.*, 18: 1229–1230.—P

Stycos, J. Mayone (1958), "Some Directions for Research on Fertility Control," *Milbank Mem. Fund. Qtrly.*, 36: 126–148.—MS

—— (1967), "Catholicism and Birth Control in the Western Hemisphere." Paper presented at the Fourth Annual Natl. Conf. of the Catholic Inter-American Coop. Prog. Washington, D.C.—MS

Sutherland, Edwin H. (1950), "The Diffu-

sion of Sexual Psychopath Laws," *Amer. J. of Soc.*, 56: 142–148.—GS

Takes, Charles A. P. (1963a), *Socio-Economic Factors Affecting Agricultural Productivity in Some Villages of Oshun Division (Western Region)*. Ibadad: Nigerian Inst. of Social and Eco. Res.—RS

Tarde, Gabriel (1903), tr. Elsie Clews Parsons, *The Laws of Imitation*. New York: Holt.—ES

Torres, Hugo B. (1961), *Estudio de Algunos Factores en la Introducción de Tecnología Agrícola en Seis Comunidades de Costa Rica (Study of Some Factors Related to the Introduction of Agriculture Technology in Six Communities of Costa Rica)*. M.S. thesis. Turrialba, Costa Rica: Instituto Interamericano de Ciencias Agricolas de la O.E.A.—EX

Tuck, Russell R., Jr. (1968), "Impact of Innovations," *Edu. Leadership*, 25: 312–315.—E

U.S. Department of Commerce (1967), "Technological Innovation: Its Environment and Management." Washington : U.S. Govt. Printing Office.—U

Valkonen, Tapani (1968), "On the Theory of Diffusion of Innovations." Helsinki, Finland: Univ. of Helsinki, unpublished paper.—GS

Van den Ban, Anne W. (1964a), "De Communicatie van Niewe Land-Bouwmethoden" (Communication of New Farm Practices), *Statistica Neerlandica*, 18: 497–506.—EX

Van Der Kroef, Justus M. (1957), "Patterns of Cultural Change in Three Primitive Societies," *Social Research*, 24: 427–456.—A

Van Es, Johannes C. and Rogers, Everett M. (1964), "Diffusion Research in Developing Societies," *J. of the Pakistan Acad. for Rural Dev.*, 4: 120–125.—RS

Wainwright, R. E. (1953), "Women's Clubs in the Central Nyanza District of Kenya," *Comm. Dev. Bul.*, 4: 77–80.—O

Waisanen, Frederick, B. (1963), "The Three-Step Flow in Communication and Change." San José, Costa Rica: Programa Interamericano de Información Popular, unpublished paper.—C

Waisanen, Frederick B. (1964), "Change Orientation and the Adoption Process." Paper presented at the First Interamer. Res. Symposium on the Role of Communication in Agri. Dev. Mexico.—C

Walker, Odell L. and others (1960), *Application of Game Theory Models to Decisions on Farm Practices and Resource Use.* Ames: Iowa Agri. Exp. Sta., Res. Bul. 488.—AE

Wasson, Chester R. (1960), "What Is 'New' About a New Product?" *J. of Mktg.,* 25: 52–56.—MR

Wilkening, Eugene A. (1958a), "An Introductory Note on the Social Aspects of Practice Adoption," *Rural Soc.,* 23: 97–102.—RS

—— (1958b), "Communication and Technological Change in Rural Society," in Alvin L. Bertrand (Ed.), *Rural Sociology: An Analysis of Contemporary Rural Life.* New York: McGraw-Hill.—RS

—— (1958c), "Process of Acceptance of Technological Innovations," in Alvin L. Bertrand (Ed.), *Rural Sociology: An Analysis of Contemporary Rural Life.* New York: McGraw-Hill.—RS

—— (1962), "The Communication of Ideas on Innovation in Agriculture," in Wilbur Schramm (Ed.), *Studies of Innovation and of Communication to the Public.*

Stanford, Cal.: Stanford Univ. Inst. for Comm. Res.—RS

—— (1964), "Some Perspectives on Change in Rural Societies," *Rural Soc.,* 29: 1–17.—RS

Woodroff, William (1963), "An Inquiry into the Origins of Invention and the Intercontinental Diffusion of Techniques of Production in the Rubber Industry," *Economic Record,* 38: 479–497.—GE

Woods, Thomas E. (1967), *The Administration of Educational Innovation.* Eugene: Univ. of Oregon, Bur. of Edu. Res., Monograph.—E

Young, Ruth (1959), "Observations on Adoption Studies Reported in June, 1958 Issue," *Rural Soc.,* 24: 272–274. —RS

Yuill, Robert S. (1964), *A Simulation Study of Barrier Effects in Spatial Diffusion Problems.* Evanston, Ill.: Northwestern Univ., Dept of Geog., Tech. Rept. 1. —G

Zald, Mayer, N. and Ash, Roberta (1966), "Social Movements and Organizations: Growth, Decay and Change," *Social Forces,* 44: 327–339.—GS

Zaltman, Gerald (1964), *Marketing: Contributions from the Behavioral Sciences,* New York: Harcourt, Brace and World. —MR

General References Cited

Agger, Robert and others (1964), *The Rulers and the Ruled: Political Power and Impotence in American Communities.* New York: Wiley.

Allport, Gordon W. and Postman, Leo (1947), *The Psychology of Rumor.* New York: Holt.

Anonymous Peace Corps Volunteer (1967), "Thrown Onto the Edge of Asia," in *The Peace Corps Reader.* New York: Quadrangle Books.

Atwood, M. S. (1964), "Small-Scale Administrative Change: Resistance to the Introduction of a High School Guidance Program," in Matthew B. Miles (ed.), *Innovation in Education.* New York: Columbia Univ., Teachers College.

Back, K. W. and others (1950), "The Methodology of Studying Rumor Transmission," *Human Relations,* 3: 307–312.

Bacon, Francis (1906), *Essays.* London: Dent.

Bagehot, Walter (1873), *Physics and Politics.* New York: Appleton-Century.

Bailey, Norman T. J. (1957), *The Mathematical Theory of Epidemics.* New York: Hafner.

Bales, Robert F. (1950), "A Set of Categories for the Analysis of Small Group Interaction," *Amer. Soc. Rev.,* 15: 257–263.

Barnes, Louis B. (1967), "Organizational Change and Field Experiment Methods," in Victor H. Vroom (Ed.), *Methods of Organizational Research.* Pittsburgh: Univ. of Pittsburgh Press.

Barnlund, Dean C. and Harland, Carroll (1963), "Propinquity and Prestige as Determinants of Communication Networks," *Sociometry,* 26: 467–479.

Bauer, Raymond A. (1963), "The Initiative of the Audience," *J. of Adv. Res.,* 3: 2–7.

Beal, George M. (1957), "How Does Social Change Occur?" in *Prospects for the Years Ahead.* Ames, Iowa: Agri. Exp. Sta., Spec. Rept. 21.

Beal, George M. and others (1964), *Social Action in Civil Defense: The Strategy of Public Involvement in County Civil Defense Educational Program.* Ames, Iowa: Agri. and Home Eco. Exp. Sta., Mimeo Rept.

Beal, George M. and others (1967c), *Vocational School Bond Issues in Iowa: Sociological Studies in Education.* Ames, Iowa: Iowa State Univ., Dept. of Soc. and Anthro., Mimeo Rept.

Bennett, Edith B. (1952), *The Relationship of Group Discussion, Decision, Commitment and Consensus to Individual Action.* Ph.D. thesis. Ann Arbor: Univ. of Michigan.

Berelson, Bernard R. and others (1954), *Voting.* Chicago: Univ. of Chicago Press.

Berlo, David K. (1960), *The Process of Communication.* New York: Holt, Rinehart and Winston.

Bible, Bond L. and Nolan, Francena L. (1960), *The Role of the Extension Committee Member in the County Extension Organization in Pennsylvania.* Univerity Park, Pa.: Agri. Exp. Sta., Bul. 665.

Blau, Peter, M. (1957), "Formal Organization: Dimensions of Analysis," *Amer. J. of Soc.,* 63: 58–69.

———— (1960), "Structural Effects," *Amer. Soc. Rev.,* 25: 178–193.

Blau, Peter M. and Scott, Richard W. (1962), *Formal Organizations: A Comparative Approach.* San Francisco: Chandler.

Boskoff, Alvin (1957), "Social Change," in Howard Becker and Alvin Boskoff (Eds.), *Modern Sociological Theory in Continuity and Change.* New York: Holt, Rinehart and Winston.

Bruner, Jerome S. and Goodman, C. C. (1947), "Value and Need as Organizing Factors in Perception," *J. of Abnor. and Soc. Psych.,* 42: 33–44.

Campbell, Ernest Q. and Alexander, C. Norman (1965), "Structural Effects and Interpersonal Relationships," *Amer. J. of Soc.,* 71: 284–289.

Carlsson, Gosta (1965), "Time and Continuity in Mass Attitude Change: The Case of Voting," *Pub. Opin. Qtrly.,* 29: 1–15.

Chandler, Alfred D., Jr. (1962), *Strategy and Structure: Chapters in the History of Industrial Enterprise.* Cambridge: Massachusetts Institute of Technology Press.

Chin, Robert (1966), "The Utility of Systems Models and Development Models for Practitioners," in Warren G. Bennis and others (Eds.), *The Planning of Change.* New York: Holt, Rinehart and Winston.

Coch, Lester and French, J. R. P., Jr. (1948), "Overcoming Resistance to Change," *Human Relations,* 1: 512–532.

Coleman, James S. (1958), "Relational Analysis: The Study of Social Organizations with Survey Methods," *Human Organization,* 14: 28–36.

———— (1962), "Analysis of Social Structures and Simulation of Social Processes with Electronic Computers," in Harold Guetzkow (Ed.), *Simulation in Social Science: Readings.* Englewood Cliffs, N.J.: Prentice-Hall.

Conner, John T. (1964), "Progress Reshapes Competition," *Printers' Ink,* 287: 36.

Cyert, R. M. and March, J. G. (1963), *A Behavioral Theory of the Firm.* Englewood Cliffs, N.J.: Prentice-Hall.

Dahl, Robert A. (1961), *Who Governs? Democracy and Power in an American City.* New Haven, Conn.: Yale Univ. Press.

Dahle, Thomas L. (1964), "An Objective and Comparative Study of Five Methods of Transmitting Information to Business and Industrial Employees," in W. C. Redding and G. A. Sanborn (Eds.), *Business and Industrial Communication: A Source Book.* New York: Harper and Row.

Davis, James A. and others (1961), "A Technique for Analyzing the Effects of Group Composition," *Amer. Soc. Rev., 26:* 215–225.

DeFleur, Melvin L. (1962c), "Mass Communication and the Study of Rumor," *Soc. Inquiry,* 32: 51–70.

DeFleur, Melvin L. (1966b), *Theories of Mass Communication.* New York: David McKay.

Edwards, Harold T. (1963), *Power Structure and Its Communication Behavior in San José, Costa Rica.* M.A. thesis. East Lansing: Michigan State Univ.

Erlich, D. and others (1957), "Post-Decision Exposure to Relevant Information," *J. of Abnor. and Soc. Psych.,* 54: 98–102.

Festinger, Leon (1949), "The Analysis of Sociograms Using Matrix Algebra," *Human Relations,* 2: 153–158.

—— (1950), "Informal Social Communication," *Psych. Rev.,* 57: 271–282.

—— (1957), *A Theory of Cognitive Dissonance.* Evanston, Ill.: Row, Peterson.

—— (1964), "Behavioral Support for Opinion Change," *Pub. Opin. Qtrly.,* 28: 404–418.

French, J. R. P. and others (1958), "Employee Participation in a Program of Industrial Change," *Personnel,* 35: 21–29.

—— (1960), "An Experiment on Participation in a Norwegian Factory," *Human Relations,* 13: 3–19.

Galbraith, Jay L. (1967), "The Use of Subordinate Participation in Decision-Making," *J. of Ind. Engrg.,* 18: 521–525.

Gans, Herbert J. (1962), *The Urban Villagers: Group and Class in the Life of Italian-Americans.* New York: The Free Press.

Gardner, John W. (1963), *Self-Renewal: The Individual and the Innovative Society.* New York: Harper and Row.

Gerard, H. B. (1957), "Some Effects of Status, Role Clarity, and Group Goal Clarity Upon the Individual's Relations to Group Process," *J. of Personnel,* 25: 475–588.

Giffin, Kim and Erlich, Larry (1963), "The Attitudinal Effects of a Group Discussion on a Proposed Change in Company Policy," *Speech Monographs,* 30: 377–379.

Gouldner, Alvin (1954), *Patterns of Industrial Bureaucracy.* New York: The Free Press.

—— (1957), "Theoretical Requirements of the Applied Social Sciences," *Amer. Soc. Rev.* 22: 92–102.

Greiner, L. E. (1965), *Organizational Change and Development.* Ph.D. thesis. Cambridge, Mass.: Harvard Univ.

Griffiths, Daniel E. (1964), "Administrative Theory and Change in Organizations," in Matthew B. Miles (Ed.), *Innovation in Education.* New York: Teachers College, Columbia Univ.

Hamblin, Robert S. and others (1961), "Group Morale and Competence of the Leader," *Sociometry,* 24: 295–311.

Hardin, Charles M. (1951), "'Natural Leaders' and the Administration of Soil Conservation Programs," *Rural Soc.,* 16: 279–281.

Heirich, Max (1964), "The Use of Time in the Study of Social Change," *Amer. Soc. Rev.,* 29: 386–397.

Hiniker, Paul J. (1968), "The Mass Media and Study Groups in Communist China," in David K. Berlo (Ed.), *Mass Communication and the Development of Nations.* East Lansing: Michigan State Univ., Intl. Communication Inst.

Homans, George, C. (1950), *The Human Group.* New York: Harcourt, Brace and World.

—— (1961), *Social Behavior: Its Elementary Forms.* New York: Harcourt, Brace and World.

Hopper, David (1957), *The Economic Organization in the Village of North Central India.* Ph.D. thesis. Ithaca, N.Y.: Cornell Univ.

Hovland, Carl I. (1959), "Reconciling Conflicting Results Derived from Experimental and Survey Studies of Attitude Change," *Amer. Psych.,* 14: 8–17.

Hovland, Carl I. and others (1953), *Communication and Persuasion: Psychological Studies of Opinion Change*. New Haven, Conn.: Yale Univ. Press.

Hubbell, Charles H. (1965), "An Input-Output Approach to Clique Identification," *Sociometry*, 28: 377–399.

Hugo, Victor M. (1893), *Historie d'un Cirme: Conclusion: LaChute*. Paris: Edition Nationale.

Humphrey, Hubert H. (1963), "The Behavioral Sciences and Survival," *Amer. Psych.,* 18: 290–294.

Hunter, Floyd (1953), *Community Power Structure*. Chapel Hill: Univ. of North Carolina Press.

Hursh, Gerald D. (forthcoming), *Survey Research Methods in Developing Nations*. Lansing: Michigan State Univ., Dept. of Communication.

Jacques, Elliott (1948), "Interpretive Group Discussion as a Method of Facilitating Social Change," *Human Relations*, 1: 533–549.

Jacobson, Eugene and Seashore, Stanley E. (1951), "Communication Practices in Complex Organizations," *J. of Social Issues*, 7: 28–40.

Jahoda, Marie and others (1951), *Research Methods in Social Relations*. New York: Holt, Rinehart and Winston.

Janowitz, Morris and Delany, William (1957), "The Bureaucrat and the Public: A Study of Information Perspectives," *Adm. Sci. Qtrly.,* 2: 141–142.

Katz, Daniel and Kahn, Robert L. (1966), *The Social Psychology of Organizations*. New York: Wiley.

Katz, Elihu and Lazarsfeld, Paul F. (1955), *Personal Influence: The Part Played by People in the Flow of Mass Communications*. New York: The Free Press.

Kelley, H. H. (1951), "Communication in Experimentally Created Hierarchies," *Human Relations*, 4: 39–56.

Kelley, H. H. and Thibaut, John W. (1954), "Experimental Studies of Group Problem-Solving and Process," in Gardner Lindzey (ed.), *Handbook of Social Psycho-*

logy, Vol. II. Reading, Mass.: Addison-Wesley.

Kelley, H. H. and Volkhart, E. H. (1952), "The Resistance to Change of Group-Anchored Attitudes," *Amer. Soc. Rev.,* 17: 453–465.

Kellstedt, Lyman (1965), "Atlanta to 'Oretown': Identifying Community Elites," *Pub. Adm. Rev.,* 25: 161–167.

Kelman, Herbert C. (1961), "Processes of Opinion Change," *Pub. Opin. Qtrly.,* 25: 57–79.

Kerr, Clark (1964), *The Uses of the University*. Cambridge, Mass.: Harvard Univ. Press.

Klapper, Joseph T. (1960), *The Effects of Mass Communication*. New York: The Free Press.

Klein, Josephine (1961), *Working with Groups: The Social Psychology of Discussion and Decision*. London: Hutchinson.

Kluckhohn, Florence R. and Strodtbeck, Fred L. (1961), *Variations in Value Orientations*. Evanston, Ill.: Row, Peterson.

Knowlton, James Q. (1965), *Studies of Patterns of Influence in the School Situation as They Affect the Use of Audio-Visual Material*. Bloomington: Indiana Univ., Div. of Edu. Media and Audio-Visual Center Rept.

Lamb, Charles (1948), "A Dissertation Upon Roast Pig," in Ernest Bernbaum (Ed.), *Anthology of Romanticism*. New York: Ronald Press.

La Piere, Richard T. (1934), "Attitudes vs. Actions," *Social Forces*, 12: 230–237.

——— (1965), *Social Change*. New York: McGraw-Hill.

Larsen, Otto N. (1964), "Social Effects of Mass Communication," in Robert E. L. Faris (Ed.), *Handbook of Modern Sociology*. Chicago: Rand McNally.

Lawrence, Lois C. and Smith, Patricia C. (1955), "Group Decision and Employee Participation," *J. of App. Psych.,* 39: 334–337.

Lazarsfeld, Paul F. and Menzel, Herbert (1963), "Mass Media and Personal Influence," in Wilbur Schramm (Ed.), *The Science of Human Communication*. New York: Basic Books.

Lazarsfeld, Paul F. and Merton, Robert K. (1964), "Friendship as Social Process: A Substantive and Methodological Analysis," in Monroe Berger and others (Eds.), *Freedom and Control in Modern Society*. New York: Octagon.

Lazarsfeld, Paul F. and others (1944), *The People's Choice*. New York: Duell, Sloan, and Pearce.

Leavitt, H. J. (1965), "Applied Organizational Change in Industry: Structural, Technological and Humanistic Approaches," in James G. March (Ed.), *Handbook of Organizations*. Chicago: Rand McNally.

Lerner, Daniel (1958), *The Passing of Traditional Society: Modernizing the Middle East*. New York: The Free Press.

—— (1963), "Toward a Communication Theory of Modernization," in Lucien W. Pye (Ed.), *Communications and Political Development*. Princeton, N.J.: Princeton Univ. Press.

Levine, Jacob and Butler, John (1952), "Lecture vs. Group Decision in Changing Behavior," *J. of App. Psych.*, 36: 29–33.

Lewin, Kurt (1936), *Principles of Topological Psychology*. New York: McGraw-Hill.

—— (1943), "Forces Behind Food Habits and Methods of Change," *Bul. of the Natl. Resource Council*, 8: 35–65.

—— (1958), "Group Decision and Social Change," in Eleanor E. Maccoby and others (Eds.), *Readings in Social Psychology*. New York: Holt, Rinehart, and Winston.

Likert, Rensis (1961), *New Patterns of Management*. New York: McGraw-Hill.

—— (1967), *The Human Organizations*. New York: McGraw-Hill.

Machiavelli, Nicola (1961), *The Prince*. George Bull (tr.). Baltimore: Penguin Books.

March, James G. and Simon, Herbert A. (1958). *Organizations*. New York: Wiley.

Marting, Elizabeth (1964), *New Products, New Profits*. New York: Amer. Mgmt. Assn.

Martyn, Howe (1964), *International Business*. New York: The Free Press.

McGrath, Joseph E. and Altman, Irwin (1966), *Small Group Research: A Synthesis and Critique of the Field*. New York: Holt, Rinehart and Winston.

McNeill, Donald R. (1957), *The Fight for Fluoridation*. New York: Oxford University Press.

Merton, Robert K. (1949), "Patterns of Influence: A Study of Interpersonal Influence and Communication Behavior in a Local Community," in Paul F. Lazarsfeld and Frank N. Stanton (Eds.), *Communication Research, 1948–49*. New York: Harper and Brothers.

—— (1957) *Social Theory and Social Structure*. New York: The Free Press.

Miles, Matthew B. (Ed.) (1964b), *Innovation in Education*. New York: Columbia University, Teachers College.

Miller, Paul A. (1953), *Community Health Action*. East Lansing: Michigan State Univ. Press.

Mills, C. Wright (1959), *The Sociological Imagination*. New York: Grove Press.

Moore, Wilbert E. (1963), *Social Change*. Englewood Cliffs, N.J.: Prentice-Hall.

Morse, Nancy C. and Reimer, Everett (1956), "The Experimental Change in a Major Organizational Variable," *J. of Abnor. and Social Psych.*, 52: 120–129.

Parsons, Talcott (1951), *The Social System*. New York: The Free Press.

Parsons, Talcott, and Shils, Edward (1951), *Toward a General Theory of Action*. Cambridge, Mass.: Harvard Univ. Press.

Pinard, Maurice (1968), "Mass Society and Political Movements: A New Formulation," *Amer. J. of Soc.*, 73: 682–690.

Polsby, Nelson W. (1963), *Community Power and Political Theory*. New Haven, Conn.: Yale Univ. Press.

Pool, Ithiel de Sola and Abelson, Robert (1961), "The Simulmatics Project," *Pub. Opin. Qtrly.*, 25: 167–183.

Popper, Karl (1961), *The Logic of Scientific Discovery*. New York: Science Editions.

President's Science Advisory Committee (1967), *The World Food Problem*, Volume I. Washington, D.C.: U.S. Govt. Printing Office.

Presthus, Robert (1964), *Men at the Top: A Study in Community Power*. New York: Oxford Univ. Press.

Pye, Lucian W. (1963), *Communications and Political Development*. Princeton, N.J.: Princeton Univ. Press.

Queenley, Mary and Street, David (1965), *Innovation in Public Education: The Impact of the Continuous Development Approach*. Chicago: Univ. of Chicago, Center for Social Organization Studies, Working Paper 45.

Radcliffe-Brown, A. R. (1957), *A Natural Science of Society*. New York: The Free Press.

Radke, Marion and Klisurich, Dana (1947), "Experiments in Changing Food Habits," *J. of the Amer. Diabetics Assn.,* 23: 403–409.

Read, W. H. (1962), "Upward Communication in Industrial Hierarchies," *Human Relations*, 15: 3–16.

Redfield, Robert (1956), *Peasant Society and Culture*. Chicago: Univ. of Chicago Press.

Rice, A. K. (1958), *Productivity and Social Organization: The Ahmedebad Experiment*. London: Tavistock Publications.

Robinson, Edward J. (1966), *Communication and Public Relations*. Columbus, Ohio: Merrill.

Rokeach, Milton (1966), "The Nature of Attitudes," in *International Encyclopedia of the Social Sciences*. New York: Macmillan.

—— (1967), "Attitude Change and Behavioral Change," *Pub. Opin. Qtrly.* 30: 529–550.

—— (1968), *Beliefs, Attitudes, and Values: A Theory of Organization and Change*. San Francisco: Jossey-Bass.

Schatzman, Leonard and Strauss, Anselm (1955), "Social Class and Modes of Communication," *Amer. J. of Soc.,* 60: 329–338.

Schorr, Burt (1961), "The Mistakes: Many New Products Fail Despite Careful Planning, Publicity," *Wall Street Journal*.

Schramm, Wilbur (1963), "Communication Development and the Development Process," in Lucien W. Pye (Ed.), *Communications and Political Development*. Princeton, N.J.: Princeton Univ. Press.

—— (1964), *Mass Media and National Development*. Stanford, Cal.: Stanford Univ. Press.

Schramm, Wilbur and others (1967), *The New Media: Memo to Educational Planners*. Paris: UNESCO.

Schultz, Theodore W. (1964), *Transforming Traditional Agriculture*. New Haven, Conn.: Yale University Press.

—— (1965), *Economic Crises in World Agriculture*. Ann Arbor: Univ. of Michigan Press.

Schwartz, Donald F. (1968), *Liaison Communication Roles in a Formal Organization*. Ph.D. thesis. East Lansing: Michigan State Univ.

Seashore, Stanley E. and Bowers, David G. (1963), *Changing the Structure and Functioning of an Organization: Report of a Field Experiment*. Ann Arbor: Univ. of Michigan, Inst. for Social Res., Survey Res. Center, Monograph 33.

Secord, Paul F. and Backman, Carl W. (1964), *Social Psychology*. New York: McGraw-Hill.

Sicniski, Andzezej (1963), "A Two-Step Flow of Communication: Verification of an Hypothesis in Poland," *Polish Soc. Bul.,* 1: 33–40.

Signorite, Vito and O'Shea, Robert M. (1965), "A Test of Significance for the Homophily Index," *Amer. J. of Soc.,* 70: 467–470.

Smith, Bruce L. and others (1946), *Propaganda, Communication and Public Opinion*. Princeton, N.J.: Princeton Univ. Press.

Sofer, C. (1961), *Organization from Within*. London: Tavistock Publications.

Sollie, Carlton R. (1966), "A Comparison of Reputational Techniques for Identifying Community Leaders," *Rural Soc.,* 31: 301–309.

Tannenbaum, Arnold S. and Bachman, Jerald G. (1964), "Structural Versus Individual Effects," *Amer. J. of Soc.,* 69: 585–595.

Tax, Sol (1963), *Penny Capitalism*. Chicago: Univ. of Chicago Press.

Taylor, F. W. (1911), *The Principles and Methods of Scientific Management*. New York: Harper and Row.

Thibaut, J. W. (1950), "An Experimental Study of the Cohesiveness of Under-Privileged Groups," *Human Relations*, 3: 251–278.

Thomas, W. I. and Znaniecki, Florian (1927), *The Polish Peasant in Europe and America.* New York: Knopf.

Thoreau, Henry D. (1899), *Walden: Or Life in the Woods.* New York: Crowell.

Vidich, Arthur J. and Bensman, Joseph (1958), *Small Town in Mass Society.* Princeton, N.J.: Princeton Univ. Press.

Vroom, Victor H. (1960), *Some Personality Determinants of the Effects of Participation.* Englewood Cliffs, N.J.: Prentice-Hall.

Wallach, M. A. and Kogan, N. (1965), "The Roles of Information, Discussion and Consensus in Group Risk Taking," *J. of Exp. Social Psych.,* 1: 1–19.

Wallach, M. A. and others (1965), "Can Group Members Recognize the Effects of Group Discussion Upon Risk Taking?" *J. of Exp. Social Psych.,* 1: 379–295.

Walton, Eugene (1959), "Communication Down the Line: How They Really Get the Word," *Personnel,* 36: 78–82.

Walton, John (1966), "Discipline, Method, and Community Power: A Note on the Sociology of Knowledge," *Amer. Soc. Rev.,* 31: 684–689.

Warneryd, Karl-Erik and Nowak, Kjell (1967), *Mass Communication and Advertising.* Stockholm: Stockholm School of Economics, Eco. Res. Inst.

Watson, James E. and Lionberger, Herbert F. (1967), *Community Leadership in a Rural Trade-Centered Community and Comparison of Methods of Identifying Leaders.* Columbia: Mo. Agri. Exp. Sta., Res. Bul. 915.

Weber, Max (1947), *The Theory of Social and Economic Organization.* A. M. Henderson and Talcott Parsons (trs.). New York: The Free Press.

Weiss, R. S. and Jacobson, Eugene (1955), "A Method for the Analysis of the Structure of Complex Organizations," *Amer. Soc. Rev.,* 20: 661–668; and in Amatai Etzioni (Ed.) (1964), *Complex Organizations: A Sociological Reader.* New York: Holt, Rinehart and Winston.

Weller, Jack E. (1965), *Yesterday's People: Life in Contemporary Appalachia.* Lexington: Univ. of Kentucky Press.

Whorf, Benjamin L. (1956), "The Relation of Habitual Thought and Behavior to Language," in *Language, Thought and Reality.* Cambridge: Massachusetts Institute of Technology Press.

Wilkening, Eugene, A. (1957), *The County Extension Agent in Wisconsin.* Madison: Wisc. Agri. Exp. Sta., Res. Bul. 203.

Wolf, Eric R. (1955), "Types of Latin American Peasantry: A Preliminary Discussion," *Amer. Anthro.,* 57: 452–471.

Wright, Charles R. and Cantor, Muriel (1967), "The Opinion Seeker and Avoider: Steps Beyond the Opinion Leader Concept," *Pacific Soc. Rev.,* 10: 33–43.

Wunderlich, Gene (1958), "Concentration of Land Ownership," *J. of Farm Eco.,* 45: 1887–1893.

Zetterberg, Hans L. (1965), *On Theory and Verification in Sociology.* Totowa, N.J.: Bedminister Press.

AUTHOR INDEX

SUBJECT INDEX